ANATOMY AND
PHYSIOLOGY
Laboratory Manual

GERARD J. TORTORA

Bergen Community College, Paramus, New Jersey

ANATOMY AND PHYSIOLOGY
Laboratory Mannual

FOURTH EDITION

MACMILLAN PUBLISHING COMPANY
NEW YORK

Maxwell Macmillan Canada
TORONTO

Editors: Sheri S. Walvoord and Gary Carlson
Production Supervisor: Dora Rizzuto
Production Manager: Paul Smolenski
Text Designer: Robert Freese
Cover Designer: Robert Freese
Cover illustration: Mary Dersch
Photo Researcher: Chris Migdol
Illustrations: Original artwork by Mary Dersch

This book was set in 10/12 Palatino by The Clarinda Company
and was printed and bound by Von Hoffmann Press.
The cover was printed by Phoenix Color Corp.

Macmillan College Publishing Company
866 Third Avenue, New York, New York 10022

Macmillan College Publishing Company is part of
the Maxwell Communication Group of Companies.

Maxwell Macmillan Canada, Inc.
1200 Eglinton Avenue East
Suite 200
Don Mills, Ontario M3C 3N1

ISBN 0-02-4211915

Printing: 1 2 3 4 5 6 7   Year: 4 5 6 7 8 9 0

Preface

Anatomy and Physiology Laboratory Manual, Fourth Edition, has been written to guide students in the laboratory study of introductory anatomy and physiology. The manual is a companion to most of the leading anatomy and physiology textbooks.

Comprehensiveness

This manual examines virtually every structure and function of the human body that is typically studied in an introductory anatomy and physiology course. Because of its detail, the need for supplemental handouts is minimized; the manual is a strong teaching device in itself.

Use of the Scientific Method

Anatomy (the science of structure) and physiology (the science of function) cannot be understood without the practical experience of laboratory work. The exercises in this manual challenge students to understand the way scientists work by asking them to make microscopic examinations and evaluations of cells and tissues, to observe and interpret chemical reactions, to record data, to make gross examinations of organs and systems, to do dissections, and to conduct physiological laboratory work and interpret and apply the results of this work.

Illustrations

The manual contains a large number and variety of illustrations. The illustrations of the human body systems have been carefully drawn to exhibit structures that are essential to students' understanding of anatomy and physiology. Numerous photographs, photomicrographs, and scanning electron micrographs are presented to show students how the structures of the body actually look. These illustrations exceed in quality of execution those found in other anatomy and physiology laboratory manuals.

Important Features

Among the key features of this manual are (1) dissection of the white rat and selected mammalian organs; (2) numerous physiological experiments; (3) emphasis on the study of anatomy through histology; (4) lists of appropriate terms accompanying drawings and photographs to be labeled; (5) inclusion of numerous scanning electron micrographs and specimen photos; (6) phonetic pronunciations and derivations for many anatomical and physiological terms; (7) diagrams of commonly used laboratory equipment; (8) laboratory report questions and reports at the end of each exercise that can be filled in, removed, and turned in for grading if the instructor so desires; (9) three appendixes dealing with units of measurement, a periodic table of elements, and eponyms used in the laboratory manual; and (10) emphasis on laboratory safety throughout the manual.

New to the Fourth Edition

Numerous changes have been made in the fourth edition of this manual in response to suggestions from instructors and students. A significant addition is the inclusion of color for the cell in Exercise 3, photomicrographs in Exercise 4, and skull bones in Exercise 7. We have also added some new physiology experiments, line drawings, photomicrographs, and phonetic pronunciations and derivations. The principal additions follow:

Exercise 2, "Introduction to the Human Body," now has a section dealing with the subdivisions of anatomy and physiology and an exercise on abdominopelvic regions.

Exercise 3, "Cells," contains revised sections on the movement of substances across and through plasma membranes and on cell division.

In Exercise 4, "Tissues," the discussions of epithelial and connective tissue have been extensively rewritten.

A revised discussion of the histology of the epidermis has been added to Exercise 5, "Integumentary System."

Exercise 6, "Bone Tissue," contains a revised discussion of the functions of bone tissue.

Exercise 8, "Articulations," has a revised discussion of joints on the basis of function rather than structure.

Exercise 10, "Skeletal Muscles," now contains phonetic pronunciations for *every skeletal muscle* discussed.

Exercise 11, "Surface Anatomy," has been expanded with several new surface anatomy features.

Exercise 12, "Nervous Tissue and Physiology," contains revised sections on the organization of the nervous system and the reflex arc.

Exercise 13, "Nervous System," now contains a section on ascending and descending pathways and tracts and a revised section on the autonomic nervous system.

In Exercise 14, "General Senses and Sensory and Motor Pathways," there is a revised discussion of the characteristics of sensations, sensory pathways, and sensory-motor integration. In addition, senses are now discussed in two exercises: general senses in Exercise 14 and special senses in Exercise 15.

Exercise 16, "Endocrine System," contains a revised section on the histology of the pituitary gland.

In Exercise 19, "Blood Vessels," new "overviews" have been prepared for *all* exhibits. These overviews orient students to the blood vessels under discussion.

In Exercise 23, "Digestive System," there is a new exercise dealing with the mouth cavity.

Exercise 25 is new. It deals with pH and acid-base balance.

Exercise 27, "Development," contains revised discussions of spermatogenesis and oogenesis and changes associated with embryonic and fetal growth.

There is a new chart on page xvi that deals with selected laboratory safety signs and labels.

Changes in Terminology

In recent years, the use of eponyms for anatomical terms has been minimized or eliminated. Anatomical eponyms are terms named after various individuals, such as Fallopian tube (after Gabriello Fallopio) and Eustachian tube (after Bartolommeo Eustachio).

Anatomical eponyms are often vague and nondescriptive and do not necessarily mean that the person whose name is applied contributed anything very original. For these reasons, we have decided to minimize their use. However, because some prevail, we provide eponyms, in parentheses, after the first reference in each chapter to the more acceptable synonym. Thus, you will see terms such as *uterine (Fallopian) tube* or *auditory (Eustachian) tube.* See Appendix C.

Instructor's Guide

A complimentary instructor's guide to accompany the manual is available from the publisher. This comprehensive guide contains (1) a listing of materials needed to complete each exercise, (2) suggested audiovisual materials, (3) answers to illustrations and questions within the exercises, and (4) answers to laboratory report questions.

Acknowledgments

We wish to thank the following reviewers for their outstanding contributions to the fourth edition of *Anatomy and Physiology Laboratory Manual.* They have shared their extensive knowledge, teaching experience, and concern for students with us, and their feedback has been invaluable: Elden W. Martin (Bowling Green State University), Rose Morgan (Minot State University), and Antonio de la Pena (Pan American University).

We also wish to thank Mary Dersch, whose outstanding medical illustrations have greatly improved the manuals pedagogic and visual appeal, and Chris Migdol at Macmillan, who located several photographs and obtained the permissions.

As the acknowledgments indicate, the participation of many individuals of diverse talent and expertise is required to produce a laboratory manual of this scope and complexity. For this reason, readers and users are invited to send their reactions and suggestions to me so that plans can be formulated for subsequent editions.

Gerard J. Tortora
Natural Sciences and Mathematics, S229
Bergen Community College
400 Paramus Road
Paramus, NJ 07652

Contents

Laboratory Safety*

In 1989, The Centers for Disease Control (CDC) published "Guidelines for Prevention of Transmission of Human Immunodeficiency Virus and Hepatitis B Virus to Health-Care and Public-Safety workers" (*MMWR*, vol. 36, no. 6S). The CDC guidelines recommend precautions to protect health care and public safety workers from exposure to human immunodeficiency virus (HIV), the causative agent of acquired immune deficiency syndrome (AIDS), and hepatitis B virus (HBV), the causative agent of hepatitis B. These guidelines are presented to reaffirm the basic principles involved in the transmission not only of the AIDS and hepatitis B viruses, but of any disease-producing organism.

Based on the CDC guidelines for health care workers, as well as on other standard laboratory precautions and procedures, the following list has been developed for your safety in the laboratory. Although specific cautions and warnings concerning laboratory safety are indicated throughout the manual, read the following *before* performing any experiments.

A. GENERAL SAFETY PRECAUTIONS AND PROCEDURES

1. Arrive on time. Laboratory directions and procedures are given at the beginning of the laboratory period.

*The authors and publisher urge consultation with each instructor's institutional policies concerning laboratory safety and first aid procedures.

2. Read all experiments before you come to class to be sure that you understand all the procedures and safety precautions. Ask the instructor about any procedure you do not understand exactly. Do not improvise any procedure.

3. Protective eyewear and laboratory coats or aprons must be worn by all students performing or observing experiments.

4. Do not perform any unauthorized experiments.

5. Do not bring any unnecessary items to the laboratory and do not place any personal items (pocketbooks, bookbags, coats, umbrellas, etc.) on the laboratory table or at your feet.

6. Make sure all apparatus is supported and squarely on the table.

7. Tie back long hair to prevent it from becoming a laboratory fire hazard.

8. Never remove equipment, chemicals, biological materials, or any other materials from the laboratory.

9. Do not operate any equipment until you are instructed in its proper use. If you are unsure of the procedures, ask the instructor.

10. Dispose of chemicals, biological materials, used apparatus, and waste materials according to your instructor's directions. Not all liquids are to be disposed of in the sink.

11. Some exercises in the laboratory manual are designed to induce some degree of cardiovascular stress. Students should not participate in these exercises if they are pregnant or have hypertension or any other known or suspected condition that might compromise

health. Before you perform any of these exercises, check with your physician.

12. Do not put anything in your mouth while in the laboratory. Never eat, drink, taste chemicals, lick labels, smoke, or store food in the laboratory.

13. Your instructor will show you the location of emergency equipment such as fire extinguishers, fire blankets, and first-aid kits, as well as eyewash stations. Memorize their locations and know how to use them.

14. Wash your hands before leaving the laboratory. Because bar soaps can become contaminated, liquid or powdered soaps should be used. Before leaving the laboratory, remove any protective clothing, such as laboratory coats or aprons, gloves, and eyewear.

B. PRECAUTIONS RELATED TO WORKING WITH BLOOD, BLOOD PRODUCTS, OR OTHER BODY FLUIDS

1. Work only with *your own* body fluids, such as blood, saliva, urine, tears, and other secretions and excretions; blood from a clinical laboratory that has been tested and certified as noninfectious; or blood from a mammal (other than a human).

2. Wear gloves when touching another person's blood or other body fluids.

3. Wear safety goggles when working with another person's blood.

4. Wear a mask and protective eyewear or a face shield during procedures that are likely to generate droplets of blood or other body fluids.

5. Wear a gown or an apron during procedures that are likely to generate splashes of blood or other body fluids.

6. Wash your hands immediately and thoroughly if contaminated with blood or other body fluids. Hands can be rapidly disinfected by using (1) a phenol disinfectant-detergent for 20 to 30 seconds (sec) and then rinsing with water, or (2) alcohol (50 to 70%) for 20 to 30 sec, followed by a soap scrub of 10 to 15 sec and rinsing with water.

7. Spills of blood, urine, or other body fluids onto bench tops can be disinfected by flooding them with a disinfectant-detergent. The spill should be covered with disinfectant for 20 min before being cleaned up.

8. Potentially infectious wastes, including human body secretions and fluids, and objects such as slides, syringes, bandages, gloves, and cotton balls contaminated with those substances, should be placed in an autoclave container. Sharp objects (including broken glass) should be placed in a puncture-proof sharps container. Contaminated glassware should be placed in a container of disinfectant and autoclaved before it is washed.

9. Use only single-use, disposable lancets, and needles. Never recap, bend, or break the lancet once it has been used. Place used lancets, needles, and other sharp instruments in a *fresh* 1:10 dilution of household bleach (sodium hypochlorite) or other disinfectant such as phenols (Amphyl), aldehydes (glutaraldehyde, 1%) and 70% ethyl alcohol and then dispose of the instruments in a puncture-proof container. These disinfectants disrupt the envelope of HIV and HBV. The fresh household bleach solution or other disinfectant should be prepared for *each* laboratory session.

10. All reusable instruments, such as hemocytometers, well slides, and reusable pipettes, should be disinfected with a *fresh* 1:10 solution of household bleach or other disinfectant and thoroughly washed with soap and hot water. The fresh household bleach solution or other disinfectant should be prepared for *each* laboratory session.

11. A laboratory disinfectant should be used to clean laboratory surfaces *before* and *after* procedures, and should be available for quick cleanup of any blood spills.

12. Mouth pipetting should never be done. Use mechanical pipetting devices for manipulating all liquids in the laboratory.

13. All procedures and manipulations that have a high potential for creating aerosols or infectious droplets (such as centrifuging, sonicating, and blending) should be performed carefully. In such instances, a biological safety cabinet or other primary containment device is required.

C. PRECAUTIONS RELATED TO WORKING WITH REAGENTS

1. Use extreme care when working with reagents. Should any reagents make contact with your eyes, flush with water for 15 minutes (min); or, if they make contact with

your skin, flush with water for 5 min. Notify your instructor immediately should a reagent make contact with your eyes or skin and seek immediate medical attention.

2. Report all accidents to your instructor, no matter how minor they may appear.

3. When you are working with chemicals or preserved specimens, the room should be well-ventilated. Avoid breathing fumes for any extended period of time.

4. Never point the opening of a test tube containing a reacting mixture (especially when heating it) toward yourself or another person.

5. Exercise care in noting the odor of fumes. Use "wafting" if you are directed to note an odor. Your instructor will demonstrate this procedure.

6. Do not force glass tubing or a thermometer into rubber stoppers. Lubricate the tubing and introduce it gradually and gently into the stopper. Protect your hands with toweling when inserting the tubing or thermometer into the stopper.

7. Never heat a flammable liquid over or near an open flame.

8. Use only glassware marked Pyrex or Kimex. Other glassware may shatter when heated. Handle hot glassware with test tube holders.

9. If you have to dilute an acid, always add acid (AAA) to water.

10. When shaking a test tube or bottle to mix its contents, do not use your fingers as a stopper.

11. Read the label on a chemical twice before using it.

12. Replace caps or stoppers on bottles immediately after using them. Return spatulas to their correct place immediately after using them and do not mix them up.

13. Mouth pipetting should never be done. Use mechanical pipetting devices for manipulating all liquids in the laboratory.

D. PRECAUTIONS RELATED TO DISSECTION

1. When you are working with chemicals or preserved specimens, the room should be well-ventilated. Avoid breathing fumes for any extended period of time.

2. Wear rubber gloves when dissecting.

3. To reduce the irritating effects of chemical preservatives to your skin, eyes, and nose, soak or wrap your specimen in a substance such as "Biostat." If this is not available, hold your specimen under running water for several minutes to wash away excess preservative and dilute what remains.

4. When dissecting, there is always the possibility of skin cuts or punctures from dissecting equipment or the specimens themselves, such as the teeth or claws of an animal. Should you sustain a cut or puncture in this manner, wash your hands with disinfectant soap, notify your instructor, and seek immediate medical attention to decrease the possibility of infection. A first aid kit should be readily available for your use.

5. When cleaning dissecting instruments, always hold the sharp edges away from you.

6. Dispose of any damaged or worn-out dissecting equipment in an appropriate container supplied by your instructor.

SELECTED LABORATORY SAFETY SIGNS/LABELS

⚠CAUTION

Protect eyes.
Wear goggles at
all times.

⚠CAUTION

Hot surface.
Do not touch.

⚠CAUTION

Cancer suspect
agent. Trained
personnel only.

⚠CAUTION

Radiation area.
Authorized
personnel only.

⚠CAUTION

Biological hazard.
Authorized
personnel only.

⚠ DANGER

Highly toxic.
Handle with care.

⚠ DANGER

Do not smoke
in this area.

⚠ DANGER

Do not smoke,
eat or drink
in this area.

⚠ DANGER

Do not pipet
liquids by mouth.

⚠ DANGER

Corrosive. Avoid
contact with eyes
and skin.

⚠ DANGER

Flammable
material. Keep
fire away.

EMERGENCY

Eye Wash
Station.
Keep area clear.

EMERGENCY

Safety Shower.
Keep area clear.

EMERGENCY

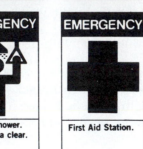

First Aid Station.

Fire extinguisher.
Remove pin and
squeeze trigger.

COMMONLY USED LABORATORY EQUIPMENT

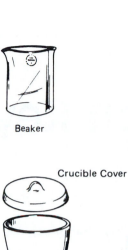
Beaker

Erlenmeyer Flask

Florence Flask

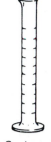
Funnel

Graduated Cylinder

Crucible Cover

Crucible

Evaporating Dish

Mortar and Pestle

Watch Glass

Stirring Rod

Test Tube

Test Tube Brush

Test Tube Holder

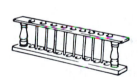
Test Tube Rack

Ring Stand and Ring

Pinch Clamp

Utility Clamp

Tripod

Clay Triangle

Wire Gauze

Crucible Tongs

Beaker Tongs

Forceps

Medicine Dropper

Nichrome Wire

Spatula

Pronunciation Key

A unique feature of this revised manual is the phonetic pronunciations given for many anatomical and physiological terms. The pronunciations are given in parentheses immediately after the particular term is introduced. The following key explains the essential features of the pronunciations.

1. The syllable with the strongest accent appears in capital letters; for example, bilateral (bī-LAT-er-al) and diagnosis (dī-ag-NŌ-sis).
2. A secondary accent is denoted by a single quotation mark ('); for example, constitution (kon'-sti-TOO-shun) and physiology (fiz'-ē-OL-ō-jē). Additional secondary accents are also noted by a single quotation mark; for example, decarboxylation (dē'-kar-bok'-si-LĀ-shun).

3. Vowels marked with a line above the letter are pronounced with the long sound, as in the following common words:

   ā as in *māke*   ī as in *ivy*
   ē as in *bē*   ō as in *pole*

4. Unmarked vowels are pronounced with the short sound, as in the following words:

   e as in *bet*   o as in *not*
   i as in *sip*   u as in *bud*

5. Other phonetic symbols are used to indicate the following sounds:

   a as in *above*   yoo as in *cute*
   oo as in *soon*   oy as in *oil*

Microscopy

Note: Before you begin any laboratory exercises in this manual, please read the section on *LABORATORY SAFETY* on page xiii.

One of the most important instruments that you will use in your anatomy and physiology course is a compound light microscope. In this instrument, the lenses are arranged so that images of objects too small to be seen with the naked eye can be highly magnified—that is, their apparent size can be increased, and their minute details can be revealed. The following discussion of some of the principles employed in light microscopy (mī-KROS-kō-pē) will be helpful before you actually learn the parts of a compound light microscopy and how to use it. Later in this exercise, some of the principles used in electron microscopy will also be discussed.

A. COMPOUND LIGHT MICROSCOPE

A *compound light microscope* uses two sets of lenses, ocular and objective, and employs light as its source of illumination. Magnification is achieved when light rays from an illuminator are passed through a condenser, which directs the light rays through the specimen under observation; then, light rays pass into the objective lens (the magnifying lens closest to the specimen), and the image of the specimen forms on a prism, where it is magnified again by the ocular lens.

A general principle of microscopy is that the shorter the wavelength of light used in the instrument, the greater the resolution. *Resolution,* or *resolving power,* is the ability of the lenses to distinguish fine detail and structure— that is, to distinguish between two points as separate objects. As an example, a microscope with a resolving power of 0.3 micrometers (mī-KROM-e-ters), symbolized μm, is capable of distinguishing two points as separate objects if they are at least 0.3 μm apart (1 μm = 0.000001 or 10^{-6} μm; see Appendix A). The light used in a compound light microscope has a relatively long wavelength and cannot resolve structures smaller than 0.3 μm. This, and practical considerations, means that even the best compound light microscopes can magnify images only about 2000 times.

A *photomicrograph* (fō-tō-MĪ-krō′-graf), a photograph of a specimen taken through a compound light microscope, is shown in Figure 4.1. In later exercises, you will be asked to examine photomicrographs of various specimens of the body before you actually view them yourself through the microscope.

1. Parts of the Microscope

Carefully carry the microscope from the cabinet to your desk by placing one hand around the arm and the other hand firmly under the base. Gently place it on your desk, directly in front of you, with the arm facing you. Locate the following parts of the microscope and, as you read about each part, label Figure 1.1 by placing the correct numbers in the spaces next to the list of terms that accompanies the figure.

1. *Base*—Bottom portion on which the microscope rests.
2. *Body tube*—The portion that receives the ocular.
3. *Arm*—The frame's angular or curved part.
4. *Inclination joint*—Movable hinge in some microscopes that allows the instrument to be tilted to a comfortable viewing position.
5. *Stage*—Platform on which slides or other objects to be studied are placed. The opening in the center, called the *stage opening*, allows light to pass from below through the specimen being examined. Some microscopes have a *mechanical stage.* An adjustor knob below the stage moves the stage forward and backward and from side to side. With a mechanical stage, the slide and the stage move simultaneously. A mechanical stage permits a smooth, precise movement of a slide. Sometimes a mechanical stage is fitted with calibrations that permit the numerical "mapping" of a specimen on a slide.
6. *Stage (spring) clips*—Two clips mounted on the stage that hold slides securely in place.
7. *Substage lamp*—Source of illumination for some light microscopes with a built-in lamp.
8. *Mirror*—A feature found in some microscopes below the stage. The mirror directs light from its source through the stage opening and through the lenses. If the light source is built in, a mirror is not necessary.
9. *Condenser*—Lens located beneath the stage opening that concentrates the light beam on the specimen.
10. *Condenser adjustment knob*—Knob that functions to raise and lower the condenser. At its highest position, it allows full illumination and thus can be used to adjust illumination.
11. *Diaphragm* (DĪ-a-fram)—Device located between the condenser and stage that regulates the intensity of the light passing through the condenser and lenses to the observer's eyes. Such regulation is needed because transparent or very thin specimens cannot be seen in bright light. One of two types of diaphragms is usually used. An *iris diaphragm,* also found in cameras, is a series of sliding leaves that vary the size of the opening and thus the amount of light entering the lenses. The leaves are moved by a *diaphragm lever* to regulate the diameter of a central opening. A *disc diaphragm* consists of a plate with a graded series of holes, any of which can be rotated into position.
12. *Coarse adjustment knob*—Usually a large knob that raises and lowers the body tube (or stage) to bring a specimen into general view.
13. *Fine adjustment knob*—Usually a small knob found below or external to the coarse adjustment knob and used for fine or final focusing. Some microscopes have both coarse and fine adjustment knobs combined into one.
14. *Nosepiece*—A plate, usually circular, at the bottom of the body tube.
15. *Revolving nosepiece*—The lower, movable part of the nosepiece that contains the various objective lenses.
16. *Scanning objective*—A lens marked 5× on most microscopes (× means "times"); it is the shortest objective and does not appear on all microscopes.
17. *Low-power objective*—A lens marked 10× on most microscopes; it is the next shortest objective.
18. *High-power objective*—A lens marked 43× or 45× on most microscopes; also called a *high-dry objective;* it is the second longest objective.
19. *Oil-immersion objective*—A lens marked 100× on most microscopes; it is distinguished by an etched colored circle (special instructions for this objective will be discussed later on page 5) and is the longest objective.
20. *Ocular (eyepiece)*—A removable lens at the top of the body tube, marked 10× on most microscopes; it is sometimes fitted with a pointer or measuring scale.

2. Rules of Microscopy

You must observe certain basic rules at all times to obtain maximum efficiency and provide proper care for your microscope.

1. Keep all parts of the microscope clean, especially the lenses of the ocular, objectives, condenser, and mirror. *Use the special lens paper provided. Never use paper towels or cloths because they tend to scratch delicate glass surfaces.* When using lens paper, use the same area on the paper only once. As you wipe the lens, change the position of the paper as you go.

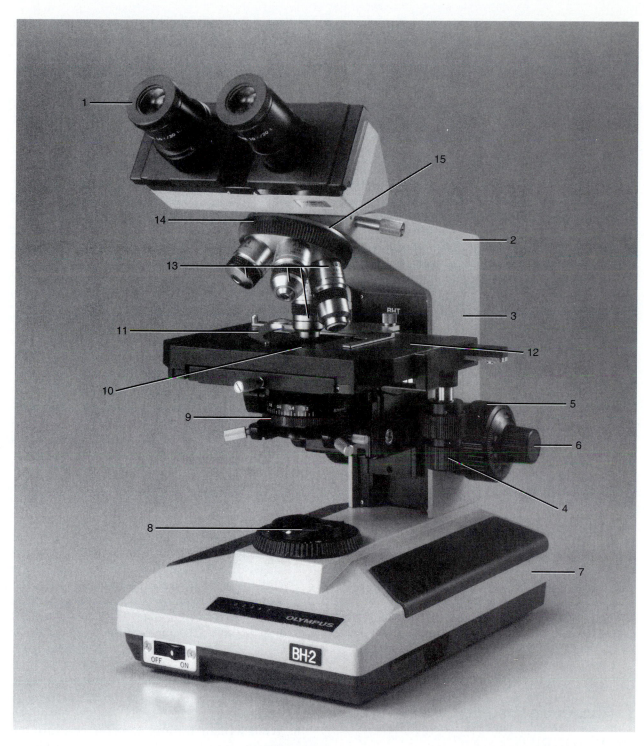

___ Arm	___ Diaphragm	___ Ocular
___ Base	___ Fine adjustment knob	___ Revolving nosepiece
___ Body tube	___ Mechanical stage knob	___ Stage
___ Coarse adjustment knob	___ Nosepiece	___ Stage clip of mechanical stage
___ Condenser	___ Objectives	___ Substage lamp

FIGURE 1.1   Olympus BH-2 microscope.

2. Do not permit the objectives to get wet, especially when observing a *wet mount.* You must use a *cover slip* when you examine a wet mount or the image becomes distorted.
3. Consult your instructor if any mechanical or optical difficulties rise. *Do not try to solve these problems yourself.*
4. Keep *both* eyes open at all times while observing objects through the microscope. This is difficult at first, but with practice becomes natural. This important technique will help you to draw and observe microscopic specimens without moving your head. Only your eyes will move.
5. Always use either the scanning or low-power objective first to locate an object; then, if necessary, switch to a higher power.
6. If you are using the high-power or oil-immersion objectives, *never focus using the coarse adjustment knob.* The distance between these objectives and the slide, called *working distance,* is very small and you may break the cover slip and the slide and scratch the lens.
7. Some microscopes have a stage that moves while focusing, others have a body tube that moves while focusing. Be sure you are familiar with the type you are using. Look at your microscope from the side (place the objective on scanning or low power) and gently turn the coarse adjustment knob. Which moves? The stage or the body tube? *Never focus downward* if the microscope's body tube moves when focusing. *Never focus upward* if the microscope's stage moves when focusing. By observing from one side you can see that the objectives do not make contact with the cover slip or slide.
8. Make sure that you raise the body tube before placing a slide on the stage or before removing a slide.

3. Setting up the Microscope

PROCEDURE

1. Place the microscope on the table with the arm toward you and with the back of the base at least 1 inch (in.) from the edge of the table.
2. Position yourself and the microscope so that you can look into the ocular comfortably.
3. Wipe the objectives, the top lens of the eyepiece, the condenser, and the mirror with lens paper. Clean the most delicate and the

least dirty lens first. Apply xylol or ethanol to the lens paper only to remove grease and oil from the lenses and microscope slides.

4. Position the low-power objective in line with the body tube. When it is in its proper place, it will click. Lower the body tube using the coarse adjustment knob until the bottom of the lens is approximately ¼ in. from the stage.
5. Admit the maximum amount of light by opening the diaphragm, if it is an iris diaphragm, or turning the disc to its largest opening if it is a disc diaphragm.
6. Place your eye to the ocular, and adjust the light. When a uniform circle (the *microscopic field*) appears without any shadows, the microscope is ready for use.

4. Using the Microscope

PROCEDURE

1. Using the coarse adjustment knob, raise the body tube to its highest fixed position.
2. Make a temporary mount using a single letter of newsprint, or use a slide that has been specially prepared with a letter, usually the letter "e." If you prepare such a slide, cut a single letter—a, b, or e—from the smallest print available and place it in the correct position to be read with the naked eye. Your instructor will provide directions for preparing the slide.
3. Place the slide on the stage, making sure that the letter is centered over the opening in the stage, directly over the condenser. Secure the slide in place with the stage clips.
4. Align the low-power objective with the body tube.
5. Lower the body tube or raise the stage as far as it will go *while you watch it from the side,* taking care not to touch the slide. The tube should reach an automatic stop that prevents the low-power objective from hitting the slide.
6. While looking through the eyepiece, turn the coarse adjustment knob counterclockwise, raising the body tube. Or, turn the coarse adjustment knob clockwise, lowering the stage. When focusing, always *raise* the body tube or *lower* the stage. Watch for the object to appear suddenly in the microscopic field. If it is in proper focus, the low-power objective is about ½ in. above the slide. When focusing, always *raise* the body tube.

7. Use the fine-adjustment knob to complete the focusing; you will usually use a counter-clockwise motion once again.

8. Compare the position of the letter with its original appearance under the microscope. Has the position of the letter been changed?

_____

9. While looking at the slide through the ocular, move the slide with your thumbs, or, if the microscope is equipped with them, the mechanical stage knobs. This exercise teaches you to move your specimen in various directions quickly and efficiently. In which direction does the letter move when you move the

slide to the left? _____
This procedure, "scanning" a slide, will be useful for examining living objects and for centering specimens so you can observe them easily.

Make a drawing of the letter as it appears under low power in the space below on the left side.

Drawing of letter   _Drawing of letter_
as seen under   _as seen under_
low power   _high power_

10. Change your magnification from low to high power by carrying out the following steps:
    a. Place the letter in the center of the field under low power. Centering is important because you are now focusing on a smaller area of the microscopic field. As you will see, _microscopic field size decreases with higher magnifications._
    b. Make sure the illumination is at its maximum. Illumination must be increased at higher magnifications because the

amount of light entering the lens decreases as the size of the objective lens increases.

c. The letter should be in focus, and if the microscope is _parfocal_ (meaning that when clear focus has been attained using any objective at random, revolving the nosepiece results in a change in magnification but leaves the specimen still in focus), the high-power objective can be switched into line with the body tube without changing focus. If it is not completely in focus after switching the lens, a slight turn of the fine-adjustment knob will focus it.

d. If your microscope is not parfocal, observe the stage from one side and carefully switch the high-power objective in line with the body tube.

e. While observing from the side and using the coarse adjustment knob, _carefully_ lower the objective or raise the stage until the objective almost touches the slide.

f. Look through the ocular and focus up slowly. Finish focusing by turning the fine-adjustment knob.

g. If your microscope has an oil-immersion objective, you must follow special procedures. Place a drop of special _immersion oil_ directly over the letter on the microscope slide, and lower the oil-immersion objective until it just contacts the oil. If your microscope is parfocal, you do not have to raise or lower the objectives. For example, if you are using the high-power objective and the specimen is in focus, just switch the high-power objective out of line with the body tube. Then add the oil and switch the oil-immersion objective into position; the specimen should be in focus. The same holds true when you switch from low power to high power. The special light-transmitting properties of the oil are such that light is refracted (bent) toward the specimen, permitting the use of powerful objectives in a relatively narrow field of vision. This objective is extremely close to the slide being examined, so when it is in position, take precautions _never to focus downward_ while you are looking through the eyepiece. Whenever you finish using immersion oil, be sure to saturate a piece of lens paper with xylol or alcohol and clean the oil-immersion objective and the slide if it is to be reused.

Is as much of the letter visible under high

power as under low power? Explain. _____

_____

Make a drawing of the letter as it appears under high power in the space next to the drawing you made for low power following Step 9.

11. Now select a prepared slide of three different-colored threads. Examination will show that a specimen mounted on a slide has depth as well as length and width. At lower magnification the amount of depth of the specimen that is clearly in focus, the depth of field, is greater than that at higher magnification. You must focus at different depths to determine the position (depth) of each thread.

    After you make your observation under low power and high power, answer the following questions about the location of the different threads:

    What color is at the bottom, closest to the

    slide? _____

    On top, closest to the cover slip? _____

    _____

    In the middle? _____

12. Your instructor might want you to prepare a wet mount as part of your introduction to microscopy. If so, the directions are given in Exercise 4, A.2.f, on page 55.

13. When you are finished using the microscope
    a. Remove the slide from the stage.
    b. Clean all lenses with lens paper.
    c. Align the mechanical stage so that it does not protrude.
    d. Leave the scanning or low-power objective in place.
    e. Lower the body tube or raise the stage as far as it will go.
    f. Wrap the cord according to your instructor's directions.
    g. Replace the dust cover or place the microscope in a cabinet.

5. Magnification

The total magnification of your microscope is calculated by multiplying the magnification of the ocular by the magnification of the objective used. Example: An ocular of $10\times$ used with an objective of $5\times$ gives a total magnification of $50\times$. Calculate the total magnification of each of the objectives on your microscope:

1. Ocular _____ × Objective _____ = _____

2. Ocular _____ × Objective _____ = _____

3. Ocular _____ × Objective _____ = _____

4. Ocular _____ × Objective _____ = _____

B. ELECTRON MICROSCOPE

Examination of specimens smaller than 0.3 μm requires an *electron microscope,* an instrument with a much greater resolving power than a compound light microscope. An electron microscope uses a beam of electrons instead of light. Electrons travel in waves just as light does. Instead of glass lenses, however, magnets are used in an electron microscope to focus a beam of electrons through a vacuum tube onto a specimen. Because the wavelength of an electron is about 1/100,000 that of visible light, the resolving power of very sophisticated transmission electron microscopes (described shortly) is close to 10 nanometers (nm). 1 nm = 0.000,000,001 m or 10^{-9}m. (See Appendix A.) Most electron microscopes have a working resolving power of about 100 nm and can magnify images up to $200,000\times$.

Two types of electron microscopes are available. In a *transmission electron microscope,* a finely focused beam of electrons passes through a specimen, usually ultrathin sections of material. The beam is then refocused, and the image reflects what the specimen has done to the transmitted electron beam. The image may be used to produce what is called a *transmission electron micrograph.* The method is extremely valuable in providing details of the interior of specimens at different layers but does not give a three-dimensional effect. Such an effect can be obtained with a *scanning electron microscope.* With this instrument, a finely focused beam of electrons is directed over the specimen and then reflected from the surface of the specimen onto a televisionlike screen or photographic plate. The photograph produced from an image generated in this manner is called a *scanning electron micrograph* (see Figure 5.2). Scanning electron micrographs commonly magnify specimens up to

10,000× and are especially useful in studying surface features of specimens.

Note: A section of LABORATORY REPORT QUESTIONS is located at the end of each exercise. The student can answer these questions and hand them in for grading at the direction of the instructor. Even if the instructor does not require them, we recommend doing them, to assess the student's level of understanding.

Some exercises also have a section of LABORATORY REPORT RESULTS, in which students can record the results of laboratory exercises, in addition to laboratory report questions. As with the laboratory report questions, the laboratory report results are located at the end of selected exercises and can be handed in as the instructor directs. Instructions in the manual tell students when and where to record laboratory results.

ANSWER THE LABORATORY REPORT QUESTIONS AT THE END OF THE EXERCISE.

Microscopy

STUDENT _____   DATE _____

LABORATORY SECTION _____   SCORE/GRADE _____

PART 1. Multiple Choice

_____ 1. The amount of light entering a microscope can be adjusted by regulating the (a) ocular (b) diaphragm (c) fine-adjustment knob (d) nosepiece

_____ 2. If the ocular on a microscope is marked 10× and the low-power objective is marked 15×, the total magnification is (a) 50× (b) 25× (c) 150× (d) 1500×

_____ 3. The size of the light beam that passes through a microscope is regulated by the (a) revolving nosepiece (b) coarse-adjustment knob (c) ocular (d) condenser

_____ 4. Parfocal means that (a) The microscope employs only one lens (b) final focusing can be done only with the fine-adjustment knob (c) changing objectives by revolving the nosepiece will still keep the specimen in focus (d) the highest magnification attainable is 1000×

_____ 5. Which of these is *not* true when changing magnification from low power to high power? (a) the specimen should be centered (b) illumination should be decreased (c) the specimen should be in clear focus (d) the high-power objective should be in line with the body tube

_____ 6. A microscope with which of the following resolving powers could distinguish the finest detail? (a) 0.01 μm (b) 10 nm (c) 0.001 μm (d) 1 dm

PART 2. Completion

7. The advantage of using immersion oil is that it has special _____ that permit the use of a powerful objective in a narrow field of vision.

8. The uniform circle of light that appears when one looks into the ocular is called the

_____.

9. In determining the position (depth) of the colored threads, the _____ (red, blue, yellow) colored thread was in the middle.

10. If you move your slide to the right, the specimen moves to the _____ you are viewing it microscopically.

11. After switching from low power to high power, _____ (more or less) of the specimen will be visible.

12. The ability of lenses to distinguish between two points at a specified distance apart is called

_____.

13. The wavelength of an electron is about _____ (what proportion?) that of visible white light.

14. A photograph of a specimen taken through a compound light microscope is called a(n) _____.

15. The type of microscope that provides three-dimensional images of nonliving specimens is _____ the microscope.

PART 3. Matching

_____ 16. Ocular

_____ 17. Stage

_____ 18. Arm

_____ 19. Condenser

_____ 20. Revolving nosepiece

_____ 21. Low-power objective

_____ 22. Fine-adjustment knob

_____ 23. Diaphragm

_____ 24. Coarse-adjustment knob

_____ 25. High-power objective

A. Platform on which slide is placed

B. Mounting for objectives

C. Lens below stage opening

D. Brings specimen into sharp focus

E. Eyepiece

F. An objective usually marked 43× or 45×

G. An objective usually marked 10×

H. Angular or curved part of frame

I. Brings specimen into general focus

J. Regulates light intensity

Introduction to the Human Body

In this exercise, you will be introduced to the organization of the human body through a study of the principal subdivisions of anatomy and physiology, levels of structural organization, principal body systems, anatomical position, regional names, directional terms, planes of the body, body cavities, abdominopelvic regions, and abdominopelvic quadrants.

A. ANATOMY AND PHYSIOLOGY

Whereas *anatomy* (a-NAT-ō-mē) refers to the study of *structure* and the relationships among structures, *physiology* (fiz-ē-OL-ō-jē) deals with the *functions* of body parts—that is, how they work. Each structure of the body is designed to carry out a particular function.

Following are selected subdivisions of anatomy and physiology. In the spaces provided, define each term.

SUBDIVISIONS OF ANATOMY

Surface anatomy. _____

_____

_____

Gross (macroscopic) anatomy. _____

_____

_____

_____

Systemic anatomy. _____

_____

_____

_____

Regional anatomy. _____

_____

_____

_____

Radiographic (rā′-dē-ō-GRAF-ik) *anatomy.* _____

_____

_____

_____

Developmental anatomy. _____

_____

_____

Embryology (em'-brē-OL-ō-jē). _____

_____

_____

Histology (hiss'-TOL-ō-jē). _____

_____

_____

Cytology (sī-TOL-ō-jē). _____

_____

_____

_____

Pathological (path'-ō-LOJ-i-kal) *anatomy.* _____

_____

_____

SUBDIVISIONS OF PHYSIOLOGY

Cell physiology. _____

_____

_____

_____

Pathophysiology (PATH-ō-fiz-ē-ol'-ō-jē). _____

_____

_____

Exercise physiology. _____

_____

_____

_____

Neurophysiology (NOO-rō-fiz-ē-ol'-ō-jē). _____

_____

_____

Endocrinology (en'-dō-kri-NOL-ō-jē). _____

_____

_____

Cardiovascular (kar-dē-ō-VAS-kyoo-lar) *physiology.* _____

_____

_____

_____

Immunology (im'-yoo-NOL-ō-jē). _____

_____

_____

_____

Respiratory (re-SPĪ-ra-tō-rē) *physiology.* _____

_____

_____

_____

B. LEVELS OF STRUCTURAL ORGANIZATION

The human body is composed of several levels of structural organization associated with one another in various ways:

1. *Chemical level*—Composed of all atoms and molecules necessary to maintain life.
2. *Cellular level*—Consists of cells, the structural and functional units of the body.
3. *Tissue level*—Formed by tissues, groups of similarly specialized cells and their intercellular material.
4. *Organ level*—Consists of organs, structures of definite form and function composed of two or more different tissues.
5. *System level*—Formed by systems, associations of organs that have a common function. The systems together constitute an *organism.*

C. SYSTEMS OF THE BODY

Using an anatomy and physiology textbook, torso, wall chart, and any other materials that might be available to you, identify the principal organs that compose the following body systems.[1] In the spaces that follow, indicate the components (organs) and functions of the systems.

Integumentary

Components. _____

_____

_____

_____

Functions. _____

_____

_____

_____

Skeletal

Components. _____

_____

_____

_____

Functions. _____

_____

_____

_____

Muscular

Components. _____

_____

_____

_____

Functions. _____

_____

_____

_____

Nervous

Components. _____

_____

_____

_____

Functions. _____

_____

_____

_____

Endocrine

Components. _____

_____

_____

Functions. _____

_____

_____

Cardiovascular

Components. _____

_____

_____

Functions. _____

_____

_____

_____

[1]You will probably need other sources, plus any aids the instructor might provide, to label many of the figures and answer some of the questions in this manual. You are encouraged to use other sources as necessary.

Lymphatic

Components. _____

_____

_____

_____

Functions. _____

_____

_____

_____

Respiratory

Components. _____

_____

_____

_____

Functions. _____

_____

_____

_____

Digestive

Components. _____

_____

_____

_____

Functions. _____

_____

_____

_____

Urinary

Components. _____

_____

_____

_____

Functions. _____

_____

_____

_____

Reproductive

Components. _____

_____

_____

_____

Functions. _____

_____

_____

_____

D. ANATOMICAL POSITION AND REGIONAL NAMES

Figure 2.1 shows anterior and posterior views of a subject in the *anatomical position*. The subject is erect (upright position) and facing the observer when in anterior view, the upper extremities (limbs) are at the sides, the palms of the hands are facing forward, and the feet are flat on the floor. The figure also shows the common names for various regions of the body. When you, as the observer, make reference to the left and right sides of the subject you are studying, this refers to the *subject's* left and right sides. In the spaces next to the list of terms in Figure 2.1, write the number of each common term next to each corresponding anatomical term. For example, the skull (29) is cranial, so write the number *29* next to the term *Cranial*.

E. EXTERNAL FEATURES OF THE BODY

Referring to your textbook and human models, identify the following external features of the body:

1. *Head (cephalic region* or *caput)*—Divided into the *cranium* (brain case) and *face.*

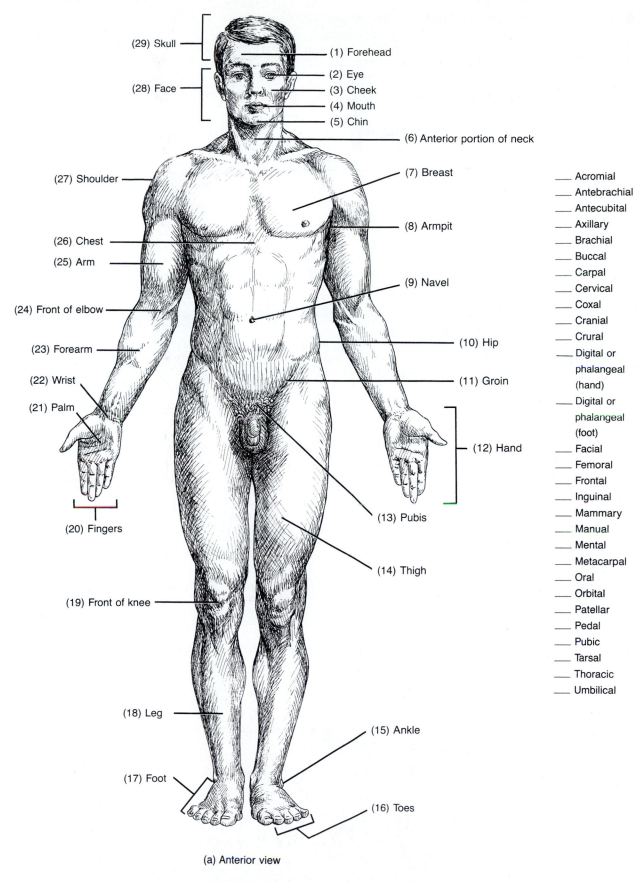

(29) Skull
(28) Face
(1) Forehead
(2) Eye
(3) Cheek
(4) Mouth
(5) Chin
(6) Anterior portion of neck
(27) Shoulder
(7) Breast
(8) Armpit
(26) Chest
(25) Arm
(9) Navel
(24) Front of elbow
(23) Forearm
(10) Hip
(22) Wrist
(11) Groin
(21) Palm
(12) Hand
(20) Fingers
(13) Pubis
(14) Thigh
(19) Front of knee
(18) Leg
(15) Ankle
(17) Foot
(16) Toes

____ Acromial
____ Antebrachial
____ Antecubital
____ Axillary
____ Brachial
____ Buccal
____ Carpal
____ Cervical
____ Coxal
____ Cranial
____ Crural
____ Digital or phalangeal (hand)
____ Digital or phalangeal (foot)
____ Facial
____ Femoral
____ Frontal
____ Inguinal
____ Mammary
____ Manual
____ Mental
____ Metacarpal
____ Oral
____ Orbital
____ Patellar
____ Pedal
____ Pubic
____ Tarsal
____ Thoracic
____ Umbilical

(a) Anterior view

FIGURE 2.1   The anatomical position.

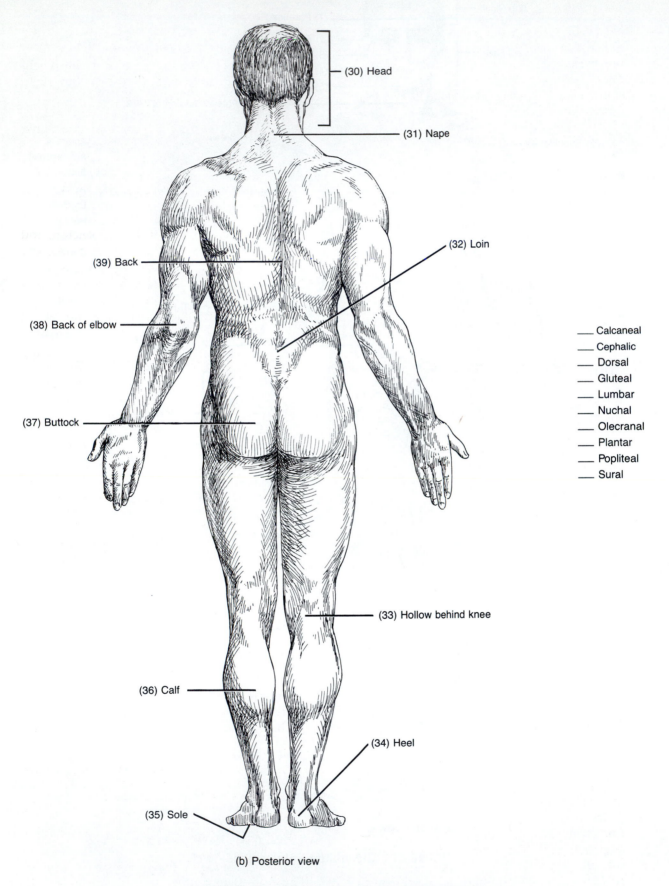

(30) Head

(31) Nape

(32) Loin

(39) Back

(38) Back of elbow

(37) Buttock

(33) Hollow behind knee

(36) Calf

(34) Heel

(35) Sole

___ Calcaneal
___ Cephalic
___ Dorsal
___ Gluteal
___ Lumbar
___ Nuchal
___ Olecranal
___ Plantar
___ Popliteal
___ Sural

(b) Posterior view

FIGURE 2.1 *(Continued)*   The anatomical position.

2. *Neck (collum)*—Region consisting of an anterior *cervix,* two lateral surfaces, and a posterior *nucha* (NOO-ka), or the nape of the neck.
3. *Trunk*—Region divided into the *back* (dorsum), *chest* (thorax), *abdomen* (venter), and *pelvis.*
4. *Upper extremity*—The *armpit* (axilla), *shoulder* (acromial region or omos), *arm* (brachium), *elbow* (cubitus), *forearm* (antebrachium), and *hand* (manus). The hand, in turn, consists of the *wrist* (carpus), *palm* (metacarpus), and *fingers* (digits). Individual bones of a digit (finger or toe) are called *phalanges. Phalanx* is the singular.
5. *Lower extremity*—The *buttocks* (gluteal region), *thigh* (femoral region), *knee* (genu), *leg* (crus), and *foot* (pes). The foot includes the *ankle* (tarsus), *sole* (metatarsus), and *toes* (digits).

F. DIRECTIONAL TERMS

To explain exactly where a structure of the body is located, it is a standard procedure to use *directional terms.* Such terms are very precise and avoid the use of unnecessary words. Commonly used directional terms for humans follow:

1. *Superior* (soó-PEER-ē-or) *(cephalic* or *cranial)*—Toward the head or the upper part of a structure; generally refers to structures in the trunk.
2. *Inferior* (in'-FEER-ē-or) *(caudal)*—Away from the head or toward the lower part of a structure; generally refers to structures in the trunk.
3. *Anterior* (an-TEER-ē-or) *(ventral)*—Nearer to or at the front or the belly surface of the body. In the *prone position,* the body lies anterior side down; in the *supine position,* the body lies anterior side up.
4. *Posterior* (pos-TEER-ē-or) *(dorsal)*—Nearer to or at the back or backbone surface of the body.
5. *Medial* (MĒ-dē-al)—Nearer the midline of the body or a structure. The *midline* is an imaginary vertical line that divides the body into equal left and right sides.
6. *Lateral* (LAT-er-al)—Farther from the midline of the body or a structure.
7. *Intermediate* (in'-ter-MĒ-dē-at)—Between two structures.
8. *Ipsilateral* (ip-si-LAT-er-al)—On the same side of the body.

9. *Contralateral* (CON-tra-lat-er-al)—On the opposite side of the body.
10. *Proximal* (PROK-si-mal)—Nearer the attachment of an extremity (limb) to the trunk; nearer to the point of origin.
11. *Distal* (DIS-tal)—Farther from the attachment of an extremity (limb) to the trunk; farther from the point of origin.
12. *Superficial* (soo'-per-FISH-al) *(external)*—Toward or on the surface of the body.
13. *Deep* (DĒP) *(internal)*—Away from the surface of the body.

Using a torso and an articulated skeleton, and consulting with your instructor as necessary, describe the location of the following structures by inserting the proper directional term.

1. The ulna is on the _____ side of the forearm.
2. The ascending colon is _____ to the urinary bladder.
3. The heart is _____ to the liver.
4. The muscles of the arm are _____ to the skin of the arm.
5. The sternum is _____ to the heart.
6. The humerus is _____ to the radius.
7. The stomach is _____ to the lungs.
8. The muscles of the thoracic wall are _____ to the viscera in the thoracic cavity.
9. The esophagus is _____ to the trachea.
10. The phalanges are _____ to the carpals.
11. The ring finger is _____ between the little (medial) and middle (lateral) fingers.
12. The ascending colon of the large intestine and the gallbladder are _____.
13. The ascending and descending colons of the large intestine are _____.

G. PLANES OF THE BODY

The structural plan of the human body can be described with respect to *planes* (imaginary flat surfaces) passing through it. Planes are fre-

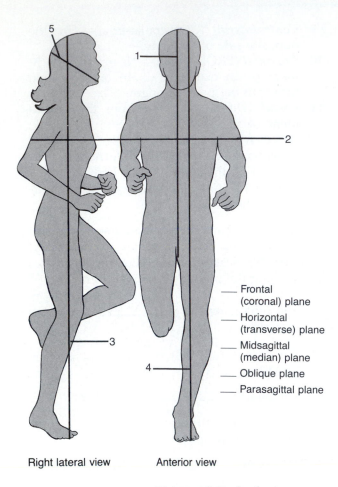

— Frontal
(coronal) plane

— Horizontal
(transverse) plane

— Midsagittal
(median) plane

— Oblique plane

— Parasagittal plane

Right lateral view   Anterior view

FIGURE 2.2   Planes of the body.

quently used to show the anatomical relationship of several structures in a region to one another. Commonly used planes follow:

1. *Midsagittal* (mid-SAJ-it-tal) or *median*—Vertical plane that passes through the midline of the body and divides the body or an organ into *equal* right and left sides.
2. *Parasagittal* (par-a-SAJ-it-tal)—*Vertical plane* that does not pass through the midline of the body; it divides the body or an organ into unequal right and left sides.
3. *Frontal (coronal;* kō-RŌ-nal)—Vertical plane at a right angle to a midsagittal (or parasagittal) plane that divides the body or an organ into anterior (front) and posterior (back) portions.
4. *Horizontal (transverse)*—Plane that is parallel to the ground (at a right angle to midsagittal, parasagittal, and frontal planes) and divides the body or an organ into superior (top) and inferior (bottom) portions.
5. *Oblique* (ō-BLĒK)—Plane that passes through the body or an organ at an angle between the

horizontal plane and either the midsagittal, parasagittal, or frontal plane.

Refer to Figure 2.2 and label the planes shown.

H. BODY CAVITIES

Spaces within the body that contain various internal organs are called *body cavities.* One way of organizing the principal body cavities follows:

> *Dorsal body cavity*
>   *Cranial cavity*
>   *Vertebral (spinal) cavity*
> *Ventral body cavity*
>   *Thoracic cavity*
>   Right pleural
>   Left pleural
>   Pericardial
>   *Abdominopelvic cavity*
>   Abdominal
>   Pelvic

The region between the pleurae of the lungs and extending from the sternum (breast bone) to the backbone is called the *mediastinum.* It contains all structures in the thoracic cavity, except the lungs themselves. Included are the heart, thymus gland, esophagus, trachea, and many large blood and lymphatic vessels.

Label the body cavities shown in Figure 2.3. Then examine a torso or wall chart, or both, and determine which organs lie within each cavity.

I. ABDOMINOPELVIC REGIONS

To describe the location of viscera more easily, the abdominopelvic cavity can be divided into *nine regions* by using four imaginary lines: (1) an upper horizontal *subcostal line* that passes just below the bottom of the rib cage through the pylorus (lower portion) of the stomach, (2) a lower horizontal line, the *transtubercular line,* that joins the iliac crests (top surfaces of the hipbones), (3) a *right midclavicular line* drawn through the midpoint of the right clavicle slightly medial to the right nipple, and (4) a *left midclavicular line* drawn through the midpoint of the left clavicle slightly medial to the left nipple.

The four imaginary lines divide the abdominopelvic cavity into the following nine regions: (1) *umbilical* (um-BIL-i-kul) *region;* which is centrally located; (2) *left lumbar (lumbus =* loin)

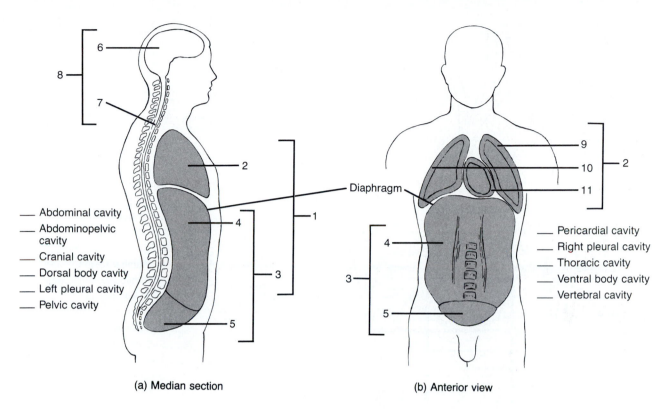

(a) Median section

— Abdominal cavity
— Abdominopelvic cavity
— Cranial cavity
— Dorsal body cavity
— Left pleural cavity
— Pelvic cavity

Diaphragm

(b) Anterior view

— Pericardial cavity
— Right pleural cavity
— Thoracic cavity
— Ventral body cavity
— Vertebral cavity

FIGURE 2.3   Body cavities.

region, to the left of the umbilical region; (3) *right lumbar,* to the right of the umbilical region; (4) *epigastric* (ep-i-GAS-trik; *epi* = above, *gaster* = stomach) *region,* directly above the umbilical region; (5) *left hypochondriac* (hī'-pō-KON-drē-ak; *hypo* = under, *chondro* = cartilage) *region,* to the left of the epigastric region; (6) *right hypochondriac region,* to the right of the epigastric region; (7) *hypogastric (pubic) region,* directly below the umbilical region; (8) *left iliac* (IL-ē-ak; *iliacus* = superior part of the hipbone) or *inguinal region,* to the left of the hypogastric (pubic) region; and (9) *right iliac (inguinal) region,* to the right of the hypogastric (pubic) region.

Label Figure 2.4 by indicating the names of the four imaginary lines and the nine abdominopelvic regions.

Examine a torso and determine which organs or parts of organs lie within each of the nine abdominopelvic regions.

In the space provided, list several organs or parts of organs found in the following abdominopelvic regions:

Right hypochrondriac. _____

_____

_____

Epigastric. _____

_____

_____

Left hypochondriac. _____

_____

_____

Right lumbar. _____

_____

_____

Umbilical. _____

_____

_____

Left lumbar. _____

_____

_____

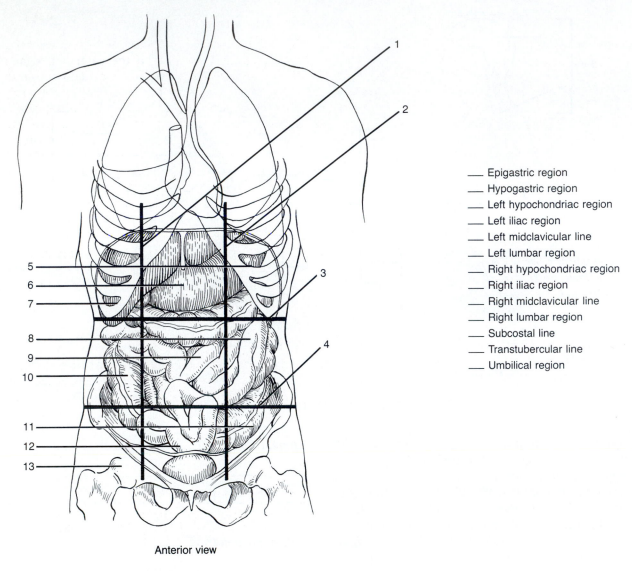

——— Epigastric region
——— Hypogastric region
——— Left hypochondriac region
——— Left iliac region
——— Left midclavicular line
——— Left lumbar region
——— Right hypochondriac region
——— Right iliac region
——— Right midclavicular line
——— Right lumbar region
——— Subcostal line
——— Transtubercular line
——— Umbilical region

Anterior view

FIGURE 2.4   Abdominopelvic regions.

Right iliac. _____

_____

_____

Hypogastric. _____

_____

_____

Left iliac. _____

_____

_____

J. ABDOMINOPELVIC QUADRANTS

Another way to divide the abdominopelvic cavity is into *quadrants* by passing one horizontal line and one vertical line through the umbilicus. The two lines thus divide the abdominopelvic cavity into a *right upper quadrant (RUQ), left upper quadrant (LUQ), right lower quadrant (RLQ),* and *left lower quadrant (LLQ).* Quadrant names are frequently used by health care professionals for locating the site of an abdominopelvic pain, tumor, or other abnormality.

Examine a torso or wall chart, or both, and determine which organs or parts of organs lie within each of the abdominopelvic quadrants.

In the space provided, list several organs or parts of organs found in the following abdominopelvic quadrants:

Right upper quadrant. _____

_____

_____

Left upper quadrant. _____

_____

_____

Right lower quadrant. _____

_____

_____

Left lower quadrant. _____

_____

_____

K. DISSECTION OF WHITE RAT

Now that you have some idea of the names of the various body systems and the principal organs that comprise each, you can actually observe some of these organs by dissecting a white rat. *Dissect* means "to separate." This dissection gives you an excellent opportunity to see the different sizes, shapes, locations, and relationships of organs and to compare their different textures and external features. In addition, this exercise will introduce you to the general procedure for dissection before you dissect in later exercises.

CAUTION! *Please reread Section D, "Precautions Related to Dissection," on Laboratory Safety at the beginning of this manual, page xv, before you begin your dissection.*

PROCEDURE

1. Place the rat on its backbone on a wax dissecting pan (tray). Using dissecting pins, anchor each of the four extremities to the wax (Figure 2.5a).

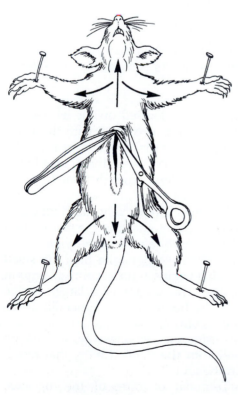

(a) Lines of incision in skin

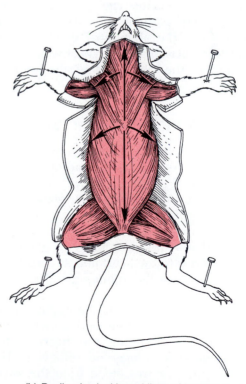

(b) Peeling back skin and lines of incision in muscles

FIGURE 2.5 Dissection procedure for exposing thoracic and abdominopelvic viscera of the white rat for examination.

2. To expose the contents of the thoracic, abdominal, and pelvic cavities, you will first have to make a midline incision. This is done by lifting the abdominal skin with a forceps to separate the skin from the underlying connective tissue and muscles. While lifting the abdominal skin, cut through it with scissors and make an incision that extends from the lower jaw to the anus (Figure 2.5a).

3. Now make four lateral incisions that extend from the midline incision into the four extremities (Figure 2.5a).

4. Peel the skin back and pin the flaps to the wax to expose the superficial muscles (Figure 2.5b).

5. Next, lift the abdominal muscles with a forceps and cut through the muscle layer, being careful not to damage any underlying organs. Keep the scissors parallel to the rat's backbone. Extend this incision from the anus to a point just below the bottom of the rib cage (Figure 2.5b). Make two lateral incisions just below the rib cage and fold back the muscle flaps to expose the abdominal and pelvic viscera (Figure 2.5b).

6. To expose the thoracic viscera, cut through the ribs on either side of the sternum. This incision should extend from the diaphragm to the neck (Figure 2.5b). The *diaphragm* is the thin muscular partition that separates the thoracic from the abdominal cavity. Again make lateral incisions in the chest wall so that you can lift the ribs to view the thoracic contents.

1. Examination of Thoracic Viscera

You will first examine the thoracic viscera (large internal organs). As you dissect and observe the various structures, palpate (feel with the hand) them so that you can compare their texture. Use Figure 2.6 as a guide.

a. *Thymus gland*—Irregular mass of glandular tissue superior to the heart and superficial to the trachea. Push the thymus gland aside or remove it.

b. *Heart*—Structure located in the midline, inferior to the thymus gland and between the lungs. The sac covering the heart is the *pericardium,* which may be removed. The large vein that returns blood from the lower regions of the body is the *inferior vena cava;* the large vein that returns blood to the heart from the upper regions of the body is the *superior vena cava.* The large artery that carries blood from the heart to most parts of the body is the *aorta.*

c. *Lungs*—Reddish, spongy structures on either side of the heart. Note that the lungs are divided into regions called *lobes.*

d. *Trachea*—Tubelike passageway superior to the heart and deep to the thymus gland. Note that the wall of the trachea consists of rings of cartilage. Identify the *larynx* (voice box) at the superior end of the trachea and the *thyroid gland,* a bilobed structure on either side of the larynx. The lobes of the thyroid gland are connected by a band of thyroid tissue, the isthmus.

e. *Bronchial tubes*—Trace the trachea inferiorly and note that it divides into bronchial tubes that enter the lungs and continue to divide within them.

f. *Esophagus*—Muscular tube posterior to the trachea that transports food from the throat into the stomach. Trace the esophagus inferiorly to see where it passes through the diaphragm to join the stomach.

2. Examination of Abdominopelvic Viscera

You will now examine the principal viscera of the abdomen and pelvis. As you do so, again refer to Figure 2.6.

a. *Stomach*—Organ located on the left side of the abdomen and in contact with the liver. The digestive organs are attached to the posterior abdominal wall by a membrane called the *mesentery.* Note the blood vessels in the mesentery.

b. *Small intestine*—Extensively coiled tube that extends from the stomach to the first portion of the large intestine called the *cecum.*

c. *Large intestine*—Wider tube than the small intestine that begins at the cecum and ends at the rectum. The cecum is a large, saclike structure. In humans, the appendix arises from the cecum.

d. *Rectum*—Muscular passageway, located on the midline in the pelvic cavity, that terminates in the anus.

e. *Anus*—Terminal opening of the digestive system to the exterior.

f. *Pancreas*—Pale gray, glandular organ posterior and inferior to the stomach.

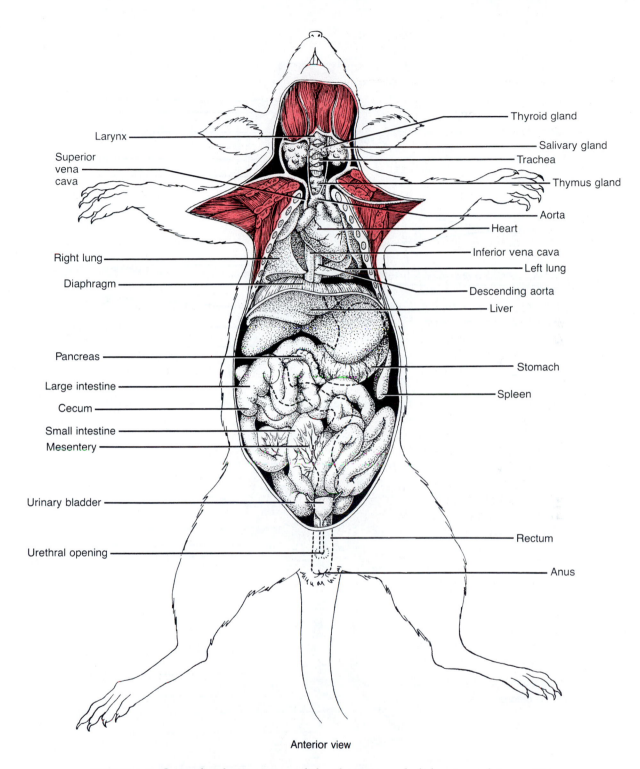

Anterior view

FIGURE 2.6 Superficial structures of the thoracic and abdominopelvic cavities of the white rat.

g. **Spleen**—Small, dark red organ lateral to the stomach.

h. **Liver**—Large, brownish-red organ directly inferior to the diaphragm. The rat does not have a gallbladder, a structure associated with the liver. To locate the remaining viscera, either move the superficial viscera aside or remove them. Use Figure 2.7 as a guide.

i. **Kidneys**—Bean-shaped organs embedded in fat and attached to the posterior abdominal

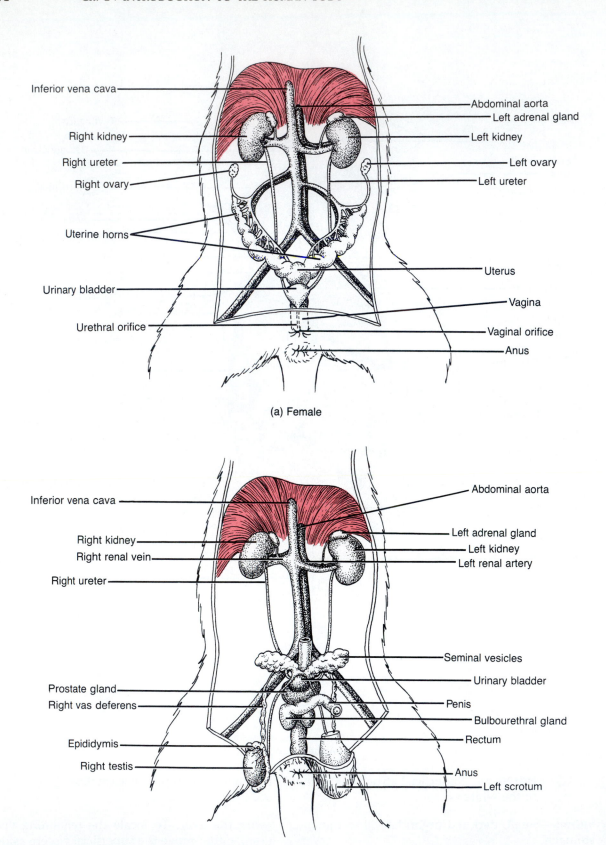

Inferior vena cava

Abdominal aorta

Left adrenal gland

Right kidney

Left kidney

Right ureter

Left ovary

Right ovary

Left ureter

Uterine horns

Uterus

Urinary bladder

Vagina

Urethral orifice

Vaginal orifice

Anus

(a) Female

Inferior vena cava

Abdominal aorta

Right kidney

Left adrenal gland

Right renal vein

Left kidney

Right ureter

Left renal artery

Seminal vesicles

Urinary bladder

Prostate gland

Penis

Right vas deferens

Bulbourethral gland

Epididymis

Rectum

Right testis

Anus

Left scrotum

(b) Male

FIGURE 2.7   Deep structures of the abdominopelvic cavity of the white rat.

wall on either side of the backbone. As will be explained later, the kidneys and a few other structures are behind the membrane that lines the abdomen *(peritoneum)*. Such structures are referred to as *retroperitoneal* and are not actually within the abdominal cavity. See if you can find the *abdominal aorta,* the large artery located along the midline behind the inferior vena cava. Also, locate the *renal arteries* branching off the abdominal aorta to enter the kidneys.

j. *Adrenal (suprarenal) glands*—Two glandular structures, one on top of each kidney.

k. *Ureters*—Tubes that extend from the medial surface of the kidney inferiorly to the urinary bladder.

l. *Urinary bladder*—Saclike structure in the pelvic cavity that stores urine.

m. *Urethra*—Tube that extends from the urinary bladder to the exterior. Its opening to the exterior is called the *urethral orifice.* In male rats, the urethra extends through the penis; in female rats, the tube is separate from the reproductive tract.

If your specimen is female (no visible scrotum anterior to the anus), identify the following:

n. *Ovaries*—Small, dark structures inferior to the kidneys.

o. *Uterus*—An organ located near the urinary bladder consisting of two sides (horns) that join separately into the vagina.

p. *Vagina*—Tube leading from the uterus to the external vaginal opening, the *vaginal orifice.* This orifice is in front of the anus and behind the urethral orifice.

If your specimen is male, identify the following:

q. *Scrotum*—Large sac anterior to the anus that contains the testes.

r. *Testes*—Egg-shaped glands in the scrotum. Make a slit into the scrotum and carefully remove one testis. See if you can find a coiled duct attached to the testis *(epididymis)* and a duct that leads from the epididymis into the abdominal cavity *(vas deferens).*

s. *Penis*—Organ of copulation medial to the testes.

When you have finished your dissection, store or dispose of your specimen per your instructor's directions. Wash your dissecting pan and dissecting instruments with laboratory detergent, dry them, and return them to their storage areas.

ANSWER THE LABORATORY REPORT QUESTIONS AT THE END OF THE EXERCISE.

Introduction to the Human Body

STUDENT _____   DATE _____

LABORATORY SECTION _____   SCORE/GRADE _____

PART 1. Multiple Choice

_____   1. The directional term that best describes the eyes in relation to the nose is (a) distal (b) superficial (c) anterior (d) lateral

_____   2. Which does *not* belong with the others? (a) right pleural cavity (b) pericardial cavity (c) vertebral cavity (d) left pleural cavity

_____   3. Which plane divides the brain into an anterior and a posterior portion? (a) frontal (b) median (c) sagittal (d) horizontal

_____   4. The urinary bladder lies in which region? (a) umbilical (b) hypogastric (c) epigastric (d) left iliac

_____   5. Which is *not* a characteristic of the anatomical position? (a) the subject is erect (b) the subject faces the observer (c) the palms face backward (d) the arms are at the sides

_____   6. The abdominopelvic region that is bordered by all four imaginary lines is the (a) hypogastric (b) epigastric (c) left hypochondriac (d) umbilical

_____   7. Which directional term best describes the position of the phalanges with respect to the carpals? (a) lateral (b) distal (c) anterior (d) proximal

_____   8. The pancreas is found in which body cavity? (a) abdominal (b) pericardial (c) pelvic (d) vertebral

_____   9. The anatomical term for the leg is (a) brachial (b) tarsal (c) crural (d) sural

_____   10. In which abdominopelvic region is the spleen located? (a) left lumbar (b) right lumbar (c) epigastric (d) left hypochondriac

_____   11. Which of the following represents the most complex level of structural organization? (a) organ (b) cellular (c) tissue (d) chemical

_____   12. Which body system is concerned with support, protection, leverage, blood-cell production, and mineral storage? (a) cardiovascular (b) lymphatic (c) skeletal (d) digestive

_____   13. The skin and structures derived from it—such as nails, hair, sweat glands, and oil glands—are components of which system? (a) respiratory (b) integumentary (c) muscular (d) lymphatic

_____   14. Hormone-producing glands belong to which body system? (a) cardiovascular (b) lymphatic (c) endocrine (d) digestive

_____   15. Which body system brings about movement, maintains posture, and produces heat? (a) skeletal (b) respiratory (c) reproductive (d) muscular

_____   16. Which abdominopelvic quadrant contains most of the liver? (a) RUQ (b) RLQ (c) LUQ (d) LLQ

_____ **17.** The physical and chemical breakdown of food for use by body cells and the elimination of solid wastes are accomplished by which body system? (a) respiratory (b) urinary (c) cardiovascular (d) digestive

PART 2. Completion

18. The tibia is _____ to the fibula.

19. The ovaries are found in the _____ body cavity.

20. The upper horizontal line that helps divide the abdominopelvic cavity into nine regions is

the _____ line.

21. The anatomical term for the hollow behind the knee is _____.

22. A plane that divides the stomach into a superior and an inferior portion is a(n)

_____ plane.

23. The wrist is described as _____ to the elbow.

24. The heart is located in the _____ cavity within the thoracic cavity.

25. The abdominopelvic region that contains the rectum is the _____ region.

26. A plane that divides the body into unequal left and right sides is the _____ plane.

27. The spinal cord is located within the _____ cavity.

28. The body system that removes carbon dioxide from body cells, delivers oxygen to body cells, helps maintain acid-base balance, helps protect against disease, helps regulate body temperature, and

prevents hemorrhage by forming clots is the _____ system.

29. The _____ quadrant contains the descending colon of the large intestine.

30. Which body system returns proteins and plasma to the cardiovascular system, transports fats from the digestive system to the cardiovascular system, filters blood, protects against disease, and

produces white blood cells? _____

PART 3. Matching

_____ **31.** Right hypochondriac region

_____ **32.** Hypogastric region

_____ **33.** Left iliac region

_____ **34.** Right lumbar region

_____ **35.** Epigastric region

_____ **36.** Left hypochondriac region

_____ **37.** Right iliac region

_____ **38.** Umbilical region

_____ **39.** Left lumbar region

A. Junction of descending and sigmoid colons of large intestine

B. Descending colon of large intestine

C. Spleen

D. Most of right lobe of liver

E. Appendix

F. Ascending colon of large intestine

G. Middle of transverse colon of large intestine

H. Adrenal (suprarenal) glands

I. Sigmoid colon of large intestine

PART 4. Matching

————— 40. Anterior

————— 41. Skull

————— 42. Transtubercular line

————— 43. Armpit

————— 44. Umbilical region

————— 45. Medial

————— 46. Cranial cavity

————— 47. Front of knee

————— 48. Breast

————— 49. Chest

————— 50. Buttock

————— 51. Superior

————— 52. Groin

————— 53. Vertebral cavity

————— 54. Cheek

————— 55. Front of neck

————— 56. Distal

————— 57. Pericardial cavity

————— 58. Forearm

————— 59. Plantar

————— 60. Mouth

A. Passes through iliac crests

B. Contains spinal cord

C. Nearer the midline

D. Thoracic

E. Cervical

E. Axillary

G. Contains the heart

H. Cranial

I. Antebrachial

J. Gluteal

K. Mammary

L. Sole of foot

M. Contains navel

N. Buccal

O. Farther from the attachment of an extremity

P. Patellar

Q. Toward the head

R. Nearer to or at the front of the body

S. Oral

T. Contains brain

U. Inguinal

3

Cells

A *cell* is the basic living structural and functional unit of the body. The study of cells is called *cytology* (sī-TOL-ō-jē; *cyto* = cell, *logos* = study of). The different kinds of cells (blood, nerve, bone, muscle, epithelial, and others) perform specific functions and differ from one another in shape, size, and structure. You will start your study of cytology by learning the important components of a theoretical, generalized cell.

A. CELL PARTS

Refer to Figure 3.1, a generalized cell based on electron micrograph studies. With the aid of your textbook and any other items made available by your instructor, label the parts of the cell indicated. In the spaces that follow, describe the function of the following cellular structures:

1. *Plasma (cell) membrane.* _____

   _____

   _____

   _____

2. *Cytoplasm* (SĪ-tō-plazm'). _____

   _____

   _____

   _____

3. *Nucleus* (NOO-klē-us). _____

   _____

   _____

   _____

4. *Endoplasmic reticulum* (en'-dō-PLAS-mik re-TIK-yoo-lum), or *ER.* _____

   _____

   _____

   _____

5. *Ribosome* (RĪ-bō-sōm). _____

   _____

   _____

   _____

6. *Golgi* (GOL-jē) *complex.* _____

   _____

   _____

   _____

7. *Mitochondria* (mī-'tō-KON-drē-a). _____

   _____

   _____

   _____

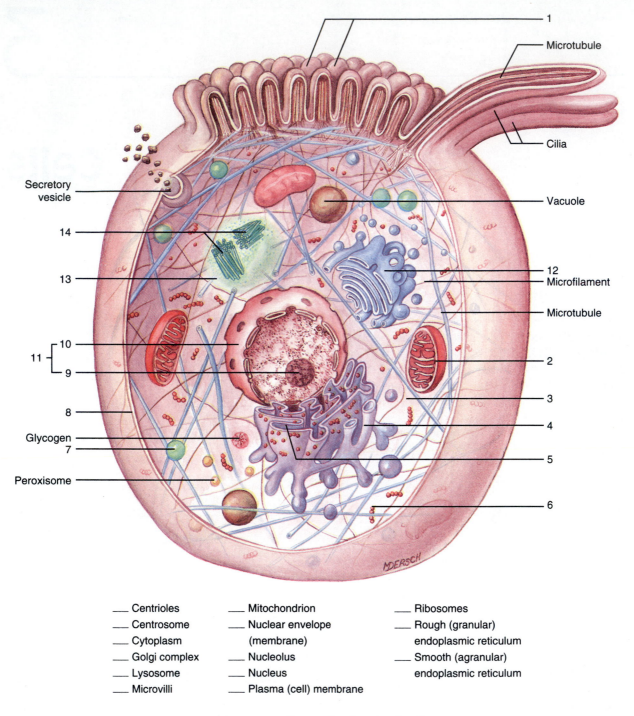

1

Microtubule

Cilia

Secretory vesicle

Vacuole

14

13

12
Microfilament

Microtubule

10

11

9

2

3

8

4

Glycogen

5

7

Peroxisome

6

MDERSCH

___ Centrioles	___ Mitochondrion	___ Ribosomes
___ Centrosome	___ Nuclear envelope	___ Rough (granular)
___ Cytoplasm	(membrane)	endoplasmic reticulum
___ Golgi complex	___ Nucleolus	___ Smooth (agranular)
___ Lysosome	___ Nucleus	endoplasmic reticulum
___ Microvilli	___ Plasma (cell) membrane	

FIGURE 3.1   Generalized animal cell.

8. *Lysosome* (LĪ-sō-sōm). _____

_____

_____

_____

9. *Peroxisome* (pe-ROKS-i-sōm). _____

_____

_____

_____

10. Cytoskeleton. _____

_____

_____

_____

11. Centrosome (SEN-trō-sōm'). _____

_____

_____

_____

12. Cilium (SIL-ē-um). _____

_____

_____

_____

13. Flagellum (fla-JEL-um). _____

_____

_____

_____

Ciliated columnar epithelial cell

Sperm cell

Nerve cell

Muscle cell

B. DIVERSITY OF CELLS

Now obtain prepared slides of the following types of cells and examine them under the magnifications suggested:

1. Ciliated columnar epithelial cells (high power)
2. Sperm cells (oil immersion)
3. Nerve cells (high power)
4. Muscle cells (high power)

After you have made your examination, draw an example of each of these kinds of cells in the spaces provided and under each cell indicate how each is adapted to its particular function.

C. MOVEMENT OF SUBSTANCES ACROSS AND THROUGH PLASMA MEMBRANES

Biological membranes serve as permeability barriers. They serve to maintain the internal integrity of the cell, as well as the maintenance of solute concentrations within the cell that are considerably different from those found in the extracellular environment. In general, most substances in the body have a high solubility in water (polar liquids) and low solubility in nonpolar liquids, such as alcohol. Because cellular membranes are composed of a combination of polar and nonpolar components, they present a formidable barrier to the movement of water-soluble substances into and out of a cell. In general, substances move across plasma membranes by two principal kinds of transport processes—passive transport, which does not require the expenditure of cellular energy, and active transport, which does require the expenditure of energy, usually via the chemical breakdown of adenosine triphosphate (ATP) into adenosine diphosphate (ADP), inorganic phosphate and energy. In *passive (physical) processes*, substances move because of differences in concentration (or pressure) from areas of higher concentration (or pressure) to areas of lower concentration (or pressure). The movement continues until an equilibrium of substances and concentration (or pressure) forces is accomplished. Passive processes are the result of the kinetic energy (energy of motion) of the substances themselves and the cell does not expend energy to move the substances. Examples of passive processes are diffusion, osmosis, filtration, and dialysis.

In *active (physiological) processes*, substances move against their concentration gradient, moving from areas of lower concentration to higher concentration. Moreover, cells must expend energy (ATP) to carry on active transport processes. Examples of active processes are active transport and endocytosis (phagocytosis and pinocytosis).

CAUTION! *Please reread Section A, "General Safety Precautions and Procedures," on page xiii, and Section C, "Precautions Related to Working with Reagents," on page xiv, at the beginning of the laboratory manual before you begin any of the following experiments. Read the experiments before you perform them, to be sure that you understand all the procedures and safety cautions.*

1. Passive Processes

a. BROWNIAN MOVEMENT

At temperatures above absolute zero ($-273°$ C or $-460°$ F), all molecules are in constant random motion because of their inherent kinetic energy. This phenomenon is called *Brownian movement.* Less energy is required to move small molecules or particles than large ones. Because all molecules are constantly bombarded by the molecules surrounding them, the smaller the particle, the greater its random motion in terms of speed and distance.

PROCEDURE

1. With a medicine dropper, place a drop of dilute detergent solution in a depression (concave) slide.
2. Using another medicine dropper, add a drop of dilute India ink to the first drop in the depression slide.

3. Stir the two solutions with a toothpick and cover the depression with a cover glass.
4. Let the slide stand for 10 min and then place it on your microscope stage and observe it under high power.
5. Using forceps, place the slide on a hot plate and warm it for 15 sec. Then remove the slide with forceps and observe it again under high power.
6. Record your observations in Section C.1.a of the LABORATORY REPORT RESULTS at the end of the exercise.

b. DIFFUSION

Diffusion is the net (greater) movement of molecules or ions from a region of higher concentration to a region of lesser concentration until they are evenly distributed (equilibrium). An example of diffusion in the human body is the movement of oxygen and carbon dioxide between body cells and blood.

The following two experiments illustrate diffusion. Either or both may be performed.

PROCEDURE

1. To demonstrate diffusion of a solid in a liquid, *using forceps,* carefully place a large crystal of potassium permanganate ($KMnO_4$) into a test tube filled with water.

CAUTION! *Avoid contact of $KMnO_4$ with your skin by using acid- or caustic-resistant gloves.*

2. Place the tube in a rack against a white background where it will not be disturbed.
3. Note the diffusion of the crystal material through the water at 10-min intervals for 1 hr.
4. Record the diffusion of the crystal in millimeters (mm) per minute at 15-min intervals in Section C.1.b of the LABORATORY REPORT RESULTS at the end of the exercise. Simply measure the distance of diffusion using a millimeter ruler.

PROCEDURE

1. To demonstrate diffusion of a solid in a solid, *using forceps,* carefully place a large crystal of methylene blue on the surface of agar in the center of a petri plate. If crystals of methylene blue are not available, use drops of methylene blue instead.
2. Note the diffusion of the crystal through the agar at 10-min intervals for 1 hr.
3. Record the diffusion of the crystal in millimeters (mm) per minute at 10-min intervals,

using a millimeter ruler, in Section C.1.b of the LABORATORY REPORT RESULTS at the end of the exercise.

c. OSMOSIS

Osmosis is the net movement of *water* through a selectively permeable membrane from a region of higher water (lower solute) concentration to a region of lower water (higher solute) concentration. In the body, fluids move between cells as a result of osmosis.

PROCEDURE

1. Refer to the osmosis apparatus in Figure 3.2.
2. Tie a knot very tightly at one end of a 4-in piece of cellophane dialysis tubing that has been soaking in water for a few minutes. Fill the dialysis tubing with a 10% sugar (sucrose) solution that has been colored with Congo red (red food coloring can also be used).
3. Close the open end of the dialysis tubing with a one-hole rubber stopper into which a glass tube has already been inserted by a laboratory assistant or your instructor.

CAUTION! *If you are unfamiliar with the procedure, do not attempt to insert the glass tube into the stopper yourself because it may break and result in serious injury.*

4. Tie a piece of string tightly around the dialysis tubing to secure it to the stopper.

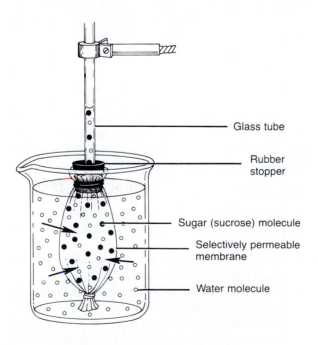

Glass tube

Rubber stopper

Sugar (sucrose) molecule

Selectively permeable membrane

Water molecule

FIGURE 3.2   Osmosis apparatus.

5. Secure and suspend the glass tube and dialysis tubing by means of a clamp attached to a ring stand.
6. Insert the dialysis tubing into a beaker or flask of water until the water comes up to the bottom of the rubber stopper.
7. As soon as the sugar solution becomes visible in the glass tube, mark it with a wax pencil and note the time.
8. Mark the height of liquid in the glass tube after 10-, 20-, and 30-min intervals by using your millimeter (mm) ruler and record your results in Section C.1.c of the LABORATORY REPORT RESULTS at the end of the exercise.

d. HEMOLYSIS AND CRENATION

Osmosis can also be understood by noting the effects of different water concentrations on red blood cells. Red blood cells maintain their normal shape when placed in an *isotonic solution* (that is, one having the same salt concentration; 0.90% solution of NaCl [sodium chloride] is isotonic to red blood cells). If, however, red blood cells are placed in a *hypotonic solution* (concentration of NaCl lower than 0.90%), a net movement of water into the cells occurs, causing the cells to swell and possibly burst. The rupture of blood cells in this manner with the loss of hemoglobin into the surrounding liquid is called *hemolysis* (hē-MOL-i-sis). If, instead, red blood cells are placed in a *hypertonic solution* (concentration of NaCl higher than 0.90%), a net movement of water out of the cells occurs, causing them to shrink. This shrinkage is known as *crenation* (kre-NĀ-shun).

PROCEDURE

CAUTION! *Please reread Section B, "Precautions Related to Working with Blood, Blood Products, or Other Body Fluids," on page xiv, at the beginning of the laboratory manual, before you begin any of the following experiments. Read the experiments before you perform them, to be sure that you understand all the procedures and safety precautions. When working with whole blood, take care to avoid any kind of contact with an open sore, cut, or wound. Wear tight-fitting surgical gloves and safety goggles.*

When you finish this part of the exercise, place the toothpicks, medicine droppers, microscope slides, and cover slips in a fresh bleach solution.

1. With a wax marking pencil, mark three microscope slides as follows: 0.90%, DW (distilled water), and 3%.
2. Using a medicine dropper, place three drops

of fresh (uncoagulated) ox blood on a microscope slide that contains 2 milliliters (ml) of a 0.90% NaCl solution (isotonic solution). Mix gently and thoroughly with a clean toothpick.

3. Using a medicine dropper, place three drops of fresh ox blood on a microscope slide that contains 2 ml of distilled water (hypotonic solution). Mix gently and thoroughly with a clean toothpick.

4. Now, using a medicine dropper, add three drops of fresh ox blood to a microscope slide that contains 2 ml of a 3% NaCl solution (hypertonic solution). Mix gently and thoroughly with a clean toothpick.

5. Using a medicine dropper, place two drops of the red blood cells in the isotonic solution on another microscope slide, cover with a cover slip, and examine the red blood cells under high power. Reduce your illumination.

What is the shape of the cells? _____

_____

Explain their shape. _____

_____

6. Using a medicine dropper, place two drops of the red blood cells in the hypotonic solution on another microscope slide, cover with a cover slip, and examine the red blood cells under high power. Reduce your illumination.

What is the shape of the cells? _____

_____

Explain their shape. _____

_____

7. Using a medicine dropper, place two drops of the red blood cells in the hypertonic solution on another microscope slide, cover with a cover slip, and examine the red blood cells under high power. Reduce your illumination.

What is the shape of the cells? _____

_____

Explain their shape. _____

_____

ALTERNATE PROCEDURE

1. Obtain three pieces of raw potato that have an *identical* weight.
2. Immerse one piece in a beaker that contains an isotonic solution; immerse a second piece in a

hypotonic solution; immerse the third piece in a hypotonic solution.

3. Continue the experiment for 1 hr. Record the time.
4. At the end of 1 hr, remove the pieces of potato and weigh them separately.

Weight of potato in isotonic solution _____

_____.

Weight of potato in hyptonic solution _____

_____.

Weight of potato in hypertonic solution _____

_____.

Explain the differences in the weights of the

three different pieces of potato. _____

_____

_____

_____

_____

_____

_____

e. FILTRATION

Filtration is the movement, under the influence of gravity, of solvents and dissolved substances across a selectively permeable membrane from regions of higher pressure to regions of lower pressure. The pressure is called **hydrostatic pressure** and is the result of the solvent, usually water. In general, any substance having a molecular weight of less than 100 is filtered because the pores in the filter paper (membrane) are larger than the molecules of the substance. Filtration is one mechanism by which the kidneys regulate the chemical composition of the blood.

PROCEDURE

1. Refer to Figure 3.3, the filtration apparatus.
2. Fold a piece of filter paper in half and then in half again.
3. Open it into a cone, place it in a funnel, and place the funnel over the beaker.
4. Shake a mixture of a few particles of powdered wood charcoal (black), 1% copper sulfate (blue), boiled starch (white), and water and pour it into the funnel until the mixture almost reaches the top of the filter paper. Gravity will

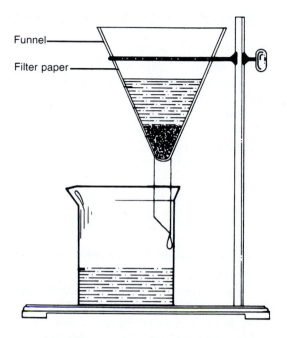

FIGURE 3.3   Filtration apparatus.

pull the particles through the pores of the membrane (that is, the filter paper).

5. Count the number of drops passing through the funnel for the following time intervals (10, 30, 60, 90, 120 sec) and record your observations in Section C.1.e of the LABORATORY REPORT RESULTS at the end of the exercise.

6. Observe which substances passed through the filter paper by noting their color in the filtered fluid in the beaker.

7. Examine the filter paper to determine whether any colored particles were not filtered.

8. To determine if any starch is in the filtrate (liquid in the beaker), add several drops of 0.01 M IKI solution. A blue-black color reaction indicates the presence of starch.

f. DIALYSIS

Dialysis (dī-AL-i-sis) is the separation of small molecules from large ones by using a selectively permeable membrane that permits diffusion of the small molecules but not the large ones. The principle of dialysis is employed in artificial kidneys.

PROCEDURE

1. Refer to the dialysis apparatus in Figure 3.4.

2. Tie off one end of a piece of dialysis tubing that has been soaking in water. Place a prepared solution containing starch, sodium chloride, 5% glucose, and albumin into the dialysis tubing.

3. Tie off the other end of the dialysis tubing and immerse it in a beaker of distilled water.

4. After 1 hr, test the solution in the beaker for the presence of each of the substances in the tubing, as follows, and record your observations in Section C.1.f of the LABORATORY REPORT RESULTS at the end of the exercise.

CAUTION! *Be extremely careful using nitric acid. It can severely damage your skin. Use acid- or caustic-resistant gloves.*

a. *Albumin*—*Carefully* add several drops of concentrated nitric acid to a test tube containing 2 ml of the solution in the beaker. Positive reaction = white coagulate.

b. *Sugar*—Test 5 ml of the solution in the beaker in a test tube with 5 ml of Benedict's solution. Place the test tube in a boiling water bath for three minutes.

CAUTION! *Make sure that the mouth of the test tube is pointed away from you and everyone else in the area.*

Using a test tube holder, remove the test tube from the water bath. Note the color. Positive reaction = green, yellow, orange, or red precipitate.

c. *Starch*—Add several drops of IKI solution to 2 ml of the solution in the beaker in a test tube. Note the color. Positive reaction = blue-black color.

d. *Sodium chloride*—Place 2 ml of the solution in the beaker in a test tube and add several drops of 1% silver nitrate. Note the color. Positive reaction = white precipitate.

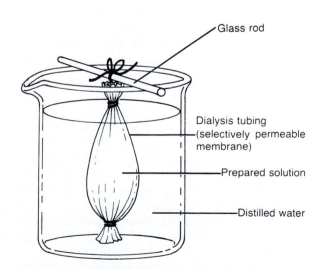

FIGURE 3.4   Dialysis apparatus.

2. Active Processes

a. ACTIVE TRANSPORT

Active transport is a process by which substances are transported across a plasma membrane via a carrier (usually termed a pump) within the cellular membrane. This pump moves the substance to be transported from an area of lower concentration, through the cellular membrane, to one of higher concentration. An example is the movement of sodium ions (Na^+) and potassium ions (K^+) across the proximal or distal convoluted tubules during the reabsorption processes involved in the formation of urine. Because reliable results are difficult to demonstrate simply, you will not be asked to demonstrate active transport.

b. ENDOCYTOSIS

Endocytosis refers to the passage of large molecules and particles across a plasma membrane, in which a segment of the membrane surrounds the substance, encloses it, and brings it into the cell. Here we will consider two types of endocytosis—phagocytosis and pinocytosis.

(1) Phagocytosis

Phagocytosis (fag'-ō-sī-TŌ-sis), or cell "eating," is the engulfment of solid particles or organisms by *pseudopods* (temporary fingerlike projections of cytoplasm) of the cell. Once the particle is surrounded by the membrane, the membrane folds inward, pinches off from the rest of the plasma membrane, and forms a vacuole around the particle. Enzymes are secreted into the vacuole or the vacuole combines with a lysosome and the particle is digested.

Phagocytosis can be demonstrated by observing the feeding of an amoeba, a unicellular organism whose movement and ingestion are similar to those of human white blood cells (leukocytes).

PROCEDURE

1. Using a medicine dropper, place a drop of culture containing amoebas that have been starved for 48 hr into the well of a depression slide and cover the well with a cover slip. Cultures containing *Chaos chaos* or *Amoeba proteus* should be used. (Your instructor may wish to use the hanging-drop method instead. If so, she or he will give you verbal instructions.)
2. Examine the amoebas under low power, and be sure that your light is reduced considerably.
3. Observe the locomotion of an amoeba for several minutes. Pay particular attention to the pseudopods that appear to flow out of the cell.
4. To observe phagocytosis, use a medicine dropper and add a drop containing small unicellular animals called *Tetrahymena pyriformis* to the culture containing the amoebas.
5. Examine under low power, and observe the ingestion of *Tetrahymena pyriformis* by an amoeba. Note the action of the pseudopods and the formation of the vesicle around the ingested organism.

(2) Pinocytosis

Pinocytosis (pi-nō-sī-TŌ-sis), or cell "drinking," is the engulfment of a liquid. The liquid is attracted to the surface of the membrane, and the membrane folds inward, surrounds the liquid, and detaches from the rest of the intact membrane.

D. CELL INCLUSIONS

Cell inclusions are a large and diverse group of mostly organic substances produced by cells that may appear or disappear at various times in the life of a cell. Using your textbook as a reference, indicate the function of the following cell inclusions:

1. *Melanin.* _____

   _____

   _____

   _____

2. *Glycogen.* _____

   _____

   _____

   _____

3. *Lipids.* _____

   _____

   _____

   _____

E. EXTRACELLULAR MATERIALS

Substances that lie outside the plasma membranes of body cells are referred to as *extracellular materials.* They include body fluids, such as interstitial fluid and plasma, which provide a medium for dissolving, mixing, and transporting substances. Extracellular materials also include special substances in which some cells are embedded.

Some extracellular materials are produced by certain cells and deposited outside their plasma membranes where they support cells, bind them together, and provide strength and elasticity. They have no definite shape and are referred to as *amorphous.* These include hyaluronic (hī-a-loo-RON-ik) acid and chondroitin (kon-DROY-tin) sulfate. Others are *fibrous* (threadlike). Examples include collagen, reticular, and elastic fibers.

Using your textbook as a reference, indicate the location and function for each of the following extracellular materials:

Hyaluronic acid

Location. _____

_____

_____

_____

Function. _____

_____

_____

_____

Chondroitin sulfate

Location. _____

_____

_____

_____

Function. _____

_____

_____

_____

Collagen fibers

Location. _____

_____

_____

_____

Function. _____

_____

_____

_____

Reticular fibers

Location. _____

_____

_____

_____

Function. _____

_____

_____

_____

Elastic fibers

Location. _____

_____

_____

_____

Function. _____

_____

_____

_____

F. CELL DIVISION

Cell division is the basic mechanism by which cells reproduce themselves. It consists of a nuclear division and a cytoplasmic division. Because nuclear division can be of two types, two kinds of

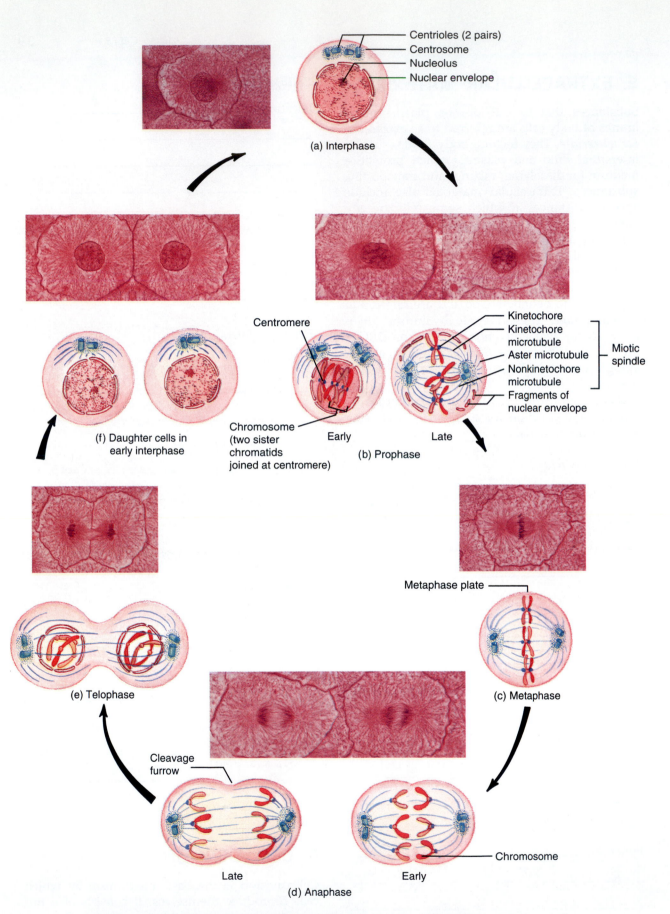

FIGURE 3.5 Cell division: mitosis and cytokinesis. Photomicrographs and diagrams of the various stages of cell division in whitefish eggs.

40

cell division are recognized. In the first type, called *somatic cell division,* a single starting cell called a *parent cell* duplicates itself and the result is two identical cells called *daughter cells.* Somatic cell division consists of a nuclear division called *mitosis* and a cytoplasmic division called *cytokinesis* (sī-tō-ki-NĒ-sis). It provides the body with a means of growth and of replacement of diseased or damaged cells (Figure 3.5). The second type of cell division is called *reproductive cell division* and is the mechanism by which sperm and eggs are produced (Exercise 25). Reproductive cell division consists of a nuclear division called *meiosis* and two cytoplasmic divisions (cytokinesis) and results in the development of four nonidentical daughter cells.

In order to study somatic cell division, obtain a prepared slide of a whitefish blastula and examine it under high power.

A cell between divisions is said to be in *interphase* of the cell cycle. Interphase is the longest part of the cell cycle and is the period of time during which a cell carries on its physiological activities. One of the most important activities of interphase is the replication of DNA so that the two daughter cells that eventually form will each have the same kind and amount of DNA as the parent cell. In addition, the proteins needed to produce structures required for doubling all cellular components, such as centrioles, are manufactured. Scan your slide and find a cell in interphase. Such a parent cell is characterized by a clearly defined nuclear envelope. Within the nucleus, look for the nucleolus (or nucleoli) and DNA (which is associated with protein in the form of a granular substance called *chromatin*). Also locate the centrioles.

Draw a labeled diagram of an interphase cell in the space provided.

Once a cell completes its interphase activities, mitosis begins. Mitosis is the distribution of two sets of chromosomes into two separate and equal nuclei after replication of the chromosomes of the parent cell, an event that takes place in the interphase preceding mitosis. Although a continuous process, mitosis is divided into four stages for purposes of study: prophase, metaphase, anaphase, and telophase.

1. *Prophase*—The first stage of mitosis is called *prophase* (*pro* = before). During early prophase, the chromatin condenses and shortens into chromosomes. Because DNA replication took place during interphase, each prophase chromosome contains a pair of identical double-stranded DNA molecules called *chromatids.* Each chromatid pair is held together by a small spherical body called a *centromere* that is required for the proper segregation of chromosomes. Attached to the outside of each centromere is a protein complex known as the *kinetochore* (ki-NET-ō-kor), whose function will be described shortly. Later in prophase, the nucleolus (or nucleoli) disperses, and the nuclear envelope breaks up. In addition, a centrosome with its pair of centrioles each moves to opposite poles (ends) of the cell. As they do so, the centrosomes start to form the *mitotic spindle,* a football-shaped assembly of microtubules that are responsible for the movement of chromosomes.

The lengthening of microtubules between centrosomes pushes the centrosomes to the poles of the cell so that the spindle extends from pole to pole. As the mitotic spindle continues to develop, three types of microtubules form: (a) *nonkinetochore microtubules,* which grow from centrosomes, extend inward, but do not bind to kinetochores; (b) *kinetochore microtubules,* which grow from centrosomes, extend inward, and attach to kinetochores; and (c) *aster microtubules,* which grow out of centrosomes but radiate outward from the mitotic spindles. Overall, the spindle is an attachment site for chromosomes. It also distributes chromosomes to opposite poles of the cell. Draw and label a cell in prophase in the space provided.

2. *Metaphase*—During **metaphase** (*meta*=after), the second stage of mitosis, the kinetochore microtubules line up the centromeres of the chromatid pairs at the exact center of the mitotic spindle. This midpoint region is called the **metaphase plate.** Draw and label a cell in metaphase in the space provided.

3. *Anaphase*—The third stage of mitosis, **anaphase** (*ana*=upward), is characterized by the splitting and separation of the centromeres (and kinetochores) and the movement of the two sister chromatids of each pair toward opposite poles of the cell. Once separated, the sister chromatids are referred to as **chromosomes.** The movement of chromosomes is the result of the shortening of kinetochore microtubules and elongation of the nonkinetochore microtubules, processes that increase the distance between separated chromosomes. As the chromosomes move during anaphase, they appear V-shaped. Draw and label a cell in anaphase in the space provided.

4. *Telophase*—The final stage of mitosis, **telophase** (*telo*=far or end), begins as soon as chromosomal movement stops. Telophase is essentially the opposite of prophase. During telophase, the identical sets of chromosomes at opposite poles of the cell uncoil and revert to their threadlike chromatin form; kinetochore microtubules disappear; nonkinetochore microtubules elongate even more; a new nuclear envelope reforms around each chro-

matin mass; new nucleoli reappear in the daughter nuclei; and eventually the mitotic spindle breaks up. Draw and label a cell in telophase in the space provided.

Cytokinesis begins during late anaphase and terminates during telophase with formation of a **cleavage furrow,** a slight indentation of the plasma membrane that extends around the center of the cell. The furrow gradually deepens until opposite surfaces of the cell make contact and the cell is split in two. The result is two separated daughter cells, each with separate portions of cytoplasm and organelles and its own set of identical chromosomes.

Following cytokinesis, each daughter cell returns to interphase. Each cell in most tissues of the body eventually grows and undergoes mitosis and cytokinesis, and a new divisional cycle begins. Examine your telophase cell again and be sure that it contains a cleavage furrow.

Using high power and starting at 12 o'clock, move around the blastula and count the number of cells in interphase and in each mitotic phase. It will be easier to do this if you imagine lines dividing the blastula into quadrants. Count the interphase cells in each quadrant, then assign each dividing cell to a specific mitotic stage. It will be hard to assign some cells to a phase—e.g., to distinguish late anaphase from early telophase. If you cannot make a decision, assign one cell to the earlier phase in question and the next cell to the later phase.

Divide the number of cells in each stage by the total number of cells counted and multiply by 100 to determine the percent of the cells in each mitotic stage at a given point in time. Record your results in Section F of the LABORATORY REPORT RESULTS at the end of the exercise.

ANSWER THE LABORATORY REPORT QUESTIONS AT THE END OF THE EXERCISE.

Cells

STUDENT _____   DATE _____

LABORATORY SECTION _____   SCORE/GRADE _____

SECTION C. MOVEMENT OF SUBSTANCES ACROSS PLASMA MEMBRANES

1. Passive Processes

a. BROWNIAN MOVEMENT

Describe the movement of the India ink particles on the unheated slide. _____

_____

How does this movement differ from that on the heated slide? _____

b. DIFFUSION

Solid in Liquid		Solid in Solid	
Time (min)	Distance (mm)	Time (min)	Distance (mm)
10	_____	10	_____
20	_____	20	_____
30	_____	30	_____
40	_____	40	_____
50	_____	50	_____
60	_____	60	_____

c. OSMOSIS

Time (min)	Height of liquid (mm)
10	_____
20	_____
30	_____

Explain what happened. _____

_____

e. FILTRATION

10 sec _____

30 sec _____

60 sec _____

90 sec _____

120 sec _____

f. DIALYSIS

Place a check in the appropriate place to indicate if the following tests are positive (+) or negative (−):

	+	(−)
Albumin	_____	_____
Sugar	_____	_____
Starch	_____	_____
Sodium chloride	_____	_____

SECTION F. CELL DIVISION

Percent of cells in interphase	_____
Percent of cells in prophase	_____
Percent of cells in metaphase	_____
Percent of cells in anaphase	_____
Percent of cells in telophase	_____

Cells

STUDENT _____   DATE _____

LABORATORY SECTION _____   SCORE/GRADE _____

PART 1. Multiple Choice

_____ 1. The portion of the cell that forms part of the mitotic spindle during division is the (a) endoplasmic reticulum (b) Golgi complex (c) cytoplasm (d) centrosome

_____ 2. Movement of molecules or ions from a region of high concentration to a region of low concentration via a process that does not require energy is called (a) phagocytosis (b) diffusion (c) active transport (d) pinocytosis

_____ 3. If red blood cells are placed in a hypertonic solution of sodium chloride, they will (a) swell (b) burst (c) shrink (d) remain the same

_____ 4. The reagent used to test for the presence of sugar is (a) silver nitrate (b) nitric acid (c) Ringer's solution (d) Benedict's solution

_____ 5. A cell that carries on a great deal of digestion also contains a large number of (a) lysosomes (b) centrioles (c) mitochondria (d) nuclei

_____ 6. Which process does *not* belong with the others? (a) active transport (b) dialysis (c) phagocytosis (d) pinocytosis

_____ 7. Movement of oxygen and carbon dioxide between blood and body cells is an example of (a) osmosis (b) active transport (c) diffusion (d) facilitated diffusion

_____ 8. Which type of solution will cause hemolysis? (a) isotonic (b) hypotonic (c) isometric (d) hypertonic

_____ 9. In addition to active transport and diffusion, the kidneys regulate the chemical composition of blood by utilizing the process of (a) phagocytosis (b) osmosis (c) filtration (d) pinocytosis

_____ 10. Engulfment of solid particles or organisms by pseudopods is called (a) active transport (b) dialysis (c) phagocytosis (d) filtration

_____ 11. Rupture of red blood cells with subsequent loss of hemoglobin into the surrounding medium is called (a) hemolysis (b) plasmolysis (c) plasmoptysis (d) hemoglobinuria

_____ 12. The area of the cell between the plasma membrane and nuclear envelope where chemical reactions occur is the (a) centrosome (b) vacuole (c) peroxisome (d) cytoplasm

_____ 13. The "powerhouses" of the cell where ATP is produced are the (a) ribosomes (b) mitochondria (c) centrioles (d) lysosomes

_____ 14. The sites of protein synthesis in the cell are (a) peroxisomes (b) flagella (c) ribosomes (d) centrosomes

_____ 15. Which process does *not* belong with the others? (a) diffusion (b) phagocytosis (c) active transport (d) pinocytosis

_____ 16. A cell inclusion that is a pigment in skin and hair is (a) glycogen (b) melanin (c) mucus (d) collagen

_____ 17. Which extracellular material is found in ligaments and tendons? (a) elastic fibers (b) chondroitin sulfate (c) collagen fibers (d) mucus

_____ 18. The organelles that contain enzymes for the metabolism of hydrogen peroxide are (a) lysosomes (b) mitochondria (c) Golgi complexes (d) peroxisomes

_____ 19. The framework of cilia, flagella, centrioles, and spindle fibers is formed by (a) endoplasmic reticulum (b) collagen fibers (c) chondroitin sulfate (d) microtubules

_____ 20. A viscous fluidlike substance that binds cells together, lubricates joints, and maintains the shape of the eyeballs is (a) elastin (b) hyaluronic acid (c) mucus (d) plasmin

PART 2. Completion

21. The external boundary of the cell through which substances enter and exit is called the

_____.

22. The cytoskeleton is formed by microtubules, intermediate filaments, and _____.

23. The portion of the cell that contains hereditary information is the _____.

24. The tail of a sperm cell is a long whiplash structure called a(n) _____.

25. Cells placed in a _____ solution will undergo hemolysis.

26. Division of cytoplasm is referred to as _____.

27. Lipid and protein secretion, carbohydrate synthesis, and assembly of glycoproteins are functions of

the _____.

28. Storage of digestive enzymes is accomplished by the _____ of a cell.

29. Projections of cells that move substances along their surfaces are called _____.

30. The _____ provides a surface area for chemical reactions, a pathway for transporting molecules, and a storage area for synthesized molecules.

31. _____ is a cell inclusion that represents stored glucose in the liver and skeletal muscles.

32. A jellylike substance that supports cartilage, bone, heart valves, and the umbilical cord

is _____.

33. The framework of many soft organs is formed by _____ fibers.

34. The net movement of water through a selectively permeable membrane from a region of higher concentration of water to a region of lower concentration of water is known as

_____.

35. The principle of _____ is employed in the operation of an artificial kidney.

36. In an interphase cell, DNA is in the form of a granular substance called _____.

37. Distribution of chromosomes into separate and equal nuclei is referred to as

_____.

38. The constant random motion of molecules due to their inherent kinetic energy is called

_____.

PART **3. Matching**

————— **39.** Anaphase

————— **40.** Metaphase

————— **41.** Interphase

————— **42.** Telophase

————— **43.** Prophase

A. Mitotic spindle appears

B. Movement of chromosome sets to opposite poles of cell

C. Centromeres line up on metaphase plate

D. Formation of two identical nuclei

E. Phase between divisions

Tissues

A *tissue* is a group of similar cells that usually have the same embryological origin and operate together to perform a specific function. The study of tissues is called *histology* (hiss-TOL-ō-jē; *histio* = tissue, *logos* = study of). The various body tissues can be categorized into four principal kinds: (1) epithelial, (2) connective, (3) muscular, and (4) nervous. In this exercise you will examine the structure and functions of epithelial and connective tissues, except for bone and blood. Other tissues will be studied later as parts of the systems to which they belong.

A. EPITHELIAL TISSUE

Epithelial (ep'-i-THĒ-lē-al) *tissue* or *epithelium* can be divided into two types: (1) covering and lining and (2) glandular. Covering and lining epithelium forms the outer layer of the skin and some internal organs; forms the inner lining of blood vessels, ducts, body cavities, and many internal organs; and helps make up special sense organs for smell, hearing, vision, and touch. Glandular epithelium constitutes the secreting portion of glands.

1. Characteristics

The general characteristics of epithelial tissue follow:

a. Epithelium consists largely or entirely of closely packed cells with little extracellular material between adjacent cells.

b. Epithelial cells are arranged in continuous sheets, in either single or multiple layers.

c. Epithelial cells have an *apical* (free) *surface* that is exposed to a body cavity, to the lining of an internal organ, or to the exterior of the body and a *basal surface* that is attached to the basement membrane (described subsequently).

d. Cell junctions are plentiful, providing secure attachments among the cells.

e. Epithelia are *avascular* (*a* = without, *vascular* = blood vessels). The vessels that supply nutrients and remove wastes are located in the adjacent connective tissue. The exchange of materials between epithelium and connective tissue is by diffusion.

f. Epithelia adhere firmly to nearby connective tissue, which holds the epithelium in position and prevents it from being torn. The attachment between the epithelium and the connective tissue is a thin extracellular layer called the *basement membrane.* It consists of two layers. The *basal lamina* contains a special type of collagen, laminin, and proteoglycans secreted by the epithelium. Cells in the connective tissue secrete the second layer, the *reticular lamina,* which contains reticular fibers, fibronectin, and glycoproteins. The basement membrane provides physical support for epithelium, provides for cell attachment, serves as a filter in the kidneys, and guides cell migration during development and tissue repair.

g. Epithelial tissue is the only tissue that makes direct contact with the environment.

h. Because epithelium is subject to a certain amount of wear and tear and injury, it has a high capacity for renewal (high mitotic rate).

i. Epithelia are diverse in origin. They are derived from all three primary germ layers (ectoderm, mesoderm, and endoderm).

j. Functions of epithelia include protection, filtration, lubrication, secretion, digestion, absorption, transportation, excretion, sensory reception, and reproduction.

2. Covering and Lining Epithelium

Before you start your microscopic examination of epithelial tissues, refer to Figure 4.1. Study the tissues carefully to familiarize yourself with their general structural characteristics. For each of the types of epithelium listed, obtain a prepared slide and, unless otherwise specified by your instructor, examine each under high power. In conjunction with your examination, consult a textbook of anatomy.

a. **Simple squamous** (SKWĀ-mus) **epithelium—** This tissue consists of a single layer of flat, scalelike cells. It is highly adapted for diffusion, osmosis, and filtration because of its thinness. Simple squamous epithelium lines the air sacs of the lungs, glomerular (Bowman's) capsules (filtering units) of the kidneys, and inner surface of the tympanic membrane (eardrum) of the ear. Simple squamous tissue that lines the heart, blood vessels, and lymphatic vessels and forms capillary walls is called **endothelium.** Simple squamous epithelium that forms the epithelial layer of a serous membrane is called **mesothe-**

lium. Serous membranes line the thoracic and abdominopelvic cavities and cover viscera within the cavities. After you make your microscopic examination, draw several cells in the space that follows and label plasma membrane, cytoplasm, and nucleus.

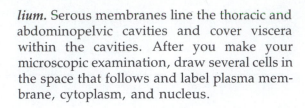

Simple squamous epithelium

b. **Simple cuboidal epithelium—**This tissue consists of a single layer of cube-shaped cells. When the tissue is sectioned at right angles, its cuboidal nature is obvious. Highly adapted for secretion and absorption, simple cuboidal epithelium covers the surface of the ovaries; lines the smaller ducts of some glands and the anterior surface of the lens capsule of the eye; and forms the pigmented epithelium of the retina of the eye, part of the tubules of the kidneys, and the secreting units of other glands, such as the thyroid gland.

After you make your microscopic examination, draw several cells in the space that follows and label plasma membrane, cyto-

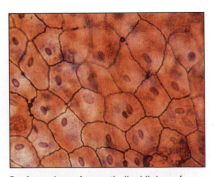

Surface view of mesothelical lining of peritoneal cavity

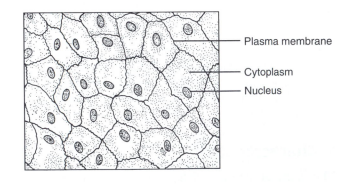

Plasma membrane

Cytoplasm

Nucleus

(a) Simple squamous epithelium

FIGURE 4.1   Epithelial tissues. Photomicrographs are labeled; the line drawing of the same tissue is unlabeled.

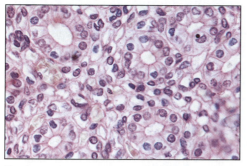

Sectional view of kidney tubules

Basement membrane
Plasma membrane
Connective tissue
Cytoplasm
Nucleus

(b) Simple cuboidal epithelium

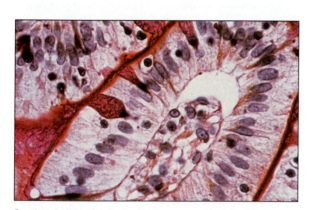

Mucus-producing goblet cell   Connective tissue   Basement membrane   Nucleus of absorptive cell

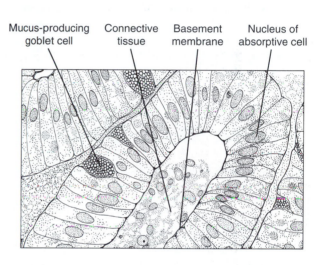

Sectional view of colonic glands

(c) Simple columnar (nonciliated) epithelium

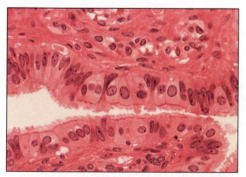

Sectional view of uterine (fallopian) tube (100 ×)

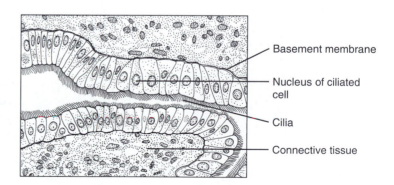

Basement membrane

Nucleus of ciliated cell

Cilia

Connective tissue

(d) Simple columnar (ciliated) epithelium

FIGURE 4.1 *(Continued)*   Epithelial tissues.

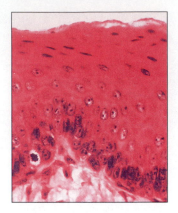

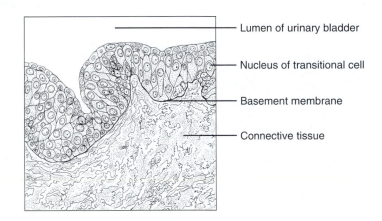

Nucleus of surface squamous cell

Basement membrane

Connective tissue
Nucleus of deep basal cell

(e) Stratified squamous epithelium

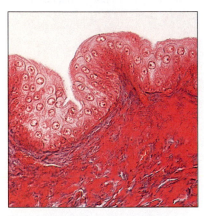

Sectional view of urinary bladder in
relaxed state (100 ×)

Lumen of urinary bladder

Nucleus of transitional cell

Basement membrane

Connective tissue

(f) Transitional epithelium

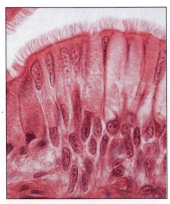

Sectional view of trachea (250 ×)

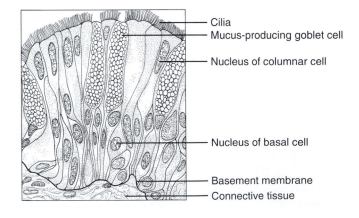

Cilia
Mucus-producing goblet cell

Nucleus of columnar cell

Nucleus of basal cell

Basement membrane
Connective tissue

(g) Pseudostratified columnar epithelium

FIGURE 4.1 *(Continued)*   Epithelial tissues.

plasm, nucleus, basement membrane, and connective tissue layer.

Simple cuboidal epithelium

c. **Simple columnar (nonciliated) epithelium—**This tissue consists of a single layer of columnar cells and, when sectioned at right angles, these cells appear as rectangles. Adapted for secretion and absorption, this tissue lines the gastrointestinal tract from the cardia of the stomach to the anus; gallbladder; and ducts of many glands. Some columnar cells are modified in that the plasma membranes are folded into fingerlike **microvilli** that increase the surface area for absorption. Other cells are modified as **goblet cells** that secrete and store mucus to protect the lining of the gastrointestinal tract. After you make your microscopic examination, draw several cells in the space that follows and label plasma membrane, cytoplasm, nucleus, goblet cell, absorptive cell, basement membrane, and connective tissue layer.

Simple columnar (nonciliated) epithelium

d. **Simple columnar (ciliated) epithelium—**This type of epithelium consists of a single layer of columnar absorptive, goblet, and ciliated cells. **Cilia** are hairlike processes that move substances over the surfaces of cells. Simple columnar (ciliated) epithelium lines some portions of the upper respiratory tract, uterine (Fallopian) tubes, uterus, some paranasal sinuses, and the central canal of the spinal cord. Mucus produced by goblet cells forms a thin film over the surface of the tissue, and movements of the cilia propel the mucus and the trapped substances over the surface of the tissue. After you make your microscopic examination, draw several cells in the space that follows and label plasma membrane, cytoplasm, nucleus, cilia, goblet cell, basement membrane, and connective tissue layer.

Simple columnar (ciliated) epithelium

e. **Stratified squamous epithelium—**This tissue consists of several layers of cells and affords considerable protection against friction. The superficial cells are flat, whereas cells of the deep layers vary in shape from cuboidal to columnar. The basal (bottom) cells continually multiply by cell division. As surface cells are sloughed off, new cells replace them from the basal layer. The surface cells of **keratinized stratified squamous** contain a waterproofing protein called **keratin** that also resists friction and bacterial invasion. The keratinized variety forms the outer layer of the skin. Surface cells of **nonkeratinized stratified squamous** do not contain keratin. The nonkeratinized variety lines wet surfaces such as the tongue, mouth, esophagus, part of the epiglottis, and vagina. After you make your microscopic examination, draw several cells in the space that follows and label plasma membrane, cytoplasm, nucleus, squamous surface cells, basal cells, basement membrane, and connective tissue layer.

Stratified squamous epithelium

f. Stratified squamous epithelium (student prepared)—Before examining the next slide, prepare a smear of cheek cells from the epithelial lining of the mouth. As noted previously, epithelium that lines the mouth is nonkeratinized stratified squamous epithelium. However, you will be examining surface cells only, and these will appear similar to simple squamous epithelium.

PROCEDURE

CAUTION! _Please reread Section B, "Precautions Related to Working with Blood, Blood Products, or Other Body Fluids," on page xiv, at the beginning of the laboratory manual, before you begin any of the following experiments. Read the experiments before you perform them, to be sure that you understand all the procedures and safety precautions. When you finish this part of the exercise, place the toothpicks, microscope slides, and cover slips in a fresh bleach solution._

a. Using the blunt end of a toothpick, _gently_ scrape the lining of your cheek several times to collect some surface cells of the stratified squamous epithelium.
b. Now move the toothpick across a clean glass microscope slide until a thin layer of scrapings is left on the slide.
c. Allow the preparation to air dry.
d. Next, cover the smear with several drops of 1% methylene blue stain. After about 1 min, gently rinse the slide in cold tap water or distilled water to remove excess stain.
e. _Gently_ blot the slide dry using a paper towel.
f. Examine the slide under low and high power. See if you can identify the plasma membrane, cytoplasm, nuclear membrane, and nucleoli.

Some bacteria are commonly found on the slide and usually appear as very small rods or spheres.

g. Transitional epithelium—This tissue resembles nonkeratinized stratified squamous, except that the superficial cells are larger and more rounded. When stretched, the surface cells are drawn out into squamouslike cells. This drawing out permits the tissue to stretch without the outer cells breaking apart from one another. The tissue lines parts of the urinary system, such as the urinary bladder, parts of the ureters, and urethra, that are subject to expansion from within. After you have made your microscopic examination, draw several cells in the space that follows and label plasma membrane, cytoplasm, nucleus, surface cells, basement membrane, and connective tissue layer.

Transitional epithelium

h. Pseudostratified columnar epithelium—Nuclei of cells in this tissue are at varying depths, and, although all the cells are attached to the basement membrane in a single layer, some do not reach the surface. This arrangement gives the impression of a multilayered tissue, thus the name _pseudostratified_. This tissue lines the large ducts of many glands, the epididymis, parts of the male urethra, and parts of the auditory (Eustachian) tubes; a special variety that lines most of the upper respiratory tract is called pseudostratified ciliated columnar epithelium. After you have made your microscopic examination, draw and label the basement membrane, a cell that reaches the surface, a cell that does not reach the surface, and the nuclei of each cell.

Pseudostratified columnar epithelium

3. Glandular Epithelium

A *gland* may consist of a single epithelial cell or a group of highly specialized epithelial cells that secrete various substances. Glands that have no ducts (ductless), secrete hormones, and release their secretions into the blood are called *endocrine glands.* Examples include the pituitary gland, thyroid gland, and adrenal glands (Exercise 15). Glands that secrete their products into ducts are called *exocrine glands.* Examples include sweat glands and salivary glands.

a. Structural Classification of Exocrine Glands

Based on the shape of the secretory portion and the degree of branching of the duct, exocrine glands can be structurally classified as follows:

1. *Unicellular*—One-celled glands that secrete mucus. An example is the goblet cell (see Figure 4.1c). These cells line portions of the respiratory and digestive systems.
2. *Multicellular*—Many-celled glands that occur in several different forms (Figure 4.2). *Simple*—Single, nonbranched duct. *Tubular*—Secretory portion is straight and tubular (intestinal glands). *Branched tubular*—Secretory portion is branched and tubular (gastric and uterine glands). *Coiled tubular*—Secretory portion is coiled (sudoriferous [soo-dor-IF-er-us], or sweat, glands). *Acinar* (AS-i-nar)—Secretory portion is flasklike (seminal vesicle glands). *Branched acinar*—Secretory portion is branched and flasklike (sebaceous [se-BĀ-shus], or oil, glands). *Compound*—Branched duct. *Tubular*—Secretory portion is tubular (bulbourethral or Cowper's glands, testes, liver). *Acinar*—Secretory portion is flasklike (sublingual and submandibular salivary glands). *Tubuloacinar*—Secretory portion is both tubular and flasklike (parotid salivary glands, pancreas).

Obtain a prepared slide of a representative of each of the types of multicellular exocrine glands just described. As you examine each slide, compare your observations to the diagrams of the glands in Figure 4.2.

b. FUNCTIONAL CLASSIFICATION OF EXOCRINE GLANDS

Functional classification is based on whether a secretion is a product of a cell or consists of entire or partial glandular cells themselves. *Holocrine glands,* such as sebaceous (oil) glands, accumulate their secretory product in their cytoplasm. The cell then dies and is discharged with its contents as the glandular secretion; the discharged cell is replaced by a new one. *Apocrine glands,* such as the sudoriferous (sweat) glands in the axilla, accumulate secretory products at the outer margins of the secreting cells. The margins pinch off as the secretion and the remaining portions of the cells are repaired so that the process can be repeated. *Merocrine (eccrine) glands,* such as the pancreas and salivary glands, produce secretions that are simply formed by the secretory cells and then discharged into a duct.

B. CONNECTIVE TISSUE

Connective tissue, the most abundant tissue in the body, functions by protecting, supporting, and separating structures (e.g., skeletal muscles) and binding structures together.

1. Characteristics

The general characteristics of connective tissue follow:

a. Connective tissue consists of three basic elements: cells, ground substance, and fibers. Together, the ground substance (fluid, gel, or

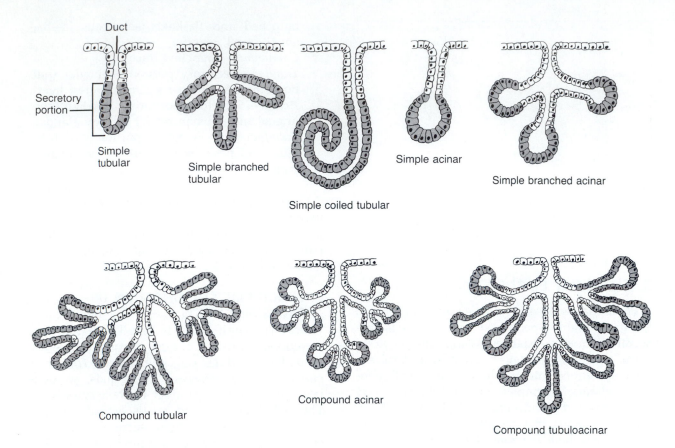

FIGURE 4.2   Structural classification of multicellular exocrine glands.

solid) and fibers, which are outside the cells, form the *matrix.* Unlike epithelial cells, connective tissue cells rarely touch one another; they are separated by a considerable amount of matrix.

b. In contrast to epithelia, connective tissues do not occur on free surfaces, such as the surfaces of a body cavity or the external surface of the body.

c. Except for cartilage, connective tissue, like epithelium, has a nerve supply.

d. Unlike epithelium, connective tissue usually is highly vascular (has a rich blood supply). Exceptions include cartilage, which is avascular, and tendons, which have a scanty blood supply.

e. The matrix of a connective tissue, which may be fluid, semifluid, gelatinous, fibrous, or calcified, is usually secreted by the connective tissue cells and adjacent cells and determines the tissue's qualities. In blood, the matrix, which is not secreted by blood cells, is fluid. In cartilage, it is firm but pliable. In bone, it is considerably harder and not pliable.

2. Connective Tissue Cells

Some of the cells contained in various types of connective tissue follow. The specific tissues to which they belong will be described subsequently.

a. *Fibroblasts* (FĪ-brō-blasts)—Large, flat, spindle-shaped cells with branching processes; they secrete the molecules that become the connective tissue fibers and ground substance of the matrix.

b. *Macrophages* (MAK-rō-fā-jez; *macro* = large, *phagein* = to eat), or *histiocytes*—Developed from *monocytes,* a type of white blood cell. Macrophages have an irregular shape with short, branching projections and are capable of engulfing bacteria and cellular debris by phagocytosis. Thus, they provide a vital defense for the body.

c. *Plasma cells*—Small and either round or irregular in shape. They develop from a type of white blood cell called a *B lymphocyte (B cell).* Plasma cells secrete specific antibodies and, accordingly, provide a defense mechanism through immunity.

d. *Mast cells*—Abundant alongside blood vessels. They produce histamine, a chemical that dilates small blood vessels and increases their permeability during inflammation.

e. *Adipocytes (fat cells)* and *white blood cells (leukocytes)*—Other cells in connective tissue.

3. Connective Tissue Ground Substance

The *ground substance* is amorphous, meaning that it has no specific shape. Connective tissue cells usually produce the ground substance and deposit it in the space between the cells.

Several examples of ground substance are as follows. *Hyaluronic* (hi-a-loo-RON-ik) *acid* is a viscous, slippery substance that binds cells together, lubricates joints, and helps maintain the shape of the eyeballs. It also appears to play a role in helping phagocytes migrate through connective tissue during development and wound repair. *Chondroitin* (kon-DROY-tin) *sulfate* is a jellylike substance that provides support and adhesiveness in cartilage, bone, the skin, and blood vessels. The skin, tendons, blood vessels, and heart valves contain *dermatan sulfate,* whereas bone, cartilage, and the cornea of the eye contain *keratan sulfate.*

The ground substance supports cells, binds them together, and provides a medium through which substances are exchanged between the blood and cells. Until recently, the ground substance was thought to function mainly as an inert scaffolding to support tissues. Now it is clear that the ground substance is quite active in functions such as influencing development, migration, proliferation, shape, and even metabolic functions.

4. Connective Tissue Fibers

Fibers in the matrix provide strength and support for tissues. Three types of fibers are embedded in the matrix between the cells of connective tissue: collagen, elastic, and reticular fibers.

a. *Collagen* (kolla = glue) *fibers*—There are at least five different types, very tough and resistant to a pulling force, yet they allow some flexibility in the tissue because they are not taut. These fibers often occur in bundles made up of many minute fibrils lying parallel

to one another. The bundle arrangement affords great strength. Chemically, collagen fibers consist of the protein *collagen.* This is the most abundant protein in the body, representing about 25% of the total protein. Collagen fibers are found in most types of connective tissues, especially bone, cartilage, tendons, and ligaments.

b. *Elastic fibers*—Smaller than collagen fibers, they freely branch and rejoin one another. They consist of a protein called *elastin.* Like collagen fibers, elastic fibers provide strength. In addition, they can be stretched to 150% of their relaxed length without breaking. Elastic fibers are plentiful in the skin, blood vessels, and lungs.

c. *Reticular* (rete = net) *fibers*—Consist of the protein collagen and a coating of glycoprotein; they provide support in the walls of blood vessels and form a network around fat cells, nerve fibers, and skeletal and smooth-muscle fibers. Produced by fibroblasts, they are much thinner than collagen fibers and form branching networks. Like collagen fibers, reticular fibers provide support and strength and also form the *stroma* (framework) of many soft-tissue organs, such as the spleen and lymph nodes. These fibers also help form the basement membrane.

5. Types

Before you start your microscopic examination of connective tissues, refer to Figure 4.3. Study the tissues carefully to familiarize yourself with their general structural characteristics. For each type of connective tissue listed, obtain a prepared slide and, unless otherwise specified by your instructor, examine each under high power.

a. LOOSE CONNECTIVE TISSUE

In this general type of connective tissue, the fibers are loosely woven and there are many cells.

1. *Areolar* (a-RĒ-ō-lar) *connective tissue*—One of the most widely distributed connective tissues in the body. It contains, at one time or another, all cells normally found in connective tissue, including fibroblasts, macrophages, plasma cells, mast cells, adipocytes, and a few white blood cells. All three types of fibers—collagen, elastic, and reticular—are present. The fluid, semifluid, or gelatinous ground substance contains hyaluronic acid, chon-

droitin sulfate, dermatan sulfate, and keratan sulfate. Areolar connective tissue is present in many mucous membranes, around blood vessels, nerves, and organs, and, together with adipose tissue, forms the *subcutaneous* (sub'-kyoo-TĀ-nē-us) *layer* or superficial fascia (FASH-ē-a). This layer is located between the skin and underlying tissues. After you make your microscopic examination, draw a small area of the tissue in the space that follows and label the elastic fibers, collagen fibers, fibroblasts, and mast cells.

Areolar connective tissue

2. *Adipose tissue*—This is fat tissue in which cells derived from fibroblasts, called *adipocytes*, are modified for fat storage. The cytoplasm and nuclei of the cells are pushed to the side. The tissue is found wherever areolar connective tissue is located and around the kidneys and heart, in the marrow of long bones, and behind the eyeball. It provides insulation, energy reserve, support, and protection. After your microscopic examination, draw several cells in the space that follows and label the fat storage area, cytoplasm, nucleus, and plasma membrane.

Adipose tissue

3. *Reticular connective tissue*—This tissue consists of fine interlacing reticular fibers in which the cells are interspersed between the fibers. It provides strength and support, forms the stroma (framework) of the liver, spleen, and lymph nodes, and binds smooth muscle fibers (cells) together. After you make your microscopic examination, draw a sample of the tissue in the space that follows and label the organ's reticular fibers and cells.

Reticular connective tissue

b. DENSE CONNECTIVE TISSUE

In this general type of connective tissue, there are more numerous and thicker fibers and fewer cells than in loose connective tissue.

1. *Dense regular connective tissue*—In this tissue, bundles of collagen fibers have an orderly, parallel arrangement that confers great strength. The tissue structure withstands pulling in one direction. Fibroblasts, which produce the fibers and ground substance, appear in rows between the fibers. The tissue is silvery white, tough, yet somewhat pliable. Because of its great strength, it is the principal component of *tendons*, which attach muscles to bones; *aponeuroses* (ap'-ō-noo-RŌ-sēz), which are sheetlike tendons connecting one muscle with another or with bone; and many *ligaments* (collagen ligaments), which hold bones together at joints. After you make your microscopic examination, draw a sample of the tissue in the space that follows and label the collagen fibers and fibroblasts.

2. *Dense irregular connective tissue*—This tissue contains collagen fibers that are interwoven without regular orientation. It is found in parts of the body where tensions are exerted in

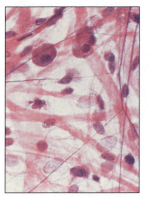

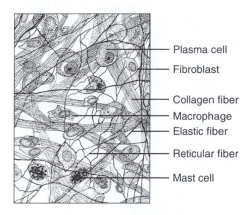

Plasma cell

Fibroblast

Collagen fiber

Macrophage

Elastic fiber

Reticular fiber

Mast cell

Sectional view of subcutaneous
tissue (160 ×)

(a) Areolar connective tissue

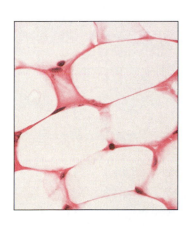

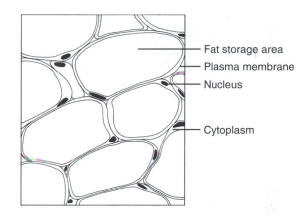

Fat storage area

Plasma membrane

Nucleus

Cytoplasm

(b) Adipose tissue

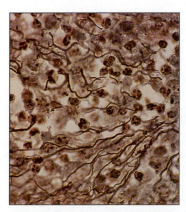

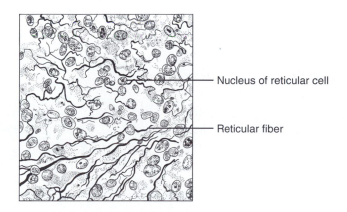

Nucleus of reticular cell

Reticular fiber

Sectional view of lymph node (250 ×)

(c) Reticular connective tissue

FIGURE 4.3 Connective tissues. Photomicrographs are labeled; the line drawing of the same tissue is unlabeled.

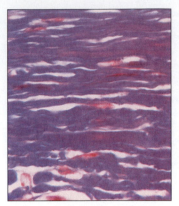

Sectional view of capsule of adrenal gland

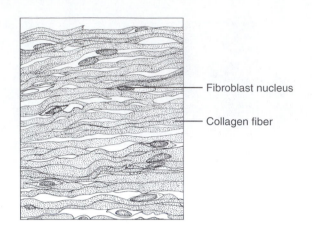

Fibroblast nucleus

Collagen fiber

(d) Dense regular connective tissue

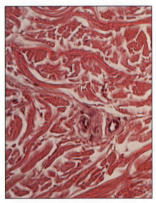

Sectional view of dermis of skin (275 ×)

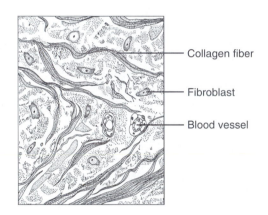

Collagen fiber

Fibroblast

Blood vessel

(e) Dense irregular connective tissue

Sectional view of ligamentum nuchae (400 ×)

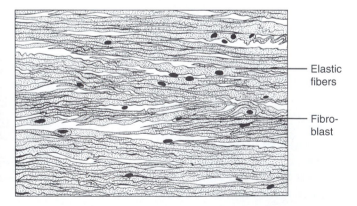

Elastic fibers

Fibro-blast

(f) Elastic connective tissue

FIGURE 4.3 *(Continued)*   Connective tissues.

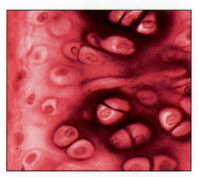

Sectional view of hyaline cartilage
from trachea (160 ×)

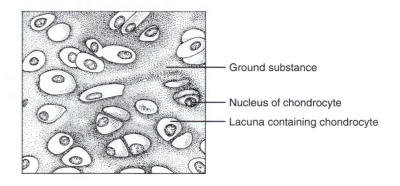

—— Ground substance

—— Nucleus of chondrocyte
—— Lacuna containing chondrocyte

(g) Hyaline cartilage

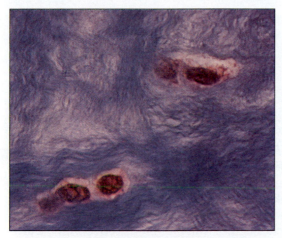

Sectional view of fibrocartilage from medial meniscus
of knee (315 ×)

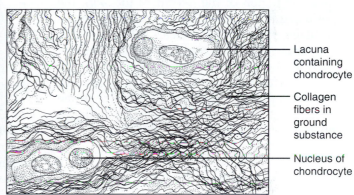

—— Lacuna
containing
chondrocyte

—— Collagen
fibers in
ground
substance

—— Nucleus of
chondrocyte

(h) Fibrocartilage

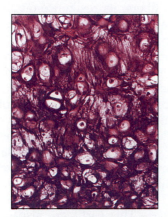

Sectional view of elastic cartilage
from auricle (pinna) of external ear

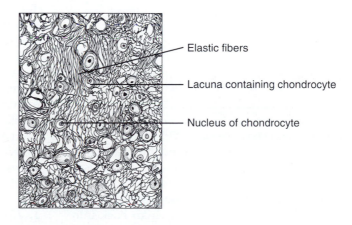

—— Elastic fibers

—— Lacuna containing chondrocyte

—— Nucleus of chondrocyte

(i) Elastic cartilage

FIGURE 4.3 *(Continued)*   Connective tissues.

Dense regular connective tissue

various directions. The tissue usually occurs in sheets. It forms some fasciae, the reticular (deeper) region of the dermis of the skin, the periosteum of bone, the perichondrium of cartilage, joint capsules, heart valves, and the *membrane (fibrous) capsules* around organs, such as the kidneys, liver, testes, and lymph nodes. After you make your microscopic examination, draw a sample of the tissue in the space that follows and label the collagen fibers.

Dense irregular connective tissue

3. *Elastic connective tissue*—This tissue has a predominance of freely branching elastic fibers. These fibers give the unstained tissue a yellowish color. Fibroblasts are present in the spaces between fibers. Elastic connective tissue can be stretched and will snap back into shape. It is a component of the walls of elastic eries, the trachea, bronchial tubes of the lungs, and the lungs themselves. Elastic connective tissue provides stretch and strength, allowing structures to perform their functions efficiently. Yellow elastic ligaments, in contrast to collagen ligaments, are composed mostly of elastic fibers; they form the ligamenta flava of the vertebrae (ligaments between successive vertebrae), the suspensory ligament of the penis, and the true vocal cords. After you make your microscopic examination, draw a sample of the tissue in the space that follows and label the elastic fibers and fibroblasts.

Elastic connective tissue

C. CARTILAGE

Cartilage is capable of enduring considerably more stress than the tissues just discussed. Unlike other connective tissues, cartilage has no blood vessels or nerves, except for those in the perichondrium (membranous covering). Cartilage consists of a dense network of collagen fibers and elastic fibers firmly embedded in chondroitin sulfate, a jellylike component of the ground substance. Whereas the strength of cartilage is due to its collagen fibers, its resilience (ability to assume its original shape after deformation) is due to chondroitin sulfate.

The cells of mature cartilage, called *chondrocytes* (KON-drō-sīts), occur singly or in groups within spaces called *lacunae* (la-KOO-nē) in the matrix. The surface of cartilage, except for fibrocartilage, is surrounded by dense irregular connective tissue called the *perichondrium* (per'-i-KON-drē-um; *peri* = around, *chondro* = cartilage). Three kinds of cartilage are recognized: hyaline cartilage, fibrocartilage, and elastic cartilage.

1. *Hyaline cartilage*—This cartilage, also called gristle, contains a resilient gel as its ground substance and appears in the body as a bluish-white, shiny substance. The fine collagen fibers, although present, are not visible with ordinary staining techniques, and the prominent chondrocytes are found in lacunae. Hyaline cartilage is the most abundant kind of

cartilage in the body. It is found at joints over the ends of the long bones and at the anterior ends of the ribs. Hyaline cartilage also helps to support the nose, larynx, trachea, bronchi, and bronchial tubes leading to the lungs. Most of the embryonic skeleton consists of hyaline cartilage, which gradually becomes calcified and develops into bone. Hyaline cartilage affords flexibility and support and, at joints, reduces friction and absorbs shock. After you make your microscopic examination, draw a sample of the tissue in the space that follows and label the perichondrium, chondrocytes, lacunae, and matrix.

Hyaline cartilage

2. *Fibrocartilage*—Chondrocytes are scattered among clearly visible bundles of collagen fibers within the matrix of this type of cartilage. Fibrocartilage forms the pubic symphysis, the point where the hipbones fuse anteriorly at the midline. It is also found in the intervertebral discs between vertebrae and in the menisci of the knee. This tissue combines strength and rigidity. After you make your microscopic examination, draw a sample of the tissue in the space that follows and label the chondrocytes, lacunae, intercellular substance, and collagen fibers.

Fibrocartilage

3. *Elastic cartilage*—In this tissue, chondrocytes are located in a threadlike network of elastic fibers within the matrix. Elastic cartilage provides strength and elasticity and maintains the shape of organs—the epiglottis of the larynx, the external part of the ear (auricle), and the auditory (Eustachian) tubes. After you make your microscopic examination, draw a sample of the tissue in the space that follows and label the perichondrium, chondrocytes, lacunae, intercellular substance, and elastic fibers.

Elastic cartilage

C. MEMBRANES

The combination of an epithelial layer and an underlying layer of connective tissue constitutes an *epithelial membrane.* Examples are mucous, serous, and cutaneous membranes. Another kind of membrane, a synovial membrane, has no epithelium. *Mucous membranes,* also called the *mucosa,* line body cavities that open directly to the exterior, such as the gastrointestinal, respiratory, urinary, and reproductive tracts. The surface tissue of a mucous membrane consists of epithelium and has a variety of functions, depending on location. Accordingly, the epithelial layer secretes mucus but may also secrete enzymes, filter dust, and have a protective and absorbent action. The underlying connective tissue layer of a mucous membrane, called the *lamina propria,* binds the epithelial layer in place, protects underlying tissues, provides the epithelium with nutrients and oxygen and removes wastes, and holds blood vessels in place.

Serous membranes, also called the *serosa,* line body cavities that do not open to the exterior and cover organs that lie within the cavities. Serous membranes consist of a surface layer of mesothelium and an underlying layer of areolar connective tissue. The mesothelium secretes a lubricating fluid. Serous membranes consist of two portions. The part attached to the cavity wall is called the *parietal layer;* the part that covers the organs in the cavity is called the *visceral layer.* Examples of serous membranes are the pleurae, pericardium, and peritoneum.

The *cutaneous membrane,* or skin, is the principal component of the integumentary system, which will be considered in the next exercise.

Synovial membranes line joint cavities. They do not contain epithelium but rather consist of areolar connective tissue, adipose tissue, and elastic fibers. Synovial membranes produce synovial fluid, which lubricates the ends of bones as they move at joints and nourishes the articular cartilage around the ends of bones.

ANSWER THE LABORATORY REPORT QUESTIONS AT THE END OF THE EXERCISE.

Tissues

STUDENT _____ DATE _____

LABORATORY SECTION _____ SCORE/GRADE _____

PART 1. Multiple Choice

_____ 1. In parts of the body such as the urinary bladder, where considerable distension (stretching) occurs, you can expect to find which epithelial tissue? (a) pseudostratified columnar (b) cuboidal (c) columnar (d) transitional

_____ 2. Stratified epithelium is usually found in areas of the body where the principal activity is (a) filtration (b) absorption (c) protection (d) diffusion

_____ 3. Ciliated epithelium destroyed by disease would cause malfunction in which system? (a) digestive (b) respiratory (c) skeletal (d) cardiovascular

_____ 4. The tissue that provides the skin with resistance to wear and tear and serves to waterproof it is (a) keratinized stratified squamous (b) pseudostratified columnar (c) transitional (d) simple columnar

_____ 5. The connective tissue cell that would most likely increase its activity during an infection is the (a) melanocyte (b) macrophage (c) adipocyte (d) fibroblast

_____ 6. Torn ligaments would involve damage to which tissue? (a) dense regular (b) reticular (c) elastic (d) areolar

_____ 7. Simple squamous tissue that lines the heart, blood vessels, and lymphatic vessels is called (a) transitional (b) adipose (c) endothelium (d) mesothelium

_____ 8. Microvilli and goblet cells are associated with which tissue? (a) hyaline cartilage (b) simple columnar nonciliated (c) transitional (d) stratified squamous

_____ 9. Superficial fascia contains which tissue? (a) elastic (b) reticular (c) fibrocartilage (d) areolar connective tissue

_____ 10. Which tissue forms articular cartilage and costal cartilage? (a) fibrocartilage (b) elastic cartilage (c) adipose (d) hyaline cartilage

_____ 11. Because the sublingual gland contains a branched duct and flasklike secretory portions, it is classified as (a) simple coiled tubular (b) compound acinar (c) simple acinar (d) compound tubular

_____ 12. Which glands discharge an entire dead cell and its contents as their secretory products? (a) merocrine (b) apocrine (c) endocrine (d) holocrine

_____ 13. Membranes that line cavities that open directly to the exterior are called (a) synovial (b) serous (c) mucous (d) cutaneous

PART 2. Completion

14. Single cells found in epithelium that secrete mucus are called _____ cells.

15. A type of epithelium that appears to consist of several layers but actually contains only one layer of

    cells is _____.

16. The cell in connective tissue that forms new fibers is the _____.

17. Histamine, a substance that dilates small blood vessels during inflammation, is secreted

    by _____ cells.

18. Cartilage cells found in lacunae are called _____.

19. The simple squamous epithelium of a serous membrane that covers viscera is called

    _____.

20. The tissue that provides insulation, support, protection, and serves as a food reserve is

    _____.

21. _____ tissue forms the stroma of organs such as the liver and spleen.

22. The cartilage that provides support for the larynx and external ear is _____.

23. The ground substance that helps lubricate joints and binds cells together is

    _____.

24. Ductless glands that secrete hormones are called _____ glands.

25. Multicellular exocrine glands that contain branching ducts are classified as

    _____ glands.

26. The mammary glands are classified as _____ glands because their secretory
    products are the pinched-off margins of cells.

27. _____ membranes consist of parietal and visceral layers and line cavities that do
    not open to the exterior.

28. If the secretory portion of a gland is flasklike, it is classified as a(n) _____ gland.

29. Membranes that line joint cavities are called _____ membranes.

30. An example of a simple branched acinar gland is a(n) _____ gland.

Integumentary System

The skin and the organs derived from it (hair, nails, and glands) and several specialized receptors constitute the ***integumentary*** (in-teg-yoo-MEN-tar-ē) ***system,*** which you will study in this exercise. An ***organ*** is an aggregation of tissues of definite form and usually recognizable shape that performs a definite function; a ***system*** is a group of organs that operate together to perform specialized functions.

A. SKIN

The ***skin*** is one of the larger organs of the body in terms of surface area, occupying a surface area of about 2 square meters (2 m^2) (22 ft^2). Among the functions performed by the skin are regulation of body temperature; protection of underlying tissues from physical abrasion, microorganisms, dehydration, and ultraviolet (UV) radiation; excretion of water and salts and several organic compounds; synthesis of vitamin D in the presence of sunlight; reception of stimuli for touch, pressure, pain, and temperature change sensations; serving as a blood reservoir; and immunity.

The skin consists of an outer, thinner ***epidermis,*** which is avascular, and an inner, thicker ***dermis,*** which is vascular. Below the dermis is the ***subcutaneous layer (superficial fascia or hypodermis)*** that attaches the skin to underlying tissues and organs.

1. Epidermis

The epidermis consists of four principal kinds of cells. ***Keratinocytes*** are the most numerous cells and they undergo keratinization—that is, newly formed cells produced in the basal layers are pushed up to the surface and in the process synthesize a waterproofing chemical, keratin. ***Melanocytes*** are pigment cells that impart color to the skin. The third type of cell in the epidermis is called a ***Langerhans cell.*** These cells are a small population of cells that arise from bone marrow and migrate to the epidermis and other stratified squamous epithelial tissue in the body. They are sensitive to UV radiation and lie above the basal layer of keratinocytes. Langerhans cells interact with cells called ***helper T cells*** to assist in the immune response. ***Merkel cells*** are found in the bottom layer of the epidermis. Their bases are in contact with flattened portions of the terminations of sensory nerves (Merkel discs) and function as receptors for touch. At this point, we will concentrate only on keratinocytes.

Obtain a prepared slide of human skin and carefully examine the epidermis. Identify the following layers from the outside inward:

a. ***Stratum corneum***—25 to 30 rows of flat, dead cells that are filled with ***keratin,*** a waterproofing protein; these cells are continuously shed and replaced by cells from deeper strata.

b. ***Stratum lucidum***—Several rows of clear, flat cells that contain ***eleidin*** (el-Ē-i-din), a precursor of keratin; they are found only in the palms and soles.

c. ***Stratum granulosum***—3 to 5 rows of flat cells that contain ***keratohyalin*** (ker′-a-tō-HĪ-a-lin), a precursor of eleidin.

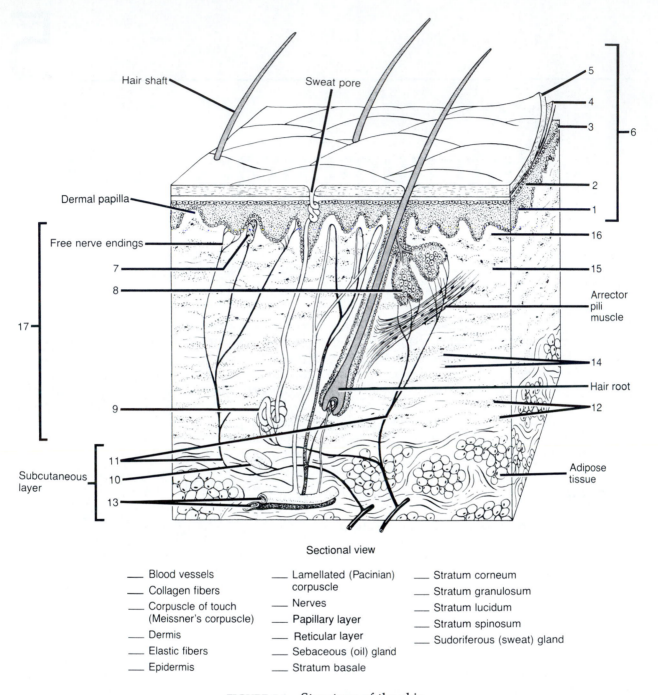

Hair shaft

Sweat pore

5
4
3
6
2
1

Dermal papilla

16

Free nerve endings

15

7

8

Arrector pili muscle

17

14

Hair root

9

12

11

10

13

Subcutaneous layer

Adipose tissue

Sectional view

___ Blood vessels
___ Collagen fibers
___ Corpuscle of touch (Meissner's corpuscle)
___ Dermis
___ Elastic fibers
___ Epidermis

___ Lamellated (Pacinian) corpuscle
___ Nerves
___ Papillary layer
___ Reticular layer
___ Sebaceous (oil) gland
___ Stratum basale

___ Stratum corneum
___ Stratum granulosum
___ Stratum lucidum
___ Stratum spinosum
___ Sudoriferous (sweat) gland

FIGURE 5.1 Structure of the skin.

d. **Stratum spinosum**—8 to 10 rows of polyhedral (many-sided) cells.

e. **Stratum basale (germinativum)**—A single layer of cuboidal to columnar cells that constantly undergo division. In hairless skin, this layer contains tactile (Merkel) discs, receptors sensitive to touch.

Label the epidermal layers in Figures 5.1 and 5.2.

2. Skin Color

The color of skin results from (1) the **hemoglobin in blood in capillaries** of the dermis (beneath the epidermis); (2) **carotene**, a yellow-orange pigment in the stratum corneum of the epidermis and fatty areas of the dermis; and (3) **melanin**, a pale yellow to black pigment found primarily in the melanocytes in the stratum basale and spinosum of the epidermis. Whereas hemoglobin in red blood

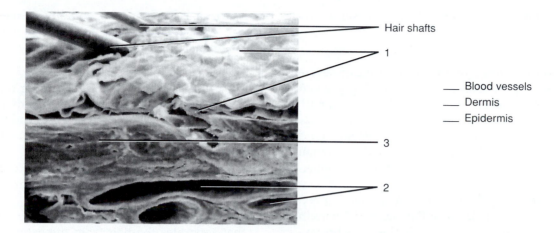

Hair shafts

1

___ Blood vessels
___ Dermis
___ Epidermis

3

2

FIGURE 5.2   Scanning electron micrograph of the skin and several hairs at a magnification of 260×. (From *Tissues and Organs: A Text-Atlas of Scanning Electron Microscopy* by Richard G. Kessel and Randy H. Kardon. W. H. Freeman and Company. Copyright © 1979.)

cells in capillaries imparts a pink color to Caucasian skin, carotene imparts the characteristic yellow color to Asian skin. Because the number of ***melanocytes*** (me-LAN-ō-sīts), or melanin-producing cells, is about the same in all races, most differences in skin color are due to the amount of melanin that the melanocytes synthesize and disperse. Exposure to ultraviolet (UV) radiation increases melanin synthesis, resulting in darkening (tanning) of the skin to protect the body against further UV radiation.

An inherited inability of a person of any race to produce melanin results in ***albinism*** (AL-binizm). The pigment is absent from the hair and eyes as well as from the skin, and the individual is referred to as an ***albino.*** In some people, melanin tends to form in patches called *freckles.* Others inherit patches of skin that lack pigment, a condition called *vitiligo* (vit-i-LĪ-gō).

3. Dermis

The dermis is divided into two regions and is composed of connective tissue containing collagen and elastic fibers and a number of other structures. The upper region of the dermis *(papillary layer)* is areolar connective tissue containing fine elastic fibers. This layer contains fingerlike projections, the ***dermal papillae*** (pa-PIL-ē). Some papillae enclose blood capillaries; others contain ***corpuscles of touch (Meissner's corpuscles),*** nerve endings sensitive to touch. The lower region of the dermis *(reticular layer)* consists of dense, irregular connective tissue with interlacing bundles of larger collagen and coarse elastic fibers. Spaces between the fibers may be occupied by ***hair follicles, sebaceous (oil) glands, bundles of smooth muscle (arrector pili muscle), sudoriferous (sweat) glands, blood vessels,*** and ***nerves.***

The reticular layer of the dermis is attached to the underlying structures (bones and muscles) by the subcutaneous layer. This layer also contains nerve endings sensitive to deep pressure called ***lamellated (Pacinian) corpuscles.***

Carefully examine the dermis and subcutaneous layer on your microscope slide. Label the following structures in Figure 5.1: papillary layer, reticular layer, corpuscle of touch, blood vessels, nerves, elastic fibers, collagen fibers, sebaceous gland, sudoriferous gland, and lamellated corpuscle. Also label the dermis and blood vessels in Figure 5.2.

If a model of the skin and subcutaneous layer is available, examine it to see the three-dimensional relationship of the structures to one another.

B. HAIR

Hairs (pili) develop from the epidermis and are variously distributed over the body. Each hair consists of a *shaft,* most of which is visible above the surface of the skin, and a *root,* the portion below the surface that penetrates deep into the dermis and even into the subcutaneous layer.

The shaft of a coarse hair consists of the following parts:

1. *Medulla*—Inner region, composed of several rows of polyhedral cells containing pigment and air spaces. In fine hairs it is poorly developed or not present.
2. *Cortex*—Several rows of dark cells surrounding the medulla; in dark hair they contain pigment and in gray hair they contain pigment and air.
3. *Cuticle of the hair*—Outermost layer, it consists of a single layer of flat, scalelike, keratinized cells arranged like shingles on a house.

The root of a hair also contains a medulla, cortex, and cuticle of the hair, along with the following associated parts:

1. *Hair follicle*—Structure surrounding the root that consists of an external root sheath and an internal root sheath. These epidermally derived layers are surrounded by a dermal layer of connective tissue.
2. *External root sheath*—Downward continuation of strata basale and spinosum of the epidermis.
3. *Internal root sheath*—Cellular tubular sheath that separates the hair from the external root

____ Bulb
____ Cortex
____ Cuticle of hair
____ Cuticle of internal root sheath
____ External root sheath
____ Granular (Huxley's) layer
____ Internal root sheath
____ Matrix
____ Medulla
____ Pallid (Henle's) layer
____ Papilla of hair

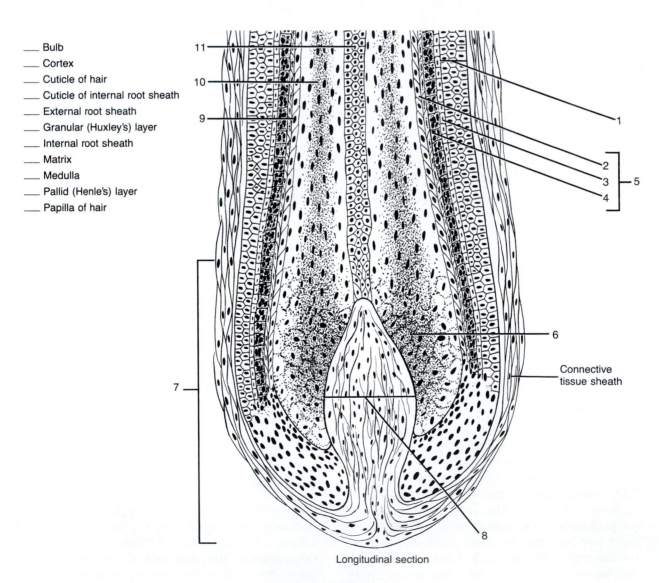

Longitudinal section

FIGURE 5.3 Hair root.

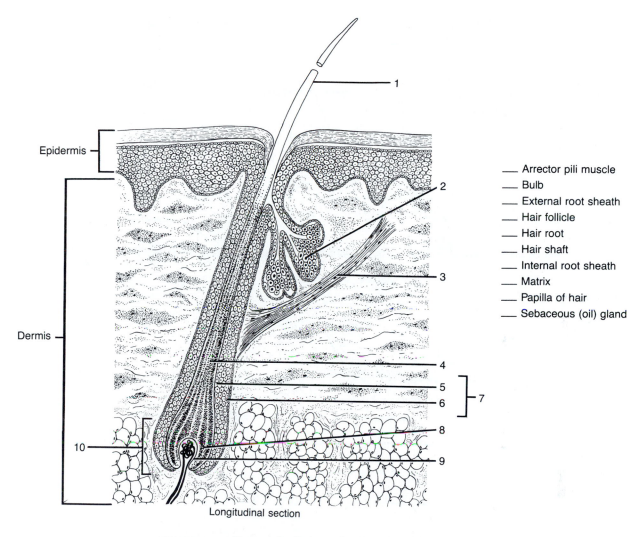

Epidermis

Dermis

— Arrector pili muscle
— Bulb
— External root sheath
— Hair follicle
— Hair root
— Hair shaft
— Internal root sheath
— Matrix
— Papilla of hair
— Sebaceous (oil) gland

1
2
3
4
5
6
7
8
9
10

Longitudinal section

FIGURE 5.4   Parts of a hair and associated structures.

sheath. It consists of (a) the *cuticle of the internal root sheath,* an inner single layer of flattened cells with atrophied nuclei, (b) a *granular (Huxley's) layer,* a middle layer of 1 to 3 rows of cells with flattened nuclei, and (c) *pallid (Henle's) layer,* an outer single layer of cuboidal cells with flattened nuclei.

4. *Bulb*—Enlarged, onion-shaped structure at the base of the hair follicle.
5. *Papilla of the hair*—Dermal indentation into the bulb, it contains areolar connective tissue and blood vessels to nourish the hair.
6. *Matrix*—Region of cells at the base of the bulb that divides to produce new hair.
7. *Arrector pili muscle*—Bundle of smooth muscle extending from the dermis of the skin to the side of the hair follicle; its contraction, under the influence of fright or cold, causes

the hair to move into a vertical position, producing "goose bumps."
8. *Hair root plexus*—Nerve endings around each hair follicle that are sensitive to touch and respond when the hair shaft is moved.

Obtain a prepared slide of a cross section and a longitudinal section of a hair root and identify as many parts as you can. Using your textbook as a reference, label Figure 5.3. Also label the parts of a hair shown in Figure 5.4.

C. GLANDS

Sebaceous (se-BĀ-shus), or *oil, glands,* with few exceptions, are connected to hair follicles (see Figures 5.1 and 5.4). They are single, branched

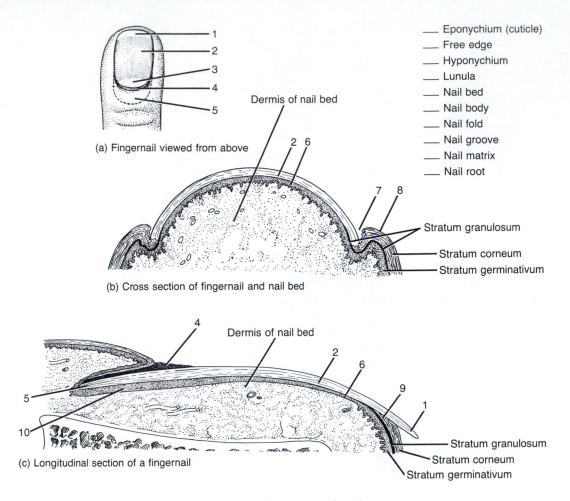

(a) Fingernail viewed from above

Dermis of nail bed

___ Eponychium (cuticle)
___ Free edge
___ Hyponychium
___ Lunula
___ Nail bed
___ Nail body
___ Nail fold
___ Nail groove
___ Nail matrix
___ Nail root

Stratum granulosum
Stratum corneum
Stratum germinativum

(b) Cross section of fingernail and nail bed

Dermis of nail bed

Stratum granulosum
Stratum corneum
Stratum germinativum

(c) Longitudinal section of a fingernail

FIGURE 5.5   Structure of nails.

acinar glands and secrete an oily substance called **sebum,** a mixture of fats, cholesterol, proteins, and salts. Sebaceous glands are absent in the skin of the palms and soles but are numerous in the skin of the face, neck, upper chest, and breasts. Sebum helps prevent hair from drying and forms a protective film over the skin that prevents excessive evaporation and keeps the skin soft and pliable.

Sudoriferous (soo'-dor-IF-er-us), or **sweat, glands** are separated into two principal types on the basis of structure, location, and secretion. **Apocrine sweat glands** are simple, branched tubular glands found primarily in the skin of the axilla, pubic region, and areolae (pigmented areas) of the breasts. Their secretory portion is located in the dermis; the excretory duct opens into hair follicles. Apocrine sweat glands begin to function at puberty and produce a more viscous secretion than the other type of sweat gland.

Eccrine sweat glands are simple, coiled tubular glands found throughout the skin, except for the margins of the lips, nail beds of the fingers and toes, glans penis, glans clitoris, and eardrums. The secretory portion of these glands is in the subcutaneous layer; the excretory duct projects upward and terminates at a pore at the surface of the epidermis (see Figure 5.1). Eccrine sweat glands function throughout life and produce a more watery secretion than the apocrine glands. Sudoriferous glands produce **perspiration,** a mixture of water, salt, urea, uric acid, amino acids, ammonia, sugar, lactic acid, and ascorbic acid. The evaporation of perspiration helps to maintain normal body temperature.

Ceruminous (se-ROO-mi-nus) **glands** are modified sudoriferous glands in the external auditory meatus. They are simple, coiled tubular glands. The combined secretion of ceruminous and sudoriferous glands is called **cerumen** (earwax). Cerumen, together with hairs in the external auditory meatus, provides a sticky barrier that prevents foreign bodies from reaching the eardrum.

D. NAILS

Nails are plates of tightly packed, hard, keratinized epidermal cells that form a clear covering over the dorsal surfaces of the terminal portions of the fingers and toes. Each nail consists of the following parts:

1. *Nail body*—Portion that is visible.
2. *Free edge*—Part that may project beyond the distal end of the digit.
3. *Nail root*—Portion hidden in the nail groove (see number 7).
4. *Lunula* (LOO-nyoo-la)—Whitish, semilunar area at the proximal end of the nail body.

5. *Nail fold*—Fold of skin that extends around the proximal end and lateral borders of the nail.
6. *Nail bed*—Strata basale and spinosum of the epidermis beneath the nail.
7. *Nail groove*—Furrow between the nail fold and the nail bed.
8. *Eponychium* (ep'-ō-NIK-ē-um)—Cuticle, a narrow band of epidermis.
9. *Hyponychium*—Thickened area of the stratum corneum below the nail's free edge.
10. *Nail matrix*—Epithelium of the proximal part of the nail bed; the division of these cells brings about the growth of nails.

Using your textbook as a reference, label the parts of a nail shown in Figure 5.5. Also, identify the parts that are visible on your own nails.

ANSWER THE LABORATORY REPORT QUESTIONS AT THE END OF THE EXERCISE.

Integumentary System

STUDENT _____   DATE _____

LABORATORY SECTION _____   SCORE/GRADE _____

PART **1.** **Multiple Choice**

_____ 1. The waterproofing quality of skin is due to the presence of (a) melanin (b) carotene (c) keratin (d) receptors

_____ 2. Sebaceous glands (a) produce a watery solution called sweat (b) produce an oily substance that prevents excessive water evaporation from the skin (c) are associated with mucous membranes (d) are part of the subcutaneous layer

_____ 3. Which of the following is the proper sequence of layering of the epidermis, going from the free surface toward the underlying tissues? (a) basale, spinosum, granulosum, corneum (b) spinosum, basale, granulosum, corneum (c) corneum, lucidum, granulosum, spinosum, basale (d) corneum, granulosum, lucidum, spinosum

_____ 4. Skin color is _not_ determined by the presence or absence of (a) melanin (b) carotene (c) keratin (d) hemoglobin in red blood cells in capillaries in the dermis

_____ 5. Destruction of what part of a single hair would result in its inability to grow? (a) sebaceous gland (b) arrector pili muscles (c) matrix (d) bulb

_____ 6. One would expect to find relatively few, if any, sebaceous glands in the skin of the (a) palms (b) face (c) neck (d) upper chest

_____ 7. Which of the following sequences, from outside to inside, is correct? (a) epidermis, reticular layer, papillary layer, subcutaneous layer (b) epidermis, subcutaneous layer, reticular layer, papillary layer (c) epidermis, reticular layer, subcutaneous layer, papillary layer (d) epidermis, papillary layer, reticular layer, subcutaneous layer

_____ 8. The attached visible portion of a nail is called the (a) nail bed (b) nail root (c) nail fold (d) nail body

_____ 9. Nerve endings sensitive to touch are called (a) corpuscles of touch (Meissner's corpuscles) (b) papillae (c) lamellated (Pacinian) corpuscles (d) follicles

_____ 10. The cuticle of a nail is referred to as the (a) matrix (b) eponychium (c) hyponychium (d) fold

_____ 11. One would _not_ expect to find sudoriferous glands associated with the (a) forehead (b) axilla (c) palms (d) nail beds

_____ 12. Fingerlike projections of the dermis that contain loops of capillaries and receptors are called (a) dermal papillae (b) nodules (c) polyps (d) pili

_____ 13. Which of the following statements about the function of skin is _not_ true? (a) It helps control body temperature (b) It prevents excessive water loss (c) It synthesizes several compounds (d) It absorbs water and salts

———— 14. Which is *not* part of the internal root sheath? (a) granular (Huxley's) layer (b) cortex (c) pallid (Henle's) layer (d) cuticle of the internal root sheath

———— 15. Growth in the length of nails is the result of the activity of the (a) eponychium (b) nail matrix (c) hyponychium (d) nail fold

PART 2. Completion

16. A group of tissues that performs a definite function is called a(n) ————————.

17. The outer, thinner layer of the skin is known as the ————————.

18. The skin is attached to underlying tissues and organs by the ————————.

19. A group of organs that operate together to perform a specialized function is called a(n)

————————.

20. The epidermal layer that contains eleidin is the stratum ————————.

21. The epidermal layers that produce new cells are the stratum spinosum and stratum

————————.

22. The smooth muscle attached to a hair follicle is called the ———————— muscle.

23. An inherited inability to produce melanin is called ————————.

24. Nerve endings sensitive to deep pressure are referred to as ———————— corpuscles.

25. The inner region of a hair shaft and root is the ————————.

26. The portion of a hair containing areolar connective tissue and blood vessels is the

————————.

27. Modified sweat glands that line the external auditory meatus are called ———————— glands.

28. The whitish, semilunar area at the proximal end of the nail body is referred to as the

————————.

29. The secretory product of sudoriferous glands is called ————————.

30. Melanin is synthesized in cells called ————————.

6

Bone Tissue

Structurally, the *skeletal system* consists of two types of connective tissue: cartilage and bone. The microscopic structure of cartilage has been discussed in Exercise 4. In this exercise, the gross structure of a typical bone and the histology of *bone (osseous) tissue* will be studied.

The skeletal system has the following basic functions:

1. *Support*—It provides a supporting framework for the soft tissues, maintaining the body's shape and posture.
2. *Protection*—It protects delicate structures such as the brain, spinal cord, heart, lungs, major blood vessels in the chest, and pelvic viscera.
3. *Movement*—When skeletal muscles contract, bones serve as levers to help produce body movements.
4. *Mineral homeostasis*—Bone tissue stores several minerals, especially calcium and phosphorus. On demand, bone releases minerals into the blood to maintain critical mineral balances and for distribution to other parts of the body.
5. *Blood cell production* or *hematopoiesis* (hē'-mat-ō-poy-Ē-sis)—*Red marrow* in certain parts of bones consists of primitive blood cells in immature stages, fat cells, and macrophages. It is responsible for producing red blood cells, platelets, and some white blood cells.
6. *Storage of energy*—Lipids stored in the cells of yellow marrow are an important source of a chemical energy reserve.

A. GROSS STRUCTURE OF A LONG BONE

Examine the external features of a fresh long bone and locate the following structures:

1. *Periosteum* (per'-ē-OS-tē-um)—Connective tissue membrane covering the surface of the bone, except for areas covered by articular cartilage. The outer *fibrous layer* is composed of dense, irregular connective tissue and contains blood vessels, lymphatic vessels, and nerves that pass into the bone. The inner *osteogenic* (os'-tē-ō-JEN-ik) *layer* contains elastic fibers; blood vessels; *osteoprogenitor* (os'-tē-ō-prō-JEN-i-tor) *cells* (stem cells derived from mesenchyme that differentiate into osteoblasts); and *osteoblasts* (OS-tē-ō-blasts), cells that secrete the organic components and mineral salts involved in bone formation. The periosteum functions in bone growth, nutrition, and repair, and as an attachment site for tendons and ligaments.
2. *Articular cartilage*—Thin layer of hyaline cartilage covering the ends of the bone where joints are formed.
3. *Epiphysis* (e-PIF-i-sis)—End or extremity of a bone; the epiphyses are referred to as proximal and distal.
4. *Diaphysis* (dī-AF-i-sis)—Elongated shaft of a bone between the epiphyses.
5. *Metaphysis* (me-TAF-i-sis)—In mature bone, the region where the diaphysis joins the epiphysis; in growing bone, the region near

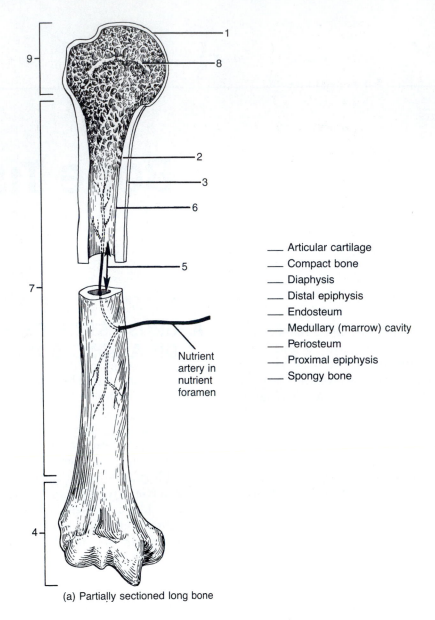

(a) Partially sectioned long bone

___ Articular cartilage
___ Compact bone
___ Diaphysis
___ Distal epiphysis
___ Endosteum
___ Medullary (marrow) cavity
___ Periosteum
___ Proximal epiphysis
___ Spongy bone

Nutrient
artery in
nutrient
foramen

FIGURE 6.1   Parts of a long bone.

the epiphyseal plate where calcified cartilage is replaced by bone as the bone lengthens.

Now examine a mature long bone that has been sectioned longitudinally and locate the following structures:

1. *Compact (dense) bone*—Bone that forms the bulk of the diaphysis and covers the epiphyses.
2. *Spongy (cancellous) bone*—Bone that forms a small portion of the diaphysis at the ends of the medullary cavity and the bulk of the epiphysis; spongy bone contains red marrow.

3. *Medullary* (MED-yoo-lar′-ē) or *marrow cavity*—A cavity within the diaphysis that contains *yellow marrow* (primarily fat cells and a few scattered cells that can produce blood cells).
4. *Endosteum* (end-OS-tē-um)—Membrane that lines the medullary cavity and contains osteoprogenitor cells, osteoblasts, and scattered *osteoclasts* (OS′-tē-ō-clasts), cells that remove bone; it separates spongy bone from marrow.

Label the parts of a long bone indicated in Figure 6.1.

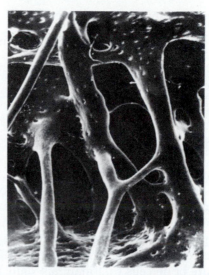

(c) Scanning electron micrograph of spongy bone trabeculae at a magnification of 25X

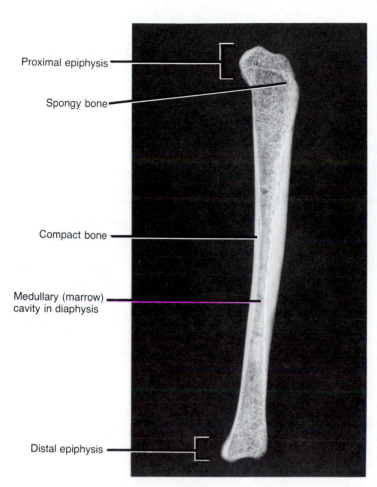

Proximal epiphysis

Spongy bone

Compact bone

Medullary (marrow) cavity in diaphysis

Distal epiphysis

(b) Photograph of a section through the tibia

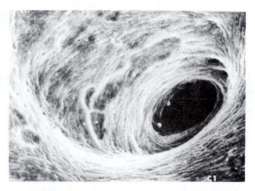

(d) Scanning electron micrograph of a view inside a central (Haversian) canal at a magnification of 250X

FIGURE 6.1 *(Continued)*   Parts of a long bone.

B. CHEMISTRY OF BONE

Unlike other connective tissues, the matrix of bone is very hard. This hardness results from the presence of mineral salts, mainly tricalcium phosphate ($Ca_3(PO_4)_2 \cdot [OH]_2$) and some calcium carbonate ($CaCO_3$). Mineral salts compose about 67% of the weight of bone. Despite its hardness, bone is also flexible, a characteristic that enables it to resist various forces. The flexibility of bone comes from organic substances in its matrix, especially collagen fibers. Organic materials compose about 33% of the weight of bone.

PROCEDURE

1. Obtain a bone that has been baked. Heating removes the organic part of bone. Apply pressure to the bone. Describe what happens.

_____

_____

_____

_____

2. Now obtain a bone that has already been soaked in nitric acid by your instructor. Acid

dissolves the minerals in bone. Apply pressure to the bone. Describe what happens. ___

_____

_____

C. HISTOLOGY OF BONE

Bone tissue, like other connective tissues, contains a large amount of matrix that surrounds widely separated cells. This matrix consists of (1) collagen fibers (2) abundant mineral salts, and (3) water. The cells include osteoprogenitor cells, osteoblasts, osteoclasts, and *osteocytes* (OS-tē-ō-sits), mature bone cells that maintain daily activities of bone tissue.

Depending on the size and distribution of spaces, bone may be categorized as spongy (cancellous) or compact (dense). *Spongy bone,* which contains many large spaces that store mainly red marrow, composes most of the tissue of short, flat, and irregularly shaped bones and most of the epiphyses of long bones. *Compact bone,* which contains few spaces and is deposited in layers over spongy bone, is thicker in the diaphysis than in the epiphyses.

Obtain a prepared slide of compact bone in which several osteons (Haversian systems) are shown in cross section. Observe them under high power. Look for the following structures:

1. *Central (Haversian) canal*—Circular canal in the center of an osteon (Haversian system) that runs longitudinally through the bone; the canal contains blood vessels, lymphatic vessels, and nerves.
2. *Lamellae* (la-MEL-ē)—Concentric layers of calcified matrix.
3. *Lacunae* (la-KOO-nē)—Spaces or cavities between lamellae that contain osteocytes.
4. *Canaliculi* (kan'-a-LIK-yoo-lē)—Minute canals that radiate in all directions from the lacunae and interconnect with each other; they contain slender processes of osteocytes; canaliculi provide routes so that nutrients can reach osteocytes and wastes can be removed from them.
5. *Osteocyte*—A mature bone cell located within a lacuna.
6. *Osteon (Haversian system)*—Microscopic structural unit of compact bone made up of a central (Haversian) canal plus its surrounding lamellae, lacunae, canaliculi, and osteocytes.

Label the parts of the osteon (Haversian system) shown in Figure 6.2.

Now obtain a prepared slide of a longitudinal section of compact bone. Examine it under high power and locate the following:

1. *Perforating (Volkmann's) canals*—Canals that extend obliquely or horizontally inward from the bone surface. They contain blood vessels, lymphatic vessels, and nerves and extend into central (Haversian) canals and the medullary cavity.
2. *Endosteum*—Layer of osteoprogenitor cells, osteoblasts, and scattered osteoclasts lining medullary (marrow) cavity and spongy bone.
3. *Medullary (marrow) cavity*—Space within diaphysis that contains yellow marrow in adults and red marrow in children.
4. *Lamellae*—Concentric rings of hard, calcified, matrix.
5. *Lacunae*—Spaces that contain osteocytes.
6. *Canaliculi*—Minute canals that radiate from lacunae and contain slender processes of osteocytes.
7. *Osteocytes*—Mature osteoblasts in lacunae that are no longer capable of producing new bone tissue but maintain daily activities of bone tissue.

Label the parts indicated in the microscopic view of bone in Figure 6.2.

D. TYPES OF BONES

The 206 named bones of the body can be classified by shape into four principal types:

1. *Long bones*—Have greater length than width, consist of a diaphysis and a variable number of epiphyses, and have a medullary (marrow) cavity; they contain more compact than spongy bone. Example: humerus.
2. *Short bones*—Somewhat cube-shaped; more spongy than compact. Example: wrist bones.
3. *Flat bones*—Generally thin and composed of two more-or-less parallel plates of compact bone enclosing a layer of spongy bone. Example: sternum.

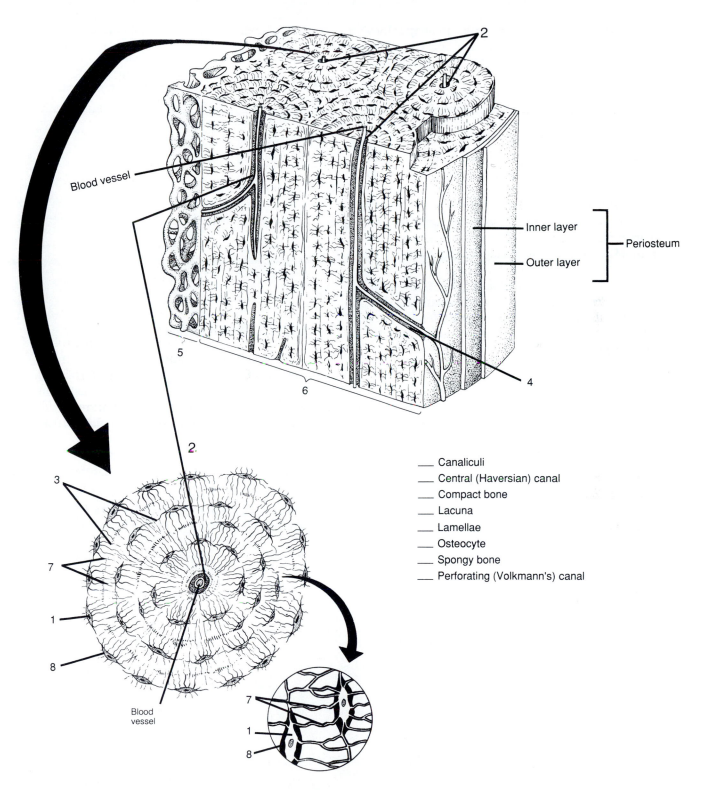

Blood vessel

Inner layer
Outer layer
Periosteum

___ Canaliculi
___ Central (Haversian) canal
___ Compact bone
___ Lacuna
___ Lamellae
___ Osteocyte
___ Spongy bone
___ Perforating (Volkmann's) canal

Blood
vessel

FIGURE 6.2   Cross section and longitudinal section of several osteons (Haversian systems).

4. *Irregular bones*—Very complex in shape, they cannot be grouped into any of the three categories just described. Example: vertebrae.

Other bones that are recognized, although not considered in the structural classification, include the following:

1. *Sutural* (SOO-chur-al) *bones*—Small bones between certain cranial bones; variable in number.
2. *Sesamoid* ("resembling a grain of sesame") *bones*—Small bones found in tendons and variable in number. The only constant sesamoid bones are the paired kneecaps.

Examine the disarticulated skeleton, Beauchene (disarticulated) skull, and articulated skeleton and find several examples of long, short, flat, and irregular bones. List examples of each type you find.

1. *Long.* _____

_____

_____

_____

2. *Short.* _____

_____

_____

_____

3. *Flat.* _____

_____

_____

_____

4. *Irregular.* _____

_____

_____

_____

E. BONE MARKINGS

The surfaces of bones contain various structural features that have specific functions. These features are called *bone markings* and are listed in Table 6.1. Knowledge of the bone markings will be very useful when you learn the bones of the body in Exercise 7.

Next to each marking listed in Table 6.1, write its definition, using your textbook as a reference.

ANSWER THE LABORATORY REPORT QUESTIONS AT THE END OF THE EXERCISE.

TABLE 6.1
Bone Markings

Marking	Description
DEPRESSIONS AND OPENINGS	
Fissure (FISH-ur)	
Foramen (fō-RĀ-men; *foramen* = hole)	
Meatus (mē-Ā-tus; *meatus* = canal)	
Paranasal sinus (*sin* = cavity)	
Grooves or sulcus (*sulcus* = ditchlike groove)	
Fossa (*fossa* = basinlike depression)	
PROCESSES (PROJECTIONS) THAT FORM JOINTS	
Condyle (KON-dīl; *condylus* = knucklelike process)	
Head	
Facet	
PROCESSES (PROJECTIONS) TO WHICH TENDONS, LIGAMENTS, AND OTHER CONNECTIVE TISSUES ATTACH	
Tubercle (TOO-ber-kul; *tube* = knob)	
Tuberosity	
Trochanter (trō-KAN-ter)	
Crest	
Line	
Spinous process of spine	
Epicondyle (*epi* = above)	

Bone Tissue

STUDENT _____   DATE _____

LABORATORY SECTION _____   SCORE/GRADE _____

PART 1. Completion

1. Small clusters of bones between certain cranial bones are referred to as _____ bones.

2. The technical name for a mature bone cell is a(n) _____.

3. Canals that extend obliquely inward or horizontally from the bone surface and contain blood vessels

   and lymphatic vessels are _____ canals.

4. The end, or extremity, of a bone is referred to as the _____.

5. Cube-shaped bones that contain more spongy than compact bone are known as

   _____ bones.

6. The cavity within the shaft of a bone that contains marrow is the _____ cavity.

7. The thin layer of hyaline cartilage covering the end of a bone where joints are formed is called

   _____ cartilage.

8. Minute canals that radiate from lacuna to lacuna are called _____.

9. The covering around the surface of a bone, except for the areas covered by cartilage, is the

   _____.

10. The shaft of a bone is referred to as the _____.

11. The _____ are concentric layers of calcified matrix.

12. The membrane that lines the medullary cavity and contains osteoblasts and a few osteoclasts is the

   _____.

13. The technical name for bone tissue is _____ tissue.

14. In a mature bone, the region where the shaft joins the extremity is called the

   _____.

15. The microscopic structural unit of compact bone is called a(n) _____.

16. The hardness of bone is primarily due to the mineral salt _____.

7

Bones

The 206 named bones of the adult skeleton are grouped into two divisions: axial and appendicular. The *axial skeleton* consists of bones that compose the axis of the body. The longitudinal axis is a straight line that runs through the center of gravity of the body, through the head, and down to the space between the feet. The *appendicular skeleton* consists of the bones of the extremities or limbs (upper and lower) and the girdles (pectoral and pelvic), that connect the extremities to the axial skeleton.

In this exercise you will study the names and locations of bones and their markings by examining various regions of the skeleton:

Region	Number of bones
AXIAL SKELETON	
Skull	
Cranium	8
Face	14
Hyoid (above the larynx)	1
*Auditory ossicles, 3 in each ear	6
Vertebral column	26
Thorax	
Sternum	1
Ribs	24
	80

Region	Number of bones
APPENDICULAR SKELETON	
Pectoral (shoulder) girdles	
Clavicle	2
Scapula	2
Upper extremities	
Humerus	2
Ulna	2
Radius	2
Carpals	16
Metacarpals	10
Phalanges	28
Pelvic (hip) girdles	
Hipbone (coxal bone)	2
Lower extremities	
Femur	2
Fibule	2
Tibia	2
Patella	2
Tarsals	14
Metatarsals	10
Phalanges	28
	126

*Not actually part of either the axial or appendicular skeleton, they are placed in the axial skeleton for convenience.

A. BONES OF ADULT SKULL

The *skull* is composed of two sets of bones— *cranial* and *facial*. The 8 cranial bones are 1 *frontal*, 2 *parietals* (pa-RĪ-i-talz), 2 *temporals*, 1 *occipital* (ok-SIP-i-tal), 1 *sphenoid* (SFĒ-noyd), and 1 *ethmoid*. The 14 facial bones include 2 *nasals*, 2 *maxillae* (mak-SIL-ē), 2 *zygo-*

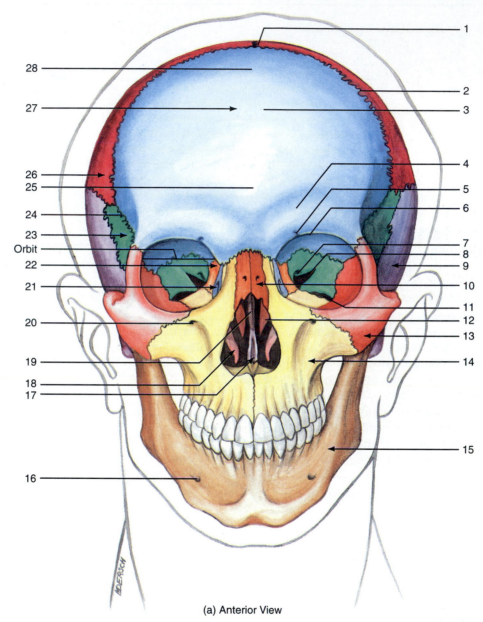

(a) Anterior View

1. Sagittal suture
2. Coronal suture
3. Frontal squama
4. Superciliary arch
5. Supraorbital foramen (notch)
6. Supraorbital margin
7. Optic foramen (canal)
8. Superior orbital fissure
9. Temporal bone
10. Nasal bone

11. Inferior orbital fissure
12. Middle nasal concha (turbinate)
13. Zygomatic bone
14. Maxilla
15. Mandible
16. Mental foramen
17. Vomer
18. Inferior nasal concha (turbinate)

19. Perpendicular plate
20. Infraorbital foramen
21. Lacrimal bone
22. Ethmoid bone
23. Sphenoid bone
24. Squamous suture
25. Glabella
26. Parietal bone
27. Frontal bone
28. Frontal eminence

FIGURE 7.1   The skull.

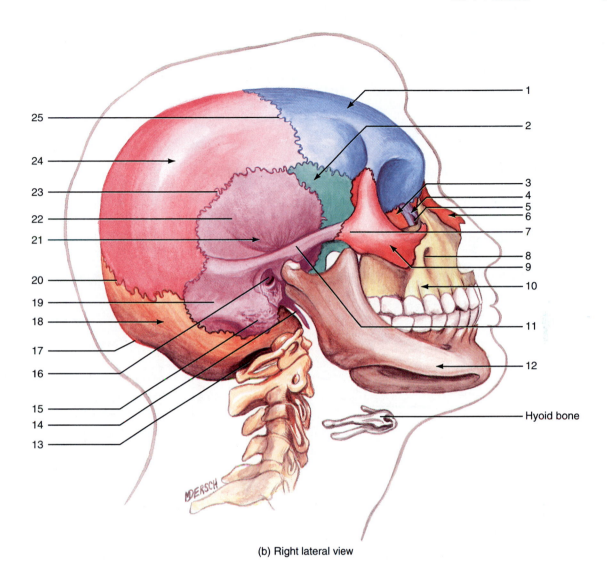

(b) Right lateral view

1. Frontal bone	11. Zygomatic process	18. Occipital bone
2. Sphenoid bone	12. Mandible	19. Mastoid portion
3. Ethmoid bone	13. Foramen magnum	20. Lambdoid suture
4. Lacrimal bone	14. Styloid process	21. Temporal bone
5. Lacrimal foramen (fossa)	15. Mastoid process	22. Temporal squama
6. Nasal bone	16. External auditory	23. Squamous suture
7. Temporal process	(acoustic) meatus	24. Parietal bone
8. Infraorbital foramen	17. External occipital	25. Coronal suture
9. Zygomatic bone	protuberance	
10. Maxilla		

FIGURE 7.1 *(Continued)*   The skull.

matics, 1 *mandible,* 2 *lacrimals* (LAK-ri-mals), 2 *palatines* (PAL-a-tīns), 2 *inferior conchae* (KONG-kē), or *turbinates,* and 1 *vomer.* These bones are indicated by *arrows* in Figure 7.1. Using the series of photographs in Figure 7.1 for reference, locate the cranial and facial bones on both a Beauchene (disarticulated) and an articulated skull.

Obtain a Beauchene skull and an articulated skull and observe them in anterior view. Using Figure 7.1a for reference, locate the parts indicated in the figure.

Turn the Beauchene skull and articulated skull so that you are looking at the right side. Using Figure 7.1b for reference, locate the parts indicated in the figure. Note also the *hyoid bone*

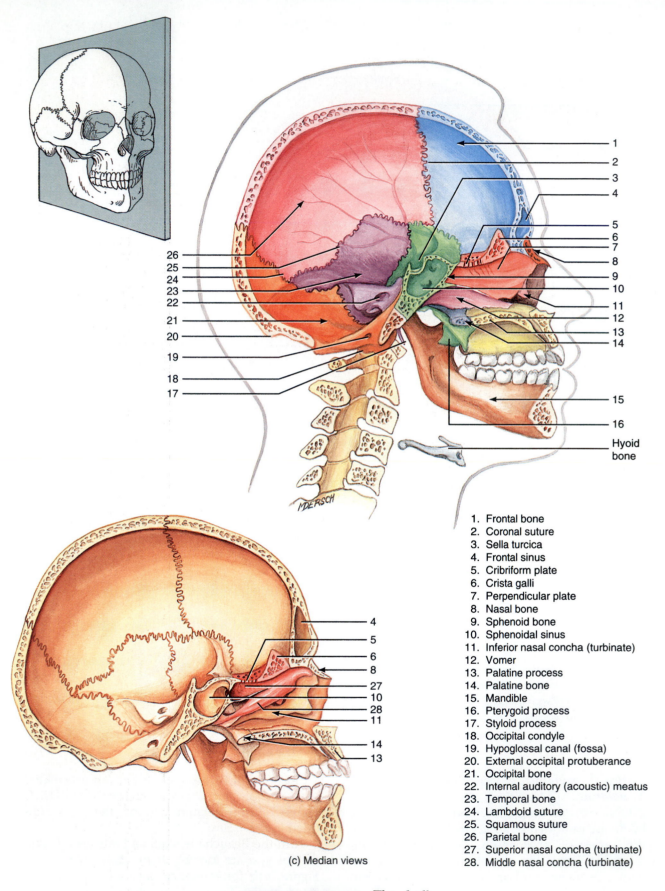

MDERSCH

Hyoid bone

1. Frontal bone
2. Coronal suture
3. Sella turcica
4. Frontal sinus
5. Cribriform plate
6. Crista galli
7. Perpendicular plate
8. Nasal bone
9. Sphenoid bone
10. Sphenoidal sinus
11. Inferior nasal concha (turbinate)
12. Vomer
13. Palatine process
14. Palatine bone
15. Mandible
16. Pterygoid process
17. Styloid process
18. Occipital condyle
19. Hypoglossal canal (fossa)
20. External occipital protuberance
21. Occipital bone
22. Internal auditory (acoustic) meatus
23. Temporal bone
24. Lambdoid suture
25. Squamous suture
26. Parietal bone
27. Superior nasal concha (turbinate)
28. Middle nasal concha (turbinate)

(c) Median views

FIGURE 7.1 *(Continued)*   The skull.

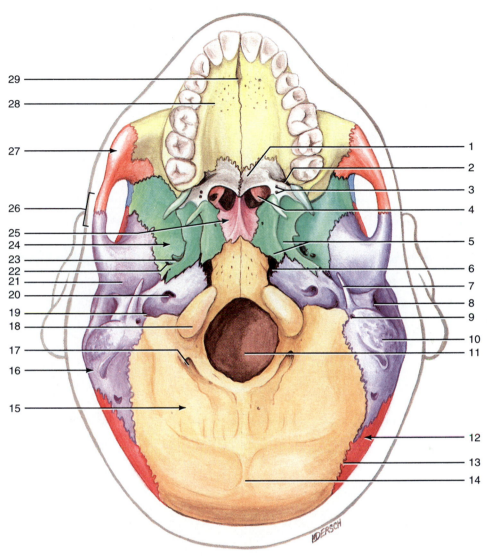

(d) Inferior view

1. Horizontal plate
2. Greater palatine foramen
3. Lesser palatine foramina
4. Middle nasal concha (turbinate)
5. Pterygoid process
6. Foramen lacerum
7. Styloid process
8. External auditory (acoustic) meatus
9. Stylomastoid foramen
10. Mastoid process
11. Foramen magnum
12. Parietal bone
13. Lambdoid suture
14. External occipital protuberance
15. Occipital bone

16. Temporal bone
17. Condylar canal
18. Occipital condyle
19. Jugular foramen
20. Carotid foramen (canal)
21. Mandibular (glenoid) fossa
22. Foramen spinosum
23. Foramen ovale
24. Sphenoid bone
25. Vomer
26. Zygomatic arch
27. Zygomatic bone
28. Palatine process
29. Incisive foramen

FIGURE 7.1 *(Continued)*   The skull.

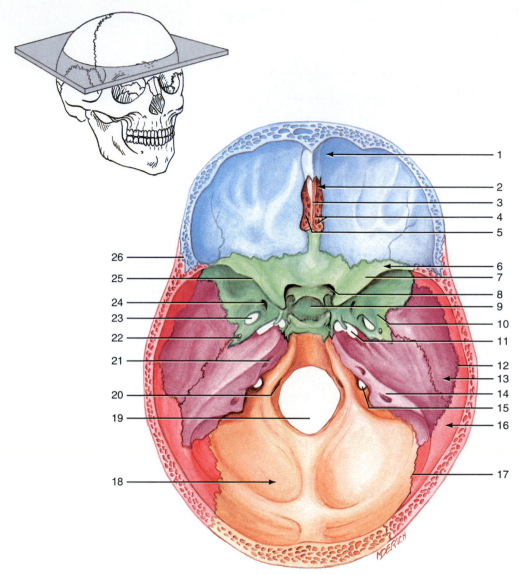

(e) Floor of cranium

1. Frontal bone
2. Ethmoid bone
3. Crista galli
4. Olfactory foramina
5. Cribriform plate
6. Sphenoid bone
7. Lesser wing
8. Optic foramen (canal)
9. Sella turcica
10. Greater wing
11. Foramen lacerum
12. Squamous suture
13. Temporal bone
14. Petrous portion
15. Jugular foramen
16. Parietal bone
17. Lambdoid suture
18. Occipital bone
19. Foramen magnum
20. Hypoglossal canal (fossa)
21. Internal auditory (acoustic) meatus
22. Foramen spinosum
23. Foramen ovale
24. Foramen rotundum
25. Superior orbital fissure
26. Coronal suture

FIGURE 7.1 *(Continued)*   The skull.

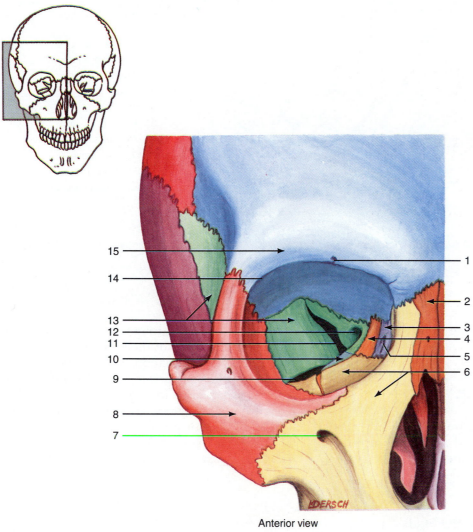

1. Supraorbital foramen (notch)
2. Nasal bone
3. Lacrimal bone
4. Ethmoid bone
5. Lacrimal foramen (fossa)
6. Maxilla
7. Infraorbital foramen
8. Zygomatic bone
9. Inferior orbital fissure
10. Palatine bone
11. Superior orbital fissure
12. Optic foramen (canal)
13. Sphenoid bone
14. Supraorbital margin
15. Frontal bone

Anterior view

FIGURE 7.2 The right orbit

below the mandible. This is not a bone of the skull; it is noted here because of its proximity to the skull.

If a skull in median section is available, use Figure 7.1c for reference and locate the parts indicated in the figure.

Take an articulated skull and turn it upside down so that you are looking at the inferior surface. Using Figure 7.1d for reference, locate the parts indicated in the figure.

Obtain an articulated skull with a removable crown. Using Figure 7.1e for reference, locate the parts indicated in the figure.

Examine the right orbit of an articulated skull. Using Figure 7.2 for reference, locate the parts indicated in the figure.

Obtain a mandible and, using Figure 7.3 for reference, identify the parts indicated in the figure.

Before you move on, examine Table 7.1, "Summary of Foramina of the Skull" on page 94. It will give you an understanding of the structures that pass through the various foramina you just studied.

B. SUTURES OF SKULL

A *suture* (SOO-chur) is an immovable joint between skull bones. The four prominent sutures are the *coronal, sagittal, lambdoid* (LAM-doyd), and *squamous* (SKWĀ-mos). Using Figures 7.1

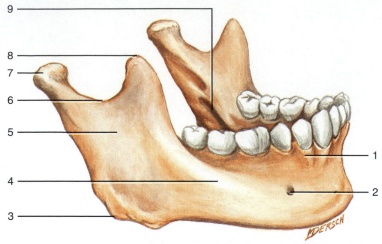

Right lateral view

1. Alveolar process
2. Mental foramen
3. Angle
4. Body
5. Ramus
6. Mandibular notch
7. Condylar process
8. Coronoid process
9. Mandibular foramen

FIGURE 7.3   Mandible

and 7.4 for reference, locate these sutures on an iculated skull.

C. FONTANELS OF SKULL

At birth, the skull bones are separated by membrane-filled spaces called *fontanels* (fon'-ta-NELZ). They (1) enable the fetal skull to compress as it passes through the birth canal, (2) permit rapid growth of the brain during infancy, (3) facilitate determination of the degree of brain development by their state of closure, (4) serve as landmarks (anterior fontanel) for withdrawal of blood from the superior sagittal sinus, and (5) aid in determining the position of the fetal head prior to birth. The principal fontanels are *anterior (frontal), posterior (occipital), anterolateral*

TABLE 7.1
Summary of Foramina of the Skull

Foramen	Structures Passing Through
Carotid	Internal carotid artery
Greater palatine	Greater palatine nerve and greater palatine vessels
Incisive	Branches of greater palatine vessels and nasopalatine nerve
Infraorbital	Infraorbital nerve and artery
Jugular	Internal jugular vein, glossopharyngeal (IX) nerve, vagus (X) nerve, and accessory (XI) nerve
Lacerum	Internal carotid artery and branch of ascending pharyngeal artery
Lesser palatine	Lesser palatine nerves and artery
Magnum	Medulla oblongata and its membranes, accessory (XI) nerve, and vertebral and spinal arteries and meninges
Mandibular	Inferior alveolar nerve and vessels
Mental	Mental nerve and vessels
Olfactory	Olfactory (I) nerve
Optic	Optic (II) nerve and ophthalmic artery
Ovale	Mandibular branch of trigeminal (V) nerve
Rotundum	Maxillary branch of trigeminal (V) nerve
Spinosum	Middle meningeal vessels
Stylomastoid	Facial (VII) nerve and stylomastoid artery
Supraorbital	Supraorbital nerve and artery

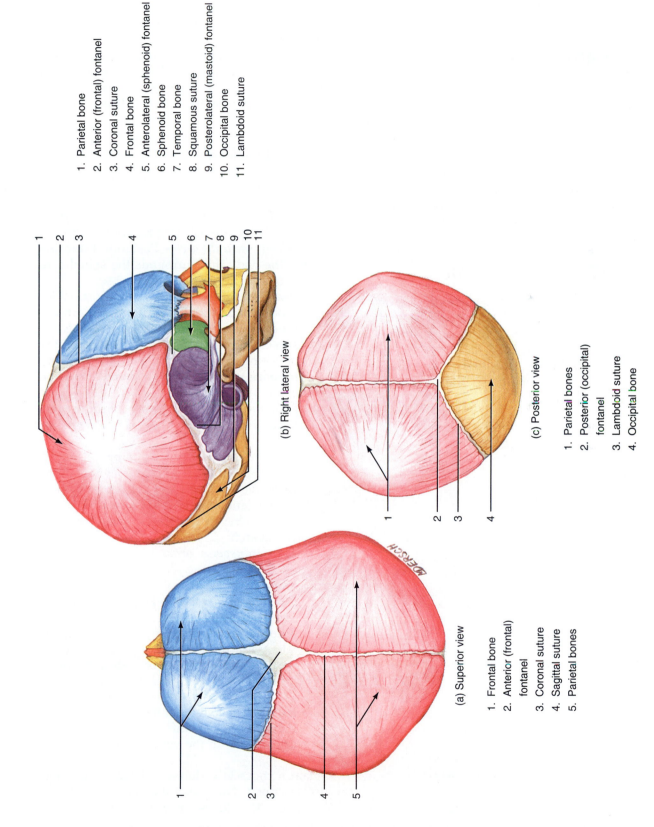

1. Parietal bone
2. Anterior (frontal) fontanel
3. Coronal suture
4. Frontal bone
5. Anterolateral (sphenoid) fontanel
6. Sphenoid bone
7. Temporal bone
8. Squamous suture
9. Posterolateral (mastoid) fontanel
10. Occipital bone
11. Lambdoid suture

(a) Superior view

1. Frontal bone
2. Anterior (frontal) fontanel
3. Coronal suture
4. Sagittal suture
5. Parietal bones

(b) Right lateral view

(c) Posterior view

1. Parietal bones
2. Posterior (occipital) fontanel
3. Lambdoid suture
4. Occipital bone

FIGURE 7.4   Fontanels.

(sphenoidal), and *posterolateral (mastoid).* Using Figure 7.4 for reference, locate the fontanels on the skull of a newborn infant.

D. PARANASAL SINUSES OF SKULL

A *paranasal (para* = beside) *sinus,* or just simply *sinus,* is a cavity in a bone located near the nasal cavity. Paired paranasal sinuses are found in the maxillae and the frontal, sphenoid, and ethmoid bones. Locate the paranasal sinuses on the Beauchene skull or other demonstration models that may be available. Label the paranasal sinuses shown in Figure 7.5.

E. VERTEBRAL COLUMN

The *vertebral column,* together with the sternum and ribs, constitutes the skeleton of the trunk. The *vertebrae* of the adult column are distributed as follows: 7 *cervical* (SER-vi-kal) (neck), 12 *thoracic* (thō-RAS-ik) (chest), 5 *lumbar* (lower back), 5 *sacral* (fused into one bone, the *sacrum* (SĀ-krum) between the hipbones), and 3, 4, or 5

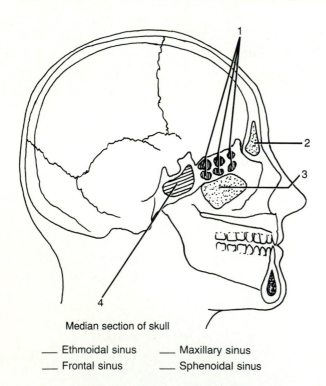

Median section of skull

___ Ethmoidal sinus   ___ Maxillary sinus
___ Frontal sinus   ___ Sphenoidal sinus

FIGURE 7.5 Paranasal sinuses.

coccygeal (kok-SIJ-ē-al) (fused into one bone, the *coccyx* (KOK-six), forming the tail of the column). Locate each of these regions on the articulated skeleton. Label the same regions in Figure 7.6, the anterior view.

Examine the vertebral column on the articulated skeleton and identify the cervical, thoracic, lumbar, and sacral (sacrococcygeal) curves. Label the curves in Figure 7.6, the right lateral view.

F. VERTEBRAE

A typical *vertebra* consists of the following portions:

1. *Body*—Thick, disc-shaped anterior portion.
2. *Vertebral (neural) arch*—Posterior extension from the body that surrounds the spinal cord and consists of the following parts:
   a. *Pedicles* (PED-i-kuls)—Two short, thick processes that project posteriorly; each has a superior and an inferior notch and, when successive vertebrae are fitted together, the opposing notches form an *intervertebral foramen* through which a spinal nerve and blood vessels pass.
   b. *Laminae* (LAM-i-nē)—Flat portions that form the posterior wall of the vertebral arch.
   c. *Vertebral foramen*—Opening through which the spinal cord passes; when all the vertebrae are fitted together, the foramina form a canal, the *vertebral canal.*
3. *Processes*—Seven processes arise from the vertebral arch:
   a. Two *transverse processes*—Lateral extensions where the laminae and pedicles join.
   b. One *spinous process*—Posterior projection of the lamina.
   c. Two *superior articular processes*—Articulate with the vertebra above. Their top surfaces, called *superior articular facets,* articulate with the vertebra above.
   d. Two *inferior articular processes*—Articulate with the vertebra below. Their bottom surfaces, called *inferior articular facets,* articulate with the vertebra below.

Obtain a thoracic vertebra and locate each part just described. Now label the vertebra in Figure 7.7a. You should be able to distinguish the

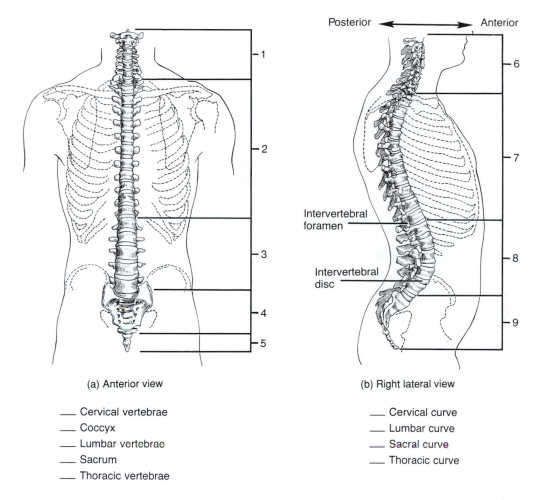

(a) Anterior view

___ Cervical vertebrae
___ Coccyx
___ Lumbar vertebrae
___ Sacrum
___ Thoracic vertebrae

(b) Right lateral view

___ Cervical curve
___ Lumbar curve
___ Sacral curve
___ Thoracic curve

FIGURE 7.6   Vertebral column.

general parts on all the different vertebrae that contain them.

Although vertebrae have the same basic design, those of a given region have special distinguishing features. Obtain examples of the following vertebrae and identify their distinguishing features.

1. *Cervical vertebrae*
   a. *Atlas*   First cervical vertebra (Figure 7.7b)
      Transverse foramen   Opening in transverse process through which an artery, a vein, and a branch of a spinal nerve pass.
      Anterior arch   Anterior wall of vertebral foramen.
      Posterior arch   Posterior wall of vertebral foramen.
      Lateral mass   Side wall of vertebral foramen.

Label the other indicated parts.

   b. *Axis*   Second cervical vertebra (Figure 7.7c)
      Dens   Superior projection of body that iculates with atlas.

Label the other indicated parts.

   c. *Cervicals 3 through 6* (Figure 7.7d)
      Bifid spinous process   Cleft in spinous processes of cervical vertebrae 2 through 6.

Label the other indicated parts.

   d. *Vertebra prominens*   Seventh cervical vertebra; contains a nonbifid and long spinous process.
2. *Thoracic vertebrae* (Figure 7.7e)
   a. *Facets*   For articulation with the tubercle

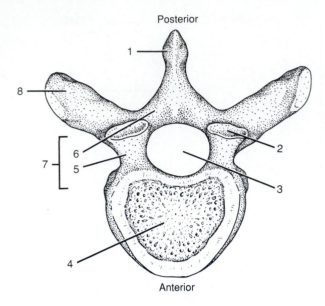

Posterior

(a) Superior view of a typical vertebra

___ Body
___ Lamina
___ Pedicle
___ Spinous process
___ Superior articular facet
___ Transverse process
___ Vertebral arch
___ Vertebral foramen

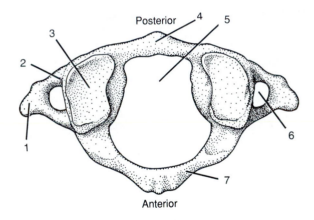

Posterior

Anterior

(b) Superior view of atlas

___ Anterior arch
___ Lateral mass
___ Posterior arch
___ Superior articular facet
___ Transverse foramen
___ Transverse process
___ Vertebral foramen

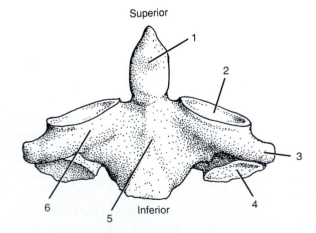

Superior

Inferior

(c) Anterior view of the axis

___ Body
___ Dens
___ Inferior articular facet
___ Lateral mass
___ Superior articular facet
___ Transverse process

FIGURE 7.7   Vertebrae.

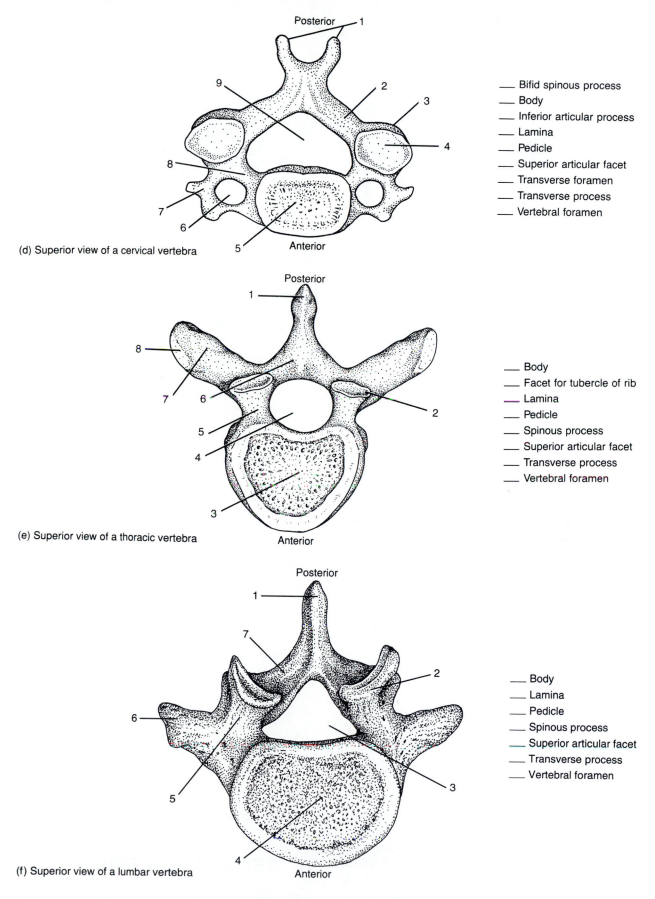

Posterior

1

9

2

3

4

8

7

6

5

Anterior

(d) Superior view of a cervical vertebra

___ Bifid spinous process
___ Body
___ Inferior articular process
___ Lamina
___ Pedicle
___ Superior articular facet
___ Transverse foramen
___ Transverse process
___ Vertebral foramen

Posterior

1

8

7

6

5

4

3

2

(e) Superior view of a thoracic vertebra

Anterior

___ Body
___ Facet for tubercle of rib
___ Lamina
___ Pedicle
___ Spinous process
___ Superior articular facet
___ Transverse process
___ Vertebral foramen

Posterior

1

7

6

2

5

3

4

(f) Superior view of a lumbar vertebra

Anterior

___ Body
___ Lamina
___ Pedicle
___ Spinous process
___ Superior articular facet
___ Transverse process
___ Vertebral foramen

FIGURE 7.7 (*Continued*)   Vertebrae.

99

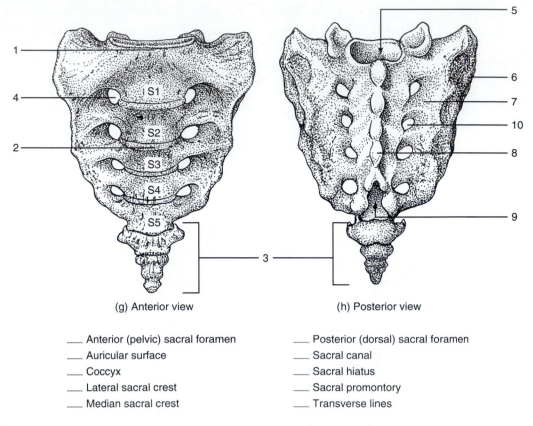

(g) Anterior view   (h) Posterior view

___ Anterior (pelvic) sacral foramen   ___ Posterior (dorsal) sacral foramen
___ Auricular surface   ___ Sacral canal
___ Coccyx   ___ Sacral hiatus
___ Lateral sacral crest   ___ Sacral promontory
___ Median sacral crest   ___ Transverse lines

FIGURE 7.7 *(Continued)*   Vertebrae. Sacrum and Coccyx.

and head of a rib; found on body and transverse processes.
  b. *Spinous processes* Long, pointed, downward projection.

Label the other indicated parts.

3. *Lumbar vertebrae* (Figure 7.7f)
  a. *Spinous processes* Broad, blunt.
  b. *Superior articular processes* Directed medially, not superiorly.
  c. *Inferior articular processes* Directed laterally, not inferiorly.

Label the other indicated parts.

4. *Sacrum* (Figure 7.7g)
  a. *Transverse lines* Areas where bodies of sacral vertebrae are joined.
  b. *Anterior (pelvic) sacral foramina* Four pairs of foramina that communicate with dorsal sacral foramina; passages for blood vessels and nerves.
  c. *Median sacral crest* Spinous processes of fused sacral vertebrae.

  d. *Lateral sacral crest* Transverse processes of fused sacral vertebrae.
  e. *Posterior (dorsal) sacral foramina* Four pairs of foramina that communicate with pelvic foramina; passages for blood vessels and nerves.
  f. *Sacral canal* Continuation of vertebral canal.
  g. *Sacral promontory* Superior, anterior projecting border.
  h. *Auricular surface* Articulates with ilium of hipbone.
  i. *Sacral hiatus* (hī-Ā-tus) Inferior entrance to sacral canal where laminae of S5, and sometimes S4, fail to meet.
5. *Coccyx* (Figure 7.7g)

Label the coccyx and the parts of the sacrum in Figure 7.7g.

G. STERNUM AND RIBS

The skeleton of the *thorax* consists of the *sternum, costal cartilages, ribs,* and *bodies of the thoracic vertebrae.*

Examine the articulated skeleton and disarticulated bones and identify the following:

1. ***Sternum (breastbone)***—Flat bone in midline of anterior thorax.
   a. *Manubrium* (ma-NOO-brē-um) Superior portion.
   b. *Body* Middle, largest portion.
   c. *Sternal angle* Junction of the manubrium and body.
   d. *Xiphoid* (ZI-foyd) *process* Inferior, smallest portion.
   e. *Suprasternal notch* Depression on the superior surface of the manubrium that can be felt.
   f. *Clavicular notches* Articular surfaces for the clavicles.

Label the parts of the sternum in Figure 7.8a.

2. ***Costal cartilage***—Strip of hyaline cartilage that attaches a rib to the sternum.
3. ***Ribs***—Parts of a typical rib (third through ninth) include
   a. *Body* Shaft, main part of rib.
   b. *Head* Posterior projection.
   c. *Neck* Constricted portion behind head.
   d. *Tubercle* (TOO-ber-kul) Knoblike elevation just below neck; consists of a *nonarticular part* that affords attachment for a ligament and an *articular part* that iculates with an inferior vertebra.
   e. *Costal groove* Depression on the inner surface containing blood vessels and a nerve.
   f. *Superior facet* Articulates with facet on superior vertebra.

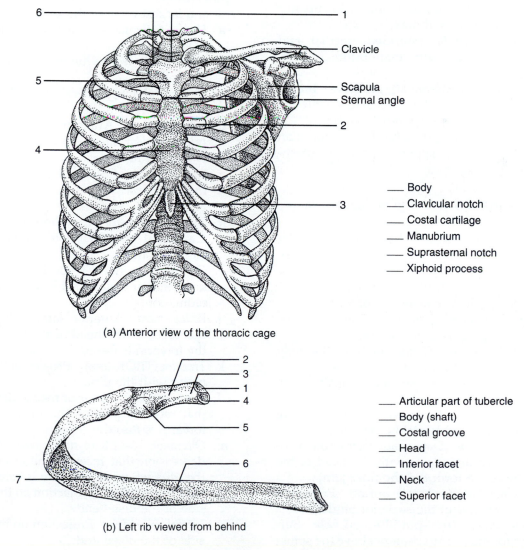

(a) Anterior view of the thoracic cage

___ Body
___ Clavicular notch
___ Costal cartilage
___ Manubrium
___ Suprasternal notch
___ Xiphoid process

(b) Left rib viewed from behind

___ Articular part of tubercle
___ Body (shaft)
___ Costal groove
___ Head
___ Inferior facet
___ Neck
___ Superior facet

FIGURE 7.8 Bones of the thorax.

g. *Inferior facet*   Articulates with facet on inferior vertebra.

Label the parts of a rib in Figure 7.8b.

H. PECTORAL (SHOULDER) GIRDLES

Each *pectoral* (PEK-tō-ral) or *shoulder girdle* consists of two bones—*clavicle* (collar bone) and *scapula* (shoulder blade). Its purpose is to attach the bones of the upper extremity to the axial skeleton.

Examine the articulated skeleton and disarticulated bones and identify the following:

1. *Clavicle* (KLAV-i-kul)—Slender bone with a double curvature; lies horizontally in superior and anterior part of thorax.
    a. *Sternal extremity*   Rounded, medial end that articulates with manubrium of sternum.
    b. *Acromial* (a-KRŌ-mē-al) *extremity*   Broad, flat, lateral end that articulates with the acromion of the scapula.
    c. *Conoid tubercle* (TOO-ber-kul)   Projection on the inferior, lateral surface for attachment of a ligament.

Label the parts of the clavicle in Figure 7.9a.

2. *Scapula* (SCAP-yoo-la)—Large, flat triangular bone in dorsal thorax between the levels of ribs 2 through 7.
    a. *Body*   Flattened, triangular portion.
    b. *Spine*   Ridge across posterior surface.
    c. *Acromion* (a-KRŌ-mē-on)   Flattened, expanded process of the spine.
    d. *Medial (vertebral) border*   Edge of the body near vertebral column.
    e. *Lateral (axillary) border*   Edge of the body near the arm.
    f. *Inferior angle*   Bottom of the body where the medial and lateral borders join.
    g. *Glenoid cavity*   Depression below the acromion that articulates with the head of the humerus to form the shoulder joint.
    h. *Coracoid* (KOR-a-koyd) *process*   Projection at lateral end of the superior border.
    i. *Supraspinous* (soo'-pra-SPĪ-nus) *fossa*   Surface for muscle attachment above the spine.

j. *Infraspinous fossa*   Surface for muscle attachment below the spine.
k. *Superior border*   Superior edge of the body.
l. *Superior angle*   Top of the body where the superior and medial borders join.

Label the parts of the scapula in Figure 7.9b, c, and d.

I. UPPER EXTREMITIES

The skeleton of the *upper extremities* consists of a humerus in each arm, an ulna and radius in each forearm, carpals in each wrist, metacarpals in each palm, and phalanges in the fingers.

Examine the articulated skeleton and disarticulated bones and identify the following:

1. *Humerus* (HYOO-mer-us)—Arm bone; longest and largest bone of the upper extremity.
    a. *Head*   Articulates with the glenoid cavity of scapula.
    b. *Anatomical neck*   Oblique groove below the head.
    c. *Greater tubercle*   Lateral projection below the anatomical neck.
    d. *Lesser tubercle*   Anterior projection.
    e. *Intertubercular sulcus (bicipital groove)*   Between the tubercles.
    f. *Surgical neck*   Constricted portion below the tubercles.
    g. *Body*   Shaft.
    h. *Deltoid tuberosity*   Roughened, V-shaped area about midway down the lateral surface of a shaft.
    i. *Capitulum* (ka-PIT-yoo-lum)   Rounded knob that articulates with the head of the radius.
    j. *Radial fossa*   Anterior lateral depression that receives the head of the radius when the forearm is flexed.
    k. *Trochlea* (TRŌK-lē-a)   Projection that articulates with the ulna.
    l. *Coronoid fossa*   Anterior medial depression that receives part of the ulna when the forearm is flexed.
    m. *Olecranon* (ō-LEK-ra-non) *fossa*   Posterior depression that receives the olecranon of the ulna when the forearm is extended.
    n. *Medial epicondyle*   Projection on the medial side of the distal end.
    o. *Lateral epicondyle*   Projection on the lateral side of the distal end.

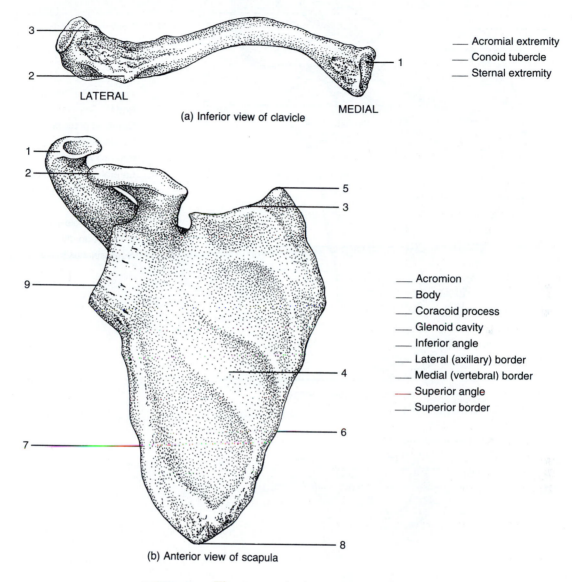

3 ———

2 ———

LATERAL

1 ——— 1

MEDIAL

(a) Inferior view of clavicle

___ Acromial extremity
___ Conoid tubercle
___ Sternal extremity

1 —
2 —

5 —
3 —

9 —

4 —

6 —

7 —

8 —

(b) Anterior view of scapula

___ Acromion
___ Body
___ Coracoid process
___ Glenoid cavity
___ Inferior angle
___ Lateral (axillary) border
___ Medial (vertebral) border
___ Superior angle
___ Superior border

FIGURE 7.9   The pectoral (shoulder) girdle.

Label the parts of the humerus in Figure 7.10a and b.

2. *Ulna*—Medial bone of the forearm.
   a. *Olecranon (olecranon process)*   Prominence of the elbow at the proximal end.
   b. *Coronoid process*   Anterior projection that, with the olecranon, receives the trochlea of the humerus.
   c. *Trochlear (semilunar) notch*   Curved area between the olecranon and the coronoid process into which the trochlea of the humerus fits.
   d. *Radial notch*   Depression lateral and inferior to the trochlear notch that receives the head of the radius.

   e. *Head*   Rounded portion at the distal end.
   f. *Styloid process*   Projection on the posterior side of the distal end.

Label the parts of the ulna in Figure 7.10c and d.

3. *Radius*—Lateral bone of the forearm.
   a. *Head*   Disc-shaped process at the proximal end.
   b. *Radial tuberosity*   Medial projection for the insertion of the biceps brachii muscle.
   c. *Styloid process*   Projection on the lateral side of the distal end.
   d. *Ulnar notch*   Medial, concave depression for articulation with the ulna.

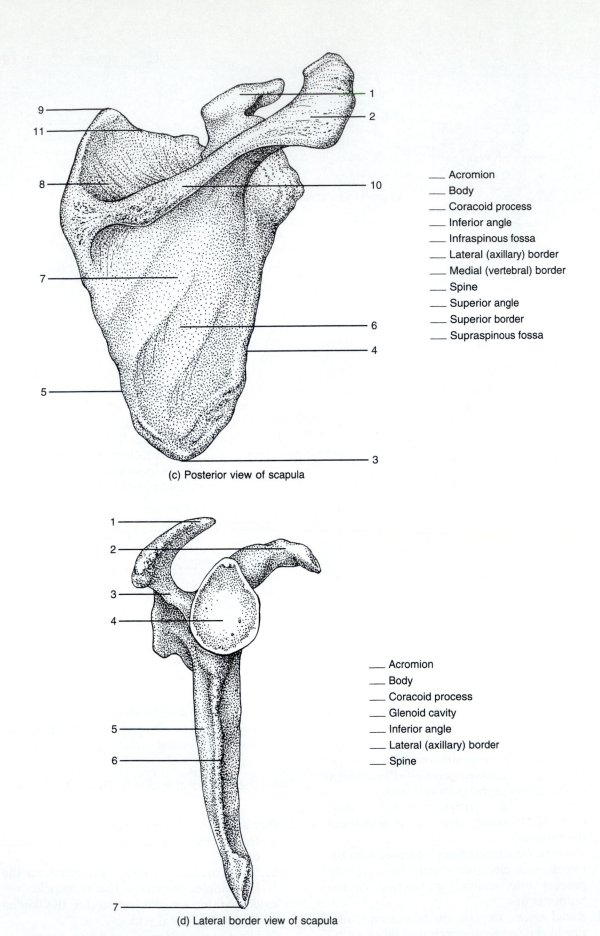

_____ Acromion
_____ Body
_____ Coracoid process
_____ Inferior angle
_____ Infraspinous fossa
_____ Lateral (axillary) border
_____ Medial (vertebral) border
_____ Spine
_____ Superior angle
_____ Superior border
_____ Supraspinous fossa

(c) Posterior view of scapula

_____ Acromion
_____ Body
_____ Coracoid process
_____ Glenoid cavity
_____ Inferior angle
_____ Lateral (axillary) border
_____ Spine

(d) Lateral border view of scapula

FIGURE 7.9 _(Continued)_ The pectoral (shoulder) girdle.

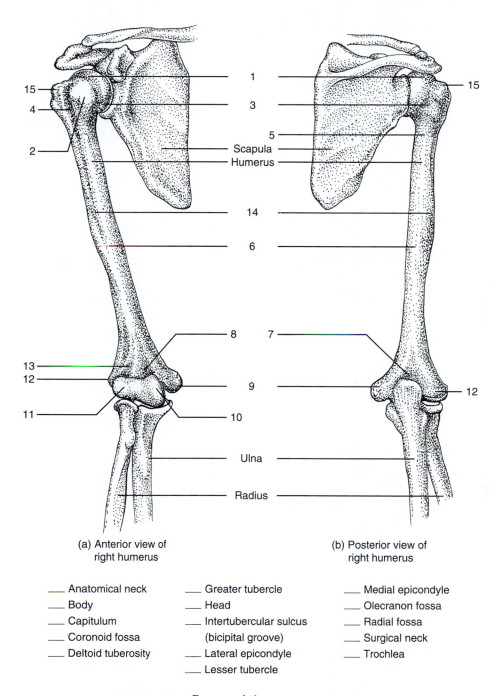

15   4   2   13   12   11

1
3
5
Scapula
Humerus
14
6
8
9
10
Ulna
Radius

7
15
12

(a) Anterior view of
right humerus

(b) Posterior view of
right humerus

____ Anatomical neck
____ Body
____ Capitulum
____ Coronoid fossa
____ Deltoid tuberosity

____ Greater tubercle
____ Head
____ Intertubercular sulcus
(bicipital groove)
____ Lateral epicondyle
____ Lesser tubercle

____ Medial epicondyle
____ Olecranon fossa
____ Radial fossa
____ Surgical neck
____ Trochlea

FIGURE 7.10 Bones of the upper extremity.

Label the parts of the radius in Figure 7.10e.

4. Carpus—Wrist, consists of eight small bones, called **carpals,** united by ligaments.
   a. *Proximal row* From medial to lateral, called *pisiform, triquetrum, lunate,* and *scaphoid.*
   b. *Distal row* From medial to lateral, called *hamate, capitate, trapezoid,* and *trapezium.*
5. Metacarpus—Five bones in the palm of the hand, numbered as follows, beginning with

the thumb side: I, II, III, IV, and V **metacarpals.**
6. Phalanges (fa-LAN-jēz)—Bones of the fingers; two in each thumb (proximal and distal) and three in each finger (proximal, middle, and distal). The singular of phalanges is **phalanx.**

Label the parts of the carpus, metacarpus, and phalanges in Figure 7.10f.

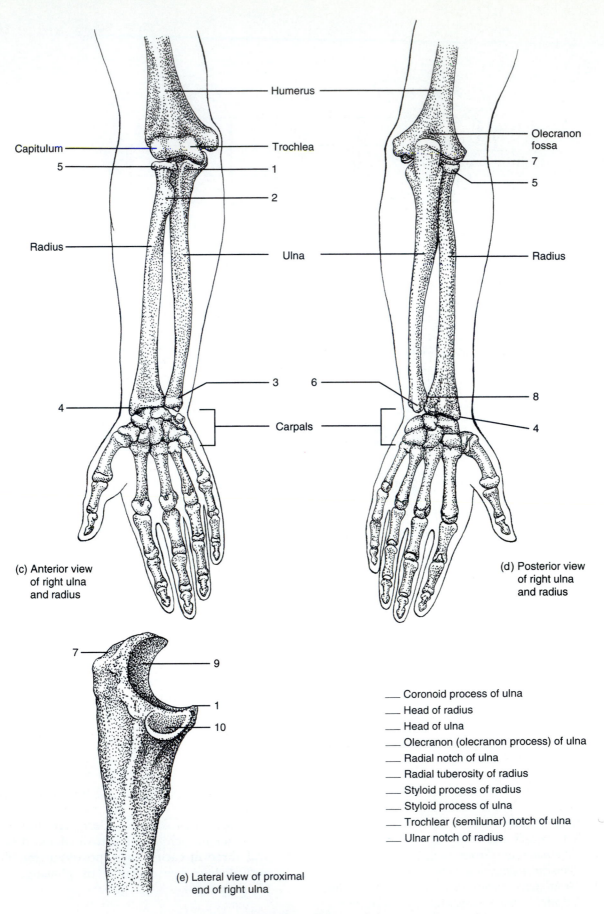

Capitulum

Trochlea

Humerus

5

1

2

Radius

Ulna

Olecranon fossa

7

5

Radius

Radius

3

6

8

4

Carpals

4

(c) Anterior view of right ulna and radius

(d) Posterior view of right ulna and radius

7

9

1

10

(e) Lateral view of proximal end of right ulna

___ Coronoid process of ulna
___ Head of radius
___ Head of ulna
___ Olecranon (olecranon process) of ulna
___ Radial notch of ulna
___ Radial tuberosity of radius
___ Styloid process of radius
___ Styloid process of ulna
___ Trochlear (semilunar) notch of ulna
___ Ulnar notch of radius

FIGURE 7.10 (Continued) Bones of the upper extremity.

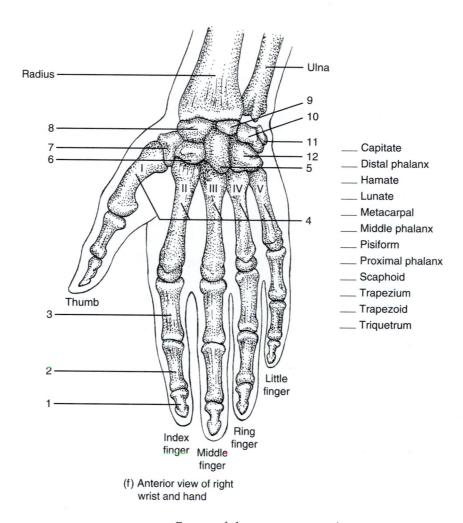

Radius — Ulna

9
10
8
7
11
6
12
5

I
II III IV V

4

___ Capitate
___ Distal phalanx
___ Hamate
___ Lunate
___ Metacarpal
___ Middle phalanx
___ Pisiform
___ Proximal phalanx
___ Scaphoid
___ Trapezium
___ Trapezoid
___ Triquetrum

Thumb
3

2

1

Little
finger

Index
finger Middle
finger Ring
finger

(f) Anterior view of right
wrist and hand

FIGURE 7.10 *(Continued)* Bones of the upper extremity.

J. PELVIC (HIP) GIRDLES

The *pelvic (hip) girdles* consist of the two *hipbones* or *coxal* (KOK-sal) *bones.* They provide a strong and stable support for the lower extremities on which the weight of the body is carried and attach the lower extremities to the axial skeleton. Examine the articulated skeleton and disarticulated bones and identify the following parts of the hipbone:

1. *Ilium*—Superior flattened portion.
   a. *Iliac crest* Superior border of the ilium.
   b. *Anterior superior iliac spine* Anterior projection of the iliac crest.
   c. *Anterior inferior iliac spine* Projection under the anterior superior iliac spine.
   d. *Posterior superior iliac spine* Posterior projection of the iliac crest.

   e. *Posterior inferior iliac spine* Projection below the posterior superior iliac spine.
   f. *Greater sciatic* (sī-AT-ik) *notch* Concavity under the posterior inferior iliac spine.
   g. *Iliac fossa* Medial concavity for the attachment of the iliacus muscle.
   h. *Iliac tuberosity* Point of attachment for the sacroiliac ligament posterior to the iliac fossa.
   i. *Auricular surface* Point of articulation with the sacrum.
2. *Ischium* (IS-kē-um)—Lower, posterior portion.
   a. *Ischial spine* Posterior projection of the ischium.
   b. *Lesser sciatic notch* Concavity under the ischial spine.
   c. *Ischial tuberosity* Roughened projection.
   d. *Ramus* Portion of the ischium that joins

the pubis and surrounds the *obturator fora-men.*

3. *Pubis*—Anterior, inferior portion.
   a. *Superior ramus*   Upper portion of the pubis.
   b. *Inferior ramus*   Lower portion of the pubis.
   c. *Pubic symphysis* (SIM-fi-sis)   Joint between the left and right hipbones.
4. *Acetabulum*   (as'-e-TAB-yoo-lum)—Socket that receives the head of the femur to form the hip joint; two-fifths of the acetabulum is formed by the ilium, two-fifths is formed by the ischium, and one-fifth is formed by the pubis.

Label Figure 7.11.

Again, examine the articulated skeleton. This time, compare the male and female *pelvis.* The pelvis consists of the two hipbones, sacrum, and coccyx. Identify the following:

1. *Greater (false) pelvis*—Expanded portion situated above the brim of the pelvis; bounded laterally by the ilia (plural of ilium) and posteriorly by the upper sacrum.
2. *Lesser (true) pelvis*—Below and behind the brim of the pelvis; constructed of parts of the ilium, pubis, sacrum, and coccyx; contains an opening above, the *pelvic inlet,* and an opening below, the *pelvic outlet.*

K. LOWER EXTREMITIES

The bones of the *lower extremities* consist of a femur in each thigh, a patella in front of each knee joint, a tibia and fibula in each leg, tarsals in each ankle, metatarsals in each foot, and phalanges in the toes.

Examine the articulated skeleton and disarticulated bones and identify the following:

1. *Femur*—Thigh bone; the longest and heaviest bone in the body.
   a. *Head*   Rounded projection at the proximal end that articulates with the acetabulum of the hipbone.
   b. *Neck*   Constricted portion below the head.
   c. *Greater trochanter* (trō-KAN-ter)   Lateral-side prominence.
   d. *Lesser trochanter*   Prominence on the dorsomedial side.
   e. *Intertrochanteric line*   Ridge on the anterior surface.

   f. *Intertrochanteric crest*   Ridge on the posterior surface.
   g. *Linea aspera*   Vertical ridge on the posterior surface.
   h. *Medial condyle*   Medial posterior projection on the distal end that articulates with the tibia.
   i. *Lateral condyle*   Lateral posterior projection on the distal end that articulates with the tibia.
   j. *Intercondylar* (in'-ter-KON-di-lar) *fossa*   Depressed area between condyles on the posterior surface.
   k. *Medial epicondyle*   Projection above the medial condyle.
   l. *Lateral epicondyle*   Projection above the lateral condyle.
   m. *Patellar surface*   Between the condyles on the anterior surface.

Label the following parts of the femur in Figure 7.12a and b.

2. *Patella*—Kneecap in front of the knee joint; develops in the tendon of the quadriceps femoris muscle.
   a. *Base*   Broad superior portion.
   b. *Apex*   Pointed inferior portion.
   c. *Articular facets*   Articulating surfaces on the posterior surface for the medial and lateral condyles of the femur.

Label the following parts of the patella in Figure 7.12c and d.

3. *Tibia*—Shinbone; medial bone of the leg.
   a. *Lateral condyle*   Articulates with the lateral condyle of the femur.
   b. *Medial condyle*   Articulates with the medial condyle of the femur.
   c. *Intercondylar eminence*   Upward projection between the condyles.
   d. *Tibial tuberosity*   Anterior projection for the attachment of the patellar ligament.
   e. *Medial malleolus* (mal-LĒ-ō-lus)   Distal projection that articulates with the talus bone of the ankle.
   f. *Fibular notch*   Distal depression that articulates with the fibula.
4. *Fibula*—Lateral bone of the leg.
   a. *Head*   Proximal projection that articulates with the tibia.
   b. *Lateral malleolus*   Projection at the distal end that articulates with the talus bone of the ankle.

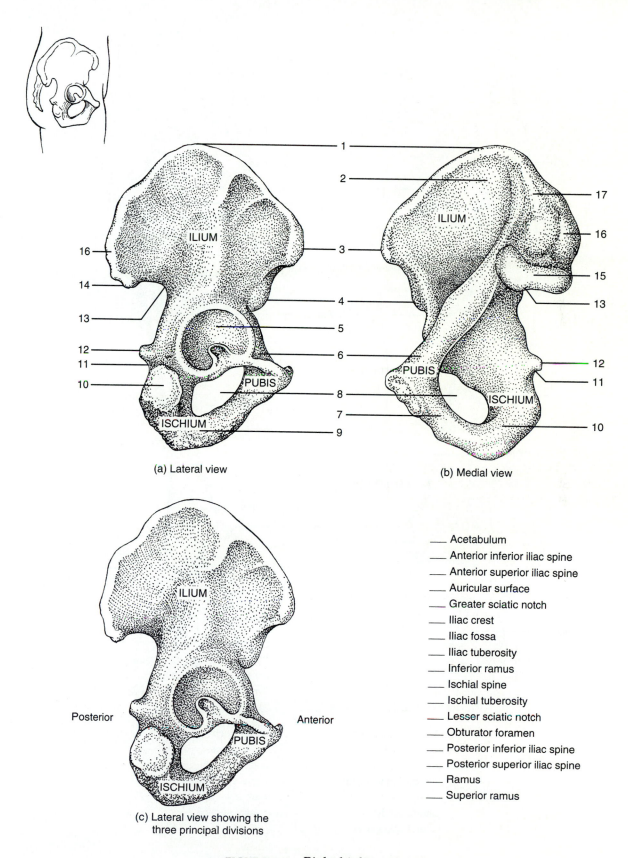

(a) Lateral view

(b) Medial view

(c) Lateral view showing the three principal divisions

Posterior

Anterior

___ Acetabulum
___ Anterior inferior iliac spine
___ Anterior superior iliac spine
___ Auricular surface
___ Greater sciatic notch
___ Iliac crest
___ Iliac fossa
___ Iliac tuberosity
___ Inferior ramus
___ Ischial spine
___ Ischial tuberosity
___ Lesser sciatic notch
___ Obturator foramen
___ Posterior inferior iliac spine
___ Posterior superior iliac spine
___ Ramus
___ Superior ramus

FIGURE 7.11   Right hipbone.

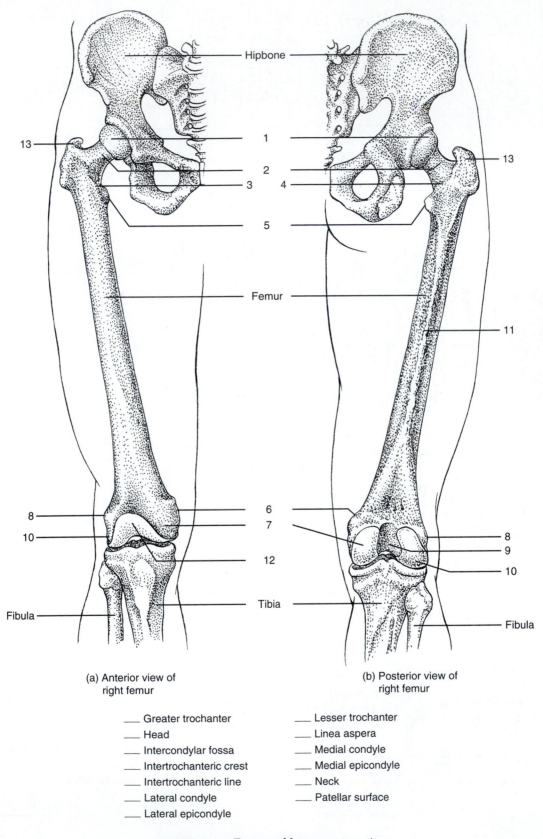

Hipbone

13

1

2

3   4

5

Femur

11

8

10

6

7

12

Tibia

Fibula

8

9

10

Fibula

(a) Anterior view of
right femur

(b) Posterior view of
right femur

___ Greater trochanter   ___ Lesser trochanter
___ Head   ___ Linea aspera
___ Intercondylar fossa   ___ Medial condyle
___ Intertrochanteric crest   ___ Medial epicondyle
___ Intertrochanteric line   ___ Neck
___ Lateral condyle   ___ Patellar surface
___ Lateral epicondyle

FIGURE 7.12 Bones of lower extremity.

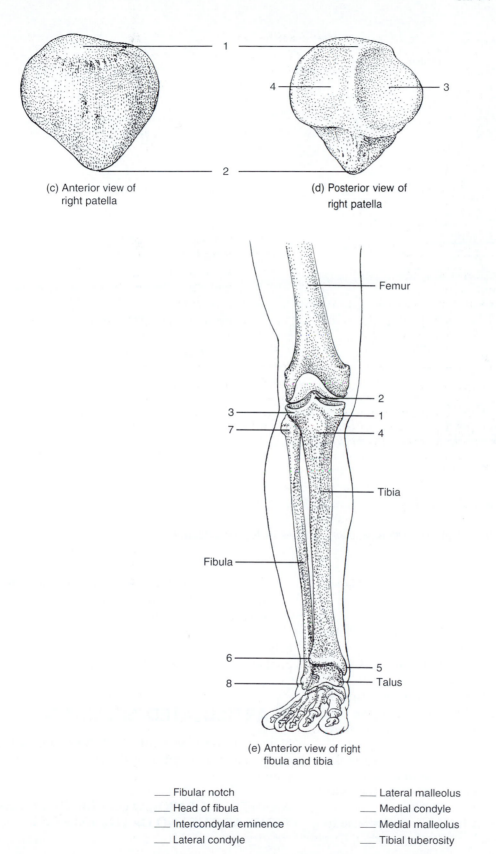

(c) Anterior view of
right patella

(d) Posterior view of
right patella

___ Apex
___ Articular facet for
lateral femoral
condyle
___ Articular facet for
medial femoral
condyle
___ Base

(e) Anterior view of right
fibula and tibia

___ Fibular notch   ___ Lateral malleolus
___ Head of fibula   ___ Medial condyle
___ Intercondylar eminence   ___ Medial malleolus
___ Lateral condyle   ___ Tibial tuberosity

FIGURE 7.12 *(Continued)*   Bones of lower extremity.

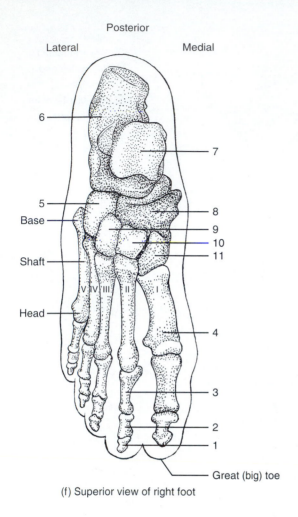

Posterior

Lateral   Medial

6

7

5

Base

8

9

Shaft   10
                                             11

V IV III   II   I

Head

4

3

2

1

Great (big) toe

(f) Superior view of right foot

___ Calcaneus
___ Cuboid
___ Distal phalanx
___ First (medial) cuneiform
___ Metatarsal
___ Middle phalanx
___ Navicular
___ Proximal phalanx
___ Second (intermediate) cuneiform
___ Talus
___ Third (lateral) cuneiform

FIGURE 7.12 *(Continued)*   Bones of lower extremity.

Label the parts of the tibia and fibula in Figure 7.12e.

5. **Tarsus**—Seven bones of the ankle called the **tarsals.**
   a. *Posterior bones* **Talus** and **calcaneus** (heel bone).
   b. *Anterior bones* **Cuboid, navicular (scaphoid)**, and three **cuneiforms** called the first (medial), second (intermediate), and third (lateral) cuneiforms.
6. **Metatarsus**—Consists of the five bones of the foot called the metatarsals, numbered as follows, beginning on the medial (large toe) side: I, II, III, IV, and V metatarsals.
7. **Phalanges**—Bones of the toes, comparable to

the phalanges of fingers; two in each large toe (proximal and distal) and three in each other toe (proximal, middle, and distal).

Label the parts of the tarsus, metatarsus, and phalanges in Figure 7.12f.

L. ARTICULATED SKELETON

Now that you have studied all the bones of the body, label the entire articulated skeleton in Figure 7.13.

ANSWER THE LABORATORY REPORT QUESTIONS AT THE END OF THE EXERCISE.

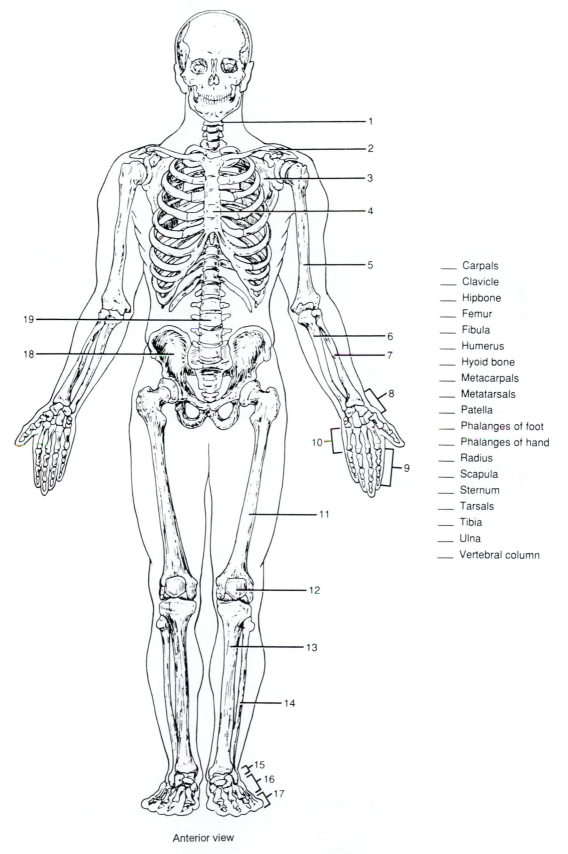

Carpals

Clavicle

Hipbone

Femur

Fibula

Humerus

Hyoid bone

Metacarpals

Metatarsals

Patella

Phalanges of foot

Phalanges of hand

Radius

Scapula

Sternum

Tarsals

Tibia

Ulna

Vertebral column

Anterior view

FIGURE 7.13   Entire skeleton.

Bones

STUDENT _____   DATE _____

LABORATORY SECTION _____   SCORE/GRADE _____

PART 1. Multiple Choice

_____ 1. The suture between the parietal and temporal bone is the (a) lambdoid (b) coronal (c) squamos (d) sagittal

_____ 2. Which bone does *not* contain a paranasal sinus? (a) ethmoid (b) maxilla (c) sphenoid (d) sacral

_____ 3. Which is the superior, concave curve in the vertebral column? (a) thoracic (b) lumbar (c) cervical (d) sacral

_____ 4. The fontanel between the parietal and occipital bones is the (a) anterolateral (b) anterior (c) posterior (d) posterolateral

_____ 5. Which is *not* a component of the upper extremity? (a) radius (b) femur (c) carpus (d) humerus

_____ 6. All are components of the appendicular skeleton *except* the (a) humerus (b) occipital bone (c) calcaneus (d) triquetral

_____ 7. Which bone does *not* belong with the others? (a) occipital (b) frontal (c) parietal (d) mandible

_____ 8. Which region of the vertebral column is closer to the skull? (a) thoracic (b) lumbar (c) cervical (d) sacral

_____ 9. Of the following bones, the one that does *not* help form part of the orbit is the (a) sphenoid (b) frontal (c) occipital (d) lacrimal

_____ 10. Which bone does *not* form a border for a fontanel? (a) maxilla (b) temporal (c) occipital (d) parietal

PART 2. Identification

For each surface marking listed, identify the skull bone to which it belongs:

11. Glabella _____

12. Mastoid process _____

13. Sella turcica _____

14. Cribriform plate _____

15. Foramen magnum _____

16. Mental foramen _____

17. Infraorbital foramen _____

18. Crista galli _____

19. Foramen ovale _____

20. Horizontal plate _____

21. Optic foramen _____

22. Superior nasal concha _____

23. Zygomatic process _____

24. Styloid process _____

25. Mandibular fossa _____

PART 3. Matching

_____ 26. Iliac crest

_____ 27. Capitulum

_____ 28. Medial malleolus

_____ 29. Laminae

_____ 30. Vertebral foramen

_____ 31. Talus

_____ 32. Olecranon

_____ 33. Pisiform

_____ 34. Acromial extremity

_____ 35. Pubic symphysis

_____ 36. Hamate

_____ 37. Costal groove

_____ 38. Xiphoid process

_____ 39. Greater trochanter

_____ 40. Transverse lines

_____ 41. Radial tuberosity

_____ 42. Greater tubercle

_____ 43. Ischium

_____ 44. Glenoid cavity

_____ 45. Lateral malleolus

A. Inferior portion of sternum

B. Medial bone of distal carpals

C. Distal projection of tibia

D. Lateral end of clavicle

E. Points where bodies of sacral vertebrae join

G. Portion of rib that contains blood vessels

G. Prominence of elbow

H. Lateral projection of humerus

I. Medial projection for insertion of biceps brachii muscle

J. Lower posterior portion of hipbone

K. Articulates with head of radius

L. Prominence on lateral side of femur

M. Distal projection of fibula

N. Superior border of ilium

O. Articulates with head of humerus

P. Anterior joint between hipbones

Q. Opening through which spinal cord passes

R. Form posterior wall of vertebral arch

S. Medial bone of proximal carpals

T. Component of tarsus

Articulations

An *articulation* (ar-tik′-yoo-LĀ-shun), or *joint,* is a point of contact between bones, cartilage and bones, or teeth and bones. Some joints permit no movement, others permit a slight degree of movement, and still others permit free movement. In this exercise you will study the structure and action of joints.

A. KINDS OF JOINTS

The joints of the body can be classified into three principal kinds on the basis of their structure and function.

The structural classification of joints is based on the presence or absence of a joint cavity (a space between the articulating bones) and the type of connective tissue that binds the bones together. Structurally, a joint is classified as

1. *Fibrous* (FĪ-brus)—There is no joint cavity and the bones are held together by fibrous connective tissue.
2. *Cartilaginous* (kar-ti-LAJ-i-nus)—There is no joint cavity and the bones are held together by cartilage.
3. *Synovial* (si-NŌ-vē-al)—There is a joint cavity and the bones forming the joint are united by a surrounding articular capsule and frequently by accessory ligaments (described in detail later).

The functional classification of joints is as follows:

1. *Synarthroses* (sin′-ar-THRŌ-sēz)—Immovable joints.
2. *Amphiarthroses* (am′-fē-ar-THRŌ-sēz)—Slightly movable joints.
3. *Diarthroses* (dī-ar-THRŌ-sēz)—Freely movable joints.

We will discuss the joints of the body on the basis of their functional classification, referring to their structural classification as well.

B. SYNARTHROSES (IMMOVABLE JOINTS)

A synarthrosis may be of three types: sutures, synchondroses, and gomphoses.

1. *Sutures* (SOO-cherz)—Composed of a thin layer of dense fibrous connective tissue; unite the bones of the skull. Example: coronal suture between the frontal and parietal bones.
2. *Gomphoses* (gom-FŌ-sēz)—Type of fibrous joint in which a cone-shaped peg fits into a socket. The substance between the bones is the periodontal ligament. Example: articulations of the roots of the teeth with the alveoli (sockets) of the maxillae and mandible.
3. *Synchondroses* (sin′-kon-DRŌ-sēz)—Cartilaginous joint in which the connecting material is hyaline cartilage. Example: epiphyseal plate between the epiphysis and diaphysis of a growing bone.

117

C. AMPHIARTHROSES (SLIGHTLY MOVABLE JOINTS)

Amphiarthroses (slightly movable joints) can be of two types: syndesmoses and symphyses.

1. *Syndesmoses* (sin'-dez-MŌ-sēz)—This is a fibrous joint in which there is considerably more fibrous connective tissue than in a suture; the fibrous connective tissue forms an interosseous membrane or ligament that permits some flexibility and movement. Example: the distal articulation of the tibia and fibula.
2. *Symphyses* (SIM-fi-sēz)—The connecting material is a broad, flat disc of fibrocartilage. Example: intervertebral discs between vertebrae and pubic symphysis between the anterior surfaces of the hipbones.

D. DIARTHROSES (FREELY MOVABLE JOINTS)

Diarthroses, or freely movable joints, have a variety of shapes and permit several different types of movements. First, we will discuss the general structure of a diarthrosis and then consider the different types.

1. Structure of a Diarthrosis

A distinguishing anatomical feature of a diarthrosis is a space, called a *synovial* (si-NŌ-vē-al), or *joint cavity* (see Figure 8.1), that separates the articulating bones. Thus, diarthroses are also called synovial joints. Another characteristic of such joints is the presence of *articular cartilage.* Articular cartilage (hyaline) covers the surfaces of the articulating bones but does not bind them.

A sleevelike *articular capsule* surrounds a synovial joint, encloses the synovial cavity, and unites the articulating bones. The articular capsule is composed of two layers. The outer layer, the *fibrous capsule,* usually consists of dense, irregular connective tissue. It attaches to the periosteum of the articulating bones at a variable distance from the edge of the articular cartilage. The flexibility of the fibrous capsule permits movement at a joint, whereas its great tensile strength resists dislocation. The fibers of some fibrous capsules are arranged in parallel bundles and are therefore highly adapted to resist recurrent strain. Such fibers are called ligaments and

are given special names. The strength of the *ligaments* is one of the principal factors in holding bone to bone. Diarthroses are freely movable joints because of the synovial cavity and the arrangement of the articular capsule and accessory ligaments. The inner layer of the articular capsule is formed by a *synovial membrane.* The synovial membrane is composed of areolar connective tissue with elastic fibers and a variable amount of adipose tissue. It secretes *synovial fluid (SF),* which lubricates the joint and provides nourishment for the articular cartilage.

Many diarthroses also contain *accessory ligaments* (*ligare* = to bind), called extracapsular ligaments and intracapsular ligaments. *Extracapsular ligaments* lie outside the articular capsule. An example is the fibular (lateral) collateral ligament of the knee joint (see Figure 8.4b). *Intracapsular ligaments* occur within the articular capsule but are excluded from the synovial cavity by folds of the synovial membrane. Examples are the cruciate ligaments of the knee joint (see Figure 8.4b).

Inside some diarthroses are pads of fibrocartilage that lie between the articular surfaces of the bones and are attached by their margins to the fibrous capsule. These pads are *articular discs (menisci).* The singular is *meniscus.* See Figure 8.4b. The discs usually subdivide the synovial cavity into two separate spaces. Articular discs allow two bones of different shapes to fit tightly, modify the shape of the joint surfaces of the articulating bones, and help to maintain the stability of the joint and direct the flow of synovial fluid to the areas of greatest friction.

Saclike structures, called *bursae,* are strategically situated to alleviate friction in some spots (see Figure 8.4f). Bursae resemble joint capsules in that their walls consist of connective tissue lined by a synovial membrane. They are also filled with a fluid similar to synovial fluid. Bursae are located between the skin and bone in places where skin rubs over bone. They are also found between tendons and bones, muscles and bones, and ligaments and bones. Such fluid-filled sacs cushion the movement of one part of the body over another.

Label Figure 8.1, the principal parts of a diarthrosis.

2. Movements at Diarthroses

Movements at diarthroses can be classified as follows:

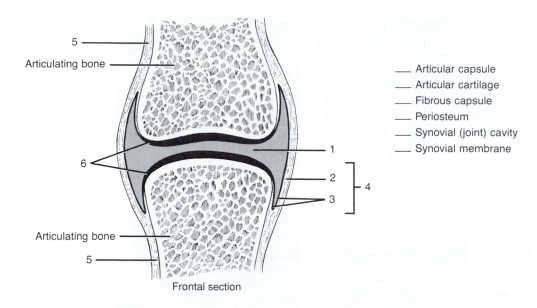

Articular capsule
Articular cartilage
Fibrous capsule
Periosteum
Synovial (joint) cavity
Synovial membrane

Frontal section

FIGURE 8.1   Parts of a diarthrosis (synovial joint).

a. **Gliding**—One surface moves back and forth and from side to side without any angular or rotary movement.

b. **Angular**—An increase or decrease in the angle between bones.
  1. *Flexion*   Decrease in the angle between the surfaces of articulating bones.
  2. *Extension*   Increase in the angle between the surfaces of articulating bones.
  3. *Hyperextension*   Continuation of the extension beyond the anatomical position.
  4. *Abduction*   Movement of a bone away from the midline.
  5. *Adduction*   Movement of a bone toward the midline.

c. **Rotation**—Movement of a bone around its own longitudinal axis.

d. **Circumduction**—Movement in which the distal end of a bone moves in a circle while the proximal end remains relatively stable; bone outlines a cone in the air.

e. **Special**—Found only at the joints indicated:
  1. *Inversion*   Movement of the sole of the foot inward (medially).
  2. *Eversion*   Movement of the sole of the foot outward (laterally).
  3. *Dorsiflexion*   Bending the foot in the direction of the dorsum (upper surface).
  4. *Plantar flexion*   Bending the foot in the direction of the plantar surface (sole).
  5. *Protraction*   Movement of the mandible or the shoulder girdle forward on a plane parallel to the ground.

  6. *Retraction*   Movement of the protracted part of the body backward on a plane parallel to the ground.
  7. *Supination*   Movement of the forearm in which the palm of the hand is turned anteriorly or superiorly.
  8. *Pronation*   Movement of the forearm in which the palm of the hand is turned posteriorly or inferiorly.
  9. *Depression*   Movement in which part of the body, such as the mandible or scapula, moves inferiorly.
  10. *Elevation*   Movement in which part of the body moves superiorly.

Label the various movements illustrated in Figure 8.2.

3. Types of Diarthroses

The principal types of diarthroses follows:

a. **Gliding**—Articulating surfaces are usually flat, permitting a gliding movement that includes side-to-side and back-and-forth. Example: between carpals, tarsals, sacrum and ilium, sternum and clavicle, scapula and clavicle, and articular processes of vertebrae.

b. **Hinge**—Convex surface of one bone fits into the concave surface of another; movement in a single plane (monoaxial movement) is usually flexion and extension. Example: elbow, knee, ankle, interphalangeal joints.

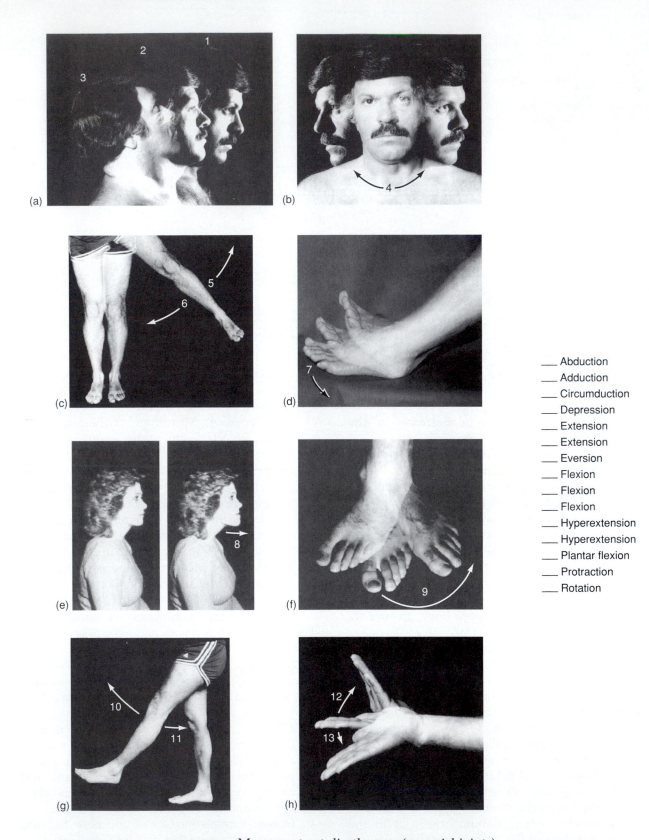

___ Abduction
___ Adduction
___ Circumduction
___ Depression
___ Extension
___ Extension
___ Eversion
___ Flexion
___ Flexion
___ Flexion
___ Hyperextension
___ Hyperextension
___ Plantar flexion
___ Protraction
___ Rotation

FIGURE 8.2   Movements at diarthroses (synovial joints).

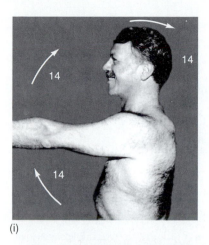

(i)

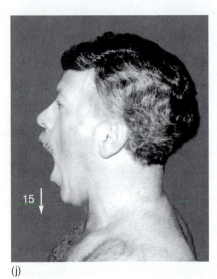

(j)

FIGURE 8.2 *(Continued)*   Movements at diarthroses (synovial joints).

c. *Pivot*—Rounded or pointed surface of one bone articulates within a ring formed partly by bone and partly by a ligament; primary movement is rotation; joint is monoaxial. Example: between atlas and axis and between proximal ends of radius and ulna.

d. *Ellipsoidal*—Oval-shaped condyle of one bone fits into an elliptical cavity of another bone; movement is in two planes (biaxial), side-to-side and back-and-forth. Example: between radius and carpals.

e. *Saddle*—Articular surface of one bone is saddle-shaped and the articular surface of the other bone is shaped like a rider sitting in the saddle; movement is similar to that of an ellipsoidal joint. Example: between trapezium of carpus and metacarpal of thumb.

f. *Ball-and-socket*—Ball-like surface of one bone fits into the cuplike depression of another; movement is in three planes (triaxial movement), flexion-extension, abduction-adduction, and rotation. Example: shoulder and hip joint.

Examine the articulated skeleton and find as many examples as you can of the joints just described. As part of your examination, be sure to note the shapes of the articular surfaces and the movements possible at each joint.

Label Figure 8.3.

E. KNEE JOINT

The knee joint is one of the largest joints in the body and illustrates the basic structure of a diarthrosis and the limitations on its movement. Some of the structures associated with the knee joint follows:

1. *Tendon of quadriceps femoris muscle*—Strengthens the joint anteriorly and externally.
2. *Gastrocnemius muscle*—Strengthens the joint posteriorly and externally.
3. *Patellar ligament*—Strengthens the anterior portion of a joint and prevents the leg from being flexed too far backward.
4. *Fibular collateral ligament*—Between the femur and fibula; strengthens the lateral side of the joint and prohibits side-to-side movement at the joint.
5. *Tibial collateral ligament*—Between the femur and tibia; strengthens the medial side of the joint and prohibits side-to-side movement at the joint.
6. *Oblique popliteal ligament*—Starts in a tendon that lies over the tibia and runs upward and laterally to the lateral side of the femur; supports the back of the knee and prevents hyperextension.
7. *Anterior cruciate ligament*—Passes posteriorly and laterally from the tibia and attaches to the femur; strengthens the joint internally and may help stabilize the knee during its movements.
8. *Posterior cruciate ligament*—Passes anteriorly and medially from the tibia and attaches to the femur; strengthens the joint internally and may help stabilize the knee during its movements.

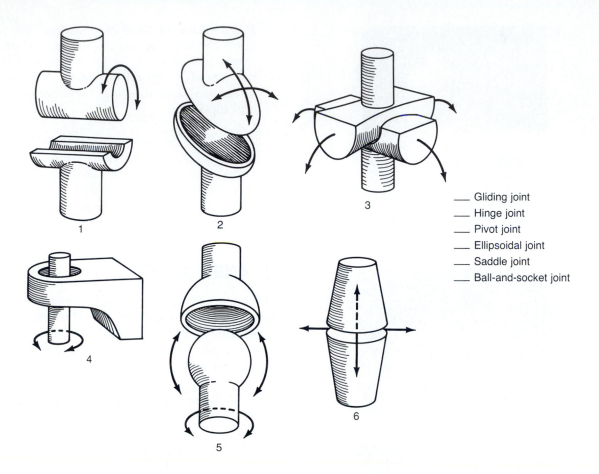

_____ Gliding joint
_____ Hinge joint
_____ Pivot joint
_____ Ellipsoidal joint
_____ Saddle joint
_____ Ball-and-socket joint

FIGURE 8.3 Types of diarthroses (synovial joints).

9. *Menisci*—Concentric wedge-shaped pieces of fibrocartilage between the femur and tibia; called *lateral meniscus* and the *medial meniscus;* provide support for the continuous weight placed on the knee joint.

Label the structures associated with a knee joint in Figure 8.4.

If a longitudinally sectioned knee joint of a cow or lamb is available, examine it and see how many structures you can identify.

ANSWER THE LABORATORY REPORT QUESTIONS AT THE END OF THE EXERCISE.

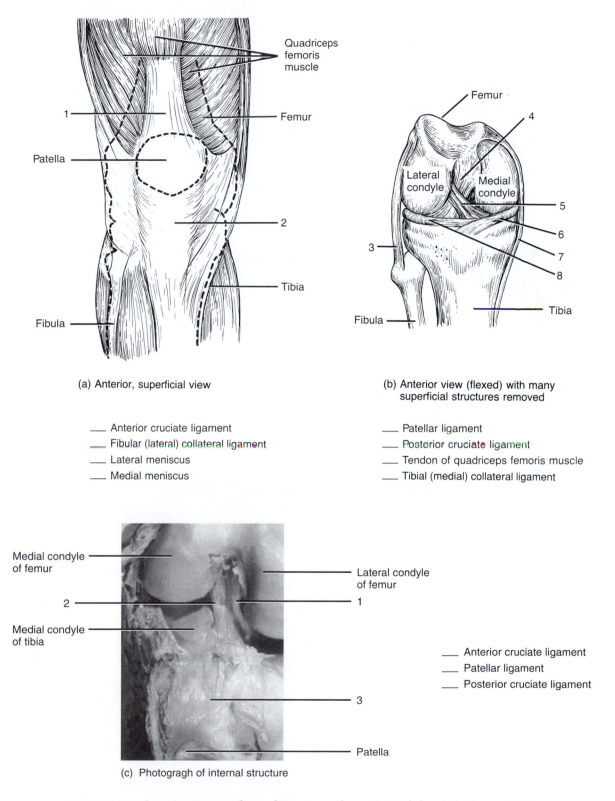

(a) Anterior, superficial view

____ Anterior cruciate ligament
____ Fibular (lateral) collateral ligament
____ Lateral meniscus
____ Medial meniscus

(b) Anterior view (flexed) with many
    superficial structures removed

____ Patellar ligament
____ Posterior cruciate ligament
____ Tendon of quadriceps femoris muscle
____ Tibial (medial) collateral ligament

(c) Photogragh of internal structure

____ Anterior cruciate ligament
____ Patellar ligament
____ Posterior cruciate ligament

FIGURE 8.4   Ligaments, tendons, bursae, and menisci of the right knee joint.

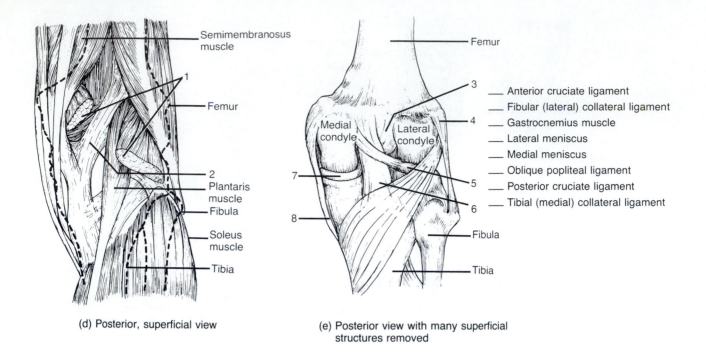

Semimembranosus muscle

Femur

1

2

Plantaris muscle

Fibula

Soleus muscle

Tibia

(d) Posterior, superficial view

Femur

Medial condyle

Lateral condyle

3

___ Anterior cruciate ligament
___ Fibular (lateral) collateral ligament
___ Gastrocnemius muscle
___ Lateral meniscus
___ Medial meniscus
___ Oblique popliteal ligament
___ Posterior cruciate ligament
___ Tibial (medial) collateral ligament

4

7

5

6

8

Fibula

Tibia

(e) Posterior view with many superficial structures removed

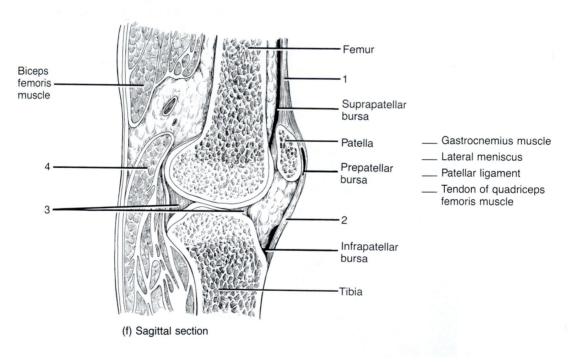

Biceps femoris muscle

Femur

1

Suprapatellar bursa

Patella

___ Gastrocnemius muscle
___ Lateral meniscus
___ Patellar ligament
___ Tendon of quadriceps femoris muscle

4

3

Prepatellar bursa

2

Infrapatellar bursa

Tibia

(f) Sagittal section

FIGURE 8.4 (Continued) Ligaments, tendons, bursae, and menisci of the right knee joint.

Articulations

STUDENT _____   DATE _____

LABORATORY SECTION _____   SCORE/GRADE _____

PART 1. Multiple Choice

_____ 1. A joint united by dense fibrous tissue that permits a slight degree of movement is a (a) suture (b) syndesmosis (c) symphysis (d) synchondrosis

_____ 2. A joint that contains a broad flat disc of fibrocartilage is classified as a (a) ball-and-socket joint (b) suture (c) symphysis (d) gliding joint

_____ 3. The following characteristics define what type of joint: presence of a synovial cavity, articular cartilage, synovial membrane, and ligaments. (a) suture (b) synchondrosis (c) syndesmosis (d) hinge

_____ 4. Which joints are slightly movable? (a) diarthroses (b) amphiarthroses (c) synovial (d) synarthroses

_____ 5. Which type of joint is immovable? (a) synarthrosis (b) syndesmosis (c) symphysis (d) diarthrosis

_____ 6. What type of joint provides triaxial movement? (a) hinge (b) ball-and-socket (c) saddle (d) ellipsoidal

_____ 7. Which ligament provides strength on the medial side of the knee joint? (a) oblique popliteal (b) posterior cruciate (c) fibular collateral (d) tibial collateral

_____ 8. On the basis of structure, which joint is fibrous? (a) symphysis (b) synchondrosis (c) pivot (d) syndesmosis

_____ 9. The elbow, knee, and interphalangeal joints are examples of which type of joint? (a) pivot (b) hinge (c) gliding (d) saddle

_____ 10. Functionally, which joint provides the greatest degree of movement? (a) diarthrosis (b) synarthrosis (c) amphiarthrosis (d) syndesmosis

PART 2. Completion

11. The thin layer of hyaline cartilage on the articulating surfaces of bones is called

_____ cartilage.

12. The synovial membrane and fibrous capsule together form the _____ capsule.

13. Pads of fibrocartilage between the articular surfaces of bones that maintain stability of the joint are

called _____.

14. Fluid-filled connective tissue sacs that cushion movements of one body part over another are referred to as _____.

15. The _____ ligament supports the back of the knee and helps to prevent hyperextension.

PART 3. Matching

_____ **16.** Circumduction

_____ **17.** Adduction

_____ **18.** Flexion

_____ **19.** Pronation

_____ **20.** Elevation

_____ **21.** Protraction

_____ **22.** Rotation

_____ **23.** Plantar flexion

_____ **24.** Dorsiflexion

_____ **25.** Inversion

A. Decrease in the angle between articulating bones

B. Moving a part upward

C. Bending the foot in the direction of the upper surface

D. Forward movement parallel to the ground

E. Movement of the sole of the foot inward at the ankle joint

F. Movement toward the midline

G. Bending the foot in the direction of the sole

H. Turning the palm posteriorly

I. Movement of a bone around its own axis

J. Distal end of a bone moves in a circle while the proximal end remains relatively stable

9

Muscle Tissue

Muscle tissue constitutes 40 to 50% of total body weight and is composed of fibers (cells) that are highly specialized with respect to four characteristics: (1) *excitability (irritability),* or ability to receive and respond to certain stimuli by producing electrical signals called action potentials (impulses); (2) *contractility,* or ability to contract (shorten and thicken); (3) *extensibility (extension),* or ability to stretch when pulled; and (4) *elasticity,* or ability to return to original shape after contraction or extension. Through contraction, muscle performs three basic functions: motion, maintenance of posture, and heat production. In this exercise you will examine the histological structure of muscle tissue and conduct exercises on the physiology of frog muscle.

A. KINDS OF MUSCLE TISSUE

Histologically, three kinds of muscle tissue are recognized:

1. *Skeletal muscle tissue*—Usually attached to bones; contains conspicuous striations when viewed microscopically; voluntary because it contracts under conscious control.
2. *Cardiac muscle tissue*—Found only in the wall of the heart; striated when viewed microscopically; involuntary because it contracts without conscious control.
3. *Smooth (visceral) muscle tissue*—Located in the walls of viscera and blood vessels; referred to as nonstriated because it lacks striations when viewed microscopically; involuntary because it contracts without conscious control.

B. SKELETAL MUSCLE TISSUE

Examine a prepared slide of skeletal muscle in longitudinal and cross section under high power. Look for the following:

1. *Sarcolemma*—Plasma membrane of the muscle fiber.
2. *Sarcoplasm*—Cytoplasm of the muscle fiber.
3. *Nuclei*—Several in each muscle fiber lying close to sarcolemma.
4. *Striations*—Cross stripes in each muscle fiber (described on page 128).
5. *Epimysium* (ep′-i-MĪZ-ē-um)—Fibrous connective tissue that surrounds the entire skeletal muscle.
6. *Perimysium* (per′-i-MĪZ-ē-um)—Fibrous connective tissue surrounding a bundle (fascicle) of muscle fibers.
7. *Endomysium* (en′-dō-MĪZ-ē-um)—Fibrous connective tissue that surrounds individual muscle fibers.

Refer to Figure 9.1 and label the structures indicated.

With the use of an electron microscope, additional details of skeletal muscle tissue may be noted. Among these are the following:

1. *Mitochondria*—Organelles that have a smooth outer membrane and a folded inner membrane in which ATP is generated.
2. *Sarcoplasmic reticulum*—(sar′-kō-PLAZ-mik re-TIK-yoo-lum), or *SR*—Network of membrane-enclosed tubules.

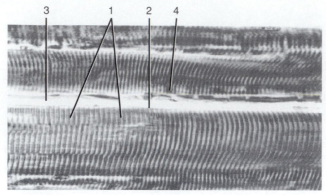

3   1   2   4

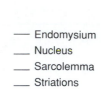

___ Endomysium
___ Nucleus
___ Sarcolemma
___ Striations

(a) Photomicrograph of several muscle fibers in longitudinal section

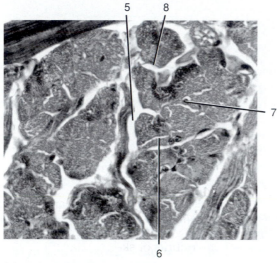

5   8

7

6

___ Endomysium
___ Nucleus
___ Perimysium
___ Sarcolemma

(b) Photomicrograph of several muscle fibers in cross section

FIGURE 9.1   Histology of skeletal muscle tissue.

3. ***Transverse (T) tubules***—Tunnel-like extensions of sarcolemma that run perpendicular to and connect with sarcoplasmic reticulum; open to the outside of muscle fiber.

4. ***Triad***—Transverse tubule and the segments of sarcoplasmic reticulum on both sides.

5. ***Myofibrils***—Threadlike structures that run lengthwise through a fiber and consist of ***thin myofilaments*** composed of the protein actin and ***thick myofilaments*** composed of the protein myosin.

6. ***Sarcomere***—Contractile unit of a muscle fiber; compartment within a muscle fiber separated from other sarcomeres by dense material called ***Z discs***.

7. ***A (anisotropic) band***—Dark region in a sarcomere represented by the length of thin myofilaments where they overlap thick myofilaments.

8. ***I (isotropic) band***—Light region in a sarcomere composed of thin myofilaments only. The combination of alternating dark A bands and light I bands gives the muscle fiber the striated (striped) appearance.

9. *H zone*—Region in the center of the A band of a sarcomere consisting of thick myofilaments only.

10. *M line*—Series of fine threads in the center of the H zone that appear to connect the middle parts of adjacent thick myofilaments.

Figure 9.2 is a diagram of skeletal muscle tissue based on electron micrographic studies. Label the structures shown.

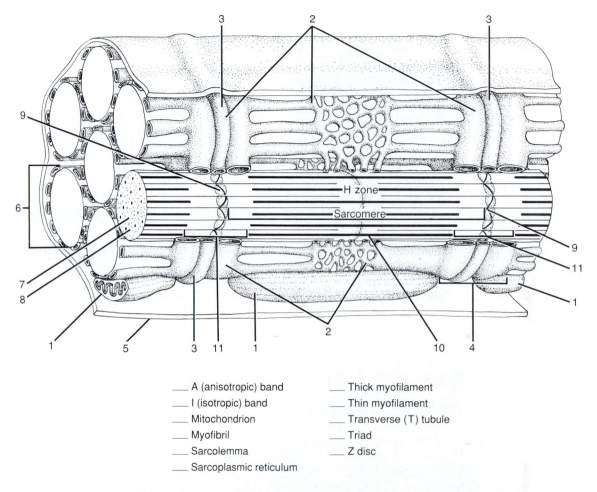

H zone

Sarcomere

___ A (anisotropic) band   ___ Thick myofilament

___ I (isotropic) band   ___ Thin myofilament

___ Mitochondrion   ___ Transverse (T) tubule

___ Myofibril   ___ Triad

___ Sarcolemma   ___ Z disc

___ Sarcoplasmic reticulum

FIGURE 9.2   Enlarged aspect of several myofibrils of skeletal muscle tissue based on an electron micrograph.

C. CARDIAC MUSCLE TISSUE

Examine a prepared slide of cardiac muscle tissue in longitudinal and cross section under high power. Locate the following structures: *sarcolemma, endomysium, nuclei, striations,* and *intercalated discs* (transverse thickenings of the sarcolemma that separate individual fibers). Label Figure 9.3.

D. SMOOTH (VISCERAL) MUSCLE TISSUE

Examine a prepared slide of smooth muscle tissue in longitudinal and cross section under high power. Locate and label the following structures in Figure 9.4: *sarcolemma, sarcoplasm, nucleus,* and *muscle fiber.*

E. PHYSIOLOGY OF SKELETAL MUSCLE CONTRACTION

The process of skeletal muscle contraction involves a series of electrical, chemical, and mechanical events. The linkage between the production of the electrical and chemical events to the ultimate mechanical contraction of skeletal muscle is termed *excitation-contraction coupling.* Muscle contraction occurs only if the muscle is stimulated with a stimulus of threshold or suprathreshold intensity. A *threshold stimulus* is defined as the minimum strength stimulus necessary to initiate a contraction. Under normal physiological conditions, this is accomplished by a nerve impulse being transmitted to the skeletal muscle cell via a nerve cell called a *motor neuron.* The portion of the *motor neuron* that extends from the nerve cell body to a muscle fiber is

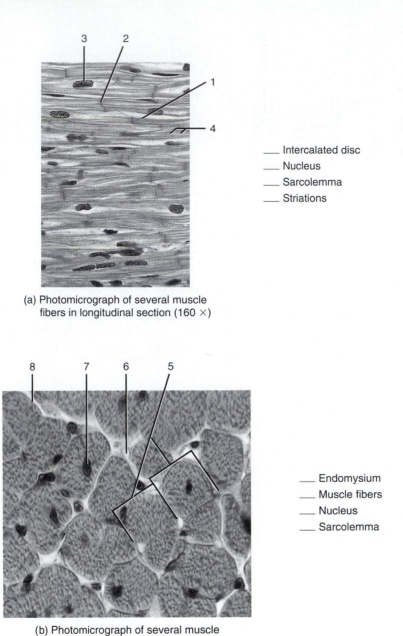

(a) Photomicrograph of several muscle
fibers in longitudinal section (160 ×)

___ Intercalated disc
___ Nucleus
___ Sarcolemma
___ Striations

(b) Photomicrograph of several muscle
fibers in cross section (325 ×)

___ Endomysium
___ Muscle fibers
___ Nucleus
___ Sarcolemma

FIGURE 9.3   Histology of cardiac muscle tissue.

called an *axon.* Upon entering the connective tissue surrounding each individual muscle fiber (termed the endomysium), an axon branches into *axon terminals* that come into close approximation to the sarcolemma of a muscle fiber. Such an area of close approximation consisting of the axon terminal of a motor neuron and the portion of the sarcolemma near it is called a *neuromuscular (myoneural) junction* (Figure 9.5). The portion of the skeletal fiber immediately beneath the axon terminal exhibits a large number of specializations, and is termed the *motor end plate.*

Close examination of a neuromuscular junction reveals that the distal ends of the axon terminals are expanded into bulblike structures called *synaptic end bulbs.* The bulbs contain membrane-enclosed sacs, called *synaptic vesicles,* that store chemicals called *neurotransmitters.* These chemicals initiate the excitation-contraction coupling process and are utilized to transmit information from the motor neuron to the skeletal muscle fiber. The invaginated area of the sarcolemma under the axon terminal is referred to as a *synaptic gutter (trough);* the space between the

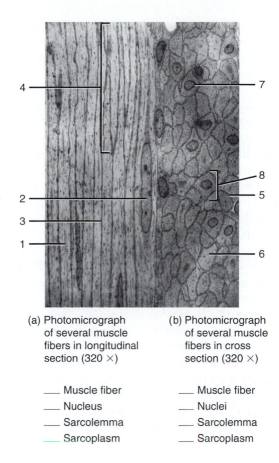

(a) Photomicrograph of several muscle fibers in longitudinal section (320 ×)

(b) Photomicrograph of several muscle fibers in cross section (320 ×)

___ Muscle fiber
___ Nucleus
___ Sarcolemma
___ Sarcoplasm

___ Muscle fiber
___ Nuclei
___ Sarcolemma
___ Sarcoplasm

FIGURE 9.4 Histology of smooth (visceral) muscle tissue.

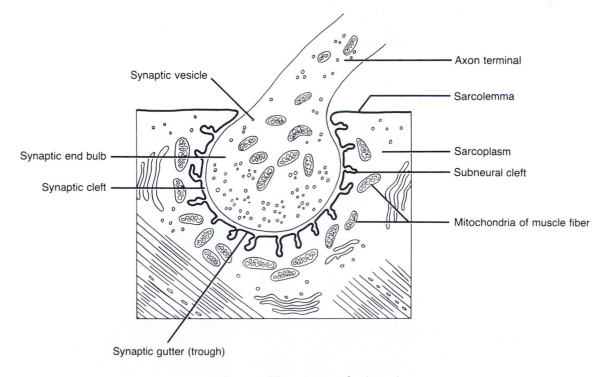

FIGURE 9.5 Neuromuscular junction.

axon terminal and sarcolemma is known as a *synaptic cleft.* Along the synaptic gutter are numerous folds in the sarcolemma, called *subneural clefts,* which greatly increase the surface area of the synaptic gutter.

In a skeletal muscle neuromuscular junction, the nerve impulse reaches the end of an axon, causing it to depolarize. This depolarization suddenly and temporarily increases the permeability of the synaptic end bulb to calcium ions (Ca^{2+}). This increase in Ca^{2+} permeability causes the synaptic vesicles to fuse with the membrane of the synaptic end bulb, thereby releasing a neurotransmitter called *acetylcholine* (as'-ē-til-KŌ-lēn), or *Ach,* which diffuses across the synaptic cleft, where it binds to receptors on the motor end plate. The binding of *Ach* to the motor end plate initiates a localized depolarization called the *end-plate potential,* which achieves threshold intensity and initiates an action potential along the sarcolemma of the muscle fiber. The action potential travels along the sarcolemma and into the interior of the muscle fiber via the transverse tubules, causing the sarcoplasmic reticulum to release Ca^{2+} among actin (thin) and myosin (thick) myofilaments. Ca^{2+} then binds to tropo-

nin (a regulatory protein found in association with the actin myofilament), causing a conformational change (change in shape) in the actin filament. This conformational change of the actin myofilament allows the cross bridges from the adjacent myosin myofilaments to attach to receptors on the actin. The attachment between the myosin cross bridges and the receptors on the actin causes ATP to be split by myosin ATPase, an enzyme found on the myosin myofilament. Energy thus released by the splitting of ATP causes the myosin cross bridges to move, and the thin myofilaments slide inward toward the H zone, causing the muscle fiber to shorten (Figure 9.6). As this is occurring, newly synthesized ATP displaces ADP from the myosin molecule. If Ca^{2+} has been taken back into the sarcoplasmic reticulum by the calcium pump, the myosin cross bridges release from the actin receptors and relaxation occurs. However, if sufficient Ca^{2+} is still present within the sarcoplasm, the entire process repeats and further contraction occurs. This theory of skeletal muscle contraction is termed the *sliding filament theory of skeletal muscle contraction.*

A contracting skeletal muscle fiber follows the

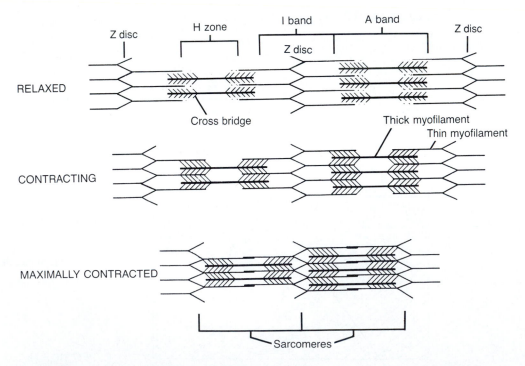

FIGURE 9.6 Sliding filament theory of a skeletal muscle contraction. The positions of the various parts of two sarcomeres in relaxed, contracting, and maximally contracted states are shown. Note the movement of thin myofilaments and the relative size of the H zone.

all-or-none principle. Simply stated, this principle says that a skeletal muscle fiber will respond maximally or not at all to a stimulus. If the stimulus is of threshold or greater intensity, the fiber will respond maximally; if the stimulus is subthreshold in intensity, the fiber will not respond. This principle, however, does not imply that the entire skeletal muscle must be either fully relaxed or fully contracted because individual fibers within a muscle possess varying thresholds for stimulation. Therefore, the muscle as a whole can have graded contractions.

F. BIOCHEMISTRY OF SKELETAL MUSCLE CONTRACTION

You will examine the effect of the following solutions on the contraction of glycerinated muscle fibers:[1] (1) ATP solution, (2) mineral ion solution, and (3) ATP plus mineral ion solution.

PROCEDURE

1. Using 7× or 10× magnification and glass needles or clean stainless steel forceps, gently tease the muscle into very thin groups of myofibers. Single cells or thin groups must be used because strands thicker than a silk thread curl when they contract.
2. Using a Pasteur pipette or medicine dropper, transfer one strand into a drop of glycerol on a clean glass slide and cover the preparation with a cover slip. Examine the strand under low and high power and note the striations. Also note that each cell has several nuclei.
3. Transfer one of the thinnest strands to a drop of glycerol on a second microscope slide. Do not add a cover slip. If the amount of glycerol on the slide is more than a small drop, soak the excess into a piece of lens paper held at the edge of the glycerol farthest from the fibers. Using a dissecting microscope and a millimeter ruler held beneath the slide, measure the length of one of the fibers.
4. Now flood the fibers with the solution containing only ATP and observe their reaction. After 30 sec or more, remeasure the same fiber.

    How much did the fiber contract? _____ mm.

5. Using clean slides and medicine droppers, and being especially sure to use clean teasing needles or forceps, transfer other fibers to a drop of glycerol on a slide. Again measure the length of one fiber. Next, flood the fibers with a solution containing mineral ions, observe their reaction, and remeasure the fiber.

    How much did the fiber contract? _____mm.
6. Repeat the exercise, this time using a solution containing a combination of ATP and ions.
7. Observe a contracted fiber under low and high power, and look for differences in appearance between muscle in a contracted state and in a relaxed state (see Figure 9.6).

G. ELECTROMYOGRAPHY

During the contraction of a single muscle fiber, an action potential is generated that lasts between 1 and 4 msec. This electrical activity is dissipated throughout the surrounding tissue. Because all the muscle fibers within a motor unit do not contract simultaneously (thereby providing a smooth muscular contraction), the electrical activity resulting from skeletal muscle contraction is prolonged considerably. By placing two electrodes on the skin or directly within the muscle an electrical recording, termed an ***electromyograph,*** or ***EMG,*** of this muscular activity can be obtained when the muscle is stimulated (Figure 9.7). EMGs are utilized clinically to distinguish between peripheral neurological and muscular diseases and for differentiating abnormalities resulting in reductions of either muscular strength or sensation that may have been caused by either dysfunction of peripheral nervous tissue or central nervous system (CNS) centers. Typically, EMGs are recorded under three different activity levels: complete inactivity, slight muscular activity, and extreme (maximal) muscular activity. Abnormal differences in records obtained under these three conditions may also indicate an abnormality in motor unit recruitment.

1. Subject Preparation

PROCEDURE

1. Prepare to place skin EMG electrodes (or ECG electrodes) on the anterior surface of the forearm superficial to the ***flexor digitorum***

---

[1]Glycerinated muscle preparation and solutions are supplied by the Carolina Biological Supply Company, Burlington, North Carolina 27215.

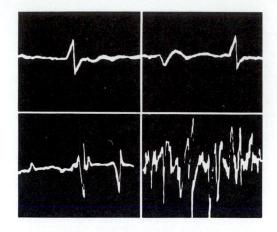

FIGURE 9.7 Diagrammatic representation of normal electromyograms. The single potential in the upper left corner has a measured amplitude of 0. mV and a duration of 7 msec.

superficialis muscle (see Figure 10.15a) and on the posterior surface of the arm superficial to the *triceps brachii* muscle (see Figure 10.14b).

2. Prior to placing the electrodes on the skin, *gently* scrub the area with a dishwashing pad in order to remove dead epithelial cells and to facilitate recording. *Make sure that the capillaries are not damaged and that no bleeding occurs.*

3. Apply a small amount of electrode gel to each electrode and apply them to the proper locations.

4. Connect the two recording electrodes (placed on the anterior surface of the forearm) and the ground (placed on the posterior surface of the arm) to the preamplifier (high-gain coupler) and set the gain to ×100, sensitivity at an appropriate setting between 20 and 100, time constant to 0.03, and paper speed at high.

2. Recording of Spontaneous Muscle Activity

PROCEDURE

1. With the subject completely relaxed and his/her arm placed horizontally on a laboratory table, record any spontaneously occurring electrical activity within the flexor digitorum superficialis over a time period of 1 to 3 min.

2. Record your observations in Section G.1 of the LABORATORY REPORT RESULTS at the end of the exercise.

3. Recruitment of Motor Units

PROCEDURE

1. Have the subject gently flex his/her digits while recording. Note the electrical activity.

2. Have the subject relax his/her digits completely, followed by a more forceful contraction. Note any change in the EMG.

3. Place a tennis ball in your subject's hand and ask him/her to squeeze the ball once or twice as strongly as possible. Note the EMG recording, and then let your subject rest thoroughly.

4. Record your observations in Section G.2 of the LABORATORY REPORT RESULTS at the end of the exercise.

4. Effect of Fatigue

PROCEDURE

1. *Without recording* an EMG on the polygraph, have your subject repeatedly squeeze the tennis ball as strongly as possible until he/she is no longer able to squeeze it.

2. When fatigue has been achieved, *quickly* start recording and ask the subject to attempt to squeeze the ball as strongly as possible four or five times in quick succession.

3. Note the EMG recording and record your observations in Section G.3 of the LABORATORY REPORT RESULTS at the end of the exercise.

ANSWER THE LABORATORY REPORT QUESTIONS AT THE END OF THE EXERCISE.

Muscle Tissue

STUDENT _____   DATE _____

LABORATORY SECTION _____   SCORE/GRADE _____

SECTION G. ELECTROMYOGRAPHY

1. Include your record of spontaneous electrical activity while the subject was resting. Give a physiological explanation. _____

_____

2. Include your record of the EMG during the recruitment exercise. Give a physiological explanation and possible function of recruitment. _____

_____

3. Include your record of the EMG following fatigue. Give a physiological explanation of your observations. _____

_____

Muscle Tissue

STUDENT _____ DATE _____

LABORATORY SECTION _____ SCORE/GRADE _____

PART 1. Multiple Choice

_____ 1. The ability of muscle tissue to return to its original shape after contraction or extension is called (a) excitability (b) elasticity (c) extension (d) tetanus

_____ 2. Which of the following is striated and voluntary? (a) skeletal muscle tissue (b) cardiac muscle tissue (c) visceral muscle tissue (d) smooth muscle tissue

_____ 3. The portion of a sarcomere composed of thin myofilaments only is the (a) H zone (b) A band (c) I band (d) Z disc

_____ 4. The area of contact between a motor axon terminal and a muscle fiber sarcolemma (motor end plate) is called the (a) synapse (b) myofilament (c) transverse tubule (d) neuromuscular junction

_____ 5. The connective tissue layer surrounding bundles of muscle fibers is called the (a) perimysium (b) endomysium (c) ectomysium (d) myomysium

_____ 6. Which of the following is striated and involuntary? (a) smooth muscle tissue (b) skeletal muscle tissue (c) cardiac muscle tissue (d) visceral muscle tissue

_____ 7. The portion of a sarcomere where thin and thick myofilaments overlap is called the (a) H zone (b) triad (c) A band (d) I band

_____ 8. Intercalated discs are characteristic of which type of muscle tissue? (a) cardiac muscle tissue (b) skeletal muscle tisssue (c) smooth muscle tissue (d) visceral muscle tissue

_____ 9. The space between an axon terminal and sarcolemma is called the (a) end plate (b) synaptic cleft (c) synaptic gutter (d) synaptic end bulb

_____ 10. The ability of muscle tissue to stretch when pulled is called (a) extensibility (b) excitability (c) elasticity (d) irritability

PART 2. Completion

11. The ability of muscle tissue to receive and respond to stimuli is called _____.

12. Fibrous connective tissue located between muscle fibers is known as _____.

13. The sections of a muscle fiber separated by Z discs are called _____.

14. The plasma membrane surrounding a muscle fiber is called the _____.

15. The region in a sarcomere consisting of thick myofilaments only is known as the

_____.

16. Muscle tissue that is nonstriated and involuntary is _____.

17. The ability of muscle tissue to stretch when pulled is called _____.

18. The phenomenon by which a muscle fiber contracts to its fullest or not at all is known as the

_____.

19. The linkage between the production of the electrical and chemical events to the contraction of

skeletal muscle is called _____.

20. A localized depolarization of a motor end plate that initiates an action potential is called an

_____.

Skeletal Muscles

In this exercise you will learn the names, locations, and actions of the principal skeletal muscles of the body.

A. HOW SKELETAL MUSCLES PRODUCE MOVEMENT

1. Origin and Insertion

Skeletal muscles produce movements by exerting force on tendons, which in turn pull on bones. Most muscles cross at least one joint and are attached to the articulating bones that form the joint. When such a muscle contracts, it draws one articulating bone toward the other. The two articulating bones usually do not move equally in response to the contraction. One is held nearly in its original position because other muscles contract to pull it in the opposite direction or because its structure makes it less movable. Ordinarily, the attachment of a muscle tendon to the stationary bone is called the *origin.* The attachment of the other muscle tendon to the movable bone is the *insertion.* A good analogy is a spring on a door. The part of the spring attached to the door represents the insertion; the part attached to the frame is the origin. The fleshy portion of the muscle between the tendons of the origin and insertion is called the *belly (gaster).* The origin is usually proximal and the insertion distal, especially in the extremities. In addition, muscles that move a body part generally do not cover the moving part. For example, although contraction of the biceps brachii muscle moves the forearm, the belly of the muscle lies over the humerus.

2. Group Actions

Most movements are coordinated by several skeletal muscles acting in groups rather than individually and most skeletal muscles are arranged in opposing pairs at joints—that is, flexors-extensors, abductors-adductors, and so on. Consider flexing the forearm at the elbow, for example. A muscle that causes a desired action is referred to as the *prime mover (agonist).* In this instance, the biceps brachii is the *prime mover* (see Figure 10.13a). Simultaneously with the contraction of the biceps brachii, another muscle, called the *antagonist,* is relaxing. In this movement, the triceps brachii serves as the *antagonist* (see Figure 10.13b). The antagonist has an effect opposite to that of the prime mover—that is, the antagonist relaxes and yields to the movement of the prime mover. You should not assume, however, that the biceps brachii is always the prime mover and the triceps brachii is always the antagonist. For example, when extending the forearm at the elbow, the triceps brachii serves as the prime mover and the biceps brachii functions as the antagonist; their roles are reversed. Note that if the prime mover and antagonist contracted simultaneously with equal force, there would be no movement.

In addition to prime movers and antagonists, most movements also involve muscles called *synergists,* which serve to steady a movement. They prevent unwanted movements and help the prime mover function more efficiently. For example, flex your hand at the wrist and then make a fist. Note how difficult this is to do. Now extend your hand at the wrist and then make a fist. Note how much easier it is to clench your fist. In this case, the extensor muscles of the wrist act as synergists in cooperation with the flexor muscles of the fingers acting as prime movers. The extensor muscles of the fingers serve as antagonists (see Figure 10.14c,d).

Some synergist muscles in a group also act as *fixators,* which stabilize the origin of the prime mover so that the prime mover can act more efficiently. For example, the scapula is a freely movable bone in the pectoral (shoulder) girdle that serves as an origin for several muscles that move the arm. However, for the scapula to serve as a firm origin for muscles that move the arm, it must be held steady. This is accomplished by fixator muscles that hold the scapula firmly against the back of the chest. In abduction of the arm, the deltoid muscle serves as the prime mover, whereas fixators (pectoralis minor, rhomboideus major, rhomboideus minor, trapezius, subclavius, and serratus anterior muscles) hold the scapula firmly (see Figure 10.11). These fixators stabilize the scapula, which serves as the attachment site for the origin of the deltoid muscle, while the insertion of the muscle pulls on the humerus to abduct the arm. Under different conditions and depending on the movement and which point is fixed, many muscles act, at various times, as prime movers, antagonists, synergists, or fixators.

B. NAMING SKELETAL MUSCLES

Most of the almost 700 skeletal muscles of the body are named on the basis of one or more distinctive criteria. If you understand these criteria, you will find it much easier to learn and remember the names of individual muscles. Some muscles are named on the basis of the **direction of the muscle fibers,** for example, *rectus* (meaning "straight"), *transverse,* and *oblique.* Rectus fibers run parallel to some imaginary line, usually the midline of the body, transverse fibers run perpendicular to the line, and oblique fibers run diagonal to the line. Muscles named according to the direction their fibers run include the rectus abdominis, transversus abdominis, and external oblique. Another criterion employed is *location.* For example, the temporalis is so named because of its proximity to the temporal

bone, and the tibialis anterior is located near the tibia. *Size* is also commonly employed. For instance, *maximus* means "largest," *minimus* means "smallest," *longus* refers to long, and *brevis* refers to short. Examples include the gluteus maximus, gluteus minimus, adductor longus, and peroneus brevis.

Some muscles—such as biceps, triceps, and quadriceps—are named on the basis of the **number of origins** they have. For instance, the biceps brachii has two origins, the triceps three, and the quadriceps four. Other muscles are named on the basis of *shape.* Common examples include the deltoid (meaning triangular) and trapezius (meaning trapezoid). Muscles may also be named after their **insertion** and their **origin.** Two such examples are the sternocleidomastoid (originates on the sternum and clavicle and inserts at the mastoid process of the temporal bone), and the stylohyoid (originates on styloid process of temporal bone and inserts at the hyoid bone).

Still another basis for naming muscles is **action.** Listed here are the principal actions of muscles, their definitions, and examples of muscles that perform the actions.

Flexor (FLEK-sor)—Decreases the angle at a joint. Example: flexor carpi radialis.

Extensor (eks-TEN-sor)—Usually increases the angle at a joint. Example: extensor carpi ulnaris.

Abductor (ab-DUK-tor)—Moves a bone away from the midline. Example: abductor pollicis brevis.

Adductor (ad-DUK-tor)—Moves a bone closer to the midline. Example: adductor longus.

Levator (le-VĀ-tor)—Produces an upward or superiorly directed movement. Example: levator scapulae.

Depressor (de-PRES-or)—Produces a downward or inferiorly directed movement. Example: depressor labii inferioris.

Supinator (soo'-pi-NĀ-tor)—Turns the palm upward or to the anterior. Example: supinator.

Pronator (prō-NŌ-tor)—Turns the palm downward or to the posterior. Example: pronator teres.

Sphincter (SFINGK-ter)—Decreases the size of an opening. Example: external anal sphincter.

Tensor (TEN-sor)—Makes a body part more rigid. Example: tensor fasciae latae.

Rotator (RŌ-tāt-or)—Moves a bone around its longitudinal axis. Example: obturator externus.

C. CONNECTIVE TISSUE COMPONENTS

Skeletal muscles are protected, strengthened, and attached to other structures by several connective tissue components. For example, the entire muscle is usually wrapped with a fibrous connective tissue called the *epimysium* (ep'-i-MĪZ-ē-um). When the muscle is cut in cross section, invaginations of the epimysium divide the muscle into bundles called fasciculi (fascicles). These invaginations of the epimysium are called the *perimysium* (per'-i-MĪZ-ē-um). In turn, invaginations of the perimysium, called *endomysium* (en'-dō-MĪZ-ē-um), penetrate into the interior of each fascicle and separate individual muscle fibers from one another. The epimysium, perimysium, and endomysium are all extensions of deep fascia and are all continuous with the connective tissue that attaches the muscle to another structure, such as bone or other muscle. All three elements may be extended beyond the muscle cells as a *tendon*—a cord of connective tissue that attaches a muscle to the periosteum of bone. The connective tissue may also extend as a broad, flat band of tendons called an *aponeurosis.* Aponeuroses also attach to the coverings of a bone or another muscle. When a muscle con-

tracts, the tendon and its corresponding bone or muscle are pulled toward the contracting muscle. In this way, skeletal muscles produce movement.

In Figure 10.1, label the epimysium, perimysium, endomysium, fasciculus, and muscle fibers.

D. PRINCIPAL MUSCLES

In the pages that follow, a series of tables has been provided for you to learn the principal skeletal muscles by region.[1] Use each table as follows:

1. Take each muscle, in sequence, and study the *learning key* that appears in parentheses after the name of the muscle. The learning key is a list of prefixes, suffixes, and definitions that explain the derivations of the muscles' names. It will help you to understand the reason for a muscle's name.
2. As you learn the name of each muscle, determine its origin, insertion, and action and write these in the spaces provided in the table. Consult your textbook if necessary.

[1]A few of the muscles listed are not illustrated in the diagrams. Please consult your textbook to locate them.

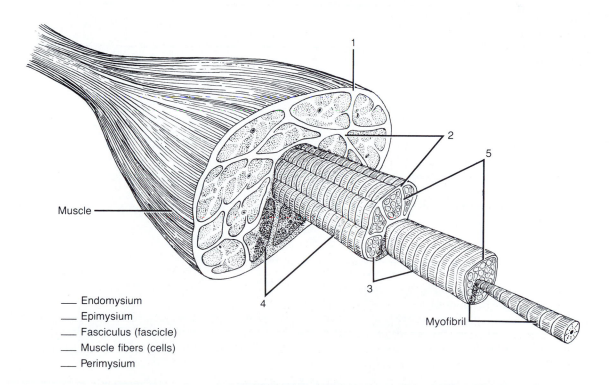

1
2
5
Muscle
3
4
Myofibril

___ Endomysium
___ Epimysium
___ Fasciculus (fascicle)
___ Muscle fibers (cells)
___ Perimysium

FIGURE 10.1   Connective tissue components of a skeletal muscle.

3. Again, using your textbook as a guide, label the diagram referred to in the table.
4. Try to visualize what happens when the muscle contracts, so that you will understand its action.
5. Do steps 1 through 4 for each muscle in the table. Before moving to the next table, examine a torso or a chart of the skeletal system so

that you can compare and approximate the positions of the muscles.
6. When possible, try to feel each muscle on your own body.

Refer to Tables 10.1 through 10.21 and Figures 10.2 through 10.20.

TABLE 10.1
Muscles of Facial Expression (After completing the table, label Figure 10.2)

OVERVIEW: The muscles in this group provide humans with the ability to express a wide variety of signals that communicate emotions, including grief, surprise, fear, and happiness. The muscles themselves lie within the layers of superficial fascia. As a rule, they arise from the fascia or bones of the skull and insert into the skin. Because of their insertions, the musclesof facial expression move the skin rather than a joint when they contract.

Muscle	Origin	Insertion	Action
Epicranius (ep-i-KRĀ-nē-us; *epi* = over, *crani* = skull)	This muscle is divisible into two portions: the frontalis, over the frontal bone, and the occipitalis, over the occipital bone. The two muscles are united by a strong aponeurosis, the galea aponeurotica, which covers the superior and lateral surfaces of the skull.		
Frontalis (fron-TA-lis; *front* = forehead)			
Occipitalis (ok-si'-pi-TA-lis; *occipito* = base of skull)			
Orbicularis oris (or-bi'-kyoo-LAR-is OR-is; *orb* = circular, *or* = mouth)			
Zygomaticus (zī-gō-MA-ti-kus) major (*zygomatic* = cheek bone, *major* = greater)			
Levator labii superioris (le-VĀ-ter LA-bē-ī soo-per'-ē-OR-is; *levator* = raises or elevates, *labii* = lip, *superioris* = upper)			
Depressor labii inferioris (de-PRE-ser LA-bē-ī-fer'-ē-OR-is; *depressor* = depresses or lowers, *inferiors* = lower)			
Buccinator (BUK-si-na'-tor; *bucc* = cheek)			

Ch. 10 / SKELETAL MUSCLES

TABLE 10.1 *(Continued)*

Muscle	Origin	Insertion	Action
Mentalis (men-TA-lis; *mentum* = chin)			
Platysma (pla-TIZ-ma; *platy* = flat, broad)			
Risorius (ri-ZOR-ē-us; *risor* = laughter)			
Orbicularis oculi (or-bi′-kyoo-LAR-is Ō-kyoo-lī; *ocul* = eye)			
Corrugator supercilii (KOR-a-gā′-tor soo-per-SI-lē-ī; *corrugo* = to wrinkle, *supercilium* = eyebrow)			
Levator palpebrae superioris (le-VĀ-tor PAL-pe-brē soo-per′-ē-OR-is; *palpebrae* = eyelids) (See Figure 10.4)			

TABLE 10.2

Muscles That Move the Lower Jaw (After completing the table, label Figure 10.3.)

OVERVIEW: Muscles that move the lower jaw are also known as muscles of mastication because they aid in biting and chewing. These muscles also assist in speech.

Muscles	Origin	Insertion	Action
Masseter (MA-se-ter; *masseter* = chewer)			
Temporalis (tem′-por-A-lis; *tempora* = temples)			
Medial pterygoid (TER-i-goid; *medial* = closer to midline, *pterygoid* = like a wing)			
Lateral pterygoid (TER-i-goid; *lateral* = farther from midline)			

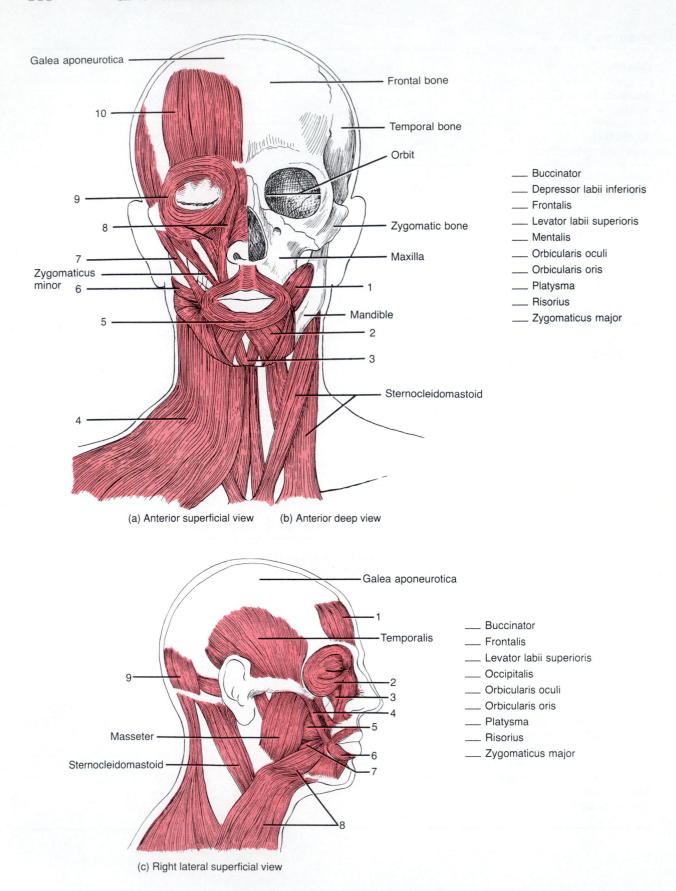

Galea aponeurotica

Frontal bone

10

Temporal bone

Orbit

9

8

Zygomatic bone

7

Maxilla

Zygomaticus
minor
6

1

5

Mandible

2

3

Sternocleidomastoid

4

(a) Anterior superficial view   (b) Anterior deep view

___ Buccinator
___ Depressor labii inferioris
___ Frontalis
___ Levator labii superioris
___ Mentalis
___ Orbicularis oculi
___ Orbicularis oris
___ Platysma
___ Risorius
___ Zygomaticus major

Galea aponeurotica

1

Temporalis

9

2

3

4

5

Masseter

6

Sternocleidomastoid

7

8

(c) Right lateral superficial view

___ Buccinator
___ Frontalis
___ Levator labii superioris
___ Occipitalis
___ Orbicularis oculi
___ Orbicularis oris
___ Platysma
___ Risorius
___ Zygomaticus major

FIGURE 10.2   Muscles of facial expression.

TABLE 10.3
Muscles That Move the Eyeball—The Extrinsic Muscles* (After completing the table, label Figure 10.4.)

OVERVIEW: Muscles associated with the eyeball are of two principal types: extrinsic and intrinsic. **Extrinsic muscles** originate outside the eyeball and are inserted on its outer surface (sclera). **Intrinsic muscles** originate and insert entirely within the eyeball.

Movements of the eyeballs are controlled by three pairs of extrinsic muscles. The two pairs of rectus muscles move the eyeball in the direction indicated by their respective names—superior, inferior, lateral, and medial. The pair of oblique muscles—superior and inferior—rotate the eyeball on its axis. The extrinsic muscles of the eyeballs are among the fastest contracting and precisely controlled skeletal muscles of the body.

Muscle	Origin	Insertion	Action
Superior rectus (REK-tus; *superior* = above, *rectus* = in this case, muscle fibers running parallel to the long axis of eyeball)			
Inferior rectus (REK-tus; *inferior* = below)			
Lateral rectus (REK-tus)			
Medial rectus (REK-tus)			
Superior oblique (ō-BLĒK; *oblique* = in this case, muscle fibers running diagonally to the long axis of the eyeball)			
Inferior oblique (ō-BLĒK)			

*Muscles situated on the outside of the eyeball.

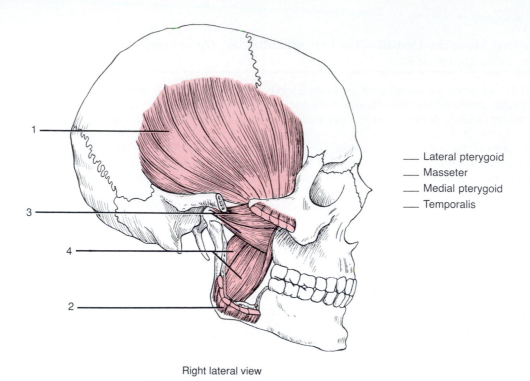

___ Lateral pterygoid
___ Masseter
___ Medial pterygoid
___ Temporalis

Right lateral view

FIGURE 10.3   Muscles that move the lower jaw.

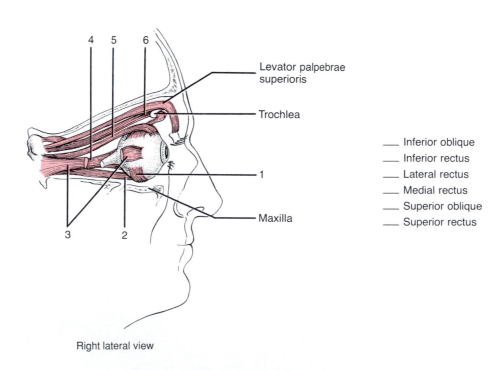

Levator palpebrae
superioris

Trochlea

1

Maxilla

___ Inferior oblique
___ Inferior rectus
___ Lateral rectus
___ Medial rectus
___ Superior oblique
___ Superior rectus

Right lateral view

FIGURE 10.4   Extrinsic muscles of the eyeball.

TABLE 10.4

Muscles That Move the Tongue—The Extrinsic Muscles (After completing the table, label Figure 10.5.)

OVERVIEW: The tongue is divided into lateral havles by a median fibrous septum. The septum extends throughout the length of the tongue and is attached inferiorly to the hyoid bone. Like the muscles of the eyeball, the muscles of the tongue are of two principal types—extrinsic and intrinsic. *Extrinsic muscles* originate outside the tongue and insert into it. *Intrinsic muscles* originate and insert within the tongue. The extrinsic and intrinsic muscles of the tongue are arranged in both lateral halves of the tongue.

Muscle	Origin	Insertion	Action
Genioglossus (jē′-nē-ō-GLOS-us; *geneion* = chin, *glossus* = tongue)			
Styloglossus (stī′-lō-GLOS-us; *stylo* = stake or pole)			
Palatoglossus (pal′-a-tō-GLOS-us; *palato* = palate)			
Hyoglossus (hī-ō-GLOS-us)			

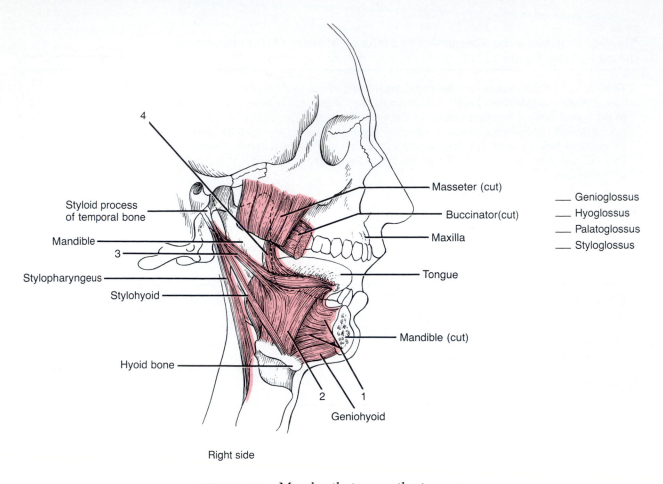

Styloid process
of temporal bone

Mandible

Stylopharyngeus

Stylohyoid

Hyoid bone

Masseter (cut)

Buccinator(cut)

Maxilla

Tongue

Mandible (cut)

Geniohyoid

___ Genioglossus
___ Hyoglossus
___ Palatoglossus
___ Styloglossus

Right side

FIGURE 10.5   Muscles that move the tongue.

TABLE 10.5
Muscles of the Floor of the Oral Cavity (Figure 10.6)

OVERVIEW: As a group, these muscles are referred to as *suprahyoid muscles.* They
lie superior to the hyoid bone and all insert into it. The digastric muscle consists of
an anterior belly and a posterior belly united by an intermediate tendon that is held
in position by a fibrous group.

Muscle	Origin	Insertion	Action
Digastric (dī'-GAS-trik; *di* = two, *gaster* = belly)			
Stylohyoid (stī'-lō-HĪ-oid; *stylo* = stake or pole, styloid process of the temporal bone, *hyoedes* = U-shaped, pertaining to the hyoid bone.)			
Mylohyoid (mī'-lō-HĪ-oid)			
Geniohyoid (je'-nē-ō-HĪ-oid; *geneion* = chin) (See Figure 10.5)			

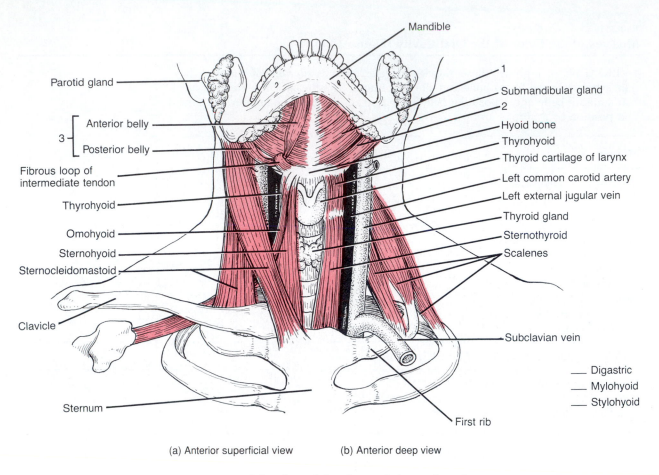

Parotid gland

Anterior belly

3

Posterior belly

Fibrous loop of
intermediate tendon

Thyrohyoid

Omohyoid

Sternohyoid

Sternocleidomastoid

Clavicle

Sternum

Mandible

1

Submandibular gland

2

Hyoid bone

Thyrohyoid

Thyroid cartilage of larynx

Left common carotid artery

Left external jugular vein

Thyroid gland

Sternothyroid

Scalenes

Subclavian vein

First rib

___ Digastric

___ Mylohyoid

___ Stylohyoid

(a) Anterior superficial view   (b) Anterior deep view

FIGURE 10.6   Muscles of the floor of the oral cavity.

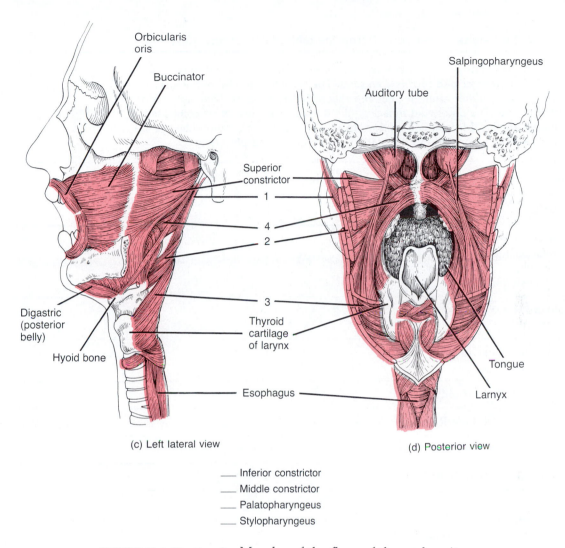

Orbicularis oris

Buccinator

Superior constrictor

1

4

2

3

Digastric (posterior belly)

Hyoid bone

Thyroid cartilage of larynx

Esophagus

(c) Left lateral view

Salpingopharyngeus

Auditory tube

Tongue

Larnyx

(d) Posterior view

___ Inferior constrictor
___ Middle constrictor
___ Palatopharyngeus
___ Stylopharyngeus

FIGURE 10.6 *(Continued)*   Muscles of the floor of the oral cavity.

TABLE 10.6

Muscles of the Larynx (After completing the table, label Figure 10.7)

OVERVIEW: The muscles of the larynx, like those of the eyeballs and tongue, are grouped into *extrinsic* and *intrinsic muscles.* The extrinsic muscles of the larynx marked (*) are together referred to as *infrahyoid (strap) muscles.* They lie inferior to the hyoid bone. The omohyoid muscles, like the digastric muscles, is composed of two bellies and an intermediate tendon. In this case, however, the two bellies are referred to as superior and inferior, rather than anterior and posterior.

Muscle	Origin	Insertion	Action
EXTRINSIC			
Omohyoid* (ō'-mō-HĪ-oid; *omo* = relationship to shoulder, *hyoedes* = U-shaped)			
Sternohyoid* (ster'-nō-HĪ-oid; *sterno* = sternum)			
Sternothyroid* (ster'-nō-THĪ-roid; *thyro*= thyroid gland)			
Thyrohyoid* (thī'-rō-HĪ-oid)			
Stylopharyngeus (stī'lō-fa-RIN-jē-us)			
Inferior constrictor (kon-STRIK-tor)			
Middle constrictor (kon-STRIK-tor)			
INTRINSIC			
Cricothyroid (kri-kō-THĪ-roid; *crico* = cricoid cartilage of larynx)			
Posterior cricoarytenoid (kri'-kō-ar'-i-TĒ-noid; *arytaina* = shaped like a jug)			
Lateral cricoarytenoid (kri'-kō-ar'-i-TĒ-noid)			
Arytenoid (ar'-i-TĒ-noid)			
Thyroartenoid (thī'-rō-ar'-i-TĒ-noid)			

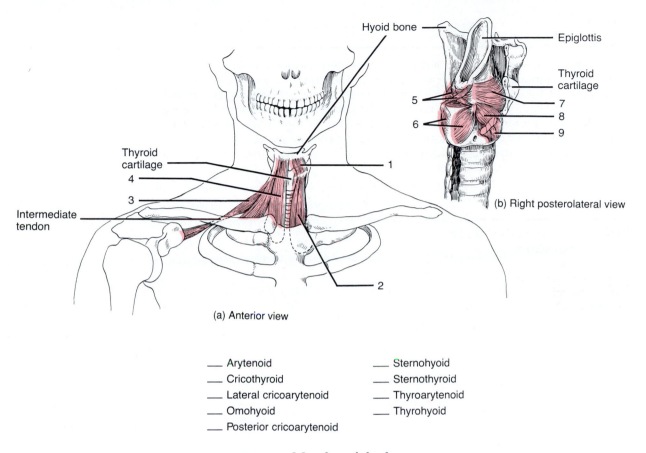

(a) Anterior view

(b) Right posterolateral view

___ Arytenoid	___ Sternohyoid
___ Cricothyroid	___ Sternothyroid
___ Lateral cricoarytenoid	___ Thyroarytenoid
___ Omohyoid	___ Thyrohyoid
___ Posterior cricoarytenoid	

FIGURE 10.7   Muscles of the larynx.

TABLE 10.7
Muscles that Move the Head

OVERVIEW: The cervical region is divided by the sternocleidomastroid muscle into two principal triangles—anterior and posterior. The *anterior triangle* is bordered superiorly by the mandible, inferiorly by the sternum, medially by the cervical midline, and laterally by the anterior border of the sternocleidomastoid muscle (see Figure 11.2). The *posterior triangle* is bordered inferiorly by the clavicle, anteriorly by the posterior border of the sternocleidomastoid muscle, and posteriorly by the anterior border of the trapezius muscle (see Figure 11.2). Subsidiary triangles exist within the two principal triangles.

Muscle	Origin	Insertion	Action
Sternocleidomastoid (ster'-nō-klī'-dō-MAS-toid; *sternum* = breastbone, *cleido* = clavicle, *mastoid* = mastoid process of the temporal bone) (Label this muscle in Figure 1.0.11.)			
Semispinalis capitis (se'-mē-spi-NA-lis KAP-i-tis; *semi* = half, *spine* = spinous process, *caput* = head) (Label this muscle in Figure 10.16.)			
Splenius capitus (SPLĒ-nē-us KAP-i-tis; *splenion* = bandage) (Label this muscle in Figure 10.16.)			
Longissimus capitus (lon-JIS-i-mus KAP-i-tis; *longissimus* = longest) (Label this muscle in Figure 10.16.)			

TABLE 10.8
Muscles That Act on the Anterior Abdominal Wall (After completing the table, label Figure 10.8.)

OVERVIEW: The anterolateral abdominal wall is composed of skin, fascia, and four pairs of flat, sheetlike muscles: rectus abdominis, external oblique, internal oblique, and transversus abdominis. The anterior surfaces of the rectus abdominis muscle are interrupted by transverse fibrous bands of tissue called *tendinous intersections,* believed to be remnants of septa that separated myotomes during embryonic development. The aponeuroses of the external oblique, internal oblique, and transversus abdominis muscles meet at the midline to form the *linea alba* (white line), a tough fibrous band that extends from the xiphoid process of the sternum to the pubic symphysis. The inferior free border of the external oblique aponeurosis, plus some collagen fibers, forms the *inguinal ligament,* which runs from the anterior superior iliac spine to the pubic tubercle (see Figure 10.17). The ligament demarcates the thigh and body wall.

   Just superior to the medial end of the inguinal ligament is a triangular slit in the aponeurosis referred to as the *superficial inguinal ring,* the outer opening of the *inguinal canal.* The canal contains the spermatic cord and ilioinguinal nerve in males and round ligament of the uterus and ilioinguinal nerve in females.

   The posterior abdominal wall is formed by the lumbar vertebrae, parts of the ilia of the hipbones, psoas major muscle (described in Table 10.18), quadratus lumborum muscle, and iliacus muscle (also described in Table 10.18). Whereas the anterolateral abdominal wall is contractile and distensible, the posterior abdominal wall is bulky and stable by comparison.

Muscle	Origin	Insertion	Action
Rectus abdominis (REK-tus ab-DO-min'-us; *rectus* = fibers parallel to midline, *abdomino* = abdomen)			
External oblique (ō-BLĒK; *external* = closer to the surface, *oblique* = fibers diagonal to the midline)			
Internal oblique (ō-BLĒK; *internal* = farther from the surface)			
Transversus abdominis (tranz-VER-sus ab-DO-min'-us; *transverse* = fibers perpendicular to the midline)			
Quadratus lumborum (kwod-RĀ-tus lum-BOR-um; *quad* = four, *lumbo* = lumbar region) (See Figure 10.6a)			

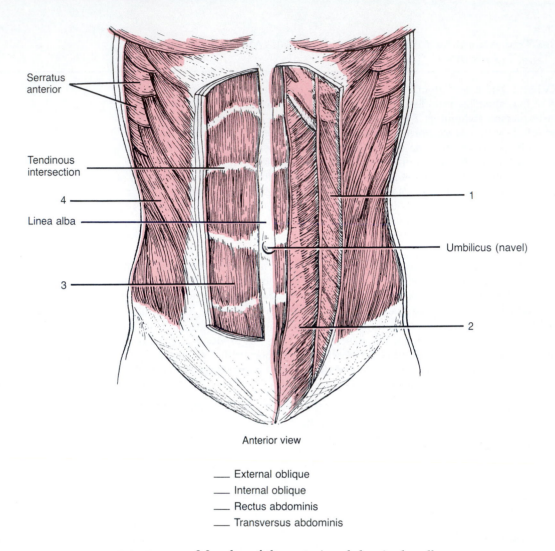

Serratus anterior

Tendinous intersection

4

Linea alba

3

1

Umbilicus (navel)

2

Anterior view

___ External oblique
___ Internal oblique
___ Rectus abdominis
___ Transversus abdominis

FIGURE 10.8   Muscles of the anterior abdominal wall.

TABLE 10.9

Muscles Used in Breathing (After completing the table, label Figure 10.9.)

OVERVIEW: The muscles described here are attached to the ribs and by their contraction and relaxation alter the size of the thoracic cavity during normal breathing.
In forced breathing, other muscles are involved as well. Essentially, inspiration occurs when the thoracic cavity increases in size. Expiration occurs when the thoracic
cavity decreases in size.

   The diaphragm, one of the muscles used in breathing, is dome-shaped and has
several openings through which various structures pass between the thorax and abdomen. These structures include the aorta along with the thoracic duct and azygos
vein, the esophagus with accompanying vagus (X) nerves, and the inferior vena
cava along with the phrenic nerve.

Muscle	Origin	Insertion	Action
Diaphragm (DĪ-a-fram; *dia* = across, between, *phragma* = wall)			
External intercostals (in'-ter-KOS-tals; *inter* = between, *costa* = rib)			
Internal intercostals (in'-ter-KOS-tals; *internal* = farther from the surface)			

TABLE 10.10

Muscles of the Pelvic Floor (After completing the table, label Figure 10.10.)

OVERVIEW: The muscles of the pelvic floor, together with the fascia covering their
external and internal surfaces, are referred to as the *pelvic diaphragm.* This diaphragm is funnel-shaped and forms the floor of the abdominopelvic cavity where it
supports the pelvic viscera. It is pierced by the anal canal and urethra in both sexes
and also by the vagina in the female.

Muscle	Muscle	Insertion	Action
Levator ani (le-VĀ-tor Ā-nē; *levator* = raises, *ani* = anus)	This muscle is divisible into two parts, the pubococcygeus muscle and the iliococcygeus muscle.		
Pubococcygeus (pu'-bo-kok-SIJ-ē-us; *pubo* = pubis, *coccygeus* = coccyx)			
Iliococcygeus (il'-ē-o-kok-SIJ-ē-us; *ilio* = ilium)			
Coccygeus* (kok-SIJ-ē-us)			

*Not illustrated

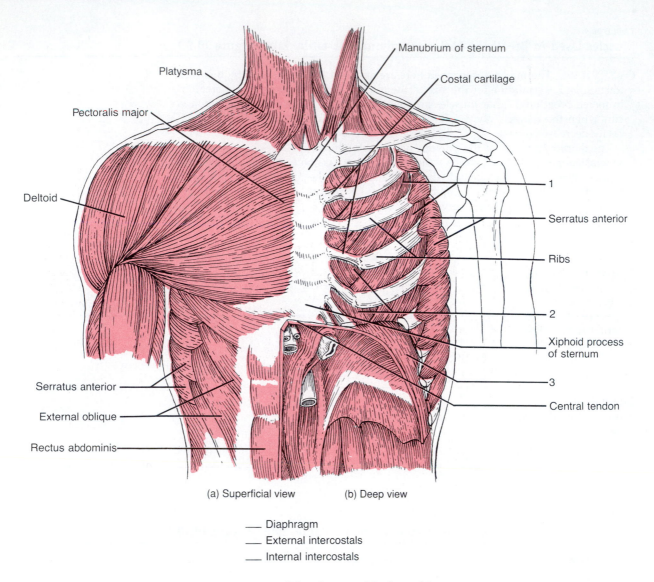

Platysma

Pectoralis major

Deltoid

Manubrium of sternum

Costal cartilage

1

Serratus anterior

Ribs

2

Xiphoid process of sternum

3

Central tendon

Serratus anterior

External oblique

Rectus abdominis

(a) Superficial view   (b) Deep view

___ Diaphragm
___ External intercostals
___ Internal intercostals

FIGURE 10.9 Muscles used in breathing.

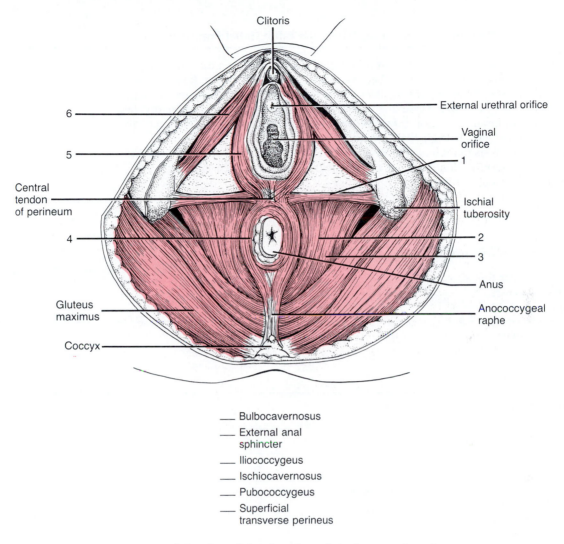

Clitoris

6

5

External urethral orifice

Vaginal orifice

1

Central tendon of perineum

Ischial tuberosity

4

2

3

Anus

Gluteus maximus

Anococcygeal raphe

Coccyx

___ Bulbocavernosus

___ External anal sphincter

___ Iliococcygeus

___ Ischiocavernosus

___ Pubococcygeus

___ Superficial transverse perineus

FIGURE 10.10 Muscles of the female pelvic floor and perineum.

TABLE 10.11
Muscles of the Perineum (After completing the table, label Figure 10.10)

OVERVIEW: The *perineum* is the entire outlet of the pelvis. It is a diamond-shaped area at the lower end of the trunk, between the thighs and buttocks. It is bordered anteriorly by the pubic symphysis, laterally by the ischial tuberosities, and posteriorly by the coccyx. A transverse line drawn between the ischial tuberosities divides the perineum into an anterior *urogenital triangle* that contains the external genitals and a posterior *anal triangle* that contains the anus.
The deep transverse perineus, the urethral sphincter, and a fibrous membrane constitute the *urogenital diaphragm.* It surrounds the urogenital ducts and helps to strengthen the pelvic floor.

Muscle	Origin	Insertion	Action
Superficial transverse perineus (per-i-NĒ-us; *superficial* = near the surface, *transverse* = across, *perineus* = perineum)			
Bulbocavernosus (bul'-bō-ka'-ver-NŌ-sus; *bulbus* = bulb, *caverna* = hollow space)			
Ischiocavernosus (is'-kē-o-ka'-ver-NŌ-sus; *ischion* = hip)			
Deep transverse perineus* (per-i-NĒ-us; *deep* = farther from the surface)			
Urethral sphincter* (yoo-RE-thral SFINGK-ter; *urethral* = pertaining to the urethra; *sphincter* = circular muscle that decreases the size of an opening)			
External anal (Ā-nal) **sphincter**			

*Not illustrated

TABLE 10.12
Muscles That Move the Pectoral (Shoulder) Girdle (After completing the table, label Figure 10.11)

OVERVIEW: Muscles that move the pectoral (shoulder) girdle originate on the axial skeleton and insert on the clavicle or scapula. The muscles can be distinguished into *anterior* and *posterior* groups. The principal action of the muscles is to stabilize the scapula so that it can function as a stable point of origin for most of the muscles that move the arm (humerus).

Muscle	Origin	Insertion	Action
ANTERIOR MUSCLES			
Subclavius (sub-KLĀ-vē-us; *sub* = under, *clavius* = clavicle)			
Pectoralis minor (pek'-tor-A-lis; *pectus* = breast, *chest* = thorax, *minor* = lesser)			
Serratus anterior (ser-Ā-tis; *serratus* = serrated, *anterior* = front)			
POSTERIOR MUSCLES			
Trapezius (tra-PĒ-zē-us; *trapezoides* = trapezoid shaped)			
Levator scapulae (le-VĀ-tor SKA-pyoo-lē; *levator* = raises, *scapulae* = scapula)			
Rhomboideus major (rom-BOID-ē-us; *rhomboides* = rhomboid-shaped or diamond-shaped)			
Rhomboideus minor (rom-BOID-ē-us)			

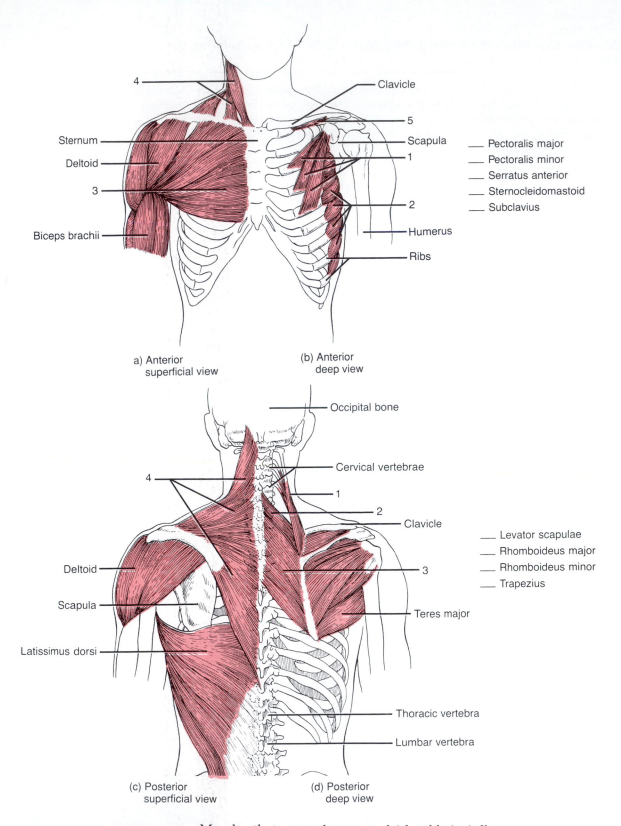

a) Anterior
superficial view

(b) Anterior
deep view

____ Pectoralis major
____ Pectoralis minor
____ Serratus anterior
____ Sternocleidomastoid
____ Subclavius

Clavicle
5
Scapula
1
2
Humerus
Ribs

Sternum
Deltoid
3
Biceps brachii
4

Occipital bone

Cervical vertebrae
1
2
Clavicle

____ Levator scapulae
____ Rhomboideus major
____ Rhomboideus minor
____ Trapezius

3
Teres major

Thoracic vertebra
Lumbar vertebra

Deltoid
Scapula
Latissimus dorsi
4

(c) Posterior
superficial view

(d) Posterior
deep view

FIGURE 10.11   Muscles that move the pectoral (shoulder) girdle.

TABLE 10.13

Muscles That Move in the Arm (After completing the table, label Figure 10.12.)

OVERVIEW: The muscles that move the arm (humerus) cross the shoulder joint. Of the nine muscles that cross the shoulder joint, only two of them (pectoralis major and latissimus dorsi) do not originate on the scapula. These two muscles are thus designated as *axial muscles* because they originate on the axial skeleton. The remaining seven muscles, the *scapular muscles,* arise from the scapula.

The strength and stability of the shoulder joint are not provided by the shape of the articulating bones or its ligaments. Instead, four deep muscles of the shoulder and their tendons—subscapularis, supraspinatus, infraspinatus, and teres minor—strengthen and stabilize the shoulder joint. The muscles and their tendons are so arranged as to form a nearly complete circle around the joint. This arrangement is referred to as the *rotator (musculotendinous) cuff.* It is a common site of injury to baseball pitchers, particularly the tearing of the supraspinatus muscle tendon. This tendon is especially predisposed to wear- and-tear changes because of its location between the head of the humerus and acromion of the scapula, which compresses the tendon during shoulder movements.

After you have studied the muscles in this table, arrange them according to the following actions: flexion, extension, abduction, adduction, medial rotation, and lateral rotation. (The same muscle can be used more than once.)

Muscle	Origin	Insertion	Action
AXIAL MUSCLES			
Pectoralis (pek'-tor-A-lis) **major** (Label this muscle in Figure 10.11.)			
Latissimus doris (la-TIS-i-mis DOR-sī; *dorsum* = back)			
SCAPULAR MUSCLES			
Deltoid (DEL-toyd; *delta* = triangular-shaped)			
Subscapularis (sub-scap'-yoo-LA-ris; *sub* = below, *scapularis* = scapula)			
Suprascapularis (sub-scap'-yoo-LA-ris; *supra* = below, *scapularis* = scapula)			
Supraspinatus (soo'-pra-spi-NĀ-tus; *supra* = above, *spinatus* = spine of the scapula)			
Infraspinatus (in'-fra-spi-NĀ-tus; *infra* = below)			
Teres major (TE-rēz) (*teres* = long and round)			
Teres (TE-rēz) **minor**			
Coracobrachialis (kor'-a-kō-BRĀ-kē-a'-lis; *coraco* = coracoid process)			

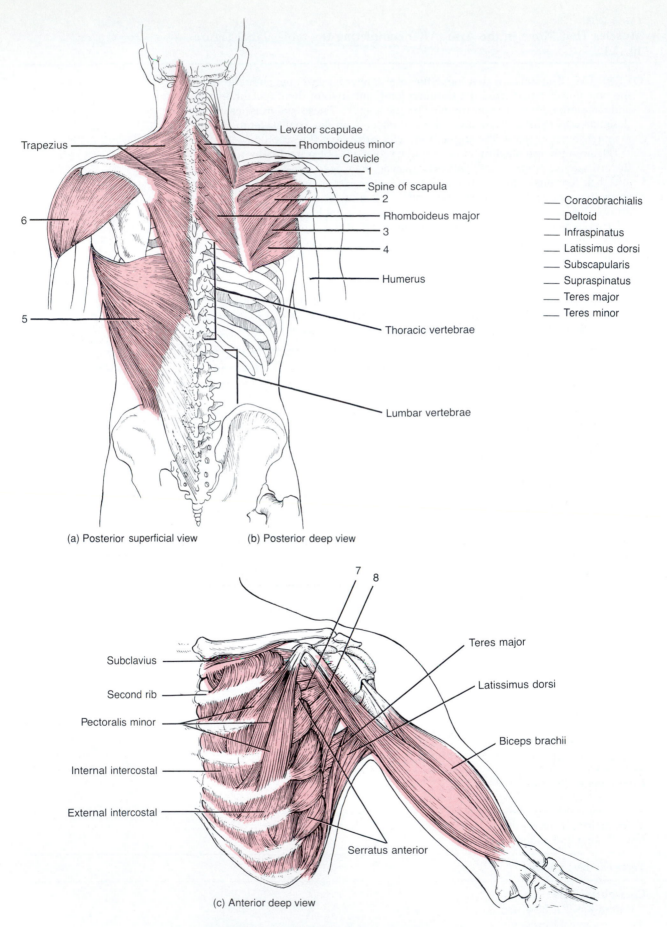

Levator scapulae

Trapezius

Rhomboideus minor

Clavicle

1

Spine of scapula

2

Rhomboideus major

3

4

Humerus

6

5

Thoracic vertebrae

Lumbar vertebrae

Coracobrachialis

Deltoid

Infraspinatus

Latissimus dorsi

Subscapularis

Supraspinatus

Teres major

Teres minor

(a) Posterior superficial view

(b) Posterior deep view

7   8

Subclavius

Second rib

Pectoralis minor

Internal intercostal

External intercostal

Teres major

Latissimus dorsi

Biceps brachii

Serratus anterior

(c) Anterior deep view

FIGURE 10.12   Muscles that move the arm.

TABLE 10.14
Muscles That Move the Forearm (After completing the table, label Figure 10.13.)

OVERVIEW: The muscles that move the forearm (radius and ulna) are divided into *flexors* and *extensors.* Recall that the elbow joint is a hinge joint, capable only of flexion and extension under normal conditions. Whereas the biceps brachii, brachialis, and brachioradialis are flexors of the elbow joint, the triceps brachii and anconeus are extensors. Other muscles that move the forearm are concerned with pronation and supination.

Muscle	Origin	Insertion	Action
FLEXORS			
Biceps brachii (BI-ceps BRA-kē-ī; *biceps* = two heads of origin, *brachion* = arm)			
Brachialis (brā′-kē-A-lis)			
Brachioradialis (bra′-kē-ō-rā′-dē-A-lis; *radialis* = radius) (Also see Figure 10.14a.)			
EXTENSORS			
Triceps brachii (TRI-ceps BRĀ-kē-ī; *triceps* = three heads of origin)			
Anconeus (an-KO-nē-us; *anconeal* = pertaining to the elbow) (Label this muscle in Figure 10.14.)			
PRONATORS			
Pronator teres (PRŌ-na′-ter TE-rēz; *pronation* = turning the palm downward or posteriorly)			
Pronator quadratus (PRŌ-na′-ter kwod-RĀ-tus; *quadratus* = squared or four-sided) (Label the muscle in Figure 10.14.)			
SUPINATOR			
Supinator (sup-pi-nā-tor; *supination* = turning the palm upward or anteriorly)			

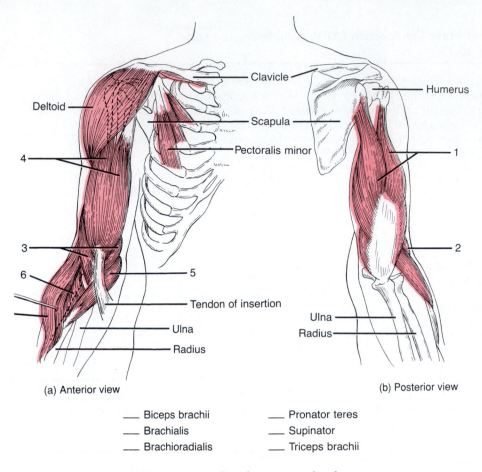

(a) Anterior view   (b) Posterior view

___ Biceps brachii   ___ Pronator teres

___ Brachialis   ___ Supinator

___ Brachioradialis   ___ Triceps brachii

FIGURE 10.13   Muscles that move the forearm.

TABLE 10.15
Muscles That Move the Wrist, Hand, and Fingers (Figure 10.14)

OVERVIEW: Muscles that move the wrist, hand, and fingers are many and varied. However, as you will see, their names for the most part give some indication of their origin, insertion, or action. On the basis of location and function, the muscles are divided into two groups—anterior and posterior. The *anterior muscles* function as flexors and pronators. They originate on the humerus and typically insert on the carpals, metacarpals, the phalanges. The bellies of these muscles form the bulk of the proximal forearm. The *posterior muscles* function as extensors and supinators. These muscles arise on the humerus and insert on the metacarpals and phalanges. Each of the two principal groups is also divided into superficial and deep muscles. The tendons of the muscles of the forearm that attach to the wrist or continue into the hand, along with blood vessels and nerves, are held close to bones by strong fascial structures. The tendons are also surrounded by tendon sheaths. At the wrist, the deep fascia is thickened into fibrous bands called retinacula. The *flexor retinaculum (transverse carpal ligament)* is located over the palmar surface of the carpal bones. Through it pass the long flexor tendons of the digits and wrist and median nerve. The *extensor retinaculum (dorsal carpal ligament)* is located over the dorsal surface of the carpal bones. Through it pass the extensor tendons of the wrist and digits.

After you have studied the muscles in the table, arrange them according to the following actions: flexion, extension, abduction, adduction, supination, and pronation. (The same muscles can be used more than once.)

Muscle	Origin	Insertion	Action
ANTERIOR GROUP (flexors and pronators) Superficial			
Flexor carpi radialis (FLEK-sor KAR-pē rā'-dē-A-lis; *flexor* = decreases the angle at a joint, *carpus* = wrist, *radialis* = radius)			
Palmaris longus (pal-MA-ris LON-gus; *palma* = palm *longus* = long)			
Flexor carpi ulnaris (FLEK-sor KAR-pē ul-NAR-is; *ulnaris* = ulna)			
Flexor digitorum superficialis (FLEK-sor di'-ji-TOR-um soo'-per-fish'-ē-A-lis; *digit* = finger or toe, *superficialis* = closer to the surface)			

TABLE 10.15 *(Continued)*

Muscle	Origin	Insertion	Action
Deep			
Flexor digitorum profundus (FLEK-sor di′-ji-TOR-um pro-FUN-dus; *profundus* = deep)			
Flexor pollicis longus (FLEK-so POL-li-kis LON-gus; *pollex* = thumb)			
POSTERIOR GROUP (extensors and supinators) Superficial			
Extensor carpi radialis longus (eks-TEN-sor KAR-pē rā′-dē-A-lis LON-gus; *extensor* = increases the angle at a joint)			
Extensor carpi radialis brevis (eks-TEN-sor KAR-pē rā′-dē-A-lis BREV-is; *brevis* = short)			
Extensor digitorum (eks-TEN-sor dī′ji-TOR-um;			
Extensor digiti minimi (eks-TEN-sor DIJ-i-tē MIN-i-mē; *mimnimi* = little finger)			
Extensor carpi ulnaris (eks-TEN-sor KAR-pē ul-NAR-is)			
Deep			
Abductor pollicis longus (ab-DUK-tor POL-li-kis LON-gus; *abductor* = moves a part away from the midline)			
Extensor pollicis brevis (eks-TEN-sor POL-li-kis BREV-is)			
Extensor pollicis longus (eks-TEN-sor POL-li-kis LON-gus)			
Extensor indicis (eks-TEN-sor IN-di-kis; *indicis* = index)			

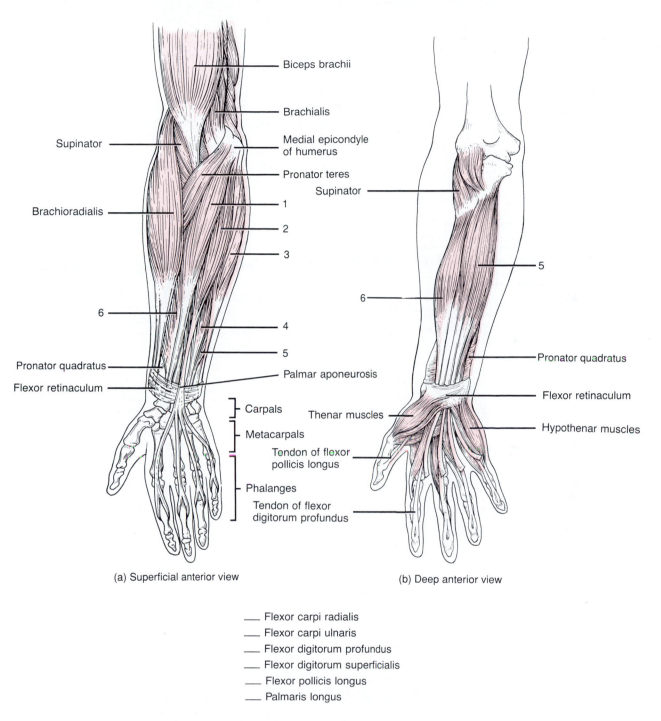

Biceps brachii

Brachialis

Medial epicondyle
of humerus

Pronator teres

Supinator

Supinator

Brachioradialis

1

2

3

6

4

5

Pronator quadratus

Flexor retinaculum

Palmar aponeurosis

Carpals

Metacarpals

Tendon of flexor
pollicis longus

Phalanges

Tendon of flexor
digitorum profundus

5

6

Pronator quadratus

Flexor retinaculum

Thenar muscles

Hypothenar muscles

(a) Superficial anterior view

(b) Deep anterior view

___ Flexor carpi radialis
___ Flexor carpi ulnaris
___ Flexor digitorum profundus
___ Flexor digitorum superficialis
___ Flexor pollicis longus
___ Palmaris longus

FIGURE 10.14 Muscles that move the wrist, hand, and fingers.

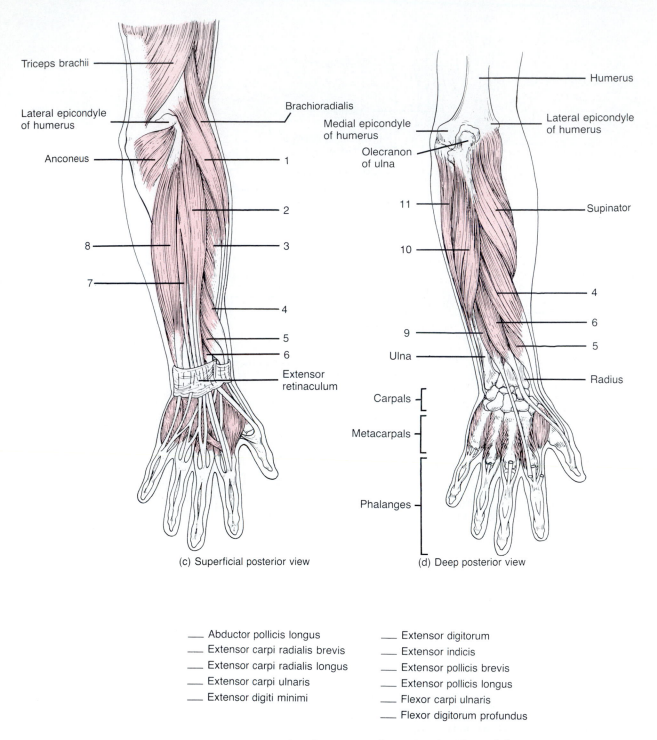

(c) Superficial posterior view

(d) Deep posterior view

___ Abductor pollicis longus
___ Extensor carpi radialis brevis
___ Extensor carpi radialis longus
___ Extensor carpi ulnaris
___ Extensor digiti minimi

___ Extensor digitorum
___ Extensor indicis
___ Extensor pollicis brevis
___ Extensor pollicis longus
___ Flexor carpi ulnaris
___ Flexor digitorum profundus

FIGURE 10.14 *(Continued)*   Muscles that move the wrist, hand, and fingers.

TABLE 10.16
Intrinsic Muscles of the Hand (Figure 10.15)

OVERVIEW: Several of the muscles discussed in Table 10.16 help to move the digits in various ways. In addition, there are muscles in the palmar surface of the hand called *intrinsic muscles* that also help to move the digits. Such muscles are so named because both their origins and insertions are within the hands. These muscles assist in the intricate and precise movements characteristic of the human hand.

The intrinsic muscles of the hand are divided into three principal groups—thenar, hypothenar, and intermediate. The four *thenar* (THĒ-nar) *muscles* act on the thumb and form the *thenar eminence* (see Figure 11.8). The four *hypothenar* (HĪ-pō-thē-nar) *muscles* act on the little finger and form the *hypothenar eminence* (see Figure 11.8). The eleven *intermediate (midpalmar) muscles* act on all the digits, except the thumb.

The functional importance of the hand is readily apparent if you consider that certain hand injuries can result in permanent disability. In fact, most of the efficiency of the hand depends on the movements of the thumb. The general activities of the hand are free motion, power grip (forcible movement of the fingers and thumb against the palm, as in squeezing), precision handling (change in position of a handled object that requires exact control of finger and thumb positions, as in winding a watch or threading a needle), and pinch (compression between the thumb and index finger or between the thumb and first two fingers).

Movement of the thumb is very important in the precise ativities of the hand. The five principal movements of the thumb, illustrated below, are flexion (movement of the thumb at right angles to the fingers), extension (movement in the opposite direction), abduction (movement of the thumb anteriorly and laterally), adduction (movement of the thumb toward the index finger), and opposition (movement of the thumb across the palm so that the tip of the thumb meets the tips of the fingers). Opposition is the single most distinctive digital movement that gives humans their characteristic tool-making and tool-using capacity.

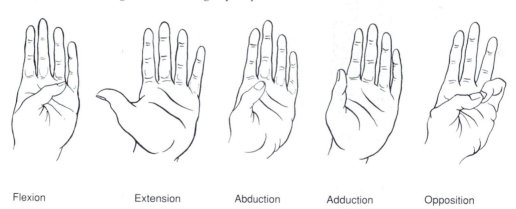

| Flexion | Extension | Abduction | Adduction | Opposition |

Muscle	Origin	Insertion	Action

THENAR (THĒ-nar) MUSCLES

Abductor pollicis brevis
(ab-DUK-tor POL-li-kis BREV-is;
abductor = moves the
part away from the midline;
pollex = thumb,
brevis = short)

TABLE 10.16 *(Continued)*

Muscle	Origin	Insertion	Action
Opponens pollicis (o-PŌ-nenz POL-li-kis; *opponens* = opposes)			
Flexor pollicis brevis (FLEK-sor POL-li-kis BREV-is; *flexor* = decreases the angle at the joint)			
Adductor pollicis (ad-DUK-tor POL-li-kis; *adductor* = moves the part toward the midline)			
HYPOTHENAR (HĪ-pō-thē′-nar) MUSCLES			
Palmaris brevis (pal-MA-ris BREV-is; *palma* = palm)			
Abductor digiti minimi (ab-DUK-tor DIJ-i-tē MIN-i-mē; *digit* = finger or toe, *minimi* = little finger)			
Flexor digiti minimi brevis (FLEK-sor DIJ-i-tē MIN-i-mē BREV-is)			
Opponens digiti minimi (o-PŌ-nenz DIJ-i-tē MIN-i-mē)			
INTERMEDIATE (MIDPALMAR) MUSCLES			
Lumbricals (LUM-bri-kals; four muscles)			
Dorsal interossei (DOR-sal in′-ter-OS-ē-ī; four muscles; *dorsal* = back surface, *inter* = between, *ossei* = bones)			
Palmar interossei (PAL-mar in′-ter-OS-ē-ī; three muscles)			

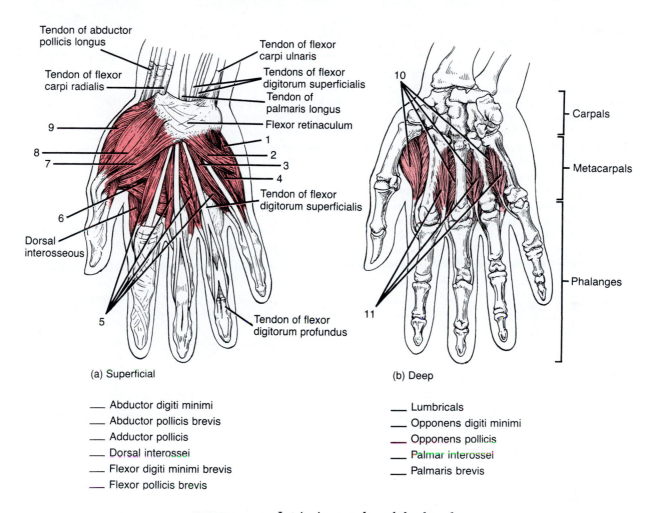

Tendon of abductor pollicis longus

Tendon of flexor carpi radialis

Tendon of flexor carpi ulnaris

Tendons of flexor digitorum superficialis

Tendon of palmaris longus

Flexor retinaculum

9

8
7

1
2
3
4

Tendon of flexor digitorum superficialis

6

Dorsal interosseous

5

Tendon of flexor digitorum profundus

(a) Superficial

10

Carpals

Metacarpals

Phalanges

11

(b) Deep

___ Abductor digiti minimi

___ Abductor pollicis brevis

___ Adductor pollicis

___ Dorsal interossei

___ Flexor digiti minimi brevis

___ Flexor pollicis brevis

___ Lumbricals

___ Opponens digiti minimi

___ Opponens pollicis

___ Palmar interossei

___ Palmaris brevis

FIGURE 10.15   Intrinsic muscles of the hand.

TABLE 10.17
Muscles That Move the Vertebral Column (Figure 10.16)

OVERVIEW: The muscles that move the vertebral column are quite complex because they have multiple origins and insertions and there is considerable overlapping among them. One way to group the muscles is on the basis of the direction of the muscle bundles and their approximate lengths. For example, the *splenius muscles* arises from the midline and run laterally and superiorly to their insertions. The *erector spinae (sacrospinalis) muscle* arises from either the midline or more laterally but usually runs almost longitudinally, with neither a marked outward nor an inward direction as it is traced superiorly. The *transversospinalis muscles* arise laterally, but run toward the midline as they are traced superiorly. Deep to these three muscle groups are small *segmental muscles* that run between spinous process or transverse processes of vertebrae. Because the scalene muscles also assist in moving the vertebral column, they are included in this table.

   Note in Table 10.8 that the rectus abdominis and quadratus lumborum muscle also assume a role in moving the vertebral column.

Muscle	Origin	Insertion	Action
SPLENIUS (SPLĒ-nē-us) MUSCLES			
Splenius capitis (SPLĒ-nē-us KAP-i-tis; *splenium* = bandage, *caput* = head)			
Splenius cervicis (SPLĒ-nē-us SER-vi-kis; *cervix* = neck)			
ERECTOR SPINAE (SACROSPINALIS)			
This is the largest muscular mass of the back and consists of three groupings—iliocostalis, longissimus, and spinalis. These groups, in turn, consist of a series of overlapping muscles. The iliocostalis group is laterally placed, the longissimus group is intermediate in placement, and the spinalis group is medially placed.			
Iliocostalis (lateral) group			
Iliocostalis lumborum (il'-ē-ō-kos-TAL-is lum-BOR-um; *ilium* = flank, *costa* = rib)			
Iliocostalis thoracis (il'-ē-ō-kos-TAL-is thō-RA-kis; *thorax* = chest)			
Iliocostalis cervicis (il'-ē-ō-kos-TAL-is SER-vi-kis)			
Longissimus (intermediate) group			
Longissimus thoracis (lon-JIS-i-mus thō-RA-kis; *longissimus* = longest)			
Longissimus cervicis (lon-JIS-i-mus SER-vi-kis)			

TABLE 10.17 *(Continued)*

Muscle	Origin	Insertion	Action
Longissimus capitis (lon-JIS-i-mus KAP-i-tis)			
Spinalis (medial) group			
Spinalis thoracis (spi-NA-lis tho-RA-kis; *spinalis* = vertebral column)			
Spinalis cervicis (spi-NA-lis SER-vi-kis)			
Spinalis capitis (spi-NA-lis KAP-i-tis)			
TRANSVERSOSPINALIS (trans-ver'-sō-spi-NA-lis) MUSCLES			
Semispinalis thoracis (sem'-ē-spi-NA-lis tho-RA-kis; *semi* = partially or one-half)			
Semispinalis cervicis (sem'-ē-spi-NA-lis SER-vi-kis)			
Semispinalis capitis (sem'ē-spi-NA-lis KAP-i-tis)			
Multifidus (mul-TIF-i-dus; *multi* = many, *findere* = to split)			
Rotatores (rō'-ta-TO-rēz; *rotate* = turn on an axis)			
SEGMENTAL MUSCLES			
Interspinales (in-ter-SPĪ-na-lēz; *inter* = between)			
Intertransversarii (in'-ter-trans-vers-AR-ē-ī;)			
SCALENE (SKĀ-lēn) MUSCLES			
Anterior scalene (SKĀ-len; *anterior* = front, *skalenos* = uneven)			
Middle scalene (SKĀ-lēn)			
Posterior scalene (SKĀ-lēn; *posterior* = back)			

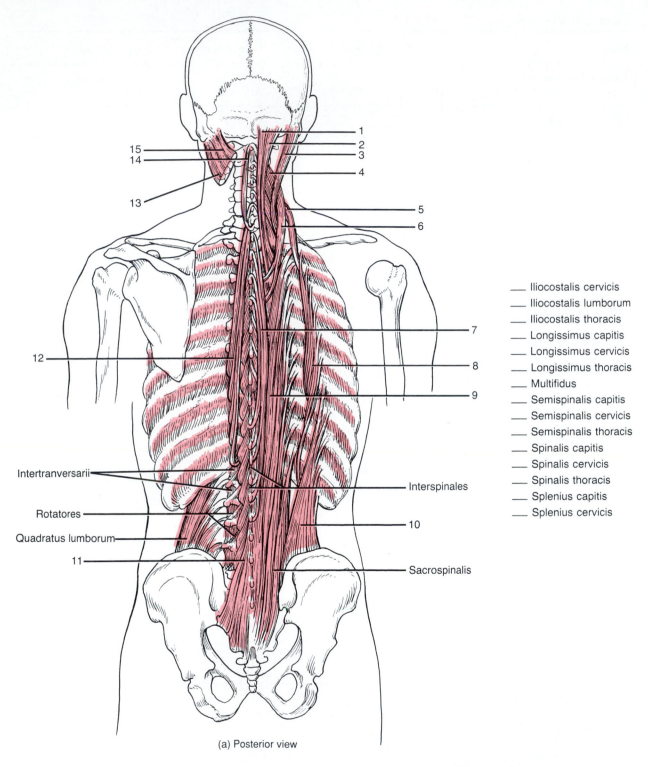

15
14
13
1
2
3
4
5
6
7
8
9
12

Iliocostalis cervicis
Iliocostalis lumborum
Iliocostalis thoracis
Longissimus capitis
Longissimus cervicis
Longissimus thoracis
Multifidus
Semispinalis capitis
Semispinalis cervicis
Semispinalis thoracis
Spinalis capitis
Spinalis cervicis
Spinalis thoracis
Splenius capitis
Splenius cervicis

Intertranversarii
Rotatores
Quadratus lumborum
11

Interspinales
10
Sacrospinalis

(a) Posterior view

FIGURE 10.16   Muscles that move the vertebral column. Many superficial muscles have been removed or cut to show deep muscles.

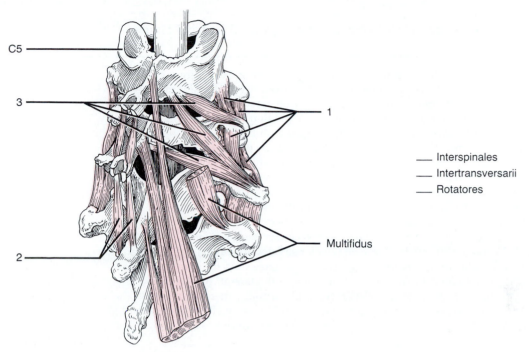

C5

3

1

2

— Interspinales
— Intertransversarii
— Rotatores

Multifidus

(b) Posterolateral view of several vertebrae
with their associated muscles

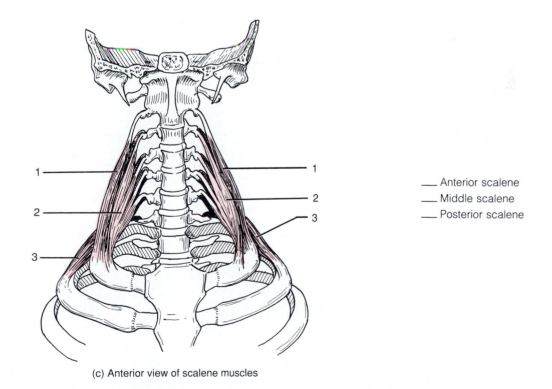

1
2
3

— Anterior scalene
— Middle scalene
— Posterior scalene

(c) Anterior view of scalene muscles

FIGURE 10.16 *(Continued)*   Muscles that move the vertebral column.

TABLE 10.18

Muscles That Move the Thigh (After completing the table, label Figure 10.17.)

OVERVIEW: As you will see, muscles of the lower extremities are larger and more powerful than those of the upper extremities because lower extremity muscles function in stability, locomotion, and maintenance of posture. Upper extremity muscles are characterized by versatility of movement. In addition, muscles of the lower extremities frequently cross two joints and act equally on both.

The majority of muscles that act on the thigh (femur) originate on the pelvic (hip) girdle and insert on the femur. The anterior muscles are the psoas major and iliacus, together referred to as the iliopsoas muscle. The remaining muscles (except for the pectineus, adductors, and tensor fasciae latae) are posterior muscles. Technically, the pectineus and adductors are components of the medial compartment of the thigh, but they are included in this table because they act on the thigh. The tensor fasciae latae muscle is laterally placed. The *fascia lata* is a deep fascia of the thigh that encircles the entire thigh. It is well developed laterally where, together with the tendons of the gluteus maximus and tensor fasciae latae muscles, it forms a structure called the *iliotibial tract.* The tract inserts into the lateral condyle of the tibia.

After you have studied the muscles in this table, arrange them according to the following actions: flexion, extension, abduction, adduction, medial rotation, and lateral rotation. (The same muscles can be used more than once.)

Muscle	Origin	Insertion	Action
Psoas (SŌ-as) **major** (*psoa* = muscle of loin)			
Iliacus (il'-ē-AK-us; *iliac* = ilium)			
Gluteus maximus (GLOO-tē-us MAK-si-mus; *glutos* = buttock, *maximus* = largest)			
Gluteus medius (GLOO-tē-us MĒ-dē-us; *media* = middle)			
Gluteus minimus (GLOO-tē-us MIN-i-mus; *minimus* = smallest)			
Tensor fasciae latae (TEN-sor FA-shē-ē LĀ-tē; *tensor* = makes tense, *fascia* = band, *latus* = wide)			
Piriformis (pir-i-FOR-mis; *pirum* = pear, *forma* = shape)			
Obturator internus* (OB-too-rā'-tor in-TER-nus; *obturator* = obturator foramen, *internus* = inside			

TABLE 10.18 *(Continued)*

Muscle	Origin	Insertion	Action
Obturator externus (OB-too-rā'-tor ex-TER-nus; *externus* = outside)			
Superior gemellus (jem-El-lus; *superior* = above, *gemellus* = twins)			
Inferior gemellus (jem-EL-lus; *inferior* = below)			
Quadratus femoris (kwod-RĀ-tus FEM-or-is; *quad* = four, *femoris* = femur)			
Adductor longus (LONG-us; *adductor* = moves the part closer to the midline, *longus* = long)			
Adductor brevis (BREV-is; *brevis* = short)			
Adductor magnus (MAG-nus; *magnus* = large)			
Pectineus (pek-TIN-ē-us; *pecten* = comb-shaped)			

*Not illustrated

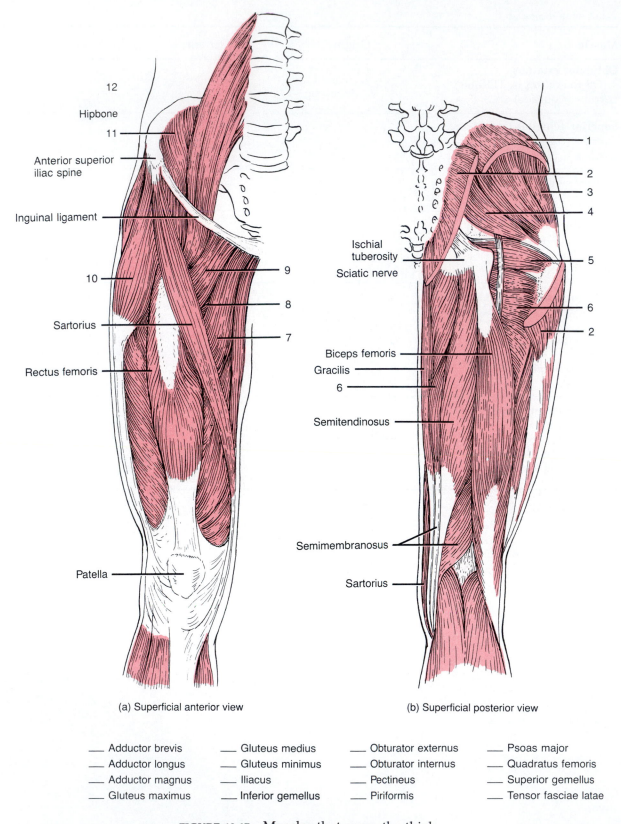

12

Hipbone

11

Anterior superior
iliac spine

Inguinal ligament

10

Sartorius

Rectus femoris

Patella

9

8

7

Ischial
tuberosity
Sciatic nerve

1

2

3

4

5

6

2

Biceps femoris

Gracilis

6

Semitendinosus

Semimembranosus

Sartorius

(a) Superficial anterior view

(b) Superficial posterior view

___ Adductor brevis   ___ Gluteus medius   ___ Obturator externus   ___ Psoas major
___ Adductor longus   ___ Gluteus minimus   ___ Obturator internus   ___ Quadratus femoris
___ Adductor magnus   ___ Iliacus   ___ Pectineus   ___ Superior gemellus
___ Gluteus maximus   ___ Inferior gemellus   ___ Piriformis   ___ Tensor fasciae latae

FIGURE 10.17   Muscles that move the thigh.

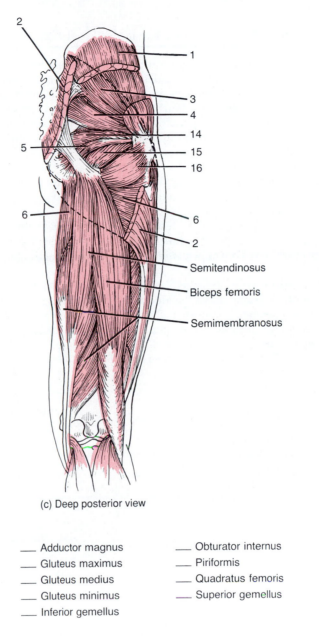

(c) Deep posterior view

Semitendinosus

Biceps femoris

Semimembranosus

___ Adductor magnus

___ Gluteus maximus

___ Gluteus medius

___ Gluteus minimus

___ Inferior gemellus

___ Obturator internus

___ Piriformis

___ Quadratus femoris

___ Superior gemellus

FIGURE 10.17 *(Continued)*   Muscles that move the thigh.

TABLE 10.19
Muscles That Act on the Leg (After completing the table, label Figure 10.18.)

OVERVIEW: The muscles that act on the leg (tibia and fibula) originate in the hip and thigh and are separated into compartments by deep fascia. The *medial (adductor) compartment* is so named because its muscles adduct the thigh. It is innervated by the obturator nerve. As noted earlier, the adductor magnus, adductor longus, adductor brevis, and pectineus muscles, components of the medial compartment, are included in Table 10.18 because they act on the femur. The gracilis, the other muscle in the medial compartment, not only adducts the thigh but also flexes the leg. For this reason, it is included in this table.

The *anterior (extensor) compartment* is so designated because its muscles act to extend the leg. It is composed of the quadriceps femoris and sartorius muscles and is innervated by the femoral nerve. The quadriceps femoris muscle is a composite muscle that includes four distinct parts, usually described as four separate muscles (rectus femoris, vastus lateralis, vastus intermedius, and vastus medialis). The common tendon for the four muscles is known as *patellar ligament* and attaches to the tibial tuberosity.

The *posterior (flexor) compartment* is so named because its muscles flex the leg. It is innervated by branches of the sciatic nerve. Included are the hamstrings (biceps femoris, semitendinosus, and semimembranosus). The hamstrings are so named because their tendons are long and stringlike in the popliteal area. The *popliteal fossa* is a diamond-shaped space on the posterior aspect of the knee bordered laterally by the biceps femoris and medially by the semitendinosus and semimembranosus muscles. (Also see Figure 11.19b.)

Muscle	Origin	Insertion	Action
MEDIAL (ADDUCTOR) COMPARTMENT			
Adductor magnus (MAG-nus)			
Adductor longus (LONG-us)			
Adductor brevis (BREV-is)			
Pectineus (pek-TIN-ē-us)			
Gracilis (gra-SIL-is; *gracilis* = slender)			
ANTERIOR (EXTENSOR) COMPARTMENT			
Quadriceps femoris (KWOD-ri-ceps FEM-or-is; *quadriceps* = four heads of origin, *femoris* = femur)			
Rectus femoris (REK-tus FEM-or-is; *rectus* = fibers parallel to the midline)			

TABLE 10.19 *(Continued)*

Muscle	Origin	Insertion	Action
Vastus lateralis (VAS-tus lat'-er-A-lis; *vastus* = large, *lateralis* = lateral)			
Vastus medialis (VAS-tus mē'-dē-A-lis; *medialis* = medial)			
Vastus intermedius (VA-tus in'-ter-MĒ-dē-us; *intermedius* = middle)			
Sartorius (sar-TOR-ē-us; *sartor* = tailor, refers to the cross-legged position of tailors)			
POSTERIOR (FLEXOR) COMPARTMENT Hamstrings			
Biceps femoris (BĪ-ceps FEM-or-is; *biceps* = two heads of origin)			
Semitendinosus (sem'-ē-TEN-di-nō-sus; *semi* = half, *tendo* = tendon)			
Semimembranosus (sem'-ē-MEM-bra-nō-sus; *membran* = membrane)			

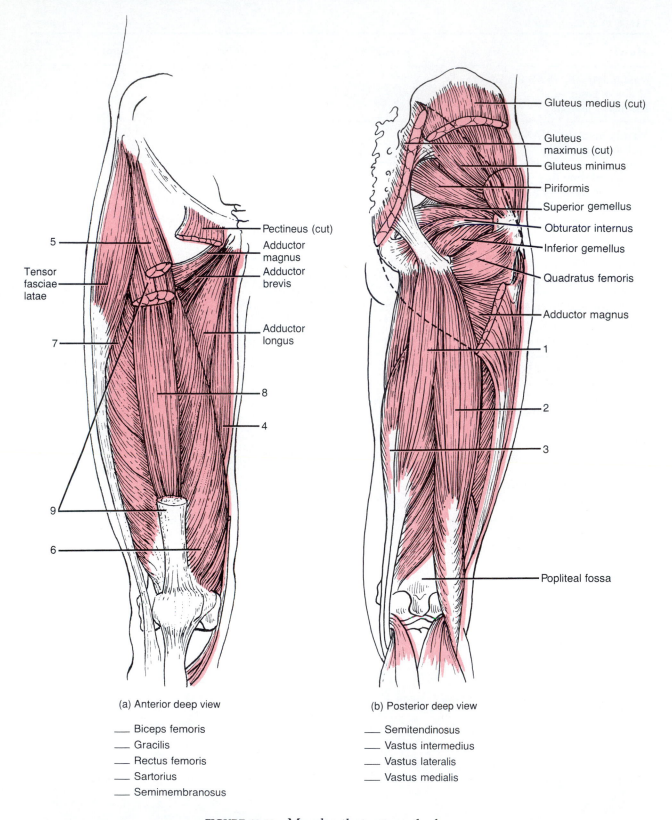

Tensor
fasciae
latae

Pectineus (cut)

Adductor
magnus

Adductor
brevis

Adductor
longus

Gluteus medius (cut)

Gluteus
maximus (cut)

Gluteus minimus

Piriformis

Superior gemellus

Obturator internus

Inferior gemellus

Quadratus femoris

Adductor magnus

Popliteal fossa

(a) Anterior deep view

(b) Posterior deep view

___ Biceps femoris

___ Gracilis

___ Rectus femoris

___ Sartorius

___ Semimembranosus

___ Semitendinosus

___ Vastus intermedius

___ Vastus lateralis

___ Vastus medialis

FIGURE 10.18   Muscles that act on the leg.

TABLE 10.20
Muscles That Move the Foot and Toes (Figure 10.19)

OVERVIEW: The musculature of the leg, like that of the thigh, can be distinguished into three compartments by deep fascia. In addition, all the muscles in the respective compartments are innervated by the same nerve. The *anterior compartment* consists of muscles that dorsiflex the foot and are innervated by the deep peroneal nerve. In a situation analogous to the wrist, the tendons of the muscles of the anterior compartment are held firmly to the ankle by thickenings of deep fascia called the *superior extensor retinaculum (transverse ligament of the ankle)* and *inferior extensor retinaculum (cruciate ligament of the ankle).*

   The *lateral (peroneal) compartment* contains two muscles that plantar flex and evert the foot. They are supplied by the superficial peroneal nerve.

   The *posterior compartment* consists of muscles that are divisible into superficial and deep groups. All are innervated by the tibial nerve. All three superficial muscles share a common tendon of insertion, the calcaneal (Achilles) tendon that inserts into the calcaneal bone of the ankle. The superficial muscles are plantar flexors of the foot. Of the four deep muscles, three plantar flex the foot.

Muscle	Origin	Insertion	Action
ANTERIOR COMPARTMENT			
Tibialis (tib'-ē-A-lis) **anterior** (*tibialis* = tibia, *anterior* = front)			
Extensor hallucis longus (HAL-a-kis LON-gus; *extensor* = increases the angle at the joint, *hallucis* = hallux, or great toe, *longus* = long)			
Extensor digitorum. longus (di'-ji-TOR-um LON-gus)			
Peroneus tertius (per'-ō-NĒ-us TER-shus; *perone* = fibula, *tertius* = third)			
LATERAL (PERONEAL) COMPARTMENT)			
Peroneus longus (per'-ō-NĒ-us LON-gus)			
Peroneus brevis (per'-ō-NĒ-us BERV-is; *brevis* = short)			

TABLE 10.20 *(Continued)*

Muscle	Origin	Insertion	Action
POSTERIOR COMPARTMENT Superficial			
Gastrocnemius (gas'-trok-NĒ-mē-us; *gaster* = belly, *kneme* = leg)			
Soleus (SŌ-lē-us; *soleus* = sole of foot)			
Plantaris (plan-TA-ris; *plantar* = sole of foot)			
Deep			
Popliteus (pop-LIT-ē-us; *poples* = posterior surface of the knee)			
Flexor hallucis longus (HAL-a-kis LON-gus; *flexor* = decreases the angle at the joint)			
Flexor digitorum longus (di'-ji-TOR-um LON-gus; *digitorum* = finger or toe)			
Tibialis (tib'-ē-A-lis) **posterior** (*posterior* = back)			

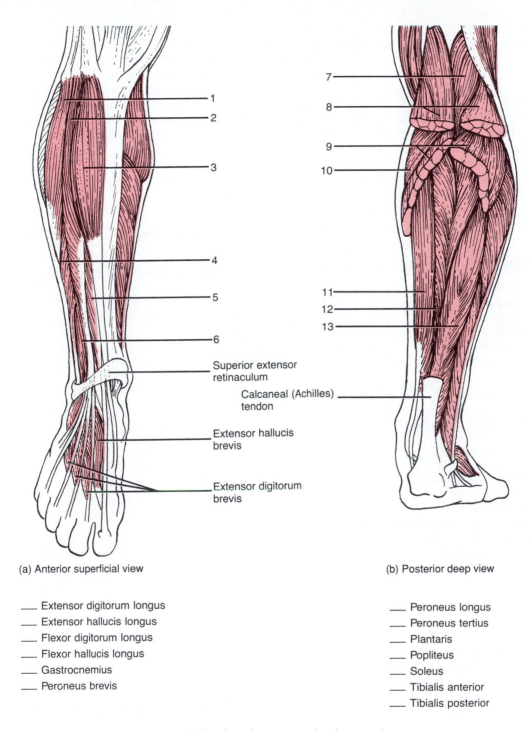

(a) Anterior superficial view

(b) Posterior deep view

___ Extensor digitorum longus
___ Extensor hallucis longus
___ Flexor digitorum longus
___ Flexor hallucis longus
___ Gastrocnemius
___ Peroneus brevis

___ Peroneus longus
___ Peroneus tertius
___ Plantaris
___ Popliteus
___ Soleus
___ Tibialis anterior
___ Tibialis posterior

FIGURE 10.19 Muscles that move the foot and toes.

TABLE 10.21
Intrinsic Muscles of the Foot (Figure 10.20)

OVERVIEW: The intrinsic muscles of the foot are comparable to those in the hand. Where the muscles of the hand are specialized for precise and intricate movements, those of the foot are limited to support and locomotion. The deep fascia of the foot forms the *plantar aponeurosis (fasica),* which extends from the calcaneus to the phalanges. The aponeurosis supports the longitudinal arch of the foot and transmits the flexor tendons of the foot.

The intrinsic musculature of the foot is divided into two groups—dorsal and plantar. There is only one *dorsal muscle.* The *plantar muscle* are arranged in four layers, the most superficial layer being referred to as the first layer.

Muscle	Origin	Insertion	Action
Extensor digitorum brevis (di'-ji-TOR-um BREV-is; *extensor* = increases the angle at the joint, *digit* = finger or toe, *brevis* = short) (Illustrated in Figure 10.19a)			
PLANTAR MUSCLES First (superficial) layer			
Abductor hallucis (HAL-a-kis; *abductor* = moves the part away from the midline, *hallucis* = hallux, or great toe)			
Flexor digitorum brevis (di'-ji-TOR-um BREV-is; *flexor* = decreases the angle at the joint)			
Abductor digiti minimi (DIJ-i-tē MIN-i-mē; *minimi* = small toe)			
Second layer			
Quadratus plantae (quod-RĀ-tus PLAN-tē; *quad* = four, *plantañ* = sole of foot)			
Lumbricals (LUM-bri-kals; four muscles)			

TABLE 10.21 (*Continued*)

Muscle	Origin	Insertion	Action
Third layer			
Flexor hallucis brevis (HAL-a-kis BREV-is)			
Adductor hallucis (HAL-a-kis; *adduct* = moves the part closer to the midline)			
Flexor digiti minimi brevis (DIJ-i-tē MIN-i-mē BREV-is)			
Fourth layer			
Dorsal interossei (in'-ter-OS-ē-ī four muscles; *dorsal* = back surface, *inter* = between, *ossei* = bone)			
Plantar interossei (in'-ter-OS-ē-ī)			

E. COMPOSITE MUSCULAR SYSTEM

Now that you have studied the muscles of the body by region, label the composite diagram shown in Figure 10.21.

ANSWER THE LABORATORY REPORT QUESTIONS AT THE END OF THE EXERCISE.

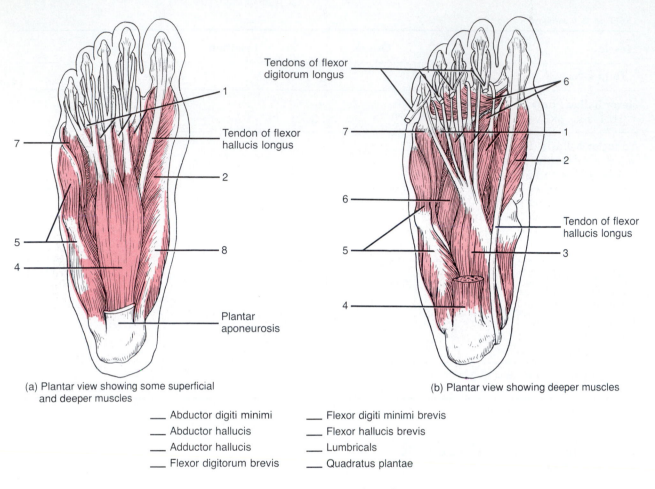

(a) Plantar view showing some superficial and deeper muscles

(b) Plantar view showing deeper muscles

___ Abductor digiti minimi   ___ Flexor digiti minimi brevis
___ Abductor hallucis   ___ Flexor hallucis brevis
___ Adductor hallucis   ___ Lumbricals
___ Flexor digitorum brevis   ___ Quadratus plantae

FIGURE 10.20   Intrinsic muscles of the foot.

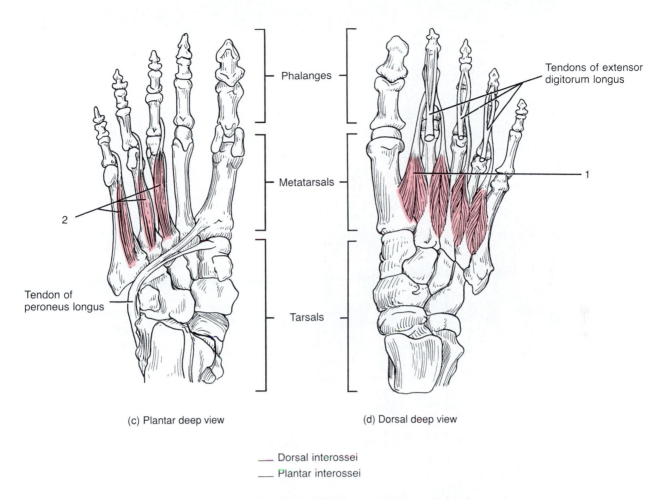

(c) Plantar deep view   (d) Dorsal deep view

—— Dorsal interossei
—— Plantar interossei

FIGURE 10.20 *(Continued)*   Intrinsic muscles of the foot.

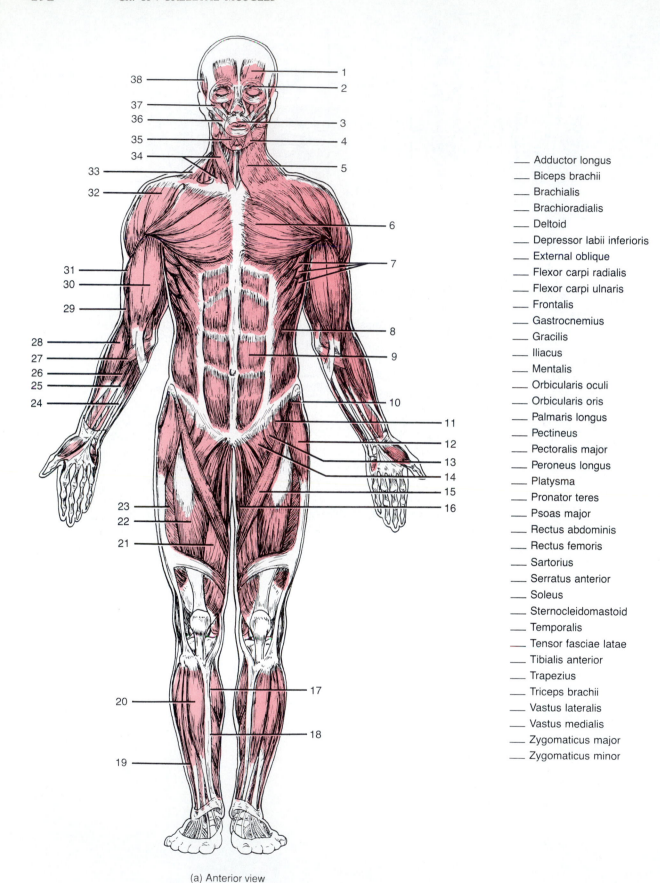

38
37
36
35
34
33
32
31
30
29
28
27
26
25
24
23
22
21
20
19

1
2
3
4
5
6
7
8
9
10
11
12
13
14
15
16
17
18

___ Adductor longus
___ Biceps brachii
___ Brachialis
___ Brachioradialis
___ Deltoid
___ Depressor labii inferioris
___ External oblique
___ Flexor carpi radialis
___ Flexor carpi ulnaris
___ Frontalis
___ Gastrocnemius
___ Gracilis
___ Iliacus
___ Mentalis
___ Orbicularis oculi
___ Orbicularis oris
___ Palmaris longus
___ Pectineus
___ Pectoralis major
___ Peroneus longus
___ Platysma
___ Pronator teres
___ Psoas major
___ Rectus abdominis
___ Rectus femoris
___ Sartorius
___ Serratus anterior
___ Soleus
___ Sternocleidomastoid
___ Temporalis
___ Tensor fasciae latae
___ Tibialis anterior
___ Trapezius
___ Triceps brachii
___ Vastus lateralis
___ Vastus medialis
___ Zygomaticus major
___ Zygomaticus minor

(a) Anterior view

FIGURE 10.21 Principal superficial muscles.

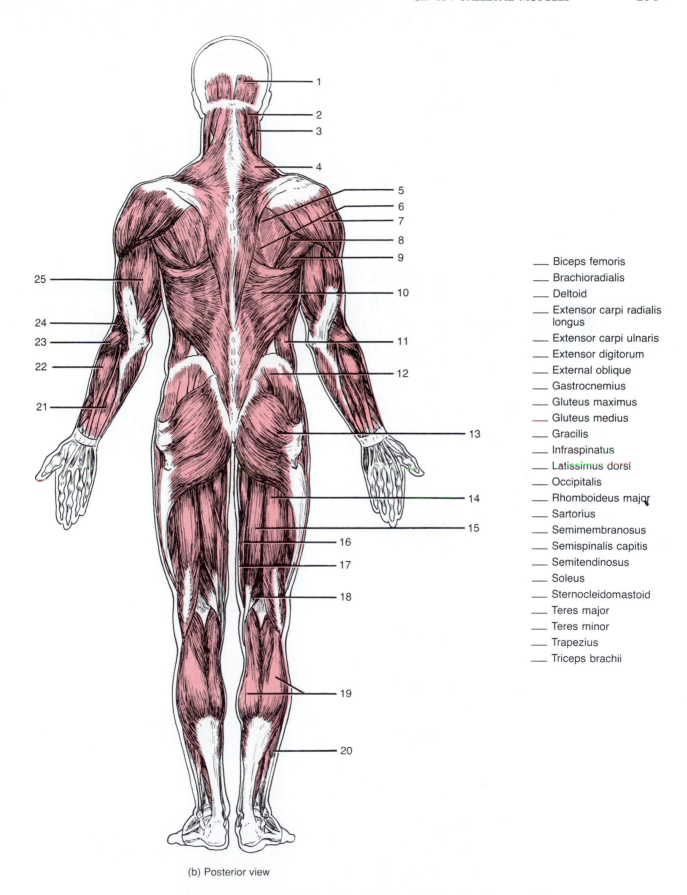

_____ Biceps femoris

_____ Brachioradialis

_____ Deltoid

_____ Extensor carpi radialis longus

_____ Extensor carpi ulnaris

_____ Extensor digitorum

_____ External oblique

_____ Gastrocnemius

_____ Gluteus maximus

_____ Gluteus medius

_____ Gracilis

_____ Infraspinatus

_____ Latissimus dorsi

_____ Occipitalis

_____ Rhomboideus major

_____ Sartorius

_____ Semimembranosus

_____ Semispinalis capitis

_____ Semitendinosus

_____ Soleus

_____ Sternocleidomastoid

_____ Teres major

_____ Teres minor

_____ Trapezius

_____ Triceps brachii

(b) Posterior view

FIGURE 10.21 _(Continued)_   Principal superficial muscles.

Skeletal Muscles

STUDENT _____   DATE _____

LABORATORY SECTION _____   SCORE/GRADE _____

PART 1. Multiple Choice

_____ **1.** The connective tissue covering that encloses the entire skeletal muscle is the (a) perimysium (b) endomysium (c) epimysium (d) mesomysium

_____ **2.** A cord of connective tissue that attaches a skeletal muscle to the periosteum of bone is called a(n) (a) ligament (b) aponeurosis (c) perichondrium (d) tendon

_____ **3.** A skeletal muscle that decreases the angle at a joint is referred to as a(n) (a) flexor (b) abductor (c) pronator (d) evertor

_____ **4.** The name *abductor* means that a muscle (a) produces a downward movement (b) moves a part away from the midline (c) elevates a body part (d) increases the angle at a joint

_____ **5.** Which connective tissue layer directly encircles the fasciculi of skeletal muscles? (a) epimysium (b) endomysium (c) perimysium (d) mesomysium

_____ **6.** Which muscle is *not* associated with a movement of the eyeball? (a) superior rectus (b) superior oblique (c) medial rectus (d) genioglossus

_____ **7.** Of the following, which muscle is involved in compression of the abdomen? (a) external oblique (b) superior oblique (c) medial rectus (d) genioglossus

_____ **8.** A muscle directly concerned with breathing is the (a) sternocleidomastoid (b) mentalis (c) brachialis (d) external intercostal

_____ **9.** Which muscle is *not* related to movement of the wrist? (a) extensor carpi ulnaris (b) flexor carpi radialis (c) supinator (d) flexor carpi ulnaris

_____ **10.** A muscle that helps move the thigh is the (a) piriformis (b) triceps brachii (c) hypoglossus (d) peroneus tertius

_____ **11.** Which muscle is *not* related to mastication? (a) temporalis (b) masseter (c) lateral rectus (d) medial pterygoid

_____ **12.** Which muscle elevates the tongue? (a) genioglossus (b) styloglossus (c) hyoglossus (d) omohyoid

_____ **13.** Which muscle does *not* belong with the others because of its relationship to the hyoid bone? (a) digastric (b) mylohyoid (c) geniohyoid (d) thyrohyoid

_____ **14.** Which muscle is *not* a component of the anterolateral abdominal wall? (a) external oblique (b) psoas major (c) rectus abdominis (d) internal oblique

_____ **15.** All are components of the pelvic diaphragm *except* the (a) anconeus (b) coccygeus (c) iliococcygeus (d) pubococcygeus

_____ **16.** Which is *not* a flexor of the forearm? (a) biceps brachii (b) brachialis (c) brachioradialis (d) triceps brachii

_____ **17.** The abductor pollicis brevis, opponeus pollicis, flexor pollicis brevis, and adductor pollicis are all components of the (a) thenar eminence (b) midpalmar muscles (c) hypothenar eminence (d) erector spinae

_____ **18.** Which muscle is *not* involved in moving the vertebral column? (a) splenius (b) longissimus (c) spinalis (d) sartorius

_____ **19.** Which muscle flexes the wrists? (a) palmaris longus (b) extensor carpi radialis longus (c) supinator (d) extensor indicis

_____ **20.** Which muscle is *not* involved in flexion of the thigh? (a) rectus femoris (b) sartorius (c) biceps femoris (d) vastus intermedius

_____ **21.** Of the muscles that move the foot and toes, the anterior compartment muscles are involved in (a) plantar flexion (b) dorsiflexion (c) abduction (d) adduction

_____ **22.** Which muscle is deepest? (a) plantar interosseous (b) adductor hallucis (c) quadratus plantae (d) abductor hallucis

PART 2. Matching

Identify the criterion (or criteria) used to name the following muscles:

_____ **23.** Supinator		A. Location
_____ **24.** Deltoid		B. Shape
_____ **25.** Stylohyoid		C. Size
_____ **26.** Flexor carpi radialis		D. Direction
_____ **27.** Gluteus maximus		E. Action
_____ **28.** External oblique		F. Number of origins
_____ **29.** Triceps brachii		G. Insertion and origin
_____ **30.** Adductor longus		
_____ **31.** Temporalis		
_____ **32.** Trapezius		

PART 3. Completion

33. The principal muscle used in compression of the cheek is the _____.

34. The muscle that protracts the tongue is the _____.

35. The eye muscle that rolls the eyeball downward is the _____.

36. The _____ muscle flexes the neck on the chest.

37. The abdominal muscle that flexes the vertebral column is the _____.

38. The muscle of the pectoral (shoulder) girdle that depresses the clavicle is the

_____.

39. Flexion, adduction, and medial rotation of the arm are accomplished by the

_____ muscle.

40. The _____ muscle is the most important extensor of the forearm.

41. The muscle that flexes and abducts the wrist is the _____.

42. The four muscles that extend the legs are the vastus lateralis, vastus medialis, vastus intermedius, and _____.

43. Muscles that lie inferior to the hyoid bone are referred to as _____ muscles.

44. The neck region is divided into two principal triangles by the _____ muscle.

45. Muscles that move the pectoral (shoulder) girdle originate on the axial skeleton and insert on the clavicle or _____.

46. The muscles that move the arm (humerus) and do not originate on the scapula are called _____ muscles.

47. Together, the subscapularis, supraspinatus, infraspinatus, and teres major muscles form the _____.

48. The posterior muscles involved in moving the wrist, hand, and fingers function in extension and _____.

49. The posterior muscles of the thigh are involved in _____ of the leg.

50. Together, the biceps femoris, semitendinosus, and semimembranosus are referred to as the _____ muscles.

11

Surface Anatomy

Now that you have studied the skeletal and muscular systems, you will be introduced to the study of *surface anatomy,* in which you will study the form and markings of the surface of the body.[1] A knowledge of surface anatomy will help you identify certain superficial structures by visual inspection or palpation (to feel with the hand).

A convenient way to study surface anatomy is first to divide the body into its principal regions: head, neck, trunk, and upper and lower extremities. These can be reviewed in Figure 2.1.

A. HEAD

The *head* (cephalic region or caput) is divisible into the cranium and face. The head has several surface features:

1. *Cranium*—(brain case or skull)
   a. *Frontal region*   Front of skull (sinciput) that includes the forehead.
   b. *Parietal region*   Crown of skull (vertex).
   c. *Temporal region*   Side of skull (tempora).
   d. *Occipital region*   Base of skull (occiput).
2. *Face*—(facies)
   a. *Orbital* or *ocular region*   Includes eyeballs (bulbi oculorum), eyebrows (supercilia), and eyelids (palpebrae).

[1]At the discretion of your instructor, surface anatomy can be studied either before or in conjunction with your study of various body systems. This exercise is an excellent review of many of the topics already studied.

b. *Nasal region*   Nose (nasus).
c. *Infraorbital region*   Inferior to orbit.
d. *Oral region*   Mouth.
e. *Mental region*   Anterior part of mandible.
f. *Buccal region*   Cheek.
g. *Parotid-masseteric region*   External to the parotid gland and masseter muscle.
h. *Zygomatic region*   Inferolateral to the orbit.
i. *Auricular region*   Ear.

Using your textbook as an aid, label Figure 11.1.

Using a mirror, examine the various features of the head just described. Working with a partner, be sure that you can identify the regions by both common *and* anatomical names.

The surface anatomy features of the eyeball and accessory structures are presented in Figure 14.10, of the ear in Figure 14.19, and of the nose in Figure 21.3.

B. NECK

The *neck* (collum) can be divided into an anterior cervical region, two lateral cervical regions, and a posterior (nuchal) region. Some of the surface features of the neck follow:

1. *Thyroid cartilage (Adam's apple)*—Triangular laryngeal cartilage in the midline of the anterior cervical region.
2. *Hyoid bone*—First resistant structure palpated in the midline below the chin, lying just above the thyroid cartilage.

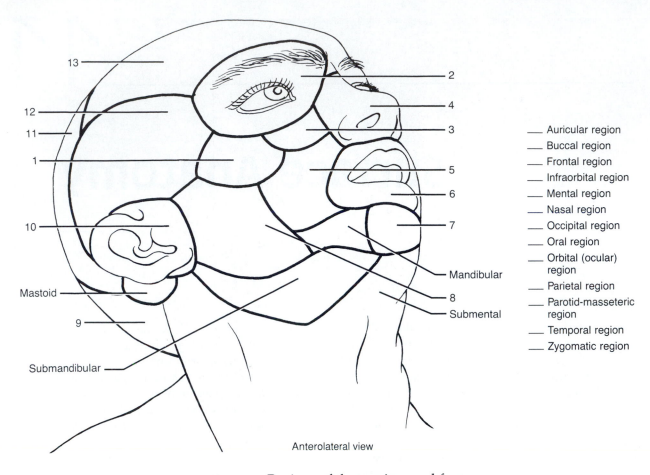

FIGURE 11.1 Regions of the cranium and face.

___ Auricular region
___ Buccal region
___ Frontal region
___ Infraorbital region
___ Mental region
___ Nasal region
___ Occipital region
___ Oral region
___ Orbital (ocular) region
___ Parietal region
___ Parotid-masseteric region
___ Temporal region
___ Zygomatic region

3. *Cricoid cartilage*—Inferior laryngeal cartilage that attaches the larynx to the trachea. This structure can be palpated by running your fingertip down from your chin over the thyroid cartilage. (After you pass the cricoid cartilage, your fingertip sinks in.) Used as a landmark for a tracheostomy.

4. *Sternocleidomastoid muscles*—Form the major portion of the lateral cervical regions, extending from the mastoid process of the temporal bone (felt as a bump behind the auricle or pinna of the ear) to the sternum and clavicle. Each muscle divides the neck into an anterior and posterior triangle. The carotid (neck) pulse is felt along its anterior border.

5. *Trapezius muscles*—Form portion of the lateral cervical region, extending downward and outward from the base of the skull. "Stiff neck" is frequently associated with inflammation of these muscles.

6. *Anterior triangle of neck*—Bordered superiorly by the mandible, inferiorly by the sternum, medially by the cervical midline, and laterally by the anterior border of the sternocleidomastoid muscle.

7. *Posterior triangle of neck*—Bordered inferiorly by the clavicle, anteriorly by the posterior border of the sternocleidomastoid muscle, and posteriorly by the anterior border of the trapezius muscle.

8. *External jugular veins* Prominent veins along the lateral cervical regions, readily seen when a person is angry or a collar fits too tightly.

Using a mirror and working with a partner, use your textbook as an aid in identifying the surface features of the neck just described. Then label Figure 11.2.

C. TRUNK

The *trunk* is divided into the back, chest, abdomen, and pelvis. Some of its surface features follow:

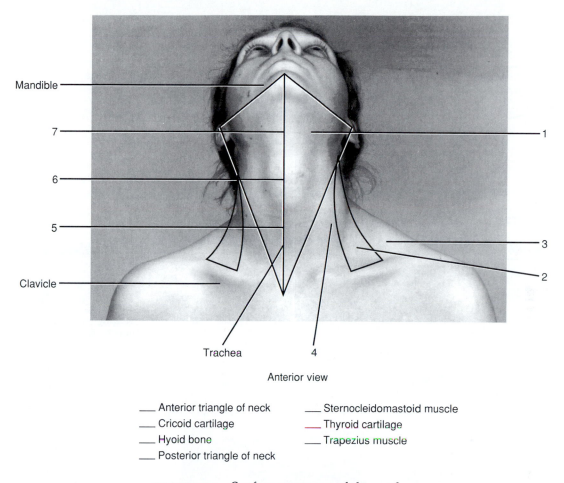

Anterior view

____ Anterior triangle of neck   ____ Sternocleidomastoid muscle
____ Cricoid cartilage   ____ Thyroid cartilage
____ Hyoid bone   ____ Trapezius muscle
____ Posterior triangle of neck

FIGURE 11.2 Surface anatomy of the neck.

1. **Back**—(dorsum)
   a. **Vertebral spines** Posteriorly pointed projections of the vertebrae. The spine of C2 is the first bony prominence encountered when the finger is drawn downward in the midline; the spine of C7 *(vertebra prominens)* is the upper of the two prominences found at the base of the neck; the spine of T1 is the lower prominence at the base of the neck; the spine of T3 is about at the same level as the spine of the scapula; the spine of T7 is about opposite the inferior angle of the scapula; and a line passing through the highest points of the iliac crests, called the supracristal line, passes through the spine of L4.
   b. **Scapula** Shoulder blade; several parts of the scapula—such as the medial (axillary) border, lateral (vertebral) border, inferior angle, spine, and acromion—can be observed or palpated.

   c. **Ribs** Visible in thin individuals.
   d. **Muscles** Among the visible superficial back muscles are the **latissimus dorsi** (covers the lower half of the back), **erector spinae** (on either side of the vertebral column), **infraspinatus** (inferior to the spine of the scapula), **trapezius,** and **teres major** (inferior to the infraspinatus).
   e. **Posterior axillary fold** Formed by the latissimus dorsi and teres major muscles, it can be palpated between the finger and thumb.
   f. **Triangle of auscultation** Triangle formed by the latissimus dorsi muscle, trapezius muscle, and vertebral border of the scapula. The space between the muscles in the region permits respiratory sounds to be heard clearly with a stethoscope.

Using your textbook as an aid, label Figure 11.3.

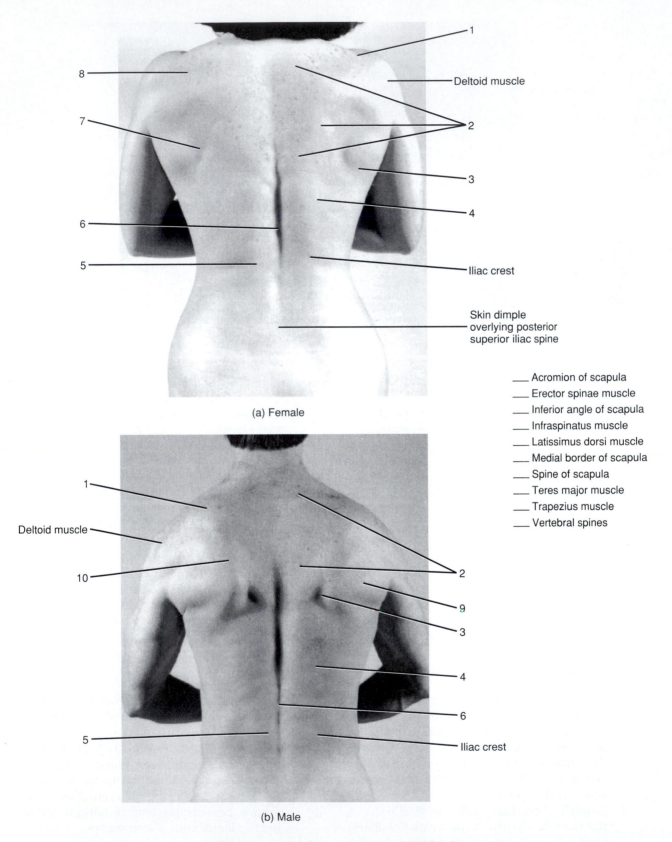

Deltoid muscle

Iliac crest

Skin dimple
overlying posterior
superior iliac spine

(a) Female

___ Acromion of scapula
___ Erector spinae muscle
___ Inferior angle of scapula
___ Infraspinatus muscle
___ Latissimus dorsi muscle
___ Medial border of scapula
___ Spine of scapula
___ Teres major muscle
___ Trapezius muscle
___ Vertebral spines

Deltoid muscle

Iliac crest

(b) Male

FIGURE 11.3   Surface anatomy of the back.

2. **Chest**—(thorax)
   a. **Clavicles** Collarbones; these lie in the superior region of the thorax and can be palpated along their entire length.
   b. **Sternum** Breastbone; lies in midline of the chest. The following parts of the sternum are important surface features:
      i. *Suprasternal notch* Depression on the superior surface of the manubrium of the sternum between the medial ends of the clavicles. The trachea can be palpated in the notch.
      ii. *Manubrium of sternum* Superior portion of the sternum at the same levels as the bodies of the third and fourth thoracic vertebrae and anterior to the arch of the aorta.
      iii. *Body of sternum* Midportion of the sternum anterior to the heart and the vertebral bodies of T5-T8.
      iv. *Sternal angle* Formed by the junction of the manubrium and the body of the sternum. This is palpable under the skin and locates the sternal end of the second rib.
      v. *Xiphoid process of sternum* Inferior portion of the sternum between the seventh costal cartilages.
   c. **Ribs** Form the bony cage of the thoracic cavity. The apex beat of the heart in adults is heard in the left fifth intercostal space, just medial to the left midclavicular line.
   d. **Costal margins** Inferior edges of the costal cartilages of ribs 7 through 10. The first costal cartilage lies below the medial end of the clavicle; the seventh costal cartilage is the lowest to articulate directly with the sternum; the tenth costal cartilage forms the lowest part of the costal margin, when viewed anteriorly.
   e. **Muscles** Among the superficial chest muscles that can be seen are the *pectoralis major* (principal upper chest muscle) and *serratus anterior* (inferior and lateral to the pectoralis major).
   f. **Mammary glands** Accessory organs of the female reproductive system located inside the breasts. They overlie the pectoralis major muscle (two-thirds) and serratus anterior muscle (one-third). After puberty, they enlarge to their hemispherical shape; in young adult females, they extend from the second through sixth ribs and from the lateral margin of the sternum to the midaxillary line.
   g. **Nipples** Superficial to fourth intercostal space, or fifth rib about 10 cm (4 in.) from the midline in males and most females. The position of the nipples in females is variable, depending on the size and pendulousness of the breasts. The right dome of the diaphragm is just inferior to the right nipple, the left dome is about 2 to 3 cm (1 in.) inferior to the left nipple, and the central tendon is at the level of the junction of the body of the sternum and xiphoid process.
   h. **Anterior axillary fold** Formed by the lateral border of the pectoralis major muscle, it can be palpated between the fingers and thumb.

Using your textbook as an aid, label Figure 11.4.

3. **Abdomen** and **Pelvis**
   a. **Umbilicus** Also called the *navel;* previous site of attachment of the umbilical cord to the fetus. It is level with the intervertebral disc between the bodies of L3 and L4 and is the most obvious surface marking on the abdomen of most individuals. The abdominal aorta bifurcates into the right and left common iliac arteries anterior to the body of vertebra L4. The inferior vena cava lies to the right of the abdominal aorta and is wider; it arises anterior to the body of vertebra L5.
   b. **Linea alba** Flat, tendonous raphe forming a furrow along the midline between the rectus abdominis muscles. The furrow extends from the xiphoid process to the pubic symphysis. It is particularly obvious in thin, muscular individuals. It is broad above the umbilicus and narrow below it. The linea alba is a frequently selected site for abdominal surgery because an incision through it severs no muscles and only a few blood vessels and nerves.
   c. **Tendinous intersections** Fibrous bands that run transversely across the rectus abdominis muscle. Three or more are visible in muscular individuals. One intersection is at the level of the umbilicus, one at the level of the xiphoid process, and one midway between.
   d. **Muscles** Among the superficial abdominal muscles are the *external oblique* (infe-

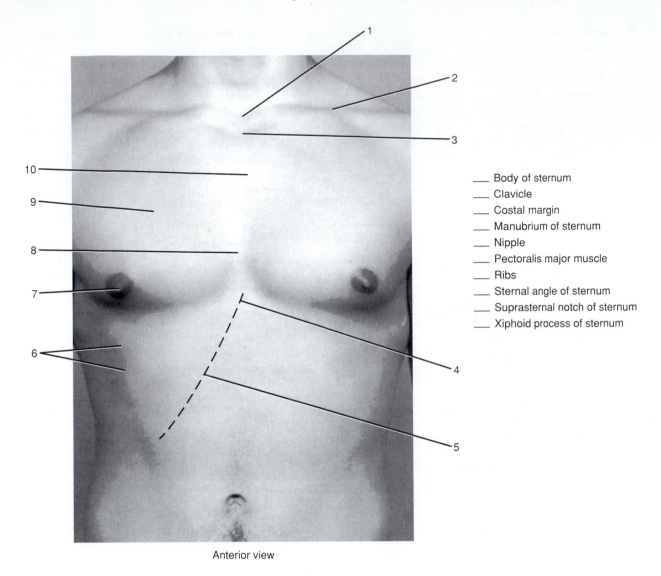

_____ Body of sternum
_____ Clavicle
_____ Costal margin
_____ Manubrium of sternum
_____ Nipple
_____ Pectoralis major muscle
_____ Ribs
_____ Sternal angle of sternum
_____ Suprasternal notch of sternum
_____ Xiphoid process of sternum

Anterior view

FIGURE 11.4 Surface anatomy of the chest. (The serratus anterior muscle is shown in Figure 11.5.)

rior to the serratus anterior) and **_rectus abdominis_** (just lateral to the midline of the abdomen).

e. **_Pubic symphysis_** Anterior joint of the hipbones. This structure is palpated as a firm resistance in the midline at the inferior portion of the anterior abdominal wall.

Using your textbook as an aid, label Figure 11.5.

D. UPPER EXTREMITY

The **_upper extremity_** consists of the armpit, arm, shoulder, elbow, forearm, wrist, and hand (palm and fingers).

1. Major surface features of the **_shoulder (acromial)_** region:
   a. **_Acromioclavicular joint_**—Slight elevation at the lateral end of the clavicle.
   b. **_Acromion_**—Expanded end of the spine of the scapula. This is clearly visible in some individuals and can be palpated about 2.5 cm (1 in.) distal to the acromioclavicular joint (see Figure 11.3).
   c. **_Deltoid muscle_**—Triangular muscle that forms the rounded prominence of the shoulder. This is a frequent site for intramuscular injections (see Figure 11.3).
2. Major surface features of the **_arm (brachium)_** and **_elbow (cubitus):_**

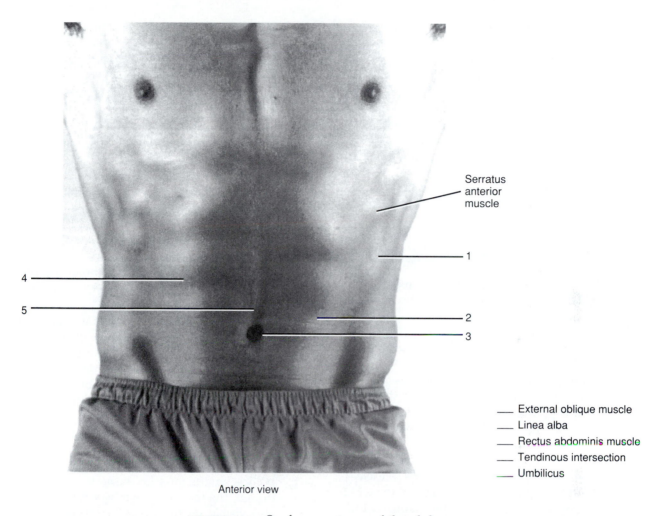

Anterior view

Serratus
anterior
muscle

1

2

3

4

5

_____ External oblique muscle
_____ Linea alba
_____ Rectus abdominis muscle
_____ Tendinous intersection
_____ Umbilicus

FIGURE 11.5   Surface anatomy of the abdomen.

a. **Biceps brachii muscle**—Forms the bulk of the anterior surface of the arm.

b. **Triceps brachii muscle**—Forms the bulk of the posterior surface of the arm.

c. **Medial epicondyle**—Medial projection at the distal end of the humerus.

d. **Ulnar nerve**—Can be palpated as a rounded cord in a groove behind the medial epicondyle.

e. **Lateral epicondyle**—Lateral projection at the distal end of the humerus.

f. **Olecranon**—Projection of the proximal end of ulna that lies between and slightly superior to the epicondyles when the forearm is extended; forms the elbow.

g. **Cubital fossa**—Triangular space in the anterior region of the elbow bounded proximally by an imaginary line between the humeral epicondyles, laterally by the medial border of the brachioradialis muscle,

and medially by the lateral border of the pronator teres muscle; contains the tendon of the biceps brachii muscle, brachial artery and its terminal branches (radial and ulnar arteries), and parts of the median and radial nerves.

h. **Median cubital vein**—Crosses the cubital fossa obliquely. This vein is frequently selected for the removal of blood.

i. **Brachial artery**—Continuation of the axillary artery that passes posterior to the coracobrachialis muscle and then medial to the biceps brachii muscle. It enters the middle of the cubital fossa and passes under the bicipital aponeurosis, which separates it from the median cubital vein. The artery is frequently used to take blood pressure. Pressure may be applied to it in cases of severe hemorrhage in the forearm and hand.

j. *Bicipital aponeurosis*—An aponeurotic band that inserts the biceps brachii muscle into the deep fascia in the medial aspect of the forearm. It can be felt when the muscle contracts.

Using your textbook as an aid, label Figure 11.6.

3. Major surface features of the *forearm (antebrachium)* and *wrist (carpus)*:
   a. *Styloid process of ulna*—Projection of the distal end of ulna at the medial side of the wrist.
   b. *Styloid process of radius*—Projection of the distal end of the radius at the lateral side of the wrist.
   c. *Brachioradialis muscle*—Located at the superior and lateral aspect of the forearm.
   d. *Flexor carpi radialis muscle*—Located along the midportion of the forearm.
   e. *Flexor carpi ulnaris muscle*—Located at the medial aspect of the forearm.
   f. *Tendon of palmaris longus muscle*—Located on the anterior surface of the wrist near the ulna. When you make a fist, you can see this tendon. The palmaris longus muscle is absent in about 13% of the population.
   g. *Tendon of flexor carpi radialis muscle*—Tendon on the anterior surface of the wrist lateral to the tendon of the palmaris longus.
   h. *Radial artery*—Can be palpated just medial to styloid process of the radius; this artery is frequently used to take the pulse.
   i. *Pisiform bone*—Medial bone of the proximal carpals. The bone is easily palpated as a projection distal to the styloid process of the ulna.
   j. *Tendon of extensor pollicis brevis muscle*—Tendon close to the styloid process of the radius along the posterior surface of the wrist; best seen when the thumb is bent backward.
   k. *Tendon of extensor pollicis longus muscle*—Tendon closer to the styloid process of the ulna along the posterior surface of the wrist; best seen when the thumb is bent backward.
   l. *"Anatomical snuffbox"*—Depression between the tendons of the extensor pollicis brevis and extensor pollicis longus muscles. Styloid process of the radius, the base of the first metacarpal, trapezium, scaphoid, and radial artery can all be palpated in the depression.
   m. *Wrist creases*—Three more or less constant lines on the anterior aspect of the wrist where the skin is firmly attached to the underlying deep fascia.

Using your textbook as an aid, label Figure 11.7.

4. Major surface features of the *hand (manus)*:
   a. *"Knuckles"*—Commonly refers to the dorsal aspects of the distal ends of metacarpals II, III, IV, and V; also includes the dorsal aspects of the metacarpophalangeal and interphalangeal joints.
   b. *Thenar eminence*—Lateral rounded contour on the anterior surface of the hand formed by the muscles of thumb.
   c. *Hypothenar eminence*—Medial rounded contour on the anterior surface of the hand formed by the muscles of the little finger.
   d. *Skin creases*—Several more or less constant lines on the anterior aspect of the palm (*palmar flexion creases*) and digits (*digital flexion creases*) where the skin is firmly attached to the underlying deep fascia.
   e. *Extensor tendons*—Besides the tendons of the extensor pollicis brevis and extensor pollicis longus muscles associated with the thumb, the following extensor tendons are also visible on the posterior aspect of the hand: *extensor digiti minimi tendon* in line with phalanx V (little finger) and *extensor digitorum* in line with phalanges II, III, and IV.
   f. *Dorsal venous arch*—Superficial veins on the posterior surface of the hand that form the cephalic vein; displayed by compressing the blood vessels at the wrist for a few minutes as the hand is opened and closed.

With the aid of your textbook, label Figure 11.8.

E. LOWER EXTREMITY

The *lower extremity* consists of the buttocks, thigh, knee, leg, ankle, and foot.

1. Major surface features of the *buttocks (gluteal region)* and *thigh (femoral region)*:
   a. *Iliac crest*—Superior margin of the ilium of the hipbone, forming the outline of the

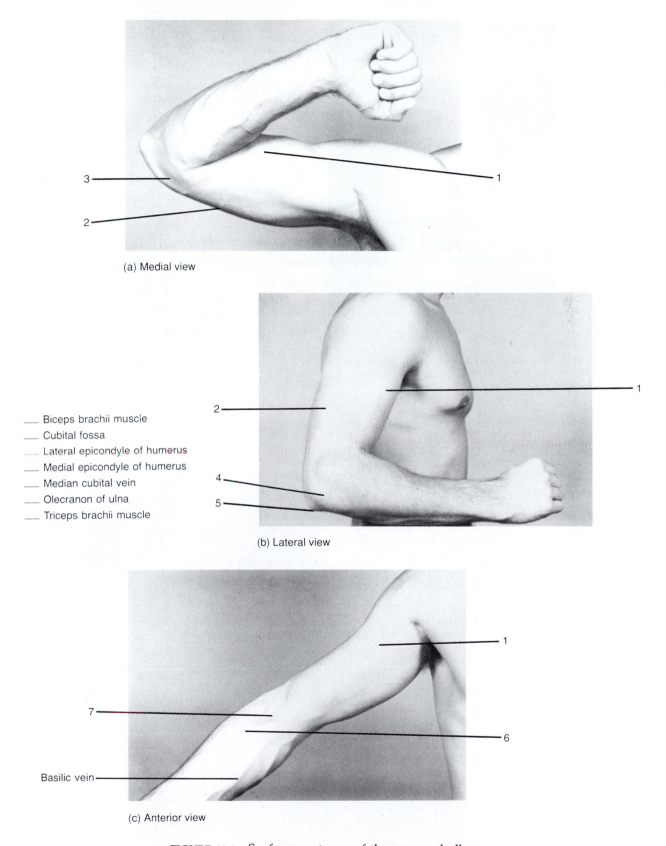

(a) Medial view

___ Biceps brachii muscle
___ Cubital fossa
___ Lateral epicondyle of humerus
___ Medial epicondyle of humerus
___ Median cubital vein
___ Olecranon of ulna
___ Triceps brachii muscle

(b) Lateral view

(c) Anterior view

FIGURE 11.6   Surface anatomy of the arm and elbow.

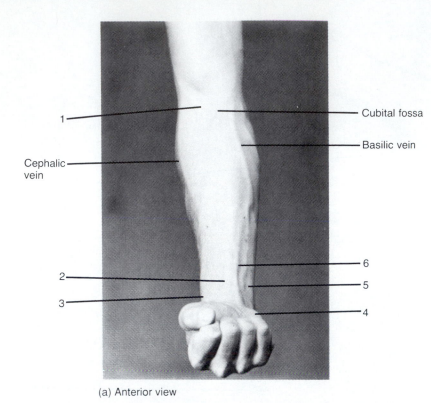

1
Cephalic vein
2
3

Cubital fossa
Basilic vein

6
5
4

(a) Anterior view

—— "Anatomical snuffbox"
—— Brachioradialis muscle
—— Pisiform bone
—— Site for palpatation of radial artery
—— Styloid process of ulna
—— Tendon of extensor pollicis brevis muscle
—— Tendon of extensor pollicis longus muscle
—— Tendon of flexor carpi radialis muscle
—— Tendon of flexor carpi ulnaris muscle
—— Tendon of palmaris longus muscle
—— Wrist creases

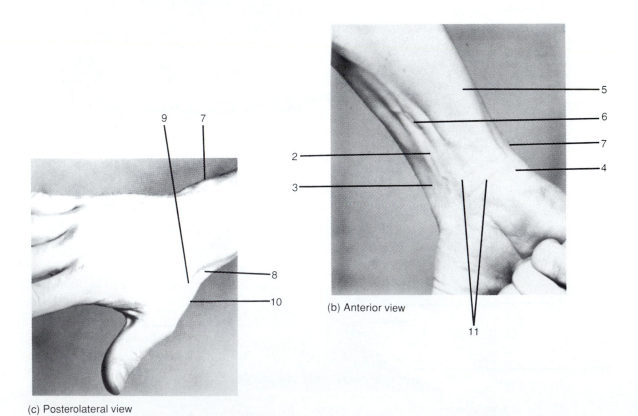

9   7

8
10

(c) Posterolateral view

5
6
7
4

2
3

11

(b) Anterior view

FIGURE 11.7   Surface anatomy of the forearm and wrist.

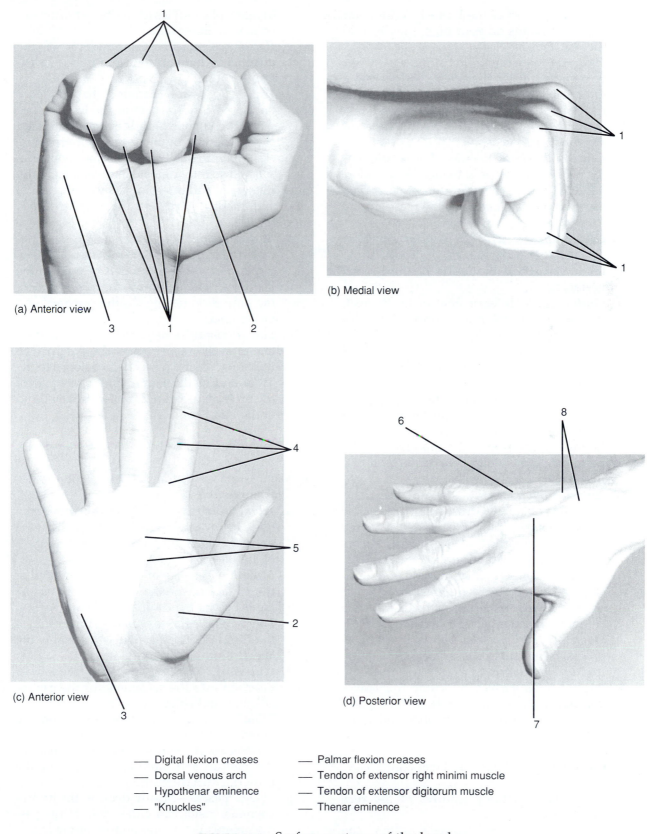

(a) Anterior view

(b) Medial view

(c) Anterior view

(d) Posterior view

— Digital flexion creases   — Palmar flexion creases
— Dorsal venous arch   — Tendon of extensor right minimi muscle
— Hypothenar eminence   — Tendon of extensor digitorum muscle
— "Knuckles"   — Thenar eminence

FIGURE 11.8   Surface anatomy of the hand.

superior border of the buttock. When you rest your hands on your hips, they rest on the iliac crests.

b. *Posterior superior iliac spine*—Posterior termination of the iliac crest; lies deep to a dimple (skin depression) about 4 cm (1.5 in.) lateral to the midline. Dimple forms because the skin and underlying fascia are attached to bone.

c. *Gluteus maximus muscle*—Forms the major portion of the prominence of buttock.

d. *Gluteus medius muscle*—Superolateral to the gluteus maximus. This is a frequent site for intramuscular injections.

e. *Gluteal (natal) cleft*—Depression along the midline that separates the buttocks; it extends as high as the fourth or third sacral vertebra.

f. *Gluteal fold*—Inferior limit of the buttock formed by the inferior margin of the gluteus maximus muscle.

g. *Ischial tuberosity*—Bony prominence of the ischium of the hipbone; bears the weight of the body when seated.

h. *Greater trochanter*—Projection of the proximal end of the femur on the lateral surface of the thigh felt and seen in front of the hollow on the side of the hip. This can be palpated about 20 cm (8 in.) inferior to the iliac crest.

i. *Anterior thigh muscles*—Include the *sartorius* (runs obliquely across thigh) and *quadriceps femoris*, which consists of *rectus femoris* (midportion of the thigh), *vastus lateralis* (anterolateral surface of the thigh), *vastus medialis* (medial inferior portion of the thigh), and *vastus intermedius* (deep to the rectus femoris). The vastus lateralis is a frequent injection site for diabetics.

j. *Posterior thigh muscles*—Include the *hamstrings (semitendinosus, semimembranosus,* and *biceps femoris).*

k. *Medial thigh muscles*—Include the *adductor magnus, adductor brevis, adductor longus, gracilis, obturator externus,* and *pectineus.*

With the aid of your textbook, label Figure 11.9.

2. Major surface anatomy features of the *knee (genu)* are follow:
   a. *Patella*—Kneecap; located within the quadriceps femoris tendon on the anterior surface of the knee along the midline.

Margins of condyles (described shortly) can be felt on either side of it.

b. *Patellar ligament*—Continuation of the quadriceps femoris tendon inferior to patella.

c. *Popliteal fossa*—Diamond-shaped space on the posterior aspect of the knee visible when the knee is flexed. Fossa is bordered superolaterally by the biceps femoris muscle, superomedially by the semimembranosus and semitendinosus muscles, and inferolaterally and inferomedially by the lateral and medial heads of the gastrocnemius muscle.

d. *Medial condyles of femur and tibia*—Medial projections just below the patella. The upper part of the projection belongs to the distal end of the femur; the lower part of the projection belongs to the proximal end of the tibia.

e. *Lateral condyles of femur and tibia*—Lateral projections just below the patella. The upper part of the projections belongs to the distal end of the femur; the lower part of the projections belongs to the proximal end of the tibia.

With the aid of your textbook, label Figure 11.10.

3. Major surface anatomy features of the *leg (crus)* and *ankle (tarsus):*
   a. *Tibial tuberosity*—Bony prominence of the tibia below the patella into which the patellar ligament inserts (see Figure 11.10a).

   b. *Medial malleolus of tibia*—Projection of the distal end of the tibia that forms the medial prominence of the ankle.

   c. *Lateral malleolus of fibula*—Projection of the distal end of the fibula that forms the lateral prominence of the ankle.

   d. *Calcaneal (Achilles) tendon*—Tendon of insertion into the calcaneus for the gastrocnemius and soleus muscles.

   e. *Tibialis anterior muscle*—Located on the anterior surface of the leg.

   f. *Gastrocnemius muscle*—Forms the bulk of the middle and superior portions of the posterior surface of the leg.

   g. *Soleus muscle*—Mostly deep to the gastrocnemius; visible on either side of the gastrocnemius below the middle of the leg.

With the aid of your textbook, label Figure 11.11.

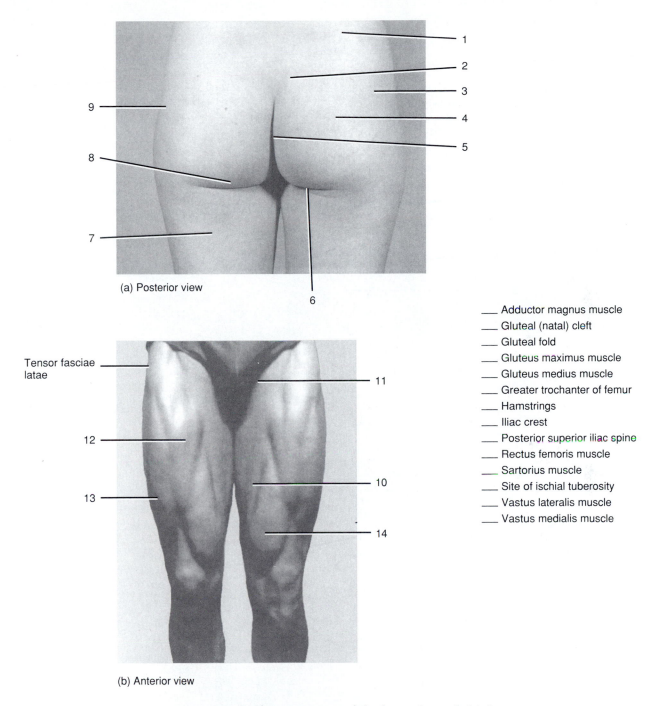

(a) Posterior view

(b) Anterior view

Tensor fasciae latae

___ Adductor magnus muscle
___ Gluteal (natal) cleft
___ Gluteal fold
___ Gluteus maximus muscle
___ Gluteus medius muscle
___ Greater trochanter of femur
___ Hamstrings
___ Iliac crest
___ Posterior superior iliac spine
___ Rectus femoris muscle
___ Sartorius muscle
___ Site of ischial tuberosity
___ Vastus lateralis muscle
___ Vastus medialis muscle

FIGURE 11.9   Surface anatomy of the buttocks and thigh.

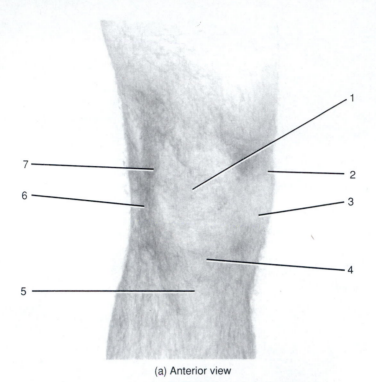

(a) Anterior view

Site of
semitendinosus and
semimembranosus
muscle

Tendon of biceps
femoris muscle

Tendon of semitendinosus
muscle

8

Gastrocnemius
muscle

(b) Posterior view

___ Lateral condyle of femur   ___ Patella

___ Lateral condyle of tibia   ___ Patellar ligament

___ Medial condyle of femur   ___ Popliteal fossa

___ Medial condyle of tibia   ___ Tibial tuberosity

FIGURE 11.10 Surface anatomy of the knee.

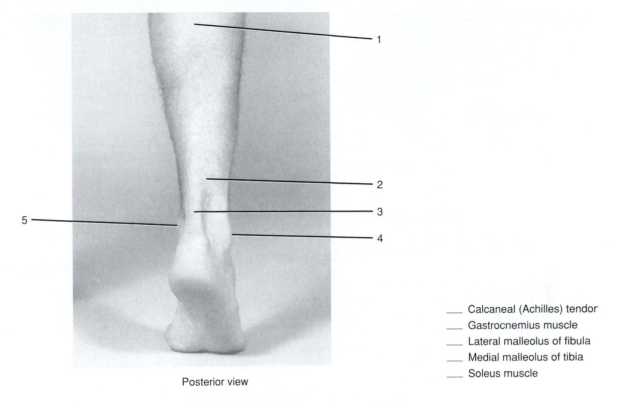

Posterior view

___ Calcaneal (Achilles) tendor
___ Gastrocnemius muscle
___ Lateral malleolus of fibula
___ Medial malleolus of tibia
___ Soleus muscle

FIGURE 11.11   Surface anatomy of the leg.

4. Major surface anatomy features of the *foot (pes):*
   a. *Calcaneus*—Heel bone into which the calcaneal (Achilles) tendon inserts.
   b. *Tendon of extensor hallucis longus muscle*—Visible in line with phalanx of the great toe. Pulsations in the dorsalis pedis artery can be felt in most people just lateral to this tendon when the blood vessel passes over the navicular and cuneiform bones.

   c. *Tendons of extensor digitorum longus muscles*—Visible in line with phalanges II through V.
   d. *Dorsal venous arch*—Superficial veins on the dorsum of the foot that unite to form the small and great saphenous veins.

With the aid of your textbook, label Figure 11.12.

ANSWER THE LABORATORY REPORT QUESTIONS AT THE END OF THE EXERCISE.

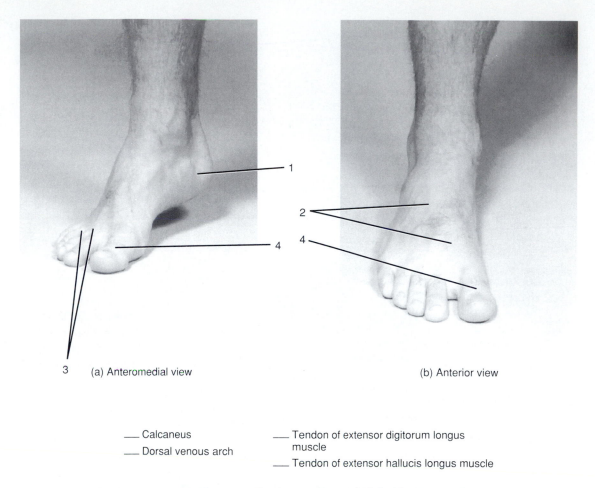

(a) Anteromedial view   (b) Anterior view

___ Calcaneus   ___ Tendon of extensor digitorum longus muscle

___ Dorsal venous arch   ___ Tendon of extensor hallucis longus muscle

FIGURE 11.12   Surface anatomy of the foot.

Surface Anatomy

STUDENT _____   DATE _____

LABORATORY SECTION _____   SCORE/GRADE _____

PART **1. Multiple Choice**

_____ **1.** The term used to refer to the crown of the skull (vertex) is (a) occipital (b) mental (c) parietal (d) zygomatic

_____ **2.** The laryngeal cartilage in the midline of the anterior cervical region known as the Adam's apple is the (a) cricoid (b) epiglottis (c) arytenoid (d) thyroid

_____ **3.** Inflammation of which muscle is associated with "stiff neck"? (a) teres major (b) trapezius (c) deltoid (d) cervicalis

_____ **4.** The skeletal muscle located directly on either side of the vertebral column is the (a) serratus anterior (b) infraspinatus (c) teres major (d) erector spinae

_____ **5.** The suprasternal notch and xiphoid process are associated with the (a) sternum (b) scapula (c) clavicle (d) ribs

_____ **6.** The expanded end of the spine of the scapula is the (a) acromion (b) linea alba (c) olecranon (d) superior angle

_____ **7.** Which nerve can be palpated as a rounded cord in a groove behind the medial epicondyle? (a) median (b) radial (c) ulnar (d) brachial

_____ **8.** Which carpal bone can be palpated as a projection distal to the styloid process of the ulna? (a) trapezoid (b) trapezium (c) hamate (d) pisiform

_____ **9.** The lateral rounded contour on the anterior surface of the hand (at the base of the thumb) formed by the muscles of the thumb is the (a) "anatomical snuffbox" (b) thenar eminence (c) hypothenar eminence (d) dorsal venous arch

_____ **10.** The superior margin of the hipbone is the (a) pubic symphysis (b) iliac spine (c) acetabulum (d) iliac crest

_____ **11.** Which bony structure bears the weight of the body when a person is seated? (a) greater trochanter (b) iliac crest (c) ischial tuberosity (d) gluteal fold

_____ **12.** Which muscle is *not* a component of the quadriceps femoris group? (a) biceps femoris (b) vastus lateralis (c) vastus medialis (d) rectus femoris

_____ **13.** The diamond-shaped space on the posterior aspect of the knee is the (a) cubital fossa (b) posterior triangle (c) popliteal fossa (d) nuchal groove

_____ **14.** The projection of the distal end of the tibia that forms the prominence on one side of the ankle is the (a) medial condyle (b) medial malleolus (c) lateral condyle (d) lateral malleolus

_____ **15.** The tendon that can be seen in line with the great toe belongs to which muscle? (a) extensor digiti minimi (b) extensor digitorum longus (c) extensor hallucis longus (d) extensor carpi radialis

215

PART 2. Completion

16. The laryngeal cartilage that connects the larynx to the trachea is the _____ cartilage.

17. The _____ triangle is bordered by the mandible, sternum, cervical midline, and sternocleidomastoid muscle.

18. The depression on the superior surface of the sternum between the medial ends of the clavicles is the _____.

19. The principal superficial chest muscle is the _____.

20. Tendinous intersections are associated with the _____ muscle.

21. The muscle that forms the rounded prominence of the shoulder is the _____ muscle.

22. The triangular space in the anterior aspect of the elbow is the _____.

23. The "anatomical snuffbox" is bordered by the tendons of the extensor pollicis brevis muscle and the _____ muscle.

24. The dorsal aspects of the distal ends of metacarpals II through V are commonly referred to as _____.

25. The tendon of the _____ muscle is in line with phalanx V.

26. The dimple that forms about 4 cm lateral to the midline just above the buttocks lies superficial to the _____.

27. The femoral projection that can be palpated about 20 cm (8 in.) inferior to the iliac crest is the _____.

28. The continuation of the quadriceps femoris tendon inferior to the patella is the _____.

29. The tendon of insertion for the gastrocnemius and soleus muscles is the _____ tendon.

30. Superficial veins in the dorsum of the foot that unite to form the small and great saphenous veins belong to the _____.

31. The prominent veins along the lateral cervical regions are the _____ veins.

32. The pronounced vertebral spine of C7 is the _____.

33. The most reliable surface anatomy feature of the chest is the _____.

34. A slight groove extending from the xiphoid process to the pubic symphysis is the _____.

35. The vein that crosses the cubital fossa and is frequently used to remove blood is the _____ vein.

PART **3. MATCHING**

_____ 36. Mental region

_____ 37. Xiphoid process

_____ 38. Arm

_____ 39. Nucha

_____ 40. Shoulder

_____ 41. Gluteus maximus muscle

_____ 42. Costal margin

_____ 43. Olecranon

_____ 44. Wrist

_____ 45. Auricular region

_____ 46. Semitendinosus muscle

_____ 47. Leg

_____ 48. Cranium

_____ 49. Ankle

_____ 50. Hand

A. Inferior edges of costalcartilages of ribs 7 through 10

B. Forms elbow

C. Inferior portion of sternum

D. Manus

E. Crus

F. Brachium

G. Tarsus

H. Anterior part of mandible

I. Brain case

J. Component of hamstrings

K. Forms main part of prominence of buttock

L. Carpus

M. Posterior neck region

N. Acromial (omos)

O. Ear

12

Nervous Tissue and Physiology

The two principal divisions of the nervous system are the *central nervous system (CNS)* and the *peripheral* (pe-RIF-er-al) *nervous system (PNS)*. The CNS consists of the brain and spinal cord; it is the control center for the entire nervous system. Within the CNS, incoming sensory information is integrated and correlated, thoughts and emotions are generated, and memories are formed. Outgoing nerve impulses may stimulate muscles to contract and glands to secrete. The PNS consists of nerves arising from the brain (cranial nerves) and from the spinal cord (spinal nerves) that carry the input to and output from the CNS. The sensory (input) component of the PNS consists of neurons (nerve cells), called *sensory* or *afferent* (AF-er-ent; *ad* = toward; *ferre* = to carry) *neurons.* They conduct nerve impulses from sensory receptors in various parts of the body to the CNS. The motor (output) component consists of neurons, called *motor* or *efferent* (EF-er-ent; *ex* = away from, *ferre* = to carry) *neurons.* They conduct nerve impulses from the CNS to effectors—that is, to muscles and glands.

On the basis of the origin of the incoming sensory information and the part of the body that responds, the PNS can be subdivided into a *somatic* (*soma* = body) *nervous system (SNS)* and an *autonomic* (*auto* = self, *nomos* = law) *nervous system (ANS).*

The somatic nervous system includes three components, two for sensory input and one for motor output:

1. *Special somatic sensory (afferent) neurons*—Relay impulses via cranial nerves for vision, hearing and equilibrium, smell, and taste.
2. *General somatic* (*soma* = body) *sensory (afferent) neurons*—Convey impulses for pain, temperature, touch, vibration, and pressure from the skin and for position from joints and muscles via spinal nerves and some cranial nerves.
3. *General somatic motor (efferent) neurons*—Conduct impulses to skeletal muscles via spinal nerves and some cranial nerves. Because these motor responses can be consciously controlled, this portion of the somatic nervous system is *voluntary.*

The autonomic nervous system includes two components: (1) *general visceral sensory (afferent) neurons,* which convey visceral information such as distention (stretching) of the organs and chemical conditions within the body into the CNS via both cranial and spinal nerves; and (2) *general visceral motor (efferent) neurons,* which conduct impulses from the CNS to smooth muscle, cardiac muscle, and glands via certain cranial and spinal nerves. Because the motor responses of the ANS are not normally under conscious control, it is *involuntary.*

The motor portion of the ANS consists of two divisions. With few exceptions, the viscera receive instructions from both the *sympathetic division* and the *parasympathetic division.* Pro-

219

cesses promoted by sympathetic neurons usually involve expenditure of energy, whereas those promoted by parasympathetic neurons restore and conserve body energy. Most often, the two divisions have opposing actions. For example, sympathetic neurons accelerate the heartbeat, whereas parasympathetic neurons slow it down.

In this exercise you will identify the parts of a neuron and the components of a reflex arc. You will also perform several experiments on reflexes in the frog.

A. HISTOLOGY OF NERVOUS TISSUE

Despite its complexity, the nervous system consists of only two principal kinds of cells: neurons and neuroglia. *Neurons* (nerve cells) constitute the nervous tissue and are highly specialized for nerve impulse conduction. Mature neurons have only limited capacity for replacement or repair. *Neuroglia* (noo-ROG-lē-a; *neuro* = nerve, *glia* = glue) can divide and multiply, support, nurture, and protect neurons and maintain homeostasis of the fluid that bathes neurons. They do not transmit nerve impulses.

A neuron consists of the following parts:

1. *Cell body*—Contains a nucleus, cytoplasm, mitochondria, Golgi apparatus, chromatophilic substance (Nissl bodies), and neurofibrils.
2. *Dendrites* (*dendro* = tree)—Usually short, highly branched extensions of the cell body that conduct nerve impulses toward it.
3. *Axon*—Single, usually relatively long process that conducts nerve impulses away from the cell body. An axon, in turn, consists of the following:
   a. *Axon hillock* Origin of an axon from the cell body; represented as a small cone-shaped region.
   b. *Axoplasm* Cytoplasm of an axon.
   c. *Axolemma* Plasma membrane around the axoplasm.
   d. *Axon collateral* Side branch of an axon.
   e. *Axon terminals* Fine, branching filaments of an axon or axon collateral.
   f. *Synaptic end bulbs* Bulblike structures at the distal end of axon terminals that contain storage sacs (*synaptic vesicles*) for neurotransmitters.
   g. *Myelin sheath* Multilayered, white phospholipid, segmented covering of many axons, especially large peripheral ones; the myelin sheath is produced by peripheral nervous system neuroglia called *neurolemmocytes (Schwann cells)*.
   h. *Neurolemma (sheath of Schwann)* Peripheral, nucleated cytoplasmic layer of the neurolemmocyte (Schwann cell) that encloses the myelin sheath.
   i. *Neurofibral node (node of Ranvier)* Unmyelinated gap between segments of the myelin sheath.

Using your textbook and models of neurons as a guide, label Figure 12.1.

Now obtain a prepared slide of an ox spinal cord (cross section), human spinal cord (cross and longitudinal sections), nerve endings in skeletal muscle, and a nerve trunk (cross and longitudinal sections). Examine each under high power and identify as many parts of the neuron as you can.

Neurons can be classified on the basis of structure and function. Structurally, neurons are *multipolar* (several dendrites and one axon), *bipolar* (one dendrite and one axon), and *unipolar* (a single process that branches into an axon and a dendrite). Functionally, neurons are classified as *sensory (afferent),* which carry nerve impulses toward the central nervous system; *motor (efferent),* which carry nerve impulses away from the central nervous system; and *association (connecting* or *interneurons),* neurons that are located in the central nervous system and carry nerve impulses between sensory and motor neurons. The neuron you have already labeled in Figure 12.1 is a motor neuron.

Using your textbook and models of neurons as a guide, label Figure 12.2.

B. HISTOLOGY OF NEUROGLIA

Among the types of neuroglial cells are the following:

1. *Astrocytes* (AS-trō-sīts; *astro* = star, *cyte* = cell)—Participate in the metabolism of neurotransmitters (glutamate and γ-aminobutyric acid) and maintain the proper balance of potassium ions (K^+) for the generation of nerve impulses by CNS neurons; participate in

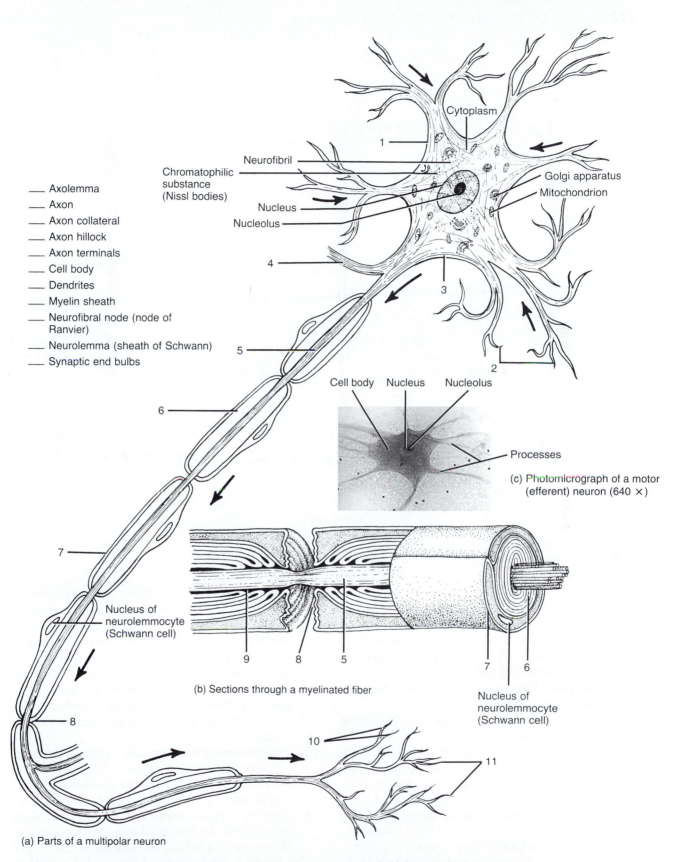

____ Axolemma

____ Axon

____ Axon collateral

____ Axon hillock

____ Axon terminals

____ Cell body

____ Dendrites

____ Myelin sheath

____ Neurofibral node (node of Ranvier)

____ Neurolemma (sheath of Schwann)

____ Synaptic end bulbs

Cytoplasm

Neurofibril

Chromatophilic substance (Nissl bodies)

Nucleus

Nucleolus

Golgi apparatus

Mitochondrion

1

4

3

2

5

6

7

Nucleus of neurolemmocyte (Schwann cell)

8

10

11

(a) Parts of a multipolar neuron

Cell body   Nucleus   Nucleolus

Processes

(c) Photomicrograph of a motor (efferent) neuron (640 ×)

9   8   5

7   6

Nucleus of neurolemmocyte (Schwann cell)

(b) Sections through a myelinated fiber

FIGURE 12.1 Structure of a neuron. In (a) arrows indicate the direction in which the nerve impulse travels. Photo © courtesy of Biophoto Associates Photo Researchers.

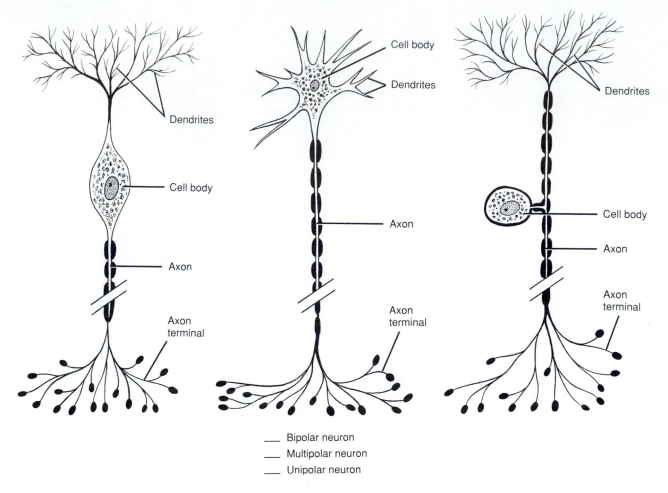

___ Bipolar neuron
___ Multipolar neuron
___ Unipolar neuron

FIGURE 12.2 Structural classification of neurons.

brain development by assisting the migration of neurons; help form the blood-brain barrier, which regulates the passage of substances into the brain; twine around neurons to form a supporting network; and provide a link between neurons and blood vessels.

a. *Protoplasmic astrocytes* are found in the gray matter of the CNS.

b. *Fibrous astrocytes* are found in the white matter of the CNS.

2. *Oligodendrocytes* (ol'-i-gō-DEN-drō-sīts; *oligo* = few, *dendro* = tree)—Resemble astrocytes but with fewer and shorter processes; they provide support in the CNS and produce a myelin sheath on axons of neurons of the CNS.

3. *Microglia* (mī-KROG-lē-a; *micro* = small)—Small cells with few processes, although normally stationary, they can migrate to damaged nervous tissue and there carry on phagocytosis; they are also called **brain macrophages.**

4. *Ependymal* (e-PEN-di-mal) *cells*—Epithelial cells arranged in a single layer that range in shape from squamous to columnar; many are ciliated; they form a continuous epithelial lining for the central canal of the spinal cord and for the ventricles of the brain, spaces that contain networks of capillaries that form cerebrospinal fluid; ependymal cells probably assist in circulating cerebrospinal fluid in these areas.

5. *Neurolemmocytes (Schwann cells)*—Flattened cells arranged around axons. They produce a phospholipid myelin sheath around the axon and dendrites of the neurons of the PNS.

6. *Satellite cells*—Flattened cells arranged around the cell bodies of ganglias (collections of neuron cell bodies outside the CNS). They support neurons in the ganglia of the PNS.

Obtain prepared slides of astrocytes (protoplasmic and fibrous), oligodendrocytes, microglia, ependymal cells, neurolemmocytes, and satellite cells. Using your textbook, Figure 12.3, and models of neuroglia as a guide, identify the various kinds of cells. In the spaces provided, draw each of the cells.

C. REFLEX ARC

A *reflex* is an involuntary response (a movement or systemic functional reaction) to a stimulus applied to the periphery and conducted either to the brain or spinal cord. Reflexes serve to restore functions to homeostasis. For a reflex to occur, a stimulus must elicit a response in one or more structural units within the body termed *reflex arcs*. A *reflex arc* consists of the following components:

1. *Receptor*—Specialized structure associated with the distal end of a neuron or the specialized distal ending of a neuron. Such a structure converts a stimulus from its particular form of energy (such as temperature, pressure, or stretch) into the electrical energy utilized by neurons. If this localized depolarization is of threshold value, an action potential will be initiated in a sensory neuron.
2. *Sensory neuron*—Nerve cell that carries the nerve impulse from the receptor to the CNS.
3. *Integrating center*—Region in the CNS where the sensory neuron makes a functional connection with one or more neurons. This CNS synapse may be with an association neuron or a motor neuron. It is here at the CNS synapse where the initial processing of sensory information occurs: the more complex the reflex involved, the greater the number of CNS synapses involved in a reflex arc.
4. *Motor neuron*—Nerve cell that transmits the action potential generated by the sensory neuron or an association neuron to the effector.
5. *Effector*—Part of the body, either a muscle or a gland, that responds to the motor neuron impulse and thus the stimulus.

Label the components of a reflex arc in Figure 12.4.

Protoplasmic astrocyte

Fibrous astrocyte

Oligodendrocyte

Microglial cell

Ependymal cell

Neurolemmocyte

Satellite cell

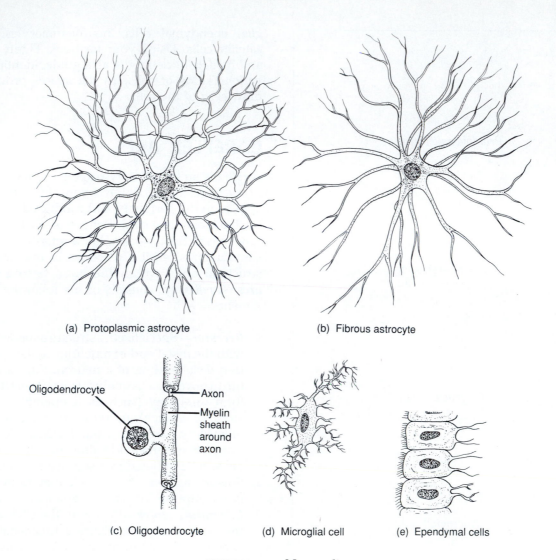

(a) Protoplasmic astrocyte

(b) Fibrous astrocyte

Oligodendrocyte

Axon

Myelin
sheath
around
axon

(c) Oligodendrocyte

(d) Microglial cell

(e) Ependymal cells

FIGURE 12.3   Neuroglia.

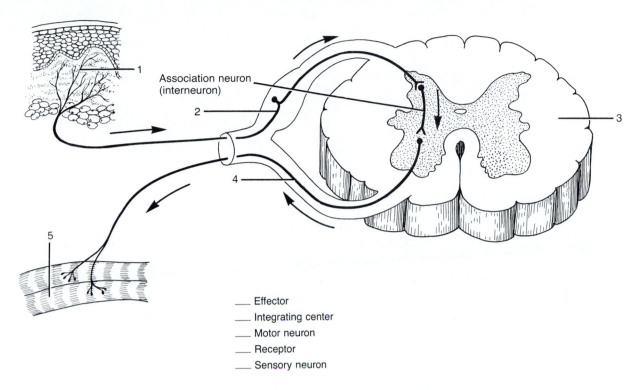

Association neuron
(interneuron)

___ Effector
___ Integrating center
___ Motor neuron
___ Receptor
___ Sensory neuron

FIGURE 12.4   Components of a reflex arc.

Nervous Tissue and Physiology

STUDENT _____   DATE _____

LABORATORY SECTION _____   SCORE/GRADE _____

PART 1. Multiple Choice

_____ 1. The portion of a neuron that conducts nerve impulses away from the cell body is the (a) dendrite (b) axon (c) receptor (d) effector

_____ 2. The fine branching filaments of an axon are called (a) myelin sheaths (b) axolemmas (c) axon terminals (d) axon hillocks

_____ 3. The component of a reflex arc that responds to a motor impulse is the (a) integrating center (b) receptor (c) sensory neuron (d) effector

_____ 4. Which type of neuron conducts nerve impulses toward the central nervous system? (a) afferent (b) association (c) internuncial (d) efferent

_____ 5. In a reflex arc, the nerve impulse is transmitted directly to the effector by the (a) sensory neuron (b) motor neuron (c) integrating center (d) receptor

_____ 6. Bulblike structures at the distal ends of axon terminals that contain storage sacs for neurotransmitters are called (a) dendrites (b) synaptic end bulbs (c) axon collaterals (d) neurofibrils

_____ 7. Which neuroglial cell is phagocytic? (a) oligodendrocyte (b) protoplasmic astrocyte (c) microglial cell (d) fibrous astrocyte

PART 2. Completion

8. A neuron that contains several dendrites and one axon is classified as _____.

9. The portion of a neuron that contains the nucleus and cytoplasm is the _____.

10. The phospholipid covering around many peripheral axons is called the _____.

11. The two types of cells that compose the nervous system are neurons and _____.

12. The peripheral, nucleated layer of the neurolemmocyte (Schwann cell) that encloses the myelin sheath is the _____.

13. The side branch of an axon is referred to as the _____.

14. The part of a neuron that conducts nerve impulses toward the cell body is the

_____.

15. Neurons that carry nerve impulses between sensory neurons and motor neurons are called

_____ neurons.

16. Neurons with one dendrite and one axon are classified as _____.

17. Unmyelinated gaps between segments of the myelin sheath are known as

_____.

18. The neuroglial cell that produces a myelin sheath around axons of neurons of the central nervous

system is called a(n) _____.

19. In a reflex arc, the muscle or gland that responds to a motor impulse is called the

_____.

Nervous System

In this exercise you will examine the principal structural features of the spinal cord and spinal nerves, perform several experiments on reflexes, identify the principal structural features of the brain, trace the course of cerebrospinal fluid, identify the cranial nerves, perform several experiments designed to test for cranial nerve function, and examine the structure and function of the autonomic nervous system. You will also dissect and study the sheep brain.

A. SPINAL CORD AND SPINAL NERVES

1. Meninges

The *meninges* (me-NIN-jēz) are coverings that run continuously around the spinal cord and brain (*meninx*, pronounced MEN-inks, is singular). They protect the central nervous system. The spinal meninges follow:

a. *Dura mater* (DYOO-ra MĀ-ter; *dura* = tough, *mater* = mother)—The outer meninx composed of dense irregular connective tissue. Between the wall of the vertebral canal and the dura mater is the *epidural space*, which is filled with fat, connective tissue, and blood vessels.

b. *Arachnoid* (a-RAK-noyd; *arachnoid* = spider) —The middle meninx composed of very delicate collagen and elastic fibers. Between the arachnoid and the dura mater is a space called the *subdural space*, which contains lymphatic fluid.

c. *Pia mater* (PĒ-a MĀ-ter)—The inner meninx is a thin, transparent connective tissue layer that adheres to the surface of the brain and spinal cord. It consists of interlacing collagen and a few elastic fibers and contains blood vessels. Between the pia mater and the arachnoid is a space called the *subarachnoid space,* where cerebrospinal fluid circulates. Extensions of the pia mater, called *denticulate* (den-TIK-yoo-lāt) *ligaments,* are attached laterally to the dura mater along the length of the spinal cord and suspend the spinal cord and afford protection against shock and sudden displacement.

Label the meninges, subarachnoid space, and denticulate ligament in Figure 13.1.

2. General Features

Obtain a model or preserved specimen of the spinal cord and identify the following general features:

a. *Cervical enlargement*—Between vertebrae C4 and T1; the origin of nerves to the upper extremities.

b. *Lumbar enlargement*—Between vertebrae T9 and T12; the origin of nerves to the lower extremities.

c. *Conus medullaris* (KŌ-nus med-yoo-LAR-is) —Tapered conical portion of the spinal cord near vertebra L1 or L2.

d. *Filum terminale* (FĪ-lum ter-mi-NAL-ē)— Nonnervous fibrous tissue arising from the conus medullaris and extending inferiorly to

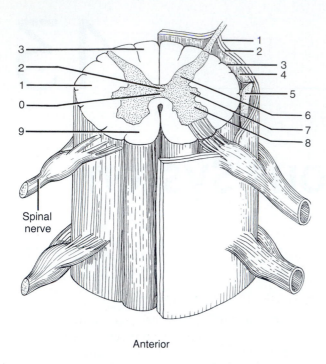

Spinal nerve

Anterior

___ Anterior gray horn   ___ Lateral gray horn
___ Anterior white column   ___ Lateral white column
___ Arachnoid   ___ Pia mater
___ Central canal   ___ Posterior gray horn
___ Denticulate ligament   ___ Posterior white column
___ Dura mater   ___ Subarachnoid space
___ Gray commissure

FIGURE 13.1   Cross section of spinal cord showing the meninges on the right side of the figure.

attach to the coccyx; consists mostly of pia mater.

e. *Cauda equina* (KAW-da ē-KWĪ-na)—Spinal nerves that angle inferiorly in the vertebral canal giving the appearance of wisps of coarse hair.

f. *Anterior median fissure*—Deep, wide groove on the anterior surface of the spinal cord.

g. *Posterior median sulcus*—Shallow, narrow groove on the posterior surface of the spinal cord.

After you have located the parts on a model or preserved specimen of the spinal cord, label the spinal cord in Figure 13.2.

3. Cross Section of Spinal Cord

Obtain a model or specimen of the spinal cord in cross section and note the gray matter, shaped like a letter H or a butterfly. Identify the following parts:

a. *Gray commissure* (KOM-mi-shur)—Cross bar of the letter H.

b. *Central canal*—Small space in the center of the gray commissure that contains cerebrospinal fluid.

c. *Anterior gray horn*—Anterior region of the upright portion of the H.

d. *Posterior gray horn*—Posterior region of the upright portion of the H.

e. *Lateral gray horn*—Intermediate region between the anterior and posterior gray horns present in the thoracic, upper lumbar, and sacral segments of the spinal cord.

f. *Anterior white column*—Anterior region of white matter.

g. *Posterior white column*—Posterior region of white matter.

h. *Lateral white column*—Intermediate region of white matter between the anterior and posterior white columns.

Label these parts of the spinal cord in cross section in Figure 13.1.

Examine a prepared slide of a spinal cord in cross section and see how many structures you can identify.

4. Spinal Nerve Attachments

Spinal nerves are the paths of communication between the spinal cord tracts and the periphery. The 31 pairs of spinal nerves are named and numbered according to the region of the spinal cord from which they emerge. The first cervical pair emerges between the atlas and occipital bone; all other spinal nerves leave the vertebral column from intervertebral foramina between adjoining vertebrae. There are 8 pairs of cervical nerves, 12 pairs of thoracic nerves, 5 pairs of lumbar nerves, 5 pairs of sacral nerves, and 1 pair of coccygeal nerves. Label the spinal nerves in Figure 13.2.

Each pair of spinal nerves is connected to the spinal cord by two points of attachment called roots. The *posterior (sensory) root* contains sensory nerve fibers only and conducts nerve impulses from the periphery to the spinal cord. Each posterior root has a swelling, the *posterior (sensory) root ganglion,* which contains the cell bodies of the sensory neurons from the periphery. Fibers extend from the ganglion into the

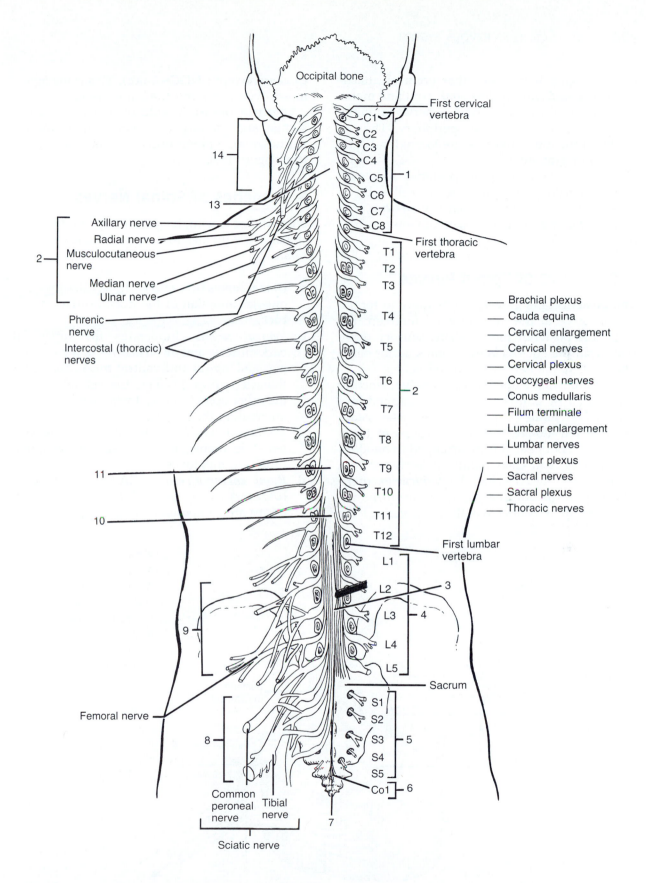

Occipital bone

First cervical vertebra

C1
C2
C3
C4
C5
C6
C7
C8

14

13

Axillary nerve
Radial nerve
Musculocutaneous nerve
Median nerve
Ulnar nerve

2

Phrenic nerve

Intercostal (thoracic) nerves

11

10

Femoral nerve

8

Common peroneal nerve

Tibial nerve

Sciatic nerve

1

First thoracic vertebra

T1
T2
T3
T4
T5
T6
T7
T8
T9
T10
T11
T12

L1
L2
L3
L4
L5

S1
S2
S3
S4
S5
Co1

2

3

4

First lumbar vertebra

Sacrum

5

6

7

9

___ Brachial plexus
___ Cauda equina
___ Cervical enlargement
___ Cervical nerves
___ Cervical plexus
___ Coccygeal nerves
___ Conus medullaris
___ Filum terminale
___ Lumbar enlargement
___ Lumbar nerves
___ Lumbar plexus
___ Sacral nerves
___ Sacral plexus
___ Thoracic nerves

Posterior view

FIGURE 13.2   Spinal cord.

231

posterior gray horn. The other point of attachment, the ***anterior (motor) root,*** contains motor nerve fibers only and conducts nerve impulses from the spinal cord to the periphery. The cell bodies of the motor neurons are located in lateral or anterior gray horns.

Label the posterior root, posterior root ganglion, anterior root, spinal nerve, cell body of sensory neuron, axon of sensory neuron, cell body of motor neuron, and axon of motor neuron in Figure 13.3.

5. Structure of Spinal Nerves

The posterior and anterior roots unite to form a spinal nerve at the intervertebral foramen. Because the posterior root contains sensory nerve fibers and the anterior root contains motor nerve fibers, all spinal nerves are mixed nerves.

Individual nerve fibers within a nerve, whether myelinated or unmyelinated, are wrapped in a connective tissue covering called the ***endoneurium*** (en'-dō-NOO-rē-um). Groups of fibers with their endoneurium are arranged in bundles, or ***fascicles,*** and each bundle is wrapped in a connective tissue covering, the ***perineurium*** (per'-i-NOO-rē-um). All the fascicles, in turn, are wrapped in a connective tissue covering, the

epineurium (ep'-i-NOO-rē-um). This is the outermost covering around the entire nerve.

Obtain a prepared slide of a nerve in cross section and identify the fibers, endoneurium, perineurium, epineurium, and fascicles. Now label Figure 13.4.

6. Branches of Spinal Nerves

Shortly after leaving its intervertebral foramen, a spinal nerve divides into several branches called ***rami:***

a. ***Dorsal ramus*** (RĀ-mus)—Innervates the deep muscles and skin of the dorsal surface of the back.
b. ***Ventral ramus***—Innervates the superficial back muscles and all structures of the extremities and lateral and ventral trunk; except for thoracic nerves T2–T11, the ventral rami of the other spinal nerves form plexuses before innervating their structures.
c. ***Meningeal branch***—Innervates vertebrae, vertebral ligaments, blood vessels of the spinal cord, and meninges.
d. ***Rami communicantes*** (RĀ-mē ko-myoo-nē-KAN-tēz)—Gray and white rami communicantes are components of the autonomic ner-

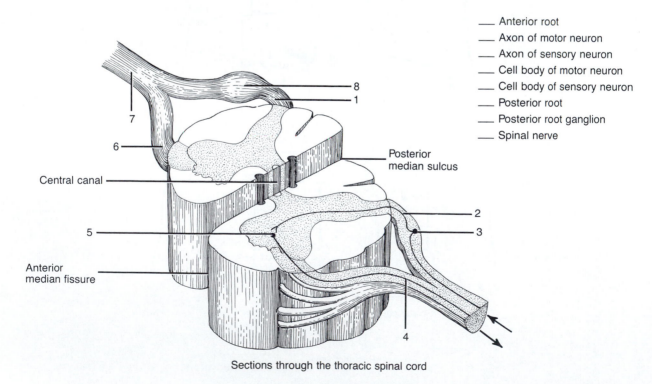

Sections through the thoracic spinal cord

FIGURE 13.3 Spinal nerve attachments.

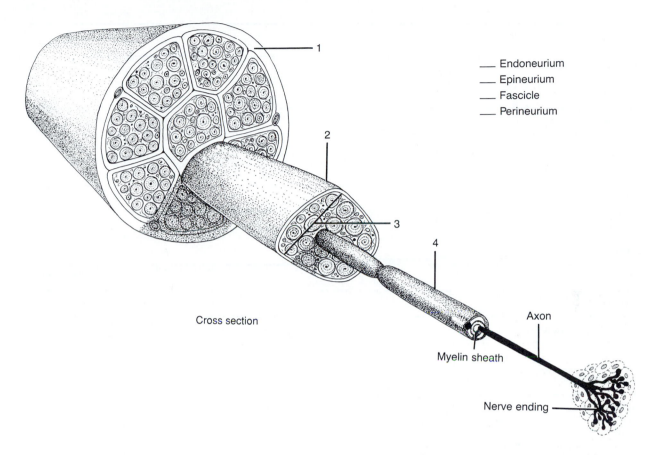

_____ Endoneurium
_____ Epineurium
_____ Fascicle
_____ Perineurium

Cross section

Axon

Myelin sheath

Nerve ending

FIGURE 13.4   Coverings of a spinal nerve.

vous system; they connect the ventral rami with sympathetic trunk ganglia.

7. Plexuses

The ventral rami of spinal nerves, except for T2–T11, join with adjacent nerves on either side of the body to form networks called *plexuses* (PLEK-sus-ēz).

a. **Cervical plexus** (PLEK-sus)—Formed by the ventral rami of the first four cervical nerves (C1–C4) with contributions from C5; one is located on each side of the neck, alongside the first four cervical vertebrae; the plexus supplies the skin and muscles of the head, neck, and upper part of the shoulders.

b. **Brachial plexus**—Formed by the ventral rami of spinal nerves C5–C8 and T1 with contributions from C4 and T2; each is located on either side of the last four cervical and first thoracic vertebrae and extends downward and laterally, over the first rib behind the clavicle, and into the axilla; the plexus constitutes the entire nerve supply for the upper extremities and shoulder region.

c. **Lumbar plexus**—Formed by the ventral rami of spinal nerves L1–L4; each is located on either side of the first four lumbar vertebrae posterior to the psoas major muscle and anterior to the quadratus lumborum muscle; the plexus supplies the anterolateral abdominal wall, external genitals, and part of the lower extremity.

d. **Sacral plexus**—Formed by the ventral rami of spinal nerves L4–L5 and S1–S4; each is located largely anterior to the sacrum; the plexus supplies the buttocks, perineum, and lower extremities.

Label the cervical, brachial, lumbar, and sacral plexuses in Figure 13.2. Also note the names of some of the major peripheral nerves that arise from the plexuses.

For each nerve listed in the following table, indicate the plexus to which it belongs and the structure(s) it innervates.

Nerve	Plexus	Innervation
Musculocutaneous		
Femoral		
Phrenic		
Pudenal		
Axillary		
Transverse cervical		
Radial		
Obturator		
Tibial		
Thoracodorsal		
Perforating cutaneous		
Ulnar		
Long thoracic		
Median		
Iliohypogastric		
Deep peroneal		

8. Spinal Cord Tracts

The vital function of conveying sensory and motor information to and from the brain is carried out by ascending (sensory) and descending (motor) tracts and pathways in the spinal cord. The names of the tracts and pathways indicate the white column in which the tract travels, where the cell bodies of the tract originate, and where the axons of the tract terminate. For example, the anterior spinothalamic tract is located in the *anterior* white column, it originates in the *spinal cord*, and it terminates in the *thalamus* of the brain. Because it conveys nerve impulses from the spinal cord upward to the brain, it is an ascending (sensory) tract.

Label the ascending and descending tracts shown in Figure 13.5.

Indicate the function of each ascending and descending tract listed in the following table:

ASCENDING PATHWAYS AND TRACTS	Function
Anterolateral (spinothalamic) pathway	
Anterior (ventral) spinothalamic (spī'-nō-THAL-am-ik) tract	
Lateral spinothalamic tract	
Posterior column-medial lemniscus pathway	
Fasciculus gracilis (fa-SIK-yoo-lus gras-I-lus) and **fasciculus cuneatus** (kū-nē-ĀT-us)	
Spinocerebellar tracts	
Posterior (dorsal) spinocerebellar (spī'-nō-ser-e-BEL-ar) tract	
Anterior (ventral) spinocerebellar tract	

DESCENDING PATHWAYS AND TRACTS	Function
Direct (pyramidal) pathway	
Lateral corticospinal (KOR'-ti-kō-spī'-nal)	
Anterior (ventral) corticospinal	
Corticobulbar (kor'-ti-kō-BUL-bar)	
Indirect (extrapyramidal) pathway	
Rubrospinal (ROO'-brō-spī'-nal)	
Tectospinal (TEK-tō-spī'-nal)	
Vestibulospinal (ves-TIB-yoo-lō-spī'-nal)	
Lateral reticulospinal (re-TIK-yoo-lō-spī'-nal)	
Anterior (ventral) or medial reticulospinal	

9. Reflex Experiments

In this section you will determine the responses obtained in various common human reflexes. While performing the procedures it is important for you to consider the various components of a reflex arc and to try to determine the receptors and effector organs involved. In addition, you should try to compare the differing levels of complexity in the reflexes studied in these procedures. Reflexes that result in the contraction of skeletal muscle are called *somatic reflexes,* while those that cause contraction of cardiac or smooth muscle, or secretion by glands, are known as *visceral (autonomic) reflexes.*

a. *Corneal reflex*—*Very slowly and carefully,* approach, but *do not touch,* the cornea of your partner's eye with lens paper.

What happens? _____

_____

What is the purpose of this reflex? _____

_____

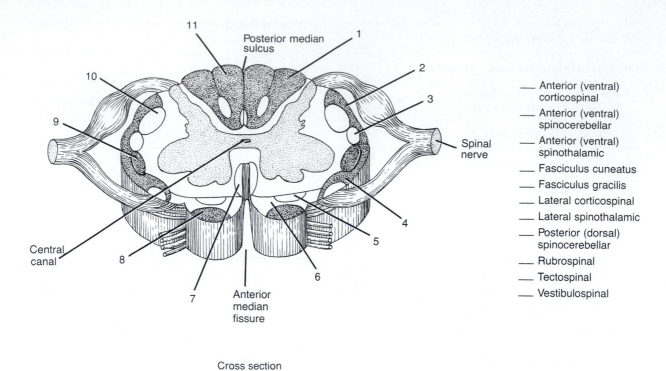

11
Posterior median sulcus
1
10
2
3
9
Spinal nerve
Central canal
8
4
5
6
7
Anterior median fissure

___ Anterior (ventral) corticospinal
___ Anterior (ventral) spinocerebellar
___ Anterior (ventral) spinothalamic
___ Fasciculus cuneatus
___ Fasciculus gracilis
___ Lateral corticospinal
___ Lateral spinothalamic
___ Posterior (dorsal) spinocerebellar
___ Rubrospinal
___ Tectospinal
___ Vestibulospinal

Cross section

FIGURE 13.5   Selected ascending and descending tracts of the spinal cord.

b. *Light (photopupil) reflex*—Have your partner close his/her eyes for about 2 min while facing a moderate (not bright) light. Now have your partner open his/her eyes and note the size of the pupils.

Describe the response. _____

_____

Now, without changing the illumination, have the subject quickly change his/her focus from the distant object to one that is only 25.4 cm (10 in.) away. (Make sure that the subject continues to focus on the close object for at least 2 min.) Note the change in the size of the pupils as the subject changes his/her focus, and then as he/she retains a focus on the closer object. Has there been any change in the size of the subject's pupil?

_____

Formulate a hypothesis as to the purpose of

this reflex. _____

_____

Which nerves and muscles act in this reflex?

_____

What is the purpose of this reflex? _____

_____

c. *Accommodation pupil reflex*—Have your partner focus on an object at a distance of 6.10 m (20 ft) or more.
Describe the size of the subject's pupils.

_____

_____

Now, without changing the illumination, have the subject quickly change his/her focus from the distant object to one that is only 25.4 cm (10 in.) away. (Make sure that the subject continues to focus on the close object for at least 2 min.) Note the change in the size of the pupils as the subject changes his or her focus, and then as he/she retains a focus on the closer object. Has there been any change in the size of the subject's pupil?

_____

Formulate a hypothesis as to the purpose of

this reflex. _____

What nerve and muscles act in this reflex?

_____

d. *Convergence reflex*—Observe the position of your partner's eyeballs while he or she again looks at a distant object observed in Procedure c. Now, without changing the illumination, have the subject quickly change his/her focus from the distant object to one that is again only 25.4 cm (10 in.) away. (Make sure that the subject continues to focus on the close object for at least 2 min.)
Note any change in the positioning of the

eyeballs. _____

_____

Formulate a hypothesis as to the purpose of

this reflex. _____

_____

e. *Swallowing reflex*—Swallow the saliva in your mouth and immediately swallow repeatedly, as fast as possible, for 20 sec. Now, repeat the procedure, only this time drink a small amount of water each time you swallow.
How do the results of the first and second

experiments compare? _____

_____

_____

Formulate a hypothesis regarding the nature of the stimulus required and the location of the receptors involved in the initiation of the

swallowing reflex. _____

_____

What muscles are involved in swallowing?

_____

_____

f. *Patellar reflex*—Have your partner sit on a table so that the knee is off of the table and the leg hangs freely. Strike the patellar ligament just inferior to the kneecap with the reflex hammer. What is the response? _____

_____

Now retest the patellar reflex while the subject is adding a column of numbers. Is there any change in the level of activity of the reflex?

_____

Test the subject while he/she interlocks fingers and pulls one hand against the other. there any change in the level of activity of the reflex?

_____

Formulate a hypothesis regarding the purpose

of the patellar reflex. _____

_____

_____

Formulate a hypothesis regarding the level of complexity of the patellar reflex as compared to any other reflex studied thus far in

this series of laboratory procedures. _____

_____

_____

g. *Achilles reflex*—Have the subject kneel on a chair and let the feet hang freely over the edge of the chair. Bend one foot to increase the tension on the gastrocnemius muscle. Tap the calcaneal (Achilles) tendon with the reflex hammer.

What is the result? _____

_____

_____

_____

h. *Plantar reflex*—Scratch the sole of your partner's foot by moving a blunt object along the sole toward the toes.

What is the response? _____

_____

_____

What is the Babinski sign? _____

_____

Why is it normal in children under the age of 18 months? _____

_____

What does the Babinski sign indicate in an adult? _____

_____

_____

B. BRAIN

1. Parts

The brain can be divided into four principal parts: (1) **brain stem,** which consists of the medulla oblongata, pons, and midbrain; (2) **diencephalon** (dī-en-SEF-a-lon), which consists primarily of the thalamus and hypothalamus; (3) **cerebellum,** which is posterior to the brain stem; and (4) **cerebrum,** which is superior to the brain stem and comprises about seven-eighths of the total weight of the brain.

Examine a model and preserved specimen of the brain and identify the parts just described. Then refer to Figure 13.6 and label the parts of the brain.

2. Meninges

As in the spinal cord, the brain is protected by **meninges.** The cranial meninges are continuous with the spinal meninges. The cranial meninges are the outer **dura mater,** the middle **arachnoid,** and the inner **pia mater.** The cranial dura mater consists of two layers called the *periosteal layer* (adheres to the cranial bones and serves as a periosteum) and the *meningeal layer* (the thinner, inner layer that corresponds to the spinal dura mater).

Refer to Figure 13.7 and label all of the meninges.

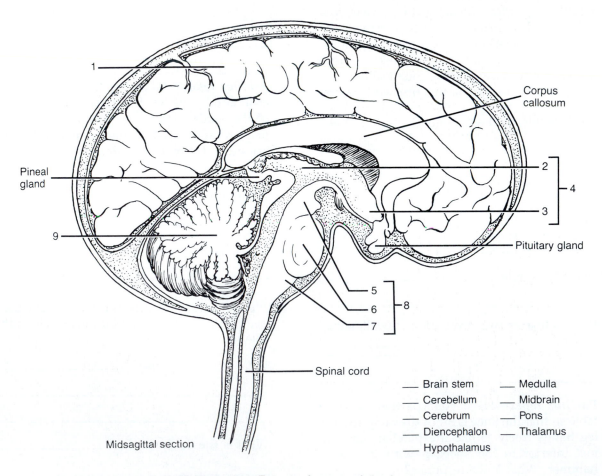

___ Brain stem	___ Medulla
___ Cerebellum	___ Midbrain
___ Cerebrum	___ Pons
___ Diencephalon	___ Thalamus
___ Hypothalamus	

Midsagittal section

FIGURE 13.6 Principal parts of the brain.

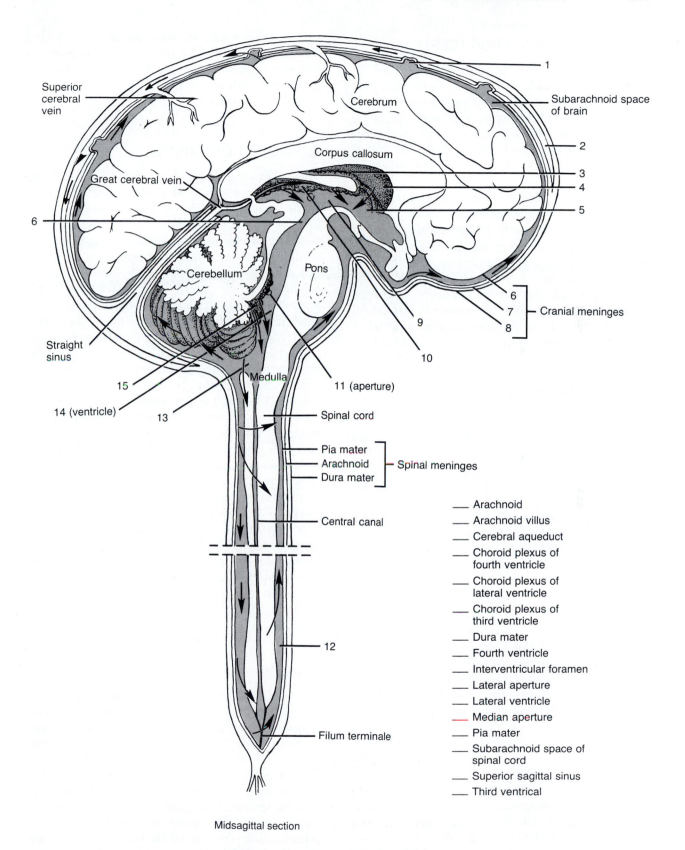

Superior cerebral vein

Cerebrum

Subarachnoid space of brain

Corpus callosum

Great cerebral vein

Cerebellum

Pons

Straight sinus

Medulla

15

14 (ventricle)   13

11 (aperture)

Spinal cord

Pia mater
Arachnoid   — Spinal meninges
Dura mater

Central canal

12

Filum terminale

Cranial meninges

9

10

___ Arachnoid
___ Arachnoid villus
___ Cerebral aqueduct
___ Choroid plexus of fourth ventricle
___ Choroid plexus of lateral ventricle
___ Choroid plexus of third ventricle
___ Dura mater
___ Fourth ventricle
___ Interventricular foramen
___ Lateral aperture
___ Lateral ventricle
___ Median aperture
___ Pia mater
___ Subarachnoid space of spinal cord
___ Superior sagittal sinus
___ Third ventrical

Midsagittal section

FIGURE 13.7   Brain and spinal cord.

3. Cerebrospinal Fluid (CSF)

The central nervous system is protected by *cerebrospinal fluid (CSF).* The fluid circulates through the subarachnoid space around the brain and spinal cord and through the ventricles of the brain. The ventricles are cavities in the brain that communicate with each other, with the central canal of the spinal cord, and with the subarachnoid space.

Cerebrospinal fluid is formed primarily by filtration and secretion from networks of capillaries and ependymal cells in the ventricles, called *choroid* (KŌ-royd) *plexuses* (see Figure 13.7). Each of the two *lateral ventricles* (VEN-tri-kuls) is located within a hemisphere (side) of the cerebrum under the corpus callosum. The fluid formed in the choroid plexuses of the lateral ventricles circulates through an opening called the *interventricular foramen* into the third ventricle. The *third ventricle* is a slit between and inferior to the right and left halves of the thalamus and between the lateral ventricles. More fluid is added by the choroid plexus of the third ventricle. Then the fluid circulates through a canal-like structure called the *cerebral aqueduct* (AK-we-dukt) into the fourth ventricle. The *fourth ventricle* lies between the inferior brain stem and the cerebellum. More fluid is added by the choroid plexus of the fourth ventricle. The roof of the fourth ventricle has three openings: one *median aperture* (AP-e-chur) and two *lateral apertures.* The fluid circulates through the apertures into the subarachnoid space around the back of the brain and downward through the central canal of the spinal cord to the subarachnoid space around the posterior surface of the spinal cord, up the anterior surface of the spinal cord, and around the anterior part of the brain. Most of the cerebrospinal fluid is absorbed into a vein called the superior sagittal sinus through its arachnoid villi. Normally, cerebrospinal fluid is absorbed as rapidly as it is formed.

Refer to Figure 13.7 and label the choroid plexus of the lateral ventricle, lateral ventricle, interventricular foramen, choroid plexus of the third ventricle, third ventricle, cerebral aqueduct, choroid plexus of the fourth ventricle, fourth ventricle, median aperture, lateral aperture, subarachnoid space of the spinal cord, superior sagittal sinus, and arachnoid villus.

Note the arrows in Figure 13.7, which indicate the path taken by cerebrospinal fluid. With the aid of your textbook, starting at the choroid plexus of the lateral ventricle and ending at the superior sagittal sinus, see if you can follow the remaining path of the fluid.

Now complete Figure 13.8.

4. Medulla Oblongata

The *medulla oblongata* (me-DULL-la ob'-long-GA-ta), or just simply *medulla,* is a continuation of the upper part of the spinal cord and forms the inferior part of the brain stem. The medulla contains all ascending and descending tracts that communicate between the spinal cord and various parts of the brain. On the ventral side of the medulla are two roughly triangular bulges called *pyramids.* They contain the largest motor tracts that run from the cerebral cortex to the spinal cord. Most fibers in the left pyramid cross to the right side of the spinal cord and most fibers in the right pyramid cross to the left side of the spinal cord. This crossing is called the *decussation* (dē'-ku-SĀ-shun) *of pyramids* and explains why motor areas of one side of the cerebral cortex control muscular movements on the opposite side of the body. The medulla contains three vital reflex centers called the *cardiac center, respiratory center,* and *vasomotor center.* The medulla also contains the nuclei of origin of several cranial nerves. These are the cochlear and vestibular branches of the vestibulocochlear (VIII) nerves, glossopharyngeal (IX) nerves, vagus (X) nerves, spinal portions of the accessory (XI) nerves, and hypoglossal (XII) nerves. Examine a model or specimen of the brain and identify the parts of the medulla. Then refer to Figure 13.6 and locate the medulla.

5. Pons

The *pons* lies directly above the medulla and anterior to the cerebellum. The pons contains fibers that connect parts of the cerebellum and medulla with the cerebrum. The pons contains the nuclei of origin of the following pairs of cranial nerves: trigeminal (V) nerves, abducens (VI) nerves, facial (VII) nerves, and vestibular branch of the vestibulocochlear (VIII) nerves. Other important nuclei in the pons are the *pneumotaxic* (noo-mō-TAK-sik) *area* and *apneustic* (ap-NOO-stik) *area* that help control respiration.

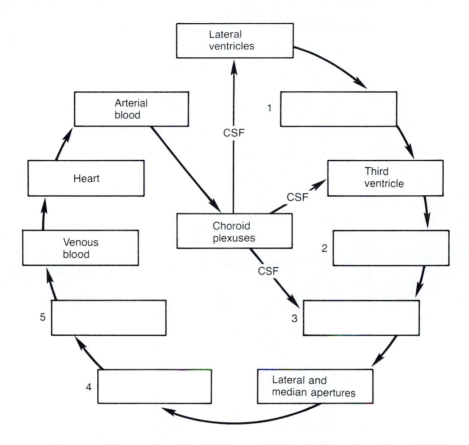

FIGURE 13.8   Formation, circulation, and absorption of cerebrospinal fluid (CSF).

Identify the pons on a model or specimen of the brain. Locate the pons in Figure 13.6.

6. Midbrain

The *midbrain* extends from the pons to the lower portion of the cerebrum. The ventral portion of the midbrain contains the paired *cerebral peduncles* (pe-DUNG-kulz), which connect the upper parts of the brain to the lower parts of the brain and spinal cord. The dorsal part of the midbrain contains four rounded elevations called the *corpora quadrigemina* (KOR-por-ra kwad-ri-JEM-in-a). Two of the elevations, the *superior colliculi* (ko-LIK-yoo-lī), serve as reflex centers for movements of the eyeballs and the head in response to visual and other stimuli. The other two elevations, the *inferior colliculi,* serve as reflex centers for movements of the head and trunk in response to auditory stimuli. The midbrain contains the nuclei of origin for two pairs of cranial nerves: oculomotor (III) and trochlear (IV).

Identify the parts of the midbrain on a model or specimen of the brain. Locate the midbrain in Figure 13.6.

7. Thalamus

The *thalamus* (THAL-a-mus; *thalamos*=inner chamber) is a large oval structure, located above the midbrain, that consists of two masses of gray matter covered by a layer of white matter. The two masses are joined by a bridge of gray matter called the *intermediate mass.* The thalamus contains numerous nuclei that serve as relay stations for all sensory impulses, except smell. The most prominent are the *medial geniculate* (je-NIK-yoo-lāt) *nuclei* (hearing), *lateral geniculate nuclei* (vision), *ventral posterior nuclei* (general sensations and taste), *ventral lateral nuclei* (voluntary motor actions), and the *ventral anterior nuclei* (voluntary motor actions and arousal).

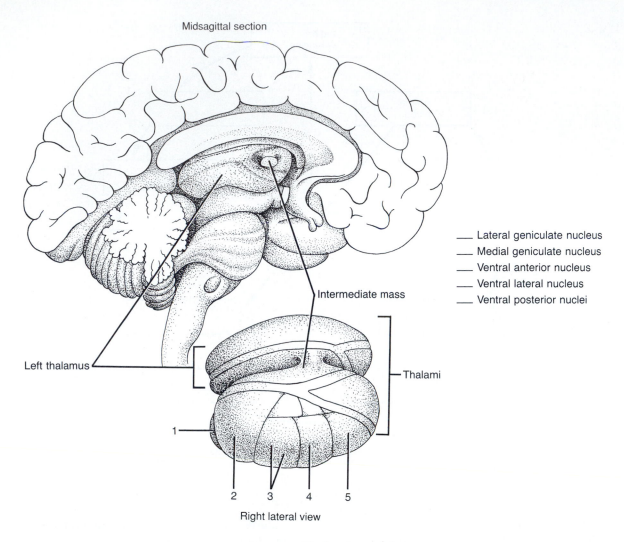

FIGURE 13.9   Thalamic nuclei.

Identify the thalamic nuclei on a model or specimen of the brain. Then refer to Figure 13.9 and label the nuclei.

Identify the hypothalamus on a model or specimen of the brain. Locate the hypothalamus in Figure 13.6.

8. Hypothalamus

The *hypothalamus* is located inferior to the thalamus and forms the floor and part of the wall of the third ventricle. Among the functions served by the hypothalamus are control and integration of the activities of the autonomic nervous system and parts of the endocrine system, reception and integration of sensory impulses from the viscera, secretion of regulating hormones, and control of body temperature. The hypothalamus also assumes a role in feelings of rage and aggression, food intake, thirst, the waking state and sleep patterns, and biological rhythms.

9. Cerebrum

The *cerebrum* is the largest portion of the brain and is supported on the brain stem. Its outer surface consists of gray matter and is called the *cerebral cortex.* Beneath the cerebral cortex is the cerebral white matter. The upfolds of the cerebral cortex are termed *gyri* (JĪ-rī), or *convolutions,* the deep downfolds are termed *fissures,* and the shallow downfolds are termed *sulci* (SUL-sī).

The *longitudinal fissure* separates the cerebrum into right and left halves called *hemispheres.* Each hemisphere is further divided into lobes by sulci or fissures. The *central sulcus* separates the *frontal lobe* from the *parietal lobe.*

The *lateral cerebral sulcus* separates the frontal lobe from the *temporal lobe.* The *parietooccipital sulcus* separates the *parietal lobe* from the *occipital lobe.* Another prominent fissure, the *transverse fissure,* separates the cerebrum from the cerebellum. Another lobe of the cerebrum, the *insula,* lies deep within the lateral cerebral fissure under the parietal, frontal, and temporal lobes. It cannot be seen in external view. Two important gyri on either side of the *central sulcus* are the *precentral gyrus* and the *postcentral gyrus.* The olfactory (I) and optic (II) cranial nerves are associated with the cerebrum.

Examine a model of the brain and identify the parts of the cerebrum just described. Refer to Figure 13.10 and label the lobes.

10. Basal Ganglia

Basal ganglia (GANG-lē-a), or *cerebral nuclei,* are paired masses of gray matter, with one member of each pair in each cerebral hemisphere. The largest of the basal ganglia is the *corpus striatum* (strī-Ā-tum; *corpus* = body, *striatum* = striped), which consists of the *caudate* (*cauda* = tail) *nucleus* and the *lentiform* (*lenticala*=shaped like a lentil or lens) or *lenticular nucleus.* The lentiform nucleus, in turn, is subdivided into a lateral *putamen* (pu-TĀ-men; *putamen*=shell) and a medial *globus pallidus* (*globus*=ball, *pallid*=pale). The basal ganglia control the large, subconscious movement of the skeletal muscles. An example is swinging the arms while walking. Other structures that are considered part of the basal ganglia

are the substantia nigra, subthalamic nuclei, and red nuclei.

Examine a model or specimen of the brain and identify the basal ganglia described. Now refer to Figure 13.11 and label the basal ganglia shown.

11. Functional Areas of Cerebral Cortex

The functions of the cerebrum are numerous and complex. In a general way, the cerebral cortex can be divided into sensory, motor, and association areas. The *sensory areas* interpret sensory impulses, the *motor areas* control muscular movement, and the *associated areas* are concerned with emotional and intellectual processes.

The principal sensory and motor areas of the cerebral cortex are indicated by numbers based on K. Brodmann's map of the cerebral cortex. His map, first published in 1909, attempts to correlate structure and function. Refer to Figure 13.12 and match the name of the sensory or motor area next to the appropriate letter (A, B, C, etc.).

12. Cerebellum

The *cerebellum* is inferior to the posterior portion of the cerebrum and separated from it by the transverse fissure. The central constricted area of the cerebellum is called the *vermis* and the lateral portions are referred to as *hemispheres.* The surface of the cerebellum, called the *cerebellar cortex,* consists of gray matter thrown into a series of slender parallel ridges called *folia.*

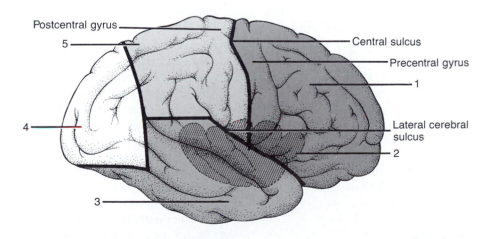

Right lateral view

FIGURE 13.10 Lobes of the cerebrum.

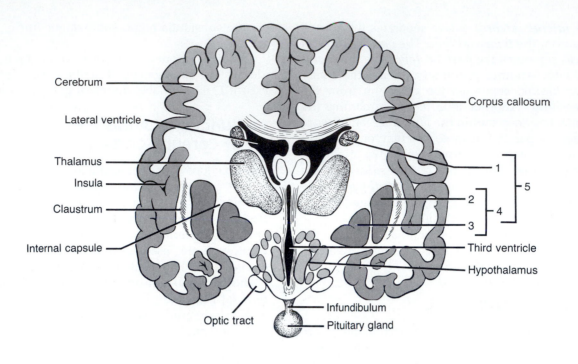

Frontal section of cerebrum

___ Caudate nucleus   ___ Lentiform nucleus
___ Corpus striatum   ___ Putamen
___ Globus pallidus

FIGURE 13.11   Basal ganglia.

Beneath the gyri are tracts of white matter called *arbor vitae.*

The cerebellum is attached to the brain stem by three paired bundles of fibers called *cerebellar peduncles.* The *inferior cerebellar peduncles* connect the cerebellum with the medulla at the base of the brain stem and with the spinal cord. The *middle cerebellar peduncles* connect the cerebellum with the pons. The *superior cerebellar peduncles* connect the cerebellum with the midbrain and thalamus.

The cerebellum is an area of the brain that coordinates certain subconscious movements in skeletal muscles required for coordination, posture, and balance.

Examine a model or specimen of the brain and locate the parts of the cerebellum. Identify the cerebellum in Figure 13.6.

13. Determination of Cerebellar Function

Working with your lab partner, perform the following tests to determine cerebellar function:

1. With your upper extremities down straight at your side, walk heel to toe for 20 ft without losing your balance.
2. Stand away from any supporting object and place your feet together and upper extremities down straight at your side; look straight ahead. Now close your eyes and stand for 2 to 3 min. Your lab partner should stand off to one side and then immediately in front of you and observe any and all body movements.
3. Stand away from any supporting object and place your feet together and upper extremities down straight at your side; look straight ahead. Now, with your eyes open raise one foot off the gound and run the heel of that foot down the shin of the other leg. Your lab partner should observe the smoothness of the movement and whether your heel remains in contact with the shin at all times.
4. Stand away from any supporting object and place your feet together and upper extremities abducted (away from midline) and palms facing anteriorly; look straight ahead. Now, gently close your eyes and touch the tip of

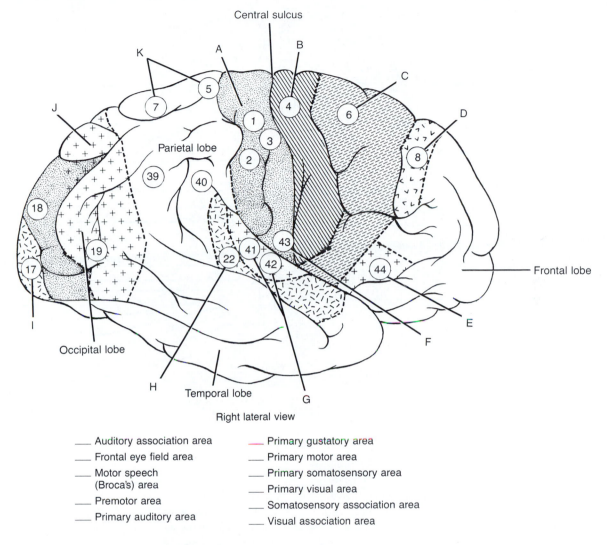

Central sulcus

Parietal lobe

Frontal lobe

Occipital lobe

Temporal lobe

Right lateral view

___ Auditory association area

___ Frontal eye field area

___ Motor speech (Broca's) area

___ Premotor area

___ Primary auditory area

___ Primary gustatory area

___ Primary motor area

___ Primary somatosensory area

___ Primary visual area

___ Somatosensory association area

___ Visual association area

FIGURE 13.12 Functional areas of the cerebrum.

your nose with your right index finger and return the right upper extremity to its abducted position. Now touch the tip of your nose with your left index finger, and return the left upper extremity to its abducted position. Slowly increase the rapidity of the activity.

5. With your eyes closed, touch your nose with the index finger of each hand.

C. CRANIAL NERVES

Of the 12 pairs of *cranial nerves,* 10 originate from the brain stem, but all pass through foramina in the base of the skull. The cranial nerves are designated by Roman numerals and names. The Roman numerals indicate the order in which the

nerves arise from the brain, from front to back. The names indicate the distribution or function of the nerves.

Obtain a model of the brain and, using your text and any other aids available, identify the 12 pairs of cranial nerves. Now refer to Figure 13.13 and label the cranial nerves.

D. TESTS OF CRANIAL NERVE FUNCTION

The following simple tests can be performed to determine cranial nerve function. Although they provide only superficial information, they will help you to understand how the various cranial nerves function. Perform each of the tests with your partner.

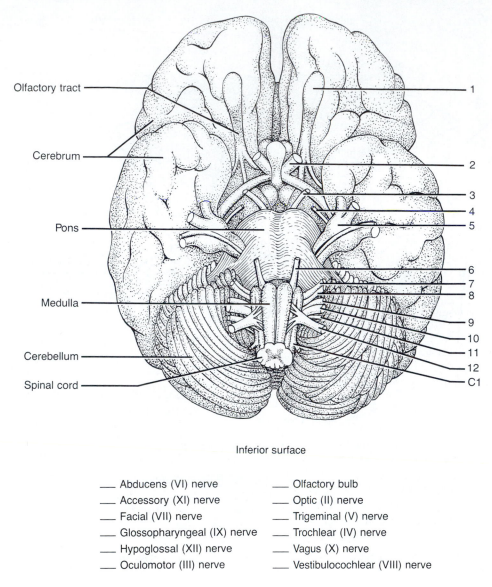

Inferior surface

____ Abducens (VI) nerve   ____ Olfactory bulb
____ Accessory (XI) nerve   ____ Optic (II) nerve
____ Facial (VII) nerve   ____ Trigeminal (V) nerve
____ Glossopharyngeal (IX) nerve ____ Trochlear (IV) nerve
____ Hypoglossal (XII) nerve   ____ Vagus (X) nerve
____ Oculomotor (III) nerve   ____ Vestibulocochlear (VIII) nerve

FIGURE 13.13   Cranial nerves of the human brain.

1. ***Olfactory (I) nerve***—Have your partner smell several familiar substances, such as spices, first using one nostril and then the other. Your partner should be able to distinguish the odors equally with each nostril.
What structures could be malfunctioning if the ability to smell is lost? _____

_____

_____

_____

2. ***Optic (II) nerve***—Have your partner read a portion of a printed page using each eye. Do the same while using a Snellen chart at a distance of 6.10 m (20 ft).
Describe the visual pathway from the optic nerve to the cerebral cortex. _____

_____

_____

3. ***Oculomotor (III)***, ***trochlear*** (TROK-lē-ar) ***(IV)***, and ***abducens*** (ab-DOO-sens) ***(VI) nerves***—To test the motor responses of these nerves, have your partner follow your finger with his/her eyes without moving the head. Move your finger up, down, medially, and laterally.

Which nerves control which movements of the eyeball? _____

_____

Look for signs of ptosis (drooping of one or both eyelids).
Which cranial nerve innervates the upper eyelid? _____
To test for pupillary light reflex, shine a small flashlight into each eye from the side.

CAUTION! *Do not make contact with the eyes.*

Observe the pupil.

What cranial nerve controls this reflex? _____

_____

4. *Trigeminal* (trī-JEM-i-nal) *(V) nerve*—To test the motor responses of this nerve, have your partner close his/her jaws tightly.

Which muscles are used? _____

_____

Now, while holding your hand under your partner's lower jaw to provide resistance, ask your partner to open his/her mouth. To test the sensory responses of this nerve, have your partner close his/her eyes. Lightly whisk a piece of dry cotton over the mandibular, maxillary, and ophthalmic areas on each side of your partner's face. Do the same with cotton that has been moistened with cold water.
Which cranial nerves bring about this response? _____

_____

5. *Facial (VII) nerve*—To test the motor responses of this nerve, ask your partner to bring the corners of the mouth straight back, smile while showing his or her teeth, whistle, puff his/her cheeks, frown, raise his/her eyebrows, and wrinkle his or her forehead.

Define Bell's palsy. _____

_____

To test the sensory responses of this nerve, touch the tip of your partner's tongue with an applicator that has been dipped into a salt solution. Have him/her rinse the mouth with water. Now place an applicator dipped into a sugar solution along the anterior surface of your partner's tongue.
What are the four basic tastes distinguished?

_____

6. *Vestibulocochlear* (ves-tib'-yoo-lō-KŌK-lē-ar) *(VIII) nerve*—To test for the functioning of the vestibular portion of this nerve, have your partner sit on a swivel stool with his/her head slightly bent forward toward the chest. Now have a student (the operator) turn your partner *slowly and very carefully* to the right about 10 times (about one turn every 2 sec). Stop the stool suddenly and look for nystagmus (nis-TAG-mus), which is rapid movement or quivering of the eyeballs. Note the direction of the nystagmus.

CAUTION! *Do not allow the subject to stand or walk until dizziness is completely gone.*

b. When the nystagmus movements have ceased, have the subject sit on the chair and, with his/her eyes open, reach forward and touch the operator's index finger with his/her finger. Now spin the subject to the right about 10 times (as before) and then suddenly stop the stool. *Immediately* after stopping the stool, have the subject again, with his/her eyes open, try to reach forward and touch the operator's index finger with his/her finger. Note the degree of ease or difficulty with which this is accomplished.

c. Again, when the nystagmus movements have ceased, repeat Procedure b. This time, however, *immediately* after stopping the chair, have the operator extend an index finger and have the subject open his/her eyes, locate the finger visually, and then *close his/her eyes and try to touch the operator's index finger with his/her eyes closed.* Note the degree of ease or difficulty with which this is accomplished, as well as the direction (right or left) in which the subject missed the operator's finger.

d. Now repeat Procedure a with the subject resting his/her head on the right shoulder. Note any changes that might occur with the nystagmus response.

_____

Formulate a hypothesis as to the differences in results obtained in procedures a through d. _____

_____

To test the functioning of the cochlear portion of the nerve, have your subject sit on a chair with his/her eyes closed and with six members of the class standing in a wide circle around the individual. Now, one at a time, and in a random order, have the students click two coins together and have the subject point in the direction from which the sound originated. Then, with the students still standing in a circle of equal radius around the chair, and with the subject's eyes still closed, hand a ticking watch to one of the individuals in the circle and have him/her *slowly* move the watch closer to the subject until the subject first hears the sound and can correctly indicate from which direction the sound is originating. Repeat this procedure until the distance at which the sound is first heard and the direction correctly identified have been determined for all six directions. Measure the distances between the subject and each encircling student. Are all six distances equal? _____

7. *Glossopharyngeal* (glos-ō-fa-RIN-jē-al) *(IX)* and *vagus (X) nerves*—The palatal (gag) reflex can be used to test the functioning of both of these nerves. Using a cotton-tipped applicator, *very slowly and gently* touch your partner's uvula.
Develop a hypothesis concerning the purpose of this reflex. _____

_____

_____

To test the sensory responses of these nerves, have your partner swallow. Does swallowing occur easily? Now *gently* hold your partner's tongue down with a tongue depressor and ask him/her to say "Ah." Does the uvula move? Are the movements on both sides of the soft palate the same? The sensory function of the glossopharyngeal nerve can be tested by *lightly* applying a cotton-tipped applicator dipped in quinine to the top, sides, and back of the tongue.

In which area of the tongue was the quinine tasted? _____

What taste sensation is located there? _____

_____

8. *Accessory (XI) nerve*—The strength and muscle tone of the sternocleidomastoid and trapezius muscles indicate the proper functioning of the accessory nerve. To ascertain the strength of the sternocleidomastoid muscle, have your partner turn his/her head from side to side against *slight* resistance that you supply by placing your hands on either side of your partner's head. To ascertain the strength of the trapezius muscle, place your hands on your partner's shoulders, and while *gently* pressing down, firmly ask him/her to shrug his/her shoulders.
Do both muscles appear to be reasonably strong? _____

9. *Hypoglossal (XII) nerve*—Have your partner protrude his/her tongue. It should protrude without deviation. Now have your partner protrude his/her tongue and move it from side to side while you attempt *gently* to resist the movements with a tongue depressor.

E. DISSECTION OF SHEEP BRAIN

CAUTION! *Please reread Section D, "Precautions Related to Dissection," at the beginning of the laboratory manual, on page xv, before you begin your dissection.*

The brains of the fetal pig, sheep, and human show many similarities. They possess the protective membranes called the **meninges,** which can easily be seen as you proceed with the dissections. The outermost layer is the **dura mater.** It is the toughest, protective one and may be missing in the preserved sheep brain. The middle membrane is the **arachnoid,** and the inner one, containing blood vessels and adhering closely to the surface of the brain itself is the **pia mater.**

PROCEDURE

1. Use Figures 13.14 through 13.17 as references for this dissection.

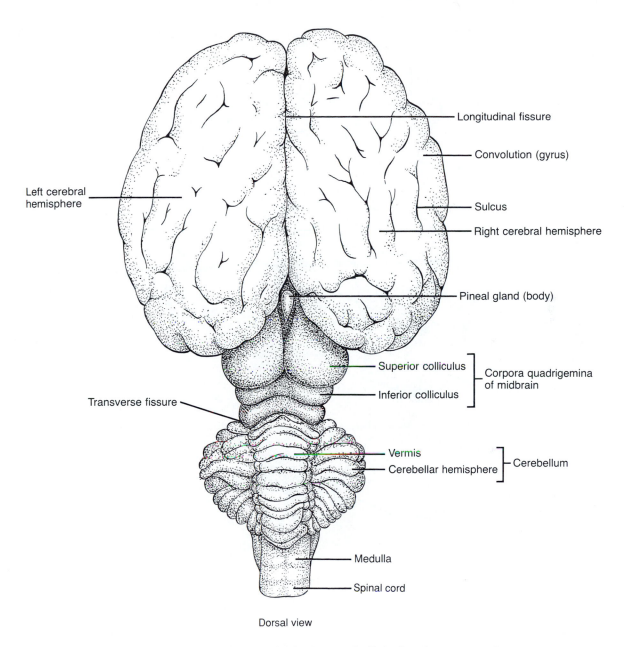

Dorsal view

FIGURE 13.14 Sheep brain in which the cerebellum has been spread apart from the cerebrum.

2. The most prominent external parts of the brain are the pair of large *cerebral hemispheres* and the posterior *cerebellum* on the dorsal surface. These large hemispheres are separated from each other by the *longitudinal fissure*. The *transverse fissure* separates the hemispheres from the cerebellum. The surfaces of these hemispheres form many *gyri,* or raised ridges, which are separated by grooves, or *sulci*.

3. If you spread the hemispheres apart gently, you can see, deep in the longitudinal fissure, thick bundles of white transverse fibers. These bundles form the *corpus callosum,* which connects the hemispheres.

4. Most of the following structures can be identified by examining a midsagittal section of the sheep brain, or by cutting an intact brain along the longitudinal fissure completely through the corpus callosum.

5. If you break through the thin ventral wall, the *septum pellucidum* of the corpus callosum, you can see part of a large chamber, the *lateral ventricle,* inside the hemisphere.

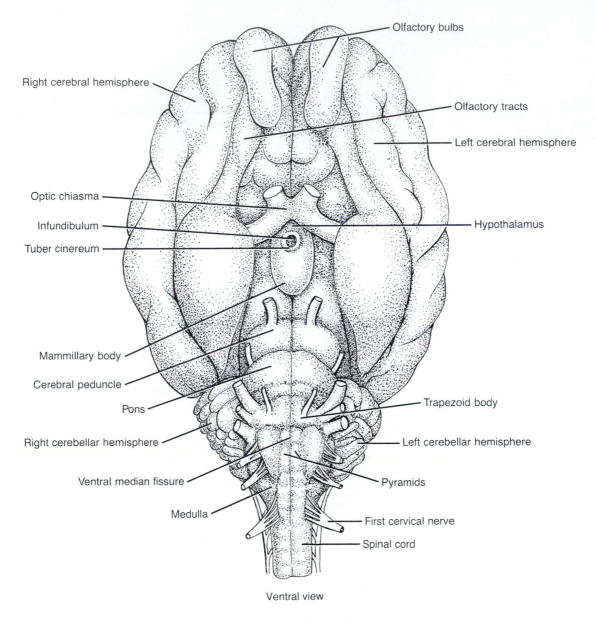

Olfactory bulbs

Right cerebral hemisphere

Olfactory tracts

Left cerebral hemisphere

Optic chiasma

Hypothalamus

Infundibulum

Tuber cinereum

Mammillary body

Cerebral peduncle

Pons

Trapezoid body

Right cerebellar hemisphere

Left cerebellar hemisphere

Ventral median fissure

Pyramids

Medulla

First cervical nerve

Spinal cord

Ventral view

FIGURE 13.15 Sheep brain.

6. Each hemisphere has one of these ventricles. Ventral to the septum pellucidum, locate a smaller band of white fibers called the *fornix*. Close to where the fornix disappears is a small, round bundle of fibers called the *anterior commissure*.

7. The *third ventricle* and the *thalamus* are located ventral to the fornix. The third ventricle is outlined by its shiny epithelial lining, and the thalamus forms the lateral walls of this ventricle. This ventricle is crossed by a large circular mass of tissue, the *intermediate mass,* which connects the two sides of the thalamus. Each lateral ventricle communi-

cates with the third ventricle through an opening, the *interventricular foramen,* which lies in a depression anterior to the intermediate mass and can be located with a dull probe.

8. Spreading the cerebral hemispheres and the cerebellum apart reveals the roof of the midbrain (mesencephalon), which is seen as two pairs of round swellings collectively called the *corpora quadrigemina.* The larger, more anterior pair are the *superior colliculi.* The smaller posterior pair are the *inferior colliculi.* The *pineal gland (body)* is seen directly between the superior colliculi. Just

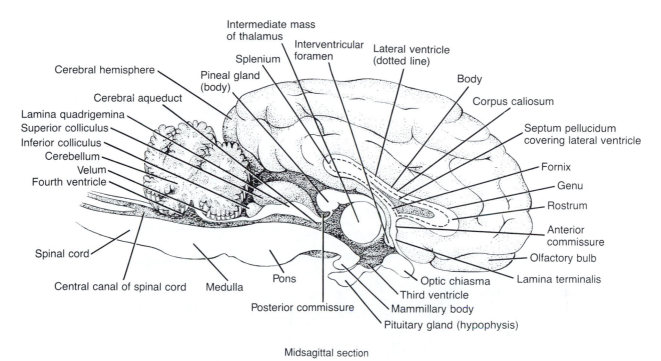

Intermediate mass of thalamus
Splenium
Interventricular foramen
Lateral ventricle (dotted line)
Pineal gland (body)
Cerebral hemisphere
Body
Cerebral aqueduct
Corpus caliosum
Lamina quadrigemina
Superior colliculus
Inferior colliculus
Septum pellucidum covering lateral ventricle
Cerebellum
Velum
Fornix
Fourth ventricle
Genu
Rostrum
Anterior commissure
Spinal cord
Olfactory bulb
Central canal of spinal cord
Medulla
Pons
Optic chiasma
Lamina terminalis
Third ventricle
Posterior commissure
Mammillary body
Pituitary gland (hypophysis)

Midsagittal section

FIGURE 13.16 Sheep brain.

posterior to the inferior colliculi, appearing as a thin white strand, is the *trochlear (IV) nerve.*

9. The cerebellum is connected to the brain stem by three prominent fiber tracts called *peduncles.* The *superior cerebellar peduncle* connects the cerebellum with the midbrain, the *inferior cerebellar peduncle* connects the cerebellum with the medulla, and the *cerebellar peduncle* connects the cerebellum with the pons.

10. Most of the following parts can be located on the ventral surface of the intact brain.

11. Just beneath the cerebral hemispheres are two *olfactory bulbs,* which continue posteriorly as two *olfactory tracts.* Posterior to these tracts, the *optic (II) nerves* undergo a crossing (decussation) known as the *optic chiasma.*

12. Locate the *pituitary gland (hypophysis)* just posterior to the chiasma. This gland is connected to the *hypothalamus* portion of the diencephalon by a stalk called the *infundibulum.* The *mammillary body* appears immediately posterior to the infundibulum.

13. Just posterior to this body are the paired *cerebral peduncles,* from which arise the large *oculomotor (III) nerves.* They may be partially covered by the pituitary gland.

14. The *pons* is a posterior extension of the hypothalamus and the *medulla oblongata* is a posterior extension of the pons.

15. The *cerebral aqueduct* dorsal to the peduncles runs posteriorly and connects the third ventricle with the *fourth ventricle,* which is located dorsal to the medulla and ventral to the cerebellum.

16. The medulla merges with the *spinal cord,* and is separated by the *ventral median fissure.* The *pyramids* are the longitudinal bands of tissue on either side of this fissure.

17. Identify the remaining *cranial nerves* on the ventral surface of the brain. They are the trigeminal (V), abducens (VI), facial (VII), vestibulocochlear (VIII), glossopharyngeal (IX), vagus (X), accessory (XI), and hypoglossal (XII). The previously identified cranial nerves are the olfactory (I), optic (II), oculomotor (III), and trochlear (IV), for a total of 12.

18. A cross section through a cerebral hemisphere reveals *gray matter* near the surface of the *cerebral cortex* and *white matter* beneath this layer.

19. A cross section through the spinal cord reveals the *central canal,* which is connected to the fourth ventricle and contains *cerebrospinal fluid.*

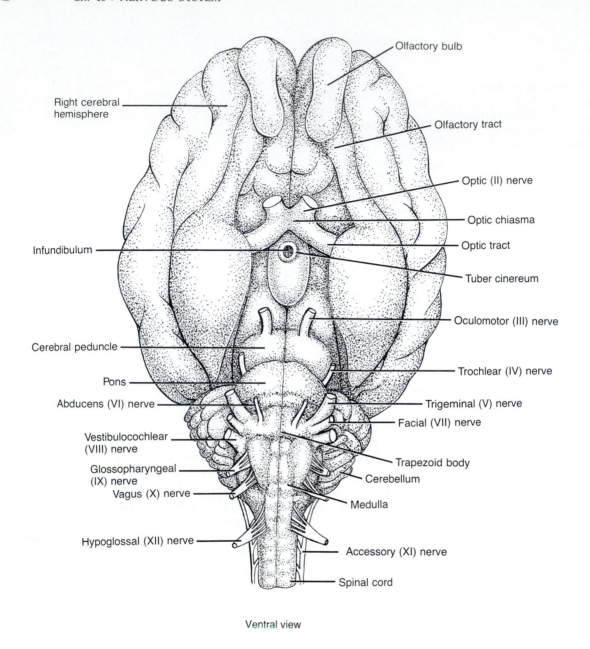

Ventral view

FIGURE 13.17   Cranial nerves of sheep brain.

20. A midsagittal section through the cerebellum reveals a treelike arrangement of gray and white matter called the *arbor vitae* (tree of life).

F. AUTONOMIC NERVOUS SYSTEM

The *autonomic nervous system (ANS)* regulates the activities of smooth muscle, cardiac muscle, and glands, usually involuntarily. In the somatic nervous system (SNS), which is voluntary, the cell bodies of the efferent (motor) neurons are in the CNS, and their axons extend all the way to skeletal muscles in spinal nerves. The ANS always has two efferent neurons in the pathway. The first efferent neuron, the *preganglionic neuron,* has its cell body in the CNS. Its axon leaves the CNS and synapses in an autonomic ganglion with the second neuron called the *postganglionic neuron.* The cell body of the postganglionic neuron is inside an autonomic ganglion, and its axon terminates in a *visceral effector* (muscle or gland).

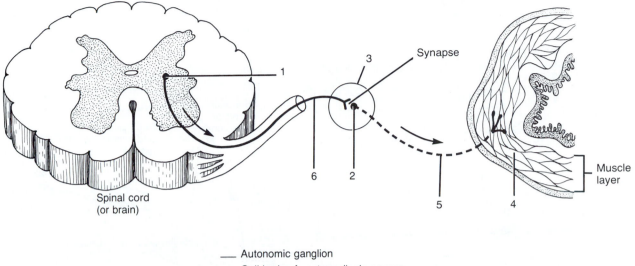

_____ Autonomic ganglion
_____ Cell body of postganglionic neuron
_____ Cell body of preganglionic neuron
_____ Axon of postganglionic neuron
_____ Axon of preganglionic neuron
_____ Visceral effector

FIGURE 13.18 Components of an autonomic pathway.

Label the components of the autonomic pathway shown in Figure 13.18.

The ANS consists of two divisions; sympathetic and parasympathetic (Figure 13.19). Most viscera are innervated by both divisions. In general, nerve impulses from one division stimulate a structure, whereas nerve impulses from the other division decrease its activity

In the *sympathetic division*, the cell bodies of the preganglionic neurons are located in the lateral gray horns of the spinal cord in the thoracic and first two lumbar segments. The axons of preganglionic neurons are myelinated and leave the spinal cord through the ventral (anterior) root of a spinal nerve. Each axon travels briefly in a ventral ramus and then through a small branch called a *white ramus communicans* to enter a sympathetic trunk ganglion. These ganglia lie in a vertical row, on either side of the vertebral column, from the base of the skull to the coccyx. In the ganglion, the axon may synapse with a postganglionic neuron, travel upward or downward through the sympathetic trunk ganglia to synapse with postganglionic neurons at different levels, or pass through the ganglion without synapsing to form part of the splanchnic nerves. If the preganglionic axon synapses in a sympathetic trunk ganglion, it reenters the ventral or dorsal ramus of a spinal nerve via a small branch called a *gray ramus communicans.* If the preganglionic axon forms part of the splanchnic nerves, it passes through the sympathetic trunk ganglion but synapses with a postganglionic neuron in a prevertebral (collateral) ganglion. These ganglia are anterior to the vertebral column close to the large abdominal arteries from which their names are derived (celiac, superior mesenteric, and inferior mesenteric).

In the *parasympathetic division,* the cell bodies of the preganglionic neurons are located in nuclei in the brain stem and lateral gray horn of the second through fourth sacral segments of the spinal cord. The axons emerge as part of cranial or spinal nerves. The preganglionic axons synapse with postganglionic neurons in terminal ganglia, near or within visceral effectors.

The sympathetic division is primarily concerned with processes that expend energy. During stress, the sympathetic division sets into operation a series of reactions collectively called the *fight-or-flight response,* designed to help the body counteract the stress and return to homeostasis. During the fight-or-flight response, the heart and breathing rates increase and the blood sugar level rises, among other things.

The parasympathetic division is primarily concerned with activities that restore and conserve energy. It is thus called the *rest-repose system.*

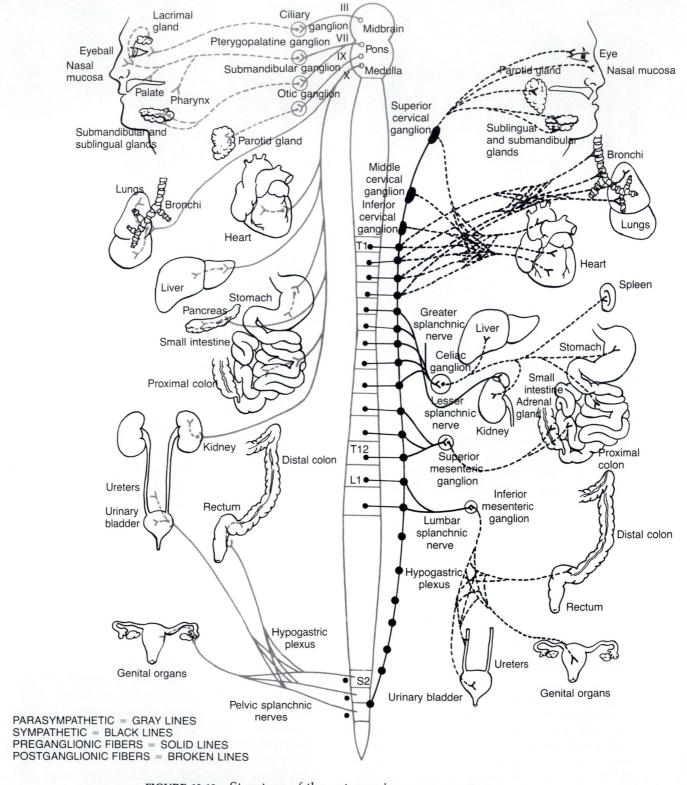

PARASYMPATHETIC = GRAY LINES
SYMPATHETIC = BLACK LINES
PREGANGLIONIC FIBERS = SOLID LINES
POSTGANGLIONIC FIBERS = BROKEN LINES

FIGURE 13.19 Structure of the autonomic nervous system.

Under normal conditions, the parasympathetic division dominates the sympathetic division in order to maintain homeostasis.

Autonomic fibers, like other axons of the nervous system, release neurotransmitters at synapses as well as at points of contact with visceral effectors *(neuroeffector junctions).* On the basis of the neurotransmitter produced, autonomic fibers may be classified as either cholinergic or adrenergic. *Cholinergic* (kō-lin-ER-jik) *fibers* release *acetylcholine (ACh)* and include the following: (1) all sympathetic and parasympathetic preganglionic axons, (2) all parasympathetic postganglionic axons, and (3) some sympathetic postganglionic axons. *Adrenergic* (ad'-ren-ER-jik) *fibers* produce *norepinephrine (NE)* or *epinephrine (adrenalin)* in the modified postganglionic sympathetic fibers of the adrenal medulla. Most sympathetic postganglionic axons are adrenergic.

The actual effects produced by ACh are determined by the type of receptor with which it interacts. The two types of cholinergic receptors are known as nicotinic receptors and muscarinic receptors. *Nicotinic receptors* are found on both sympathetic and parasympathetic postganglionic neurons. These receptors are so named because the action of ACh on them are similar to those produced by nicotine. *Muscarinic receptors* are found on effectors innervated by parasympathetic postganglionic axons. These receptors are so named because the actions of ACh on them are similar to those produced by muscarine, a toxin produced by a mushroom. The effects of NE and epinephrine, like those of ACh, are also determined by the type of receptor with which they interact. Such receptors are found on visceral effectors innervated by most sympathetic postganglionic axons and are referred to as *alpha receptors* and *beta receptors.* In general, alpha receptors are excitatory. Some beta receptors are excitatory, while others are inhibitory. Although cells of most effectors contain either alpha or beta receptors, some effector cells contain both. NE, in general, stimulates alpha receptors to a greater extent than beta receptors, and epinephrine, in general, stimulates both alpha and beta receptors about equally.

Using your textbook as a reference, write in the effects of sympathetic and parasympathetic stimulation for the visceral effectors listed in Table 13.1.

ANSWER THE LABORATORY REPORT QUESTIONS AT THE END OF THE EXERCISE.

TABLE 13.1
Activities of the Autonomic Nervous System

Visceral effector	Effect of sympathetic stimulation	Effect of parasympathetic stimulation
Glands		
Sweat		
Lacrimal (tear)		
Adrenal medulla		
Liver		
Adipose (fat) cells		
Kidney		
Pancreas		
Smooth muscle		
Iris, radial muscle		
Iris, sphincter muscle		
Ciliary muscle of eye		
Salivary glands		
Gastric glands		
Intestinal glands		
Lungs, bronchial muscle		
Heart arterioles		
Skin and mucosa arterioles		
Skeletal muscle arterioles		
Abdominal viscera arterials		
Brain arterioles		
Systemic veins		
Gallbladder and ducts		
Stomach and intestines		
Kidney		
Ureter		
Spleen		
Urinary bladder		
Uterus		

TABLE 13.1 *(Continued)*

Visceral effector	Effect of sympathetic stimulation	Effect of parasympathetic stimulation
Sex organs		
Hair follicles, arrector pili muscle		
Cardiac muscle (heart)		

Nervous System

STUDENT _____   DATE _____

LABORATORY SECTION _____   SCORE/GRADE _____

PART 1. Multiple Choice

_____   1. The tapered, conical portion of the spinal cord is the (a) filum terminale (b) conus medullaris (c) cauda equina (d) lumbar enlargement

_____   2. The outermost meninx composed of dense fibrous connective tissue is the (a) pia mater (b) arachnoid (c) dura mater (d) denticulate

_____   3. The portion of a spinal nerve that contains motor nerve fibers only is the (a) posterior root (b) posterior root ganglion (c) lateral root (d) anterior root

_____   4. The connective tissue covering around individual nerve fibers is the (a) endoneurium (b) epineurium (c) perineurium (d) ectoneurium

_____   5. On the basis of organization, which does *not* belong with the others? (a) pons (b) medulla (c) thalamus (d) midbrain

_____   6. The lateral ventricles are connected to the third ventricle by the (a) interventricular foramen (b) cerebral aqueduct (c) median aperture (d) lateral aperture

_____   7. The vital centers for heartbeat, respiration, and blood vessel diameter regulation are found in the (a) pons (b) cerebrum (c) cerebellum (d) medulla

_____   8. The reflex centers for movements of the head and trunk in response to auditory stimuli are located in the (a) inferior colliculi (b) medial geniculate nucleus (c) superior colliculi (d) ventral posterior nucleus

_____   9. Which thalamic nucleus controls general sensations and taste? (a) medial geniculate (b) ventral posterior (c) ventral lateral (d) ventral anterior

_____   10. Integration of the autonomic nervous system, secretion of regulating factors, control of body temperature, and the regulation of food intake and thirst are functions of the (a) pons (b) thalamus (c) cerebrum (d) hypothalamus

_____   11. The left and right cerebral hemispheres are separated from each other by the (a) central sulcus (b) transverse fissure (c) longitudinal fissure (d) insula

_____   12. Which structure does *not* belong with the others? (a) putamen (b) caudate nucleus (c) insula (d) globus pallidus

_____   13. Which peduncles connect the cerebellum with the midbrain? (a) superior (b) inferior (c) middle (d) lateral

_____   14. Which cranial nerve has the most anterior origin? (a) XI (b) IX (c) VII (d) IV

_____   15. Extensions of the pia mater that suspend the spinal cord and protect against shock are the (a) choroid plexuses (b) pyramids (c) denticulate ligaments (d) superior colliculi

_____ **16.** Which branch of a spinal nerve enters into formation of plexuses? (a) meningeal (b) dorsal (c) rami communicantes (d) ventral

_____ **17.** Which plexus innervates the upper extremities and shoulders? (a) sacral (b) brachial (c) lumbar (d) cervical

_____ **18.** How many pairs of thoracic spinal nerves are there? (a) 1 (b) 5 (c) 7 (d) 12

PART 2. Completion

19. The narrow, shallow groove on the posterior surface of the spinal cord is the

_____.

20. The space between the dura mater and wall of the vertebral canal is called the

_____.

21. In a spinal nerve, the cell bodies of sensory neurons are found in the _____.

22. The outermost connective tissue covering around a spinal nerve is the _____.

23. The middle meninx is referred to as the _____.

24. The nuclei of origin for cranial nerves IX, X, XI, and XII are found in the _____.

25. The portion of the brain containing the cerebral peduncles is the _____.

26. Cranial nerves V, VI, VII, and VIII have their nuclei of origin in the _____.

27. A shallow downfold of the cerebral cortex is called a(n) _____.

28. The _____ separates the frontal lobe of the cerebrum from the parietal lobe.

29. White matter tracts of the cerebellum are called _____.

30. The space between the dura mater and the arachnoid is referred to as the _____.

31. Together, the thalamus and hypothalamus constitute the _____.

32. Cerebrospinal fluid passes from the third ventricle into the fourth ventricle through the

_____.

33. The cerebrum is separated from the cerebellum by the _____ fissure.

34. The branches of a spinal nerve that are components of the autonomic nervous system are known as

_____.

35. The plexus that innervates the buttocks, perineum, and lower extremities is the

_____ plexus.

36. There are _____ pairs of spinal nerves.

37. The part of the brain that coordinates subconscious movements in skeletal muscles is the

_____.

38. The cell bodies of _____ neurons of the ANS are found inside autonomic ganglia.

39. The portion of the ANS concerned with the fight-or-flight response is the _____ division.

40. The autonomic ganglia that are anterior to the vertebral column and close to large abdominal arteries

are called _____ ganglia.

41. The _____ nerve innervates the diaphragm.

42. The _____ tract conveys nerve impulses for touch and pressure.

43. The flexor muscles of the thigh and extensor muscles of the leg are innervated by the _____ nerve.

44. The _____ tract conveys nerve impulses related to muscle tone and posture.

45. The _____ nerve supplies the extensor muscles of the arm and forearm.

46. The _____ area of the cerebral cortex receives sensations from cutaneous, muscular, and visceral receptors in various parts of the body.

47. The portion of the cerebral cortex that translates thoughts into speech is the area.

14

General Senses and Sensory and Motor Pathways

Now that you have studied the nervous system, you will study sensations, which are closely related to and integrated with this system. If you could not "sense" your environment and make the necessary homeostatic adjustments, you could not survive on your own. In its broadest sense, the term *sensation* refers to the conscious or subconscious awareness of external or internal conditions of the body. Sensations are picked up by *sensory receptors* that change mechanical, thermal, or chemical energy into electrical energy termed *receptor potentials.* If the receptor potential reaches threshold value, a nerve impulse will be transmitted to the CNS for decoding.

A. CHARACTERISTICS OF SENSATIONS

Conscious sensations are characterized into three types: superficial sensations, deep sensations, and combined sensations. Touch, temperature, two-point discrimination, and pain are termed *superficial sensations. Deep sensations* include muscle and joint pain, vibratory sense, and proprioception (muscle and joint position or location). *Combined sensations* are involved in a process termed *stereognosis,* which is the process of recognizing objects by the sense of touch while the eyes are closed, and *topognosis,* which is the ability to localize cutaneous sensations.

For a sensation to arise, four events must occur:

1. *Stimulation*—A *stimulus,* or change in the environment, capable of activating certain sensory neurons must be present.
2. *Transduction*—A *sensory unit* (sensory receptor or sense organ) must receive the stimulus and *transduce* (convert) it to a receptor potential. A sensory receptor or sense organ is a specialized type of neuron that is very sensitive to certain types of changes in internal or external conditions. Such changes are termed *stimuli.*
3. *Conduction*—The receptor potential ultimately elicits nerve impulses that are conducted along a neural pathway into the CNS. The sensory neuron that conveys such impulses into the CNS is called the *first-order neuron.*
4. *Translation*—A region of the CNS must translate the nerve impulses into a sensation. Most conscious sensations or perceptions occur in the cortical regions of the brain. In other words, you see, hear, and feel in the brain. You see with your eyes, hear with your ears, and feel pain in an injured part of your body because sensory impulses from each part of the body project to specific regions in the cortex and the cortex interprets the sensation as coming from the stimulated sensory receptors.

A *sensory unit* consists of a single peripheral neuron located in either a posterior spinal ganglion or within a cranial nerve ganglion. The peripheral processes may terminate either as free nerve endings or in association with a sensory receptor. Sensory receptors may be very simple or quite complex, containing highly specialized neurons, epithelium, and connective tissue components. All sensory receptors contain the terminal processes of a neuron, exhibit a high degree of excitability, and possess a specific threshold for a particular type of stimulus, termed a *sensory modality.* In addition, a sensory unit possesses a peripheral *receptive field,* which is an area within which a stimulus of appropriate quality and strength will cause an afferent sensory neuron to initiate an action potential (nerve impulse). The majority of sensory impulses are conducted to the sensory areas of the cerebral cortex, for it is in this region of the brain that a stimulus produces conscious feeling. Different sets of sensory nerve fibers, when activated, will elicit different sensations by virtue of their unique CNS connections. Therefore, a particular sensory nerve fiber will provoke an identical sensation *regardless of how it is excited.* This interpretation of sensory perception is termed *Muller's Doctrine of Specific Energies.*

One characteristic of sensations, that of *projection,* describes the process by which the brain refers sensations to their point of *learned origin* of the stimulation. A second characteristic of many sensations is *adaptation*—that is, a change in sensitivity, usually a decrease, even though a stimulus is still being applied. For example, when you first get into a tub of hot water, you might feel an intense burning sensation. After a brief period of time, however, the sensation decreases to one of comfortable warmth, even though the stimulus (hot water) is still present. Another characteristic is that of *afterimages*—that is, the persistence of a sensation after the stimulus has been removed. One common example of an afterimage occurs when you look at a bright light and then look away. You will still see the light for several seconds afterward. The fourth characteristic of sensations is that of *modality.* Modality is the possession of distinct properties by which one sensation can be distinguished from another. For example, pain, pressure, touch, body position, equilibrium, hearing, vision, smell, and taste are all distinctive because the brain perceives each differently.

B. CLASSIFICATION OF RECEPTORS

No single classification format has proved adequate in describing the types of receptors present within the body, or in describing the method of receptor stimulation or the information conveyed to the CNS by the afferent nervous connections. One convenient method of classifying receptors is by their location:

1. *Exteroceptors* (EKS'-ter-ō-sep'-tors)—Located near the body surface, they provide information about the external environment. They transmit sensations such as hearing, sight, smell, taste, touch, pressure, temperature, and pain.
2. *Visceroceptors* (VIS-er-ō-sep'-tors), or *interoceptors*—Located in blood vessels and viscera, they provide information about the internal environment. These stimuli are perceived as pain, taste, fatigue, hunger, thirst, and nausea. Two specific types of visceroceptors are *baroreceptors (pressoreceptors),* stimulated by changes in the hydrostatic pressure of blood, and *osmoreceptors,* which are stimulated by changes in the solute concentration of extracellular fluid.
3. *Proprioceptors* (PRŌ-prē-ō-sep'-tors), or stretch receptors—Located in muscles, tendons, or joints, they are stimulated by stretching or movement. They provide information about the position and activity of our joints and equilibrium.

Another classification of sensations is based on the form of energy to which the receptors respond at *the lowest stimulus intensity.*

1. *Mechanoreceptors*—Detect mechanical deformation of the receptor itself or in an adjacent tissue or fluid. They provide sensations of touch, pressure, vibration, proprioception (awareness of the location and movement of body parts), hearing, and equilibrium. Mechanoreceptors are also triggered by changes in blood pressure.
2. *Thermoreceptors*—Detect changes in temperature.
3. *Nociceptors* (NŌ-sē-sep'-tors)—detect pain.
4. *Photoreceptors*—Detect light that strikes the retina of the eye.

5. *Chemoreceptors*—Detect taste in the mouth, smell in the nose, and chemicals in body fluids such as water, oxygen, carbon dioxide, hydrogen ions and certain other electrolytes, hormones, and glucose.

Sensations derived from receptors can also be classified into general sensory systems and special sensory systems. *General sensory systems* involve reflex responses through pathways in the spinal cord and the brain stem. Impulses from some of these receptors may even reach the cerebellum. *Special sensory systems* are related to one of the organs of sight, hearing and equilibrium, smell, and taste.

1. *General senses*—Involve simple receptors and neural pathways. Taken together, a, b, and c below are referred to as *cutaneous sensations.*
   a. *Tactile* (TAK-tīl; *tact*=touch) *sensations* (touch, pressure, vibration).
   b. *Thermoreceptive sensations* (hot and cold).
   c. *Pain sensations.*
   d. *Proprioceptive* (prō'-prē-ō-SEP-tiv) *sensations* (awareness of the activities of muscles, tendons, and joints and equilibrium).
2. *Special senses*—Involve complex receptors and neural pathways.
   a. *Olfactory* (ōl-FAK-tō-rē) *sensations* (smell).
   b. *Gustatory* (GUS-ta-tō-rē) *sensations* (taste).
   c. *Visual sensations* (sight).
   d. *Auditory sensations* (hearing).
   e. *Equilibrium* (ē'-kwi-LIB-rē-um) *sensations* (orientation of the body).

C. RECEPTORS FOR GENERAL SENSES

1. Tactile Receptors

Although touch, pressure, and vibration are classified as separate sensations, all are detected by mechanoreceptors.

Touch sensations generally result from stimulation of the tactile receptors in the skin or the in tissues immediately beneath the skin. *Crude touch* refers to the ability to perceive that something has touched the skin, although neither the size nor texture can be determined. *Discriminative touch* refers to the ability to recognize exactly at what point the body is touched. Receptors for touch include the corpuscles of touch, tactile discs, hair

root plexuses, and free (naked) nerve endings. *Corpuscles of touch (Meissner's corpuscles)* are receptors for discriminative touch that are found in dermal papillae. They have already been discussed in Exercise 5. *Tactile discs,* also called *Merkel* (MER-kel) *discs,* consist of disclike formations of dendrites attached to the deeper layers of the epidermis. They also function in discriminative touch. *Hair root plexuses* are dendrites arranged in networks around hair follicles that detect movement when hairs are disturbed. *Free (naked) nerve endings* are found everywhere in the skin and in many other tissues (see Figure 5.1). They are important pain receptors and also respond to objects, such as clothing, in contact with the skin.

Receptors for touch, like other cutaneous receptors, are not randomly distributed; some areas contain many receptors, others contain few. Such a clustering of receptors is called *punctate distribution.* Touch receptors are most numerous in fingertips, the palms of the hands, and the soles of the feet. They are also abundant in the eyelids, tip of the tongue, lips, nipples, clitoris, and tip of the penis.

Examine prepared slides of corpuscles of touch (Meissner's corpuscles), tactile (Merkel) discs, hair root plexuses, and free nerve endings. With the aid of your textbook, draw each of the receptors in the spaces that follow.

Corpuscles of touch (Meissner's corpuscles)

Tactile (Merkel) discs

Hair root plexuses

Free (naked) nerve endings

Pressure sensations generally result from deformation of deeper tissues and are longer lasting and have less variation in intensity than touch sensations. Pressure is really sustained touch. Moreover, whereas touch is felt in a small "pinprick" area, pressure is felt over a much larger area. Receptors for pressure sensations include free nerve endings (discussed in the section on touch sensations), lamellated corpuscles, and Type-II cutaneous mechanoreceptors. *Lamellated (Pacinian) corpuscles* are found in the subcutaneous layer and have already been discussed in Exercise 5. *Type-II cutaneous mechanoreceptors (end organs of Ruffini)* are deeply embedded in the dermis.

Pressure receptors are found in the subcutaneous tissue under the skin, in the deep subcutaneous tissues that lie under mucous membranes, around joints and tendons, in the perimysium of muscles, in the mammary glands, in the external genitals of both sexes, and in some viscera.

Examine prepared slides of lamellated (Pacinian) corpuscles and Type-II cutaneous mechanoreceptors (end organs of Ruffini). With the aid of your textbook, draw the receptors in the spaces that follow.

Lamellated (Pacinian) corpuscles

Type-II cutaneous mechanoreceptors (end organs of Ruffini)

Vibration sensations result from rapidly repetitive sensory signals from tactile receptors. Receptors for vibration include corpuscles of touch (Meissner's corpuscles) that detect low-frequency vibration and lamellated (Pacinian) corpuscles that detect high-frequency vibration.

2. Thermoreceptors

The cutaneous receptors for the sensation of heat and cold are free (naked) nerve endings that are widely distributed in the dermis and subcutaneous tissue. They are also located in the cornea of the eye, tip of the tongue, and external genitals.

3. Pain Receptors

Receptors for *pain,* called *nociceptors* (nō'-sē-SEP-tors; *noci*=harmful), are free (naked) nerve endings (see Figure 5.1). Pain receptors are found in practically every tissue of the body and adapt only slightly or not at all. They can be excited by any type of stimulus. Excessive stimulation of any sense organ causes pain. For example, when stimuli for other sensations such as touch, pressure, heat, and cold reach a certain threshold, they stimulate pain receptors as well. Pain receptors, because of their sensitivity to all stimuli, have a general protective function of informing us of changes that could be potentially dangerous to health or life. Adaptation to pain does not readily occur. This low level of adaptation is important because pain indicates disorder or disease. If we became used to it and ignored it, irreparable damage could result.

4. Proprioceptive Receptors

An awareness of the activities of muscles, tendons, and joints and equilibrium is provided by the *proprioceptive (kinesthetic) sense.* It informs us of the degree to which tendons are tensed and muscles are contracted. The proprioceptive sense enables us to recognize the location and rate of movement of one part of the body in relation to other parts. It also allows us to estimate weight and to determine the muscular work necessary to perform a task. With the proprioceptive sense, we can judge the position and movements of our limbs without using our eyes when we walk, type, play a musical instrument, or dress in the dark.

Proprioceptive receptors are located in skeletal muscles and tendons, in and around joints, and in the internal ear. Proprioceptors adapt only slightly. This slight adaptation is beneficial because the brain must be apprised of the status of different parts of the body at all times so that adjustments can be made to ensure coordination.

Receptors for proprioception follow. The *joint kinesthetic* (kin'-es-THET-ik) *receptors* are located in the articular capsules of joints and ligaments about joints. These receptors provide feedback information on the degree and rate of angulation (change of position) of a joint. *Muscle spindles* consist of the endings of sensory neurons that are wrapped around specialized muscle fibers (cells). They are located in nearly all skeletal muscles and are more numerous in the muscles of the extremities. Muscle spindles provide feedback information on the degree of muscle stretch. This information is relayed to the central nervous system to assist in the coordination and efficiency of muscle contraction. *Tendon organs (Golgi tendon organs)* are located at the junction of skeletal muscle and tendon. They function by sensing the tension applied to a tendon. The degree of tension is related to the degree of muscle contraction and is translated by the central nervous system.

Proprioceptors in the internal ear are the maculae and cristae that function in equilibrium. These are discussed at the end of the exercise.

Examine prepared slides of joint kinesthetic receptors, muscle spindles, and tendon organs (Golgi tendon organs). With the aid of your textbook, draw the receptors in the spaces that follow.

Muscle spindles

Tendon organs (Golgi tendon organs)

D. TESTS FOR GENERAL SENSES

Areas of the body that have few cutaneous receptors are relatively insensitive, whereas those regions that contain large numbers of cutaneous receptors are quite sensitive. This difference can be demonstrated by the *two-point discrimination test* for touch. In the following tests, students can work in pairs, with one acting as subject and the other as experimenter. The subject will keep his/her eyes closed during the experiments.

1. Two-Point Discrimination Test

In this test, the two points of a measuring compass are applied to the skin and the distance in millimeters (mm) between the two points is varied. The subject indicates when he/she feels two points and when he/she feels only one.

PROCEDURE

CAUTION! *Wash the compass in a fresh bleach solution before using it on another subject to prevent the transfer of saliva.*

1. *Very gently,* place the compass on the tip of the tongue, an area where receptors are very densely packed.
2. Narrow the distance between the two points to 1.4 mm. At this distance, the points are able to stimulate two different receptors, and the subject feels that he/she is being touched by two objects.

Joint kinesthetic receptors

3. Decrease the distance to less than 1.4 mm. The subject feels only one point, even though both points are touching the tongue, because the points are so close together that they reach only one receptor.
4. Now *gently* place the compass on the back of the neck, where receptors are relatively few and far between. Here the subject feels two distinctly different points only if the distance between them is 36.2 mm or more.
5. The two-point discrimination test shows that the more sensitive the area, the closer the compass points can be placed and still be felt separately.
6. The following order, from greatest to least sensitivity, has been established from the test: tip of tongue, tip of finger, side of nose, back of hand, and back of neck.
7. Test the tip of finger, side of nose, and back of hand and record your results in Section D.1 of the LABORATORY REPORT RESULTS at the end of the exercise.

2. Identifying Touch Receptors

PROCEDURE

1. Using a water-soluble colored felt marking pen, draw a 1-in. square on the back of the subject's forearm and divide the square into 16 smaller squares.
2. With the subject's eyes closed, press a Von Frey hair or bristle against the skin, just enough to cause the hair to bend, once in each of the 16 squares. The pressure should be applied in the same manner each time.
3. The subject should indicate when he/she experiences the sensation of touch, and the experimenter should make dots in Square 1 at the places corresponding to the points at which the subject feels the sensations.
4. The subject and the experimenter should switch roles and repeat the test.

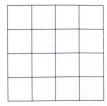

The pair of students working as a team should examine their 1-in. squares after the test is done. They should compare the number of positive and negative responses in each of the 16 small squares, to see how uniformly the touch receptors are distributed throughout the entire 1-in. square. Other general areas used for locating touch receptors are the arm and the back of the hand.

3. Identifying Pressure Receptors

PROCEDURE

1. The experimenter touches the skin of the subject (whose eyes are closed) with the point of a piece of colored chalk.
2. With eyes still closed, the subject then tries to touch the same spot with a piece of differently colored chalk. The distance between the two points is then measured.
3. Proceed using various parts of the body, such as the palm of the hand, arm, forearm, and back the of neck.
4. Record your results in Section D.3 of the LABORATORY REPORT RESULTS at the end of the exercise.

4. Identifying Thermoreceptors

PROCEDURE

1. Draw a 1-in. square on the back of the subject's wrist.
2. Place a forceps or other metal probe in ice-cold water for a minute, dry it quickly, and, with the *dull* point, explore the area in the square for the presence of cold spots.
3. Keep the probe cold and, using ink, mark the position of each spot that you find.
4. Mark each corresponding place in square 2 with the letter *c*.
5. Immerse the forceps in hot water so that it will give a sensation of warmth when it is removed and applied to the skin, but *avoid having it so hot that it causes pain.*
6. Proceeding as before, locate the position of the warm spots in the same area of the skin.
7. Mark these spots with ink of a different color, and then mark each corresponding place in square 2 with the letter *h*.
8. Repeat the entire procedure, using both cold forceps and warm forceps on the back of the hand and the palm of the hand, respectively, and mark squares 3 and 4 as you did Square 2.

9. In order to demonstrate the adaptation of thermoreceptors, fill three 1-liter beakers with 700 ml of (1) ice water, (2) water at room temperature, and (3) water at 45° C. Immerse your left hand in the ice water and your right hand in the water at 45° C for 1 min. Now move your left hand to the beaker with water at room temperature and record the sensation.

_____

Move your right hand to the beaker with the water at room temperature and record the sensation. _____

How do you explain the experienced difference in the temperature of the water in the beaker with the water at room temperature?

_____

_____

5. Identifying Pain Receptors

PROCEDURE

1. Using the same 1-in. square of the forearm previously used for the touch test in Section D.2, perform the following experiment.
2. Apply a piece of absorbent cotton soaked with water to the area of the forearm for 5 min, to soften the skin.
3. Add water to the cotton as needed.
4. Place the blunt end of a probe on the surface of the skin and press enough to produce a sensation of pain. Explore the marked area systematically.
5. Using dots, mark the places in Square 5 that correspond to the points that give the sensation of pain when stimulated.
6. Distinguish between the sensations of pain and touch. Are the areas for touch and pain identical _____ ?
7. At the end of the test, compare your squares as you did in Section D.2.

Perform the following test to demonstrate the phenomenon of **referred pain.** Place your

elbow in a large shallow pan of ice water, and note the progression of sensation you experience. At first, you will feel some discomfort in the region of the elbow. Later, pain sensations will be felt elsewhere.

Where do you feel the referred pain? _____

_____

Label the areas of referred pain indicated in Figure 14.1.

6. Identifying Proprioceptors

PROCEDURE

1. Face a blackboard close enough so that you can easily reach to mark it. Mark a small "X" on the board in front of you and keep the chalk on the X for a moment. Now close your eyes and raise your right hand above your head. With your eyes still closed, mark a dot as near as possible to the X. Repeat the procedure by placing your chalk on the X, closing your eyes, raising your arm above your head, and then marking another dot as close as possible to the X. Repeat the procedure a third time. Record your results by estimating or measuring how far you missed the X for each trial in Section D.6 of the LABORATORY REPORT RESULTS at the end of the exercise.
2. Write the word _physiology_ on the left line that follows. Now, with your _eyes closed_, write the same word immediately to the right. How do the samples of writing compare?

_____ _____

Explain your results. _____

_____

_____

_____

3. The following experiments demonstrate that kinesthetic sensations facilitate the repetition

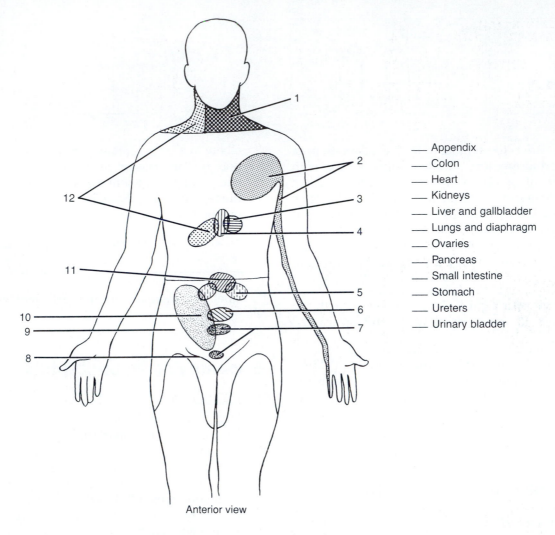

___ Appendix
___ Colon
___ Heart
___ Kidneys
___ Liver and gallbladder
___ Lungs and diaphragm
___ Ovaries
___ Pancreas
___ Small intestine
___ Stomach
___ Ureters
___ Urinary bladder

Anterior view

FIGURE 14.1   Referred pain.

of certain acts involving muscular coordination.

Students work in pairs for these experiments.

a. The experimenter asks the subject to carry out certain movements with his/her eyes closed: for example, to point to the middle finger of the subject's left hand with the index finger of the subject's right hand.

b. With his/her eyes closed, the subject extends the right arm as far as possible behind the body and then brings the index finger quickly to the tip of his/her nose. How accurate is the subject in doing this?

_____

c. Ask the subject, with his/her eyes shut, to touch the named fingers of one hand with the index finger of the other hand.

How well does the subject carry out the directions? _____

_____

_____

_____

E. SENSORY PATHWAYS

Most input from somatic receptors on one side of the body crosses over to the opposite side in the spinal cord or brain stem before ascending to the thalamus. It then projects from the thalamus to the somatosensory cortex (primary somatosensory or general sensory area), where conscious sensations result. Axon collaterals (branches) of somatic sensory neurons also carry signals into the cerebellum and the reticular formation of the brain stem. Two general pathways lead from sensory receptors to the cortex: the posterior column--medial lemniscus pathway and the anterolateral (spinothalamic) pathway.

1. Posterior Column-Medial Lemniscus Pathway

Nerve impulses for conscious proprioception and most tactile sensations ascend to the cortex along a common pathway formed by three-neuron sets. *First-order neurons* extend from sensory receptors into the spinal cord and up to the medulla on the same side of the body. The cell body of a first-order neuron is in the posterior (dorsal) root ganglion of a spinal nerve. Its axon is part of the *posterior column (fasciculus gracilis* and *fasciculus cuneatus)* in the spinal cord. The axon terminals form synapses with *second-order neurons* in the medulla. The cell body of a second-order neuron is located in the nucleus cuneatus (which receives input conducted along axons in the fasciculus cuneatus from the neck, arms, and upper chest) or nucleus gracilis (which receives input conducted along axons in the fasciculus gracilis from the trunk and legs). The axon of the second-order neuron crosses to the opposite side of the medulla and enters the *medial lemniscus,* a projection tract that extends from the medulla to the thalamus. In the thalamus, the axon terminals of second-order neurons synapse with *third-order neurons,* which project their axons to the somatosensory area of the cerebral cortex.

Impulses conducted along the posterior column–medial lemniscus pathway give rise to several highly evolved and refined sensations:

a. *Discriminative touch*—Ability to recognize the exact location of light touch stimuli and to make two-point discriminations.
b. *Stereognosis*—Ability to recognize by "feel" the size, shape, and texture of an object. Examples are identifying (with closed eyes) a paperclip put into your hand or reading Braile.
c. *Proprioception*—Awareness of the precise position of body parts, and *kinesthesia,* the awareness of directions of movement.
d. *Weight discrimination*—Ability to assess the weight of an object.
e. *Vibratory sensations*—Ability to sense rapidly fluctuating touch.

Label the components of the posterior column-medial lemniscus pathway in Figure 14.2.

2. Anterolateral (Spinothalamic) Pathways

The *anterolateral (spinothalamic) pathways* carry mainly pain and temperature impulses. In addition, they relay tickle, itch, and some tactile impulses, which give rise to a very crude, not well-localized touch or pressure sensation. Like the posterior column–medial lemniscus pathway, the anterolateral pathways are also composed of three-neuron sets. The first-order neuron connects a receptor of the neck, trunk, or extremities with the spinal cord. The cell body of the first-order neuron is in the posterior root ganglion. The axon of the first-order neuron synapses with the second-order neuron, which is located in the posterior gray horn of the spinal cord. The axon of the second-order neuron extends to the opposite side of the spinal cord and passes upward to the brain stem in either the *lateral spinothalamic tract* or *anterior spinothalamic tract.* The axon from the second-order neuron terminates in the thalamus, where it synapses with the third-order neuron. The axon of the third-order neuron extends through the *internal capsule* and projects to the somatosensory area of the cerebral cortex. The lateral spinothalamic tract conveys sensory impulses for pain and temperature, whereas the anterior spinothalamic tract conveys tickle, itch, and crude touch and pressure impulses.

Label the components of the lateral spinothalamic tract in Figure 14.3 and the anterior spinothalamic tract in Figure 14.4.

F. SENSORY-MOTOR INTEGRATION

Sensory systems provide the input that keeps the CNS informed of changes in the external and internal environment. Responses to this information are conveyed to motor systems, which enable us to move about, alter glandular secretions, and change our relationship to the world around us. As sensory information reaches the CNS, it becomes part of a large pool of sensory

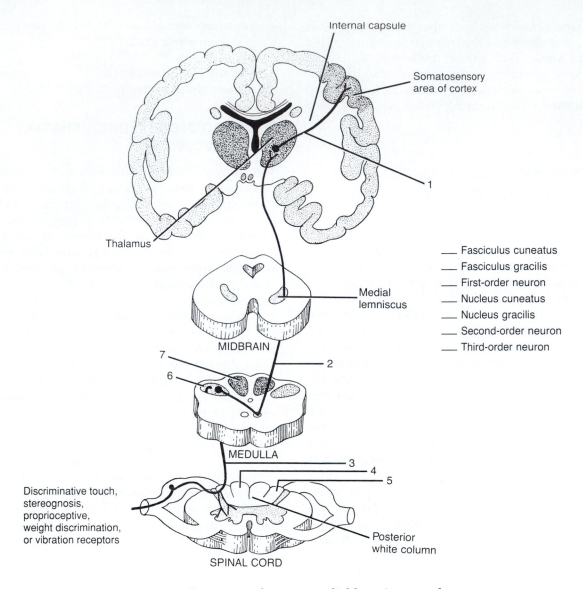

Internal capsule

Somatosensory
area of cortex

1

Thalamus

Medial
lemniscus

___ Fasciculus cuneatus
___ Fasciculus gracilis
___ First-order neuron
___ Nucleus cuneatus
___ Nucleus gracilis
___ Second-order neuron
___ Third-order neuron

MIDBRAIN

7

6

2

MEDULLA

3
4
5

Discriminative touch,
stereognosis,
proprioceptive,
weight discrimination,
or vibration receptors

Posterior
white column

SPINAL CORD

FIGURE 14.2   Posterior column—medial lemniscus pathway.

input. We do not actively respond to every bit of input the CNS receives. Rather, the incoming information is integrated with other information arriving from all other operating sensory receptors. The integration process occurs not just once, but at many stations along the pathways of the CNS and at both conscious and subconscious levels. It occurs within the spinal cord, brain stem, cerebellum, basal ganglia, and cerebral cortex. As a result, a motor response to make a muscle contract or a gland secrete can be modified at any of these levels. The cerebral motor cortex plays the major role in initiating and

controlling precise, discrete muscular movements. The basal ganglia largely integrate semivoluntary, automatic movements like walking, swimming, and laughing. The cerebellum assists the motor cortex and basal ganglia by making body movements smooth and coordinated and by contributing significantly to maintaining normal posture and balance.

After receiving and interpreting sensory information, the CNS generates nerve impulses to direct responses to that sensory input. The nerve impulses are sent down the spinal cord in two major descending motor pathways: the direct

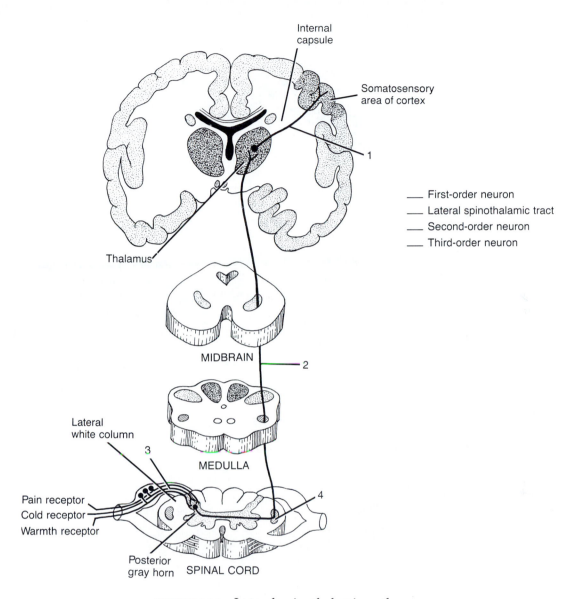

Internal capsule

Somatosensory area of cortex

1

___ First-order neuron
___ Lateral spinothalamic tract
___ Second-order neuron
___ Third-order neuron

Thalamus

MIDBRAIN

2

Lateral white column

3

MEDULLA

Pain receptor
Cold receptor
Warmth receptor

4

Posterior gray horn   SPINAL CORD

FIGURE 14.3   Lateral spinothalamic pathway.

(pyramidal) pathways and the indirect (extrapyramidal) pathways.

1. Direct (Pyramidal) Pathways

Voluntary motor impulses propagate from the motor cortex to somatic efferent neurons (voluntary motor neurons) that innervate skeletal muscles via the *direct* or *pyramidal* (pi-RAM-i-dal) *pathways.* The simplest of these pathways consists of sets of two neurons: upper and lower motor neurons. About one million cell bodies of the direct-pathway *upper motor neurons* are in the cerebral cortex. Their axons descend through the internal capsule of the cerebrum. In the medulla, the axon bundles form the ventral bulges known as the pyramids; this is the reason for the name *pyramidal pathway.* Most of these axons also cross to the opposite side in the medulla. They terminate in nuclei of cranial nerves that innervate voluntary muscles in the head or in the anterior gray horn of the spinal cord. *Lower motor neurons* extend from the cranial nerve motor nuclei or spinal cord anterior horns to the skeletal muscle fibers. Close to their termination point, most upper motor neurons synapse with a short-association neuron, which in turn synapses with a lower motor neuron.

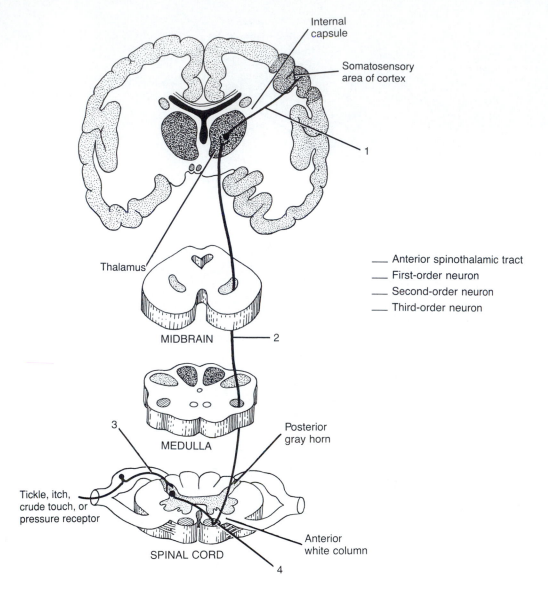

Internal
capsule

Somatosensory
area of cortex

1

Thalamus

___ Anterior spinothalamic tract
___ First-order neuron
___ Second-order neuron
___ Third-order neuron

MIDBRAIN   — 2

MEDULLA

3

Posterior
gray horn

Tickle, itch,
crude touch, or
pressure receptor

Anterior
white column

SPINAL CORD

4

FIGURE 14.4   Anterior spinothalamic pathway.

Some upper motor neurons synapse directly with lower motor neurons.

The direct pathways convey impulses from the cortex that result in precise, voluntary movements and include three tracts:

a. *Lateral corticospinal tracts (pyramidal tracts proper)*—These pathways begin in the right and left motor cortex and descend through the internal capsule of the cerebrum, the cerebral peduncle of the midbrain, and then to the pons on the same side as the point of origin. About 85 to 90% of the upper motor neurons decussate (cross) in the medulla. The axons that have decussated form the lateral

corticospinal tracts in the right and left lateral white columns of the spinal cord. Thus, the motor cortex of the right side of the brain controls muscles on the left side of the body, and vice versa. The lower motor neurons then receive input from synapses with both upper motor neurons and association neurons. Axons of lower motor neurons (somatic motor neurons) exit all levels of the cord via the anterior roots of spinal nerves and terminate in skeletal muscles. Movement controlled by these motor neurons includes precise contraction of muscles in the distal extremities. Label the components of the lateral corticospinal tracts in Figure 14.5.

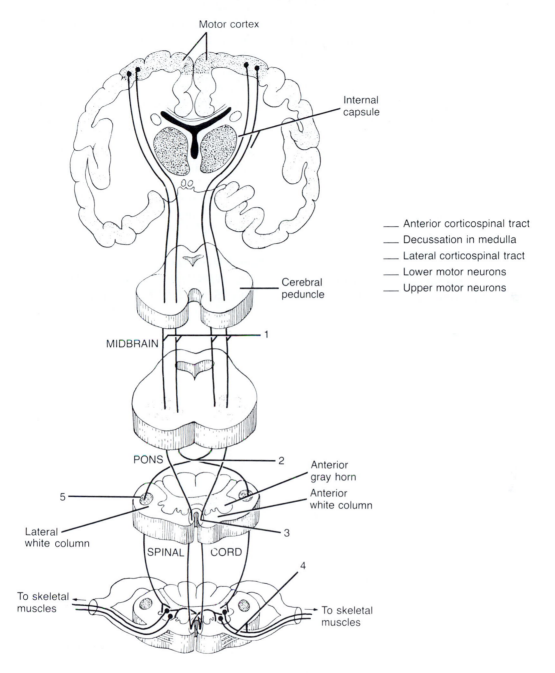

Motor cortex

Internal capsule

___ Anterior corticospinal tract
___ Decussation in medulla
___ Lateral corticospinal tract
___ Lower motor neurons
___ Upper motor neurons

Cerebral peduncle

MIDBRAIN   1

PONS   2

Anterior gray horn

Anterior white column

5

Lateral white column   3

SPINAL CORD

4

To skeletal muscles

To skeletal muscles

FIGURE 14.5   Direct (pyrimidal) pathways: lateral and anterior corticospinal tracts.

b. *Anterior corticospinal tracts*—About 10 to 15 the axons of upper motor neurons do not cross in the medulla. They pass through the medulla, descend on the same side, and form the anterior corticospinal tracts in the right and left anterior white columns. The axons of these upper motor neurons eventually decussate. They cross to the opposite side and synapse with association or lower motor neu- rons in the anterior gray horn of the spinal cord. Axons of these lower motor neurons exit the cervical and upper thoracic segments of the cord via the anterior roots of spinal nerves. They terminate in skeletal muscles that control muscles of the neck and part of the trunk, coordinating movements of the axial skeleton. Label the components of the anterior corticospinal tracts in Figure 14.5.

c. *Corticobulbar tracts*—The axons of these tracts accompany the corticospinal tracts from the motor cortex, through the internal capsule to the brain stem. There they decussate and terminate in the nuclei of nine pairs of cranial nerves in the pons and medulla. These cranial nerves include the oculomotor (III), trochlear (IV), trigeminal (V), abducens (VI), facial (VII), glossopharyngeal (IX), vagus (X), accessory (XI), and hypoglossal (XII). The corticobulbar tracts convey impulses that control voluntary movements of the head and neck.

2. Indirect (Extrapyramidal) Pathways

The *indirect,* or *extrapyramidal, pathways* include all descending (motor) tracts other than the corticospinal and corticobulbar tracts (see p. 235). Nerve impulses conducted along the indirect pathways follow complex circuits that involve the motor cortex, basal ganglia, thalamus, cerebellum, reticular formation, and nuclei in the brain stem. Upper motor neurons of the indirect pathways all begin in various nuclei of the brain stem. Their axons extend into the spinal cord along several tracts and finally synapse with association neurons or lower motor neurons. The lower motor neurons are the same ones activated by the direct pathways. For this reason, lower motor neurons are also called the *final common pathway.*

ANSWER THE LABORATORY REPORT QUESTIONS AT THE END OF THE EXERCISE.

General Senses and Sensory and Motor Pathways

STUDENT _____   DATE _____

LABORATORY SECTION _____   SCORE/GRADE _____

SECTION D. TESTS FOR GENERAL SENSES

1. Two-Point Discrimination Test

Part of body	Least distance at which two points can be detected
Tip of tongue	1.4 mm
Tip of finger	
Side of nose	
Back of hand	
Back of neck	36.2 mm

3. Identifying Pressure Receptors

Part of body	Distance between points touched by chalk
Palm of hand	
Arm	
Forearm	
Back of Neck	

6. Identifying Proprioceptors

First trial   _____

Second trial   _____

Third trial   _____

General Senses and Sensory and Motor Pathways

STUDENT _____   DATE _____

LABORATORY SECTION _____   SCORE/GRADE _____

PART 1. Multiple Choice

_____ 1. The process by which the brain refers sensations to their point of stimulation is referred to as (a) modality (b) projection (c) accommodation (d) convergence

_____ 2. An awareness of the activities of muscles, tendons, and joints is known as (a) referred pain (b) adaptation (c) refraction (d) proprioception

_____ 3. The inability to feel a sensation consciously, even though a stimulus is still being applied, is called (a) modality (b) projection (c) adaptation (d) afterimage formation

_____ 4. Which receptor does *not* belong with the others? (a) muscle spindle (b) tendon organ (Golgi tendon organ) (c) joint kinesthetic receptor (d) lamellated (Pacinian) corpuscle

_____ 5. A characteristic of sensations by which one sensation may be distinguished from another is called (a) modality (b) projection (c) adaptation (d) afterimage formation

_____ 6. Which are *not* cutaneous receptors? (a) tactile (Merkel) discs (b) muscle spindles (c) lamellated (Pacinian) corpuscles (d) corpuscles of touch (Meissner's corpuscles)

_____ 7. Which is *not* a deep sensation? (a) muscle pain (b) vibratory sense (c) two-point discrimination (d) proprioception

_____ 8. The ability to localize cutaneous sensations is called (a) topognosis (b) steneognosis (c) translation (d) transduction

_____ 9. The persistence of a sensation after a stimulus has been removed is called (a) adaptation (b) afterimage (c) translation (d) learned origin

_____ 10. Receptors that provide information about the external environment are called (a) visceroceptors (b) proprioceptors (c) baroreceptors (d) exteroceptors

PART 2. Completion

11. Receptors found in blood vessels and viscera are classified as _____.

12. Tactile sensations include touch, pressure, and _____.

13. Receptors for pressure are free (naked) nerve endings, type II cutaneous mechanoreceptors (end

organs of Ruffini), and _____ corpuscles.

14. Proprioceptive receptors that provide information about the degree and rate of angulations of joints are _____.

15. The _____ tract conveys sensory impulses for pain and temperature.

16. The _____ conveys motor impulses for precise contraction of muscles in the distal extremities.

17. Pain receptors are called _____.

18. The ability to recognize exactly which point of the body is touched is called _____.

19. _____ neurons extend from cranial nerve motor nuclei or spinal cord anterior horns to skeletal muscle fibers.

20. _____ tracts convey impulses that control voluntary movements of the head and neck.

15

Special Senses

In this exercise, we will consider *special senses* that involve complex receptors and neural pathways. These include *olfactory* (ōl-FAK-tō-rē) *sensations* (smell), *gustatory* (GUS-ta-tō-rē) *sensations* (taste), *visual sensations* (sight), *auditory sensations* (hearing), and *equilibrium* (ē'-kwi-LIB-rē-um) *sensations* (orientation of the body).

A. OLFACTORY SENSATIONS

1. Olfactory Receptors

The receptors for the *olfactory* (ol-FAK-tō-rē; *olfactus*=smell) *sense* are found in the nasal epithelium in the superior portion of the nasal cavity on either side of the nasal septum. The nasal epithelium consists of three principal kinds of cells: olfactory receptors, supporting cells, and basal cells. The *olfactory receptors (cells)* are biopolar neurons. Their cell bodies lie between the supporting cells. The distal (free) end of each olfactory cell contains a knob-shaped dendrite from which six to eight cilia, called *olfactory hairs,* protrude. The *supporting (sustentacular) cells* are columnar epithelial cells of the mucous membrane that lines the nose. *Basal cells* lie between the bases of the supporting cells and produce new supporting cells. Within the connective tissue beneath the olfactory epithelium are *olfactory (Bowman's) glands* that secrete mucus.

The unmyelinated axons of the olfactory cells unite to form the *olfactory (I) nerves,* which pass through foramina in the cribriform plate of the ethmoid bone. The olfactory nerves terminate in paired masses of gray matter, the *olfactory bulbs* that lie beneath the frontal lobes of the cerebrum on either side of the crista galli of the ethmoid bone. The first synapse of the olfactory neural pathway occurs in the olfactory bulbs between the axons of the olfactory (I) nerves and the dendrites of neurons inside the olfactory bulbs. Axons of these neurons run posteriorly to form the *olfactory tract.* From here, nerve impulses are conveyed to the olfactory area of the cerebral cortex. In the cortex, the nerve impulses are interpreted as odor and give rise to the sensation of smell.

Adaptation happens quickly, especially adaptation to odors. For this reason, we become accustomed to some odors and are also able to endure unpleasant ones. Rapid adaptation also accounts for the failure of a person to detect gas that accumulates slowly in a room.

Label the structures associated with olfaction in Figure 15.1.

Now examine a slide of the olfactory epithelium under high power. Identify the olfactory receptors and supporting cells and label Figure 15.2.

2. Olfactory Adaptation

PROCEDURE

1. The subject should close his/her eyes after plugging one nostril with cotton.

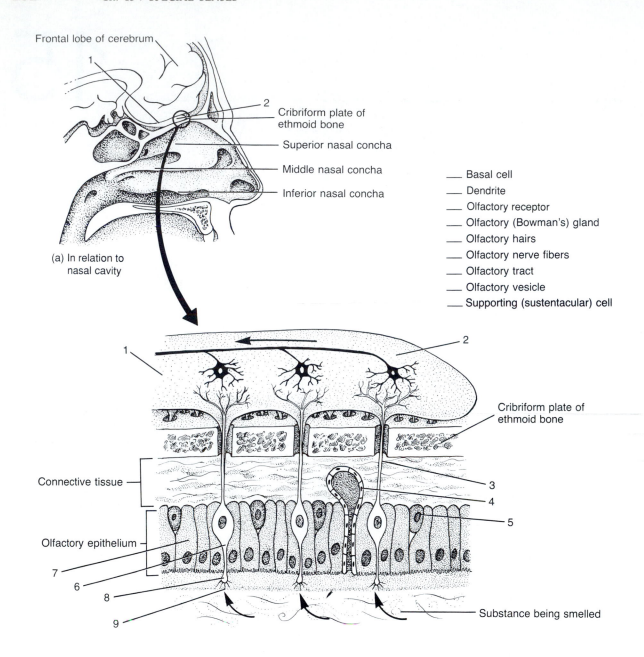

Frontal lobe of cerebrum

1

2 — Cribriform plate of ethmoid bone

— Superior nasal concha

— Middle nasal concha

— Inferior nasal concha

(a) In relation to nasal cavity

___ Basal cell
___ Dendrite
___ Olfactory receptor
___ Olfactory (Bowman's) gland
___ Olfactory hairs
___ Olfactory nerve fibers
___ Olfactory tract
___ Olfactory vesicle
___ Supporting (sustentacular) cell

1

2

Cribriform plate of ethmoid bone

3

4

5

Connective tissue

Olfactory epithelium

7

6

8

9

Substance being smelled

(b) Enlarged aspect

FIGURE 15.1   Olfactory receptors

2. Hold a bottle of oil of cloves, or other substance having a distinct odor, under the open nostril.

3. The subject breathes in through the open nostril and exhales through the mouth. Note the time required for the odor to disappear, and repeat with the other nostril.

4. As soon as olfactory adaptation has occurred, test an entirely different substance.

5. Compare results for the various materials tested.

Olfactory stimuli, such as pepper, onions, ammonia, ether, and chloroform, are irritating and may cause tearing because they stimulate the receptors of the trigeminal (V) nerve as well as the olfactory neurons.

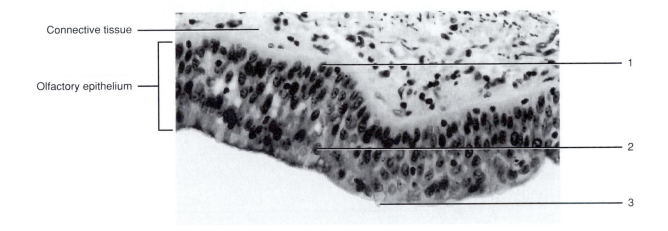

Connective tissue

Olfactory epithelium

1

2

3

___ Basal cell
___ Olfactory receptor
___ Supporting (sustentacular) cell

FIGURE 15.2 Photomicrograph of olfactory epithelium.

B. GUSTATORY SENSATIONS

1. Gustatory Receptors

The receptors for *gustatory* (GUS-ta-tō-rē; *gusto*=taste) *sensations,* or sensations of taste, are located in the taste buds. Although taste buds are most numerous on the tongue, they are also found on the soft palate and in the pharynx. *Taste buds* are oval bodies consisting of three kinds of cells: supporting cells, gustatory receptors, and basal cells. The *supporting (sustentacular) cells* are a specialized epithelium that forms a capsule. Inside each capsule are 4 to 20 *gustatory receptors (cells).* Each gustatory receptor contains a hairlike process *(gustatory hair)* that projects to the surface through an opening in the taste bud called the *taste pore.* Gustatory cells make contact with taste stimuli through the taste pore. *Basal cells* are found at the periphery of the taste bud and produce new supporting cells, which then develop into gustatory receptors. Examine a slide of taste buds and label the structures associated with gustation in Figure 15.3.

Taste buds are located in some connective tissue elevations on the tongue called *papillae* (pa-PILL-ē). They give the upper surface of the tongue its rough texture and appearance. *Circumvallate* (ser-kum-VAL-āt) *papillae* are circu-

lar and form an inverted V-shaped row at the posterior portion of the tongue. *Fungiform* (FUN-ji-form) *papillae* are knoblike elevations found primarily on the tip and sides of the tongue. All circumvallate and most fungiform papillae contain taste buds. *Filiform* (FIL-i-form) *papillae* are threadlike structures that cover the anterior two thirds of the tongue.

Have your partner protrude his/her tongue and examine its surface with a hand lens to identify the shape and position of the papillae.

2. Identifying Taste Zones

For gustatory cells to be stimulated, substances tasted must be in solution in the saliva in order to enter the taste pores in the taste buds. Despite the many substances tasted, there are basically only four taste sensations: sour, salty, bitter, and sweet. Each taste is due to a different response to different chemicals. Some regions of the tongue react more strongly than others to particular taste sensations.

To identify the taste zones for the four taste sensations, perform the following steps and record the results in Section B.2 of the LABORATORY REPORT RESULTS at the end of the exercise by inserting a "+" (taste detected) or a "−" (taste not detected) where appropriate.

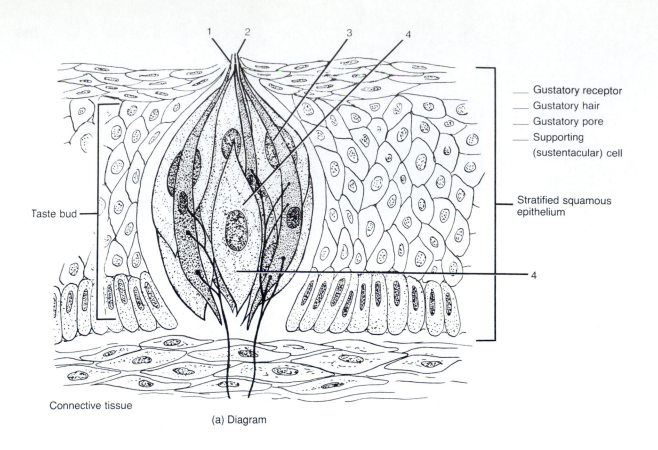

_____ Gustatory receptor
_____ Gustatory hair
_____ Gustatory pore
_____ Supporting
(sustentacular) cell

Stratified squamous
epithelium

Taste bud

4

Connective tissue

(a) Diagram

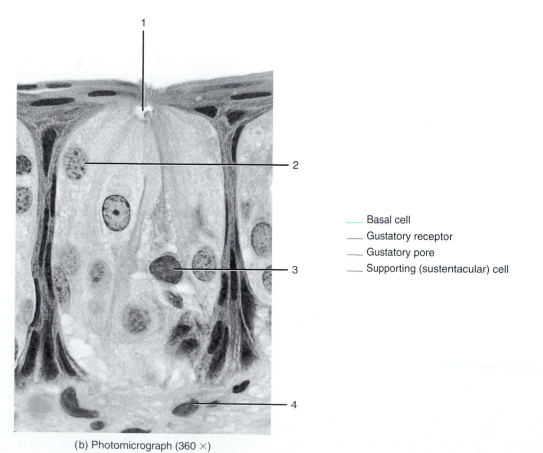

1

2

_____ Basal cell
_____ Gustatory receptor
_____ Gustatory pore
_____ Supporting (sustentacular) cell

3

4

(b) Photomicrograph (360 ×)

FIGURE 15.3 Structure of a taste bud.

PROCEDURE

1. The subject thoroughly dries his/her tongue (use a clean paper towel). The experimenter places some granulated sugar on the tip of the tongue and notes the time. The subject indicates when he/she tastes sugar by raising a hand. The experimenter notes the time again and records how long it takes for the subject to taste the sugar.

2. Repeat the experiment, but this time use a drop of sugar solution. Again record how long it takes for the subject to taste the sugar. How do you explain the difference in time periods?

3. The subject rinses his/her mouth again. The experiment is then repeated using the quinine solution (bitter taste), and then the salt solution.

4. After rinsing yet again, the experiment is repeated using the acetic acid solution or vinegar (sour) placed on the tip and *sides* of the tongue.

3. Taste and Inheritance

Taste for certain substances is inherited, and geneticists for many years have been using the chemical *phenylthiocarbamide (PTC)* to test taste. To some individuals this substance tastes bitter; to others it is sweet; some cannot taste it at all.

PROCEDURE

1. Place a few crystals of PTC or PTC paper on the subject's tongue. Does he/she taste it? If so, have the subject describe the taste.

2. Special paper flavored with this chemical can be chewed, mixed with saliva, and tested in the same manner.

3. Record on the blackboard your response to the PTC test. Usually about 70% of the people tested can taste this compound; 30% cannot. Compare this percentage with the class results.

4. Taste and Smell

This test combines the effect of smell on the sense of taste.

PROCEDURE

1. Obtain small cubes of carrot, onion, potato, and apple.

2. The subject dries the tongue with a clean paper towel, closes his/her eyes, and pinches the nostrils shut. The experimenter places the cubes, one by one, on the subject's tongue.

3. The subject attempts to identify each cube in the following sequences: (1) immediately, (2) after chewing (nostrils closed), and (3) after opening the nostrils.

4. Record your results in Section B.4 of the LABORATORY REPORT RESULTS at the end of the exercise.

C. VISUAL SENSATIONS

Structures related to *vision* are the eyeball (the receptor organ for visual sensations), optic (II) nerve, brain, and accessory structures. The extrinsic muscles of the eyeball can be reviewed in Figure 10.4.

1. Accessory Structures

Among the *accessory structures* are the eyebrows, eyelids, eyelashes, lacrimal (tearing) apparatus, and extrinsic eye muscles. *Eyebrows* protect the eyeball from falling objects, prevent perspiration from getting into the eye, and shade the eye from the direct rays of the sun. *Eyelids (palpebrae)* (PAL-pe-brē) consist primarily of muscle covered externally by skin. The underside of the muscle is lined by a mucous membrane called the *conjunctiva* (kon-junk-TĪ-va) which also covers the surface of the eyeball. Also within eyelids are *tarsal (Meibomian) glands,* modified sebaceous glands whose oily secretion keeps the eyelids from adhering to each other. Infection of these glands produces a *chalazion* (cyst) in the eyelid. Eyelids shade the eye during sleep, protect the eye from light rays and foreign objects, and spread lubricating secretions over the surface of the eyeball. Projecting from the border of each eyelid is a row of short, thick hairs, the *eyelashes.* Sebaceous glands at the base of the hair follicles of the eyelashes, called *sebaceous ciliary glands (glands of Zeis),* pour a lubricating fluid into the follicles. An infection of these glands is called a *sty.*

The *lacrimal* (LAK-ri-mal; *lacrima* = tear) *apparatus* consists of a group of structures that manufacture and drain tears. Each *lacrimal gland* is located at the superior lateral portion of both orbits. Leading from the lacrimal glands are 6 to 12 *excretory lacrimal ducts* that empty tears onto the surface of the conjunctiva of the upper lid. From here, the tears pass medially and enter two small openings called *lacrimal puncta* that appear as two small pores, one in each papilla of the eyelid, at the medial commissure of the eye. The tears then pass into two ducts, the *lacrimal canals,* and are next conveyed into the lacrimal sac. The *lacrimal sac* is the superior expanded portion of the *nasolacrimal duct,* a canal that carries the tears into the inferior meatus of the nasal cavity. Tears clean, lubricate, and moisten the external surface of the eyeball.

Label the parts of the lacrimal apparatus in Figure 15.4.

2. Structure of the Eyeball

The eyeball can be divided into three principal layers: (1) fibrous tunic, (2) vascular tunic, and (3) retina (nervous tunic). See Figure 15.5.

a. FIBROUS TUNIC

The *fibrous tunic* is the outer coat. It is divided into the posterior sclera and the anterior cornea. The *sclera* (SKLE-ra), called the white of the eye, is a white coat of dense connective tissue that covers all the eyeball except the most anterior

portion (cornea). The sclera gives shape to the eyeball and protects its inner parts. The anterior portion of the fibrous tunic is known as the *cornea* (KOR-nē-a). This nonvascular, transparent fibrous coat covers the colored iris. The outer surface of the cornea contains epithelium that is continuous with the epithelium of the bulbar conjunctiva. At the junction of the sclera and cornea is the *scleral venous sinus (canal of Schlemm).*

b. VASCULAR TUNIC

The *vascular tunic* is the middle layer of the eyeball and consists of three portions: posterior choroid, anterior ciliary body, and iris. The *choroid* (KŌ-royd) is a thin, dark brown membrane that lines most of the internal surface of the sclera and contains blood vessels and melanin. The choroid absorbs light rays so they are not reflected back out of the eyeball; it also maintains the nutrition of the retina. The anterior portion of the choroid is the *ciliary* (SIL-ē-ar'-ē) *body,* the thickest portion of the vascular tunic. It extends from the *ora serrata* (Ō-ra ser-RĀ-ta) of the retina (inner tunic) to a point just behind the sclerocor-

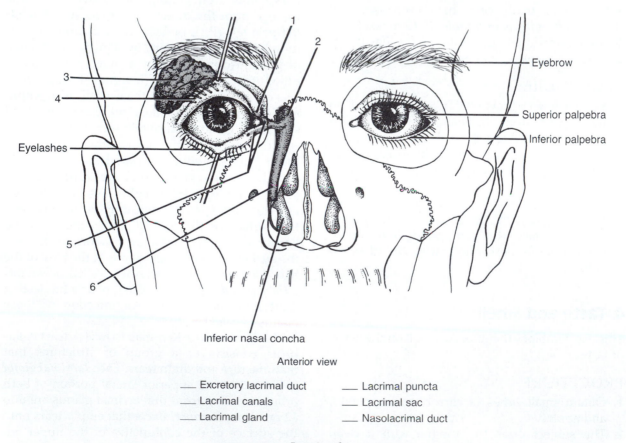

Anterior view

___ Excretory lacrimal duct   ___ Lacrimal puncta
___ Lacrimal canals   ___ Lacrimal sac
___ Lacrimal gland   ___ Nasolacrimal duct

FIGURE 15.4 Lacrimal apparatus.

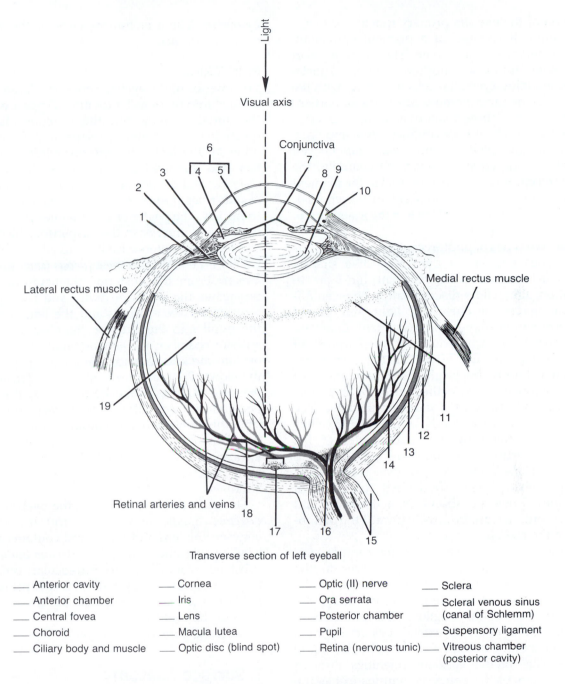

Light

Visual axis

Conjunctiva

6

4 5

7

8 9

3

10

2

1

Medial rectus muscle

Lateral rectus muscle

19

11

12

13

14

15

16

17

18

Retinal arteries and veins

Transverse section of left eyeball

___ Anterior cavity   ___ Cornea   ___ Optic (II) nerve   ___ Sclera

___ Anterior chamber   ___ Iris   ___ Ora serrata   ___ Scleral venous sinus
                                                                                                (canal of Schlemm)

___ Central fovea   ___ Lens   ___ Posterior chamber   ___ Suspensory ligament

___ Choroid   ___ Macula lutea   ___ Pupil   ___ Vitreous chamber
                                                                                                (posterior cavity)

___ Ciliary body and muscle   ___ Optic disc (blind spot)   ___ Retina (nervous tunic)

FIGURE 15.5 Parts of the eyeball.

neal junction. The *ora serrata* is the jagged margin of the retina. The ciliary body consists of the *ciliary processes* (folds of the ciliary body that secrete aqueous humor) and the *ciliary muscle* (a smooth muscle that alters the shape of the lens for near or far vision). The *iris* (*irid*=colored circle), the third portion of the vascular tunic, is the colored portion of the eyeball and consists of circular and radial smooth-muscle fibers arranged

to form a doughnut-shaped structure. The black hole in the center of the iris is the *pupil,* through which light enters the eyeball. One function of the iris is to regulate the amount of light entering the eyeball.

C. RETINA
The third and inner coat of the eye, the *retina (nervous tunic),* is found only in the posterior

portion of the eye. Its primary function is image formation. It consists of a pigment epithelium and a nervous tissue layer. The outer *pigment epithelium* (nonvisual portion) consists of melanin-containing epithelial cells in contact with the choroid. The inner *nervous layer* (visual portion) is composed of three zones of neurons. Named in the order in which they conduct nerve impulses, these are the *photoreceptor layer, bipolar cell layer,* and *ganglion cell layer.* Structurally, the photoreceptor layer is just internal to the pigment epithelium, which lies adjacent to the choroid. The zone of ganglion cell layer is the innermost of the neuronal layers.

The two types of photoreceptors are called rods and cones because of their respective shapes. *Rods* are specialized for vision in dim light. In addition, they allow discrimination between different shades of dark and light and permit discernment of shapes and movement. *Cones* are specialized for color vision and for sharpness of vision—that is, for *visual acuity.* Cones are stimulated only by bright light and are most densely concentrated in the *central fovea,* a small depression in the center of the macula lutea. The *macula lutea* (MAK-yoo-la LOO-tē-a), or yellow spot, is situated in the exact center of the posterior portion of the retina and corresponds to the visual axis of the eye. The fovea is the area of sharpest vision because of the high concentration of cones. Rods are absent from the fovea and macula but increase in density toward the periphery of the retina.

When light stimulates photoreceptors, impulses are conducted across synapses to the bipolar neurons in the intermediate zone of the nervous layer of the retina. From there, the impulses pass to the ganglion cell layer. Axons of the ganglion neurons extend posteriorly to a small area of the retina called the *optic disc (blind spot).* This region contains openings through which fibers of the ganglion neurons exit as the *optic (II) nerve.* Because this area contains neither rods nor cones, and only nerve fibers, no image is formed on it. For this reason it is called the blind spot.

d. LENS

The eyeball itself also contains the lens, just behind the pupil and iris. The *lens* is constructed of numerous layers of protein fibers arranged like the layers of an onion. Normally, the lens is perfectly transparent and is enclosed by a clear capsule and held in position by the *suspensory ligament.* A loss of transparency of the lens is called a *cataract.*

e. INTERIOR

The interior of the eyeball contains a large cavity divided into two smaller cavities. These are called the anterior cavity and the vitreous chamber (posterior cavity) and are separated from each other by the lens. The *anterior cavity,* in turn, has two subdivisions known as the anterior chamber and the posterior chamber. The *anterior chamber* lies posterior to the cornea and anterior to the iris. The *posterior chamber* lies posterior to the iris and anterior to the suspensory ligaments and lens. The anterior cavity is filled with a clear, watery fluid known as the *aqueous* (aqua = water) *humor.* From the posterior chamber, the fluid permeates the posterior cavity and then passes forward between the iris and the lens, through the pupil into the anterior chamber. From the anterior chamber, the aqueous humor is drained off into the scleral venous sinus and passes into the blood. Pressure in the eye, called *intraocular pressure (IOP),* is produced mainly by the aqueous humor. Intraocular pressure keeps the retina smoothly applied to the choroid so that the retina can form clear images. Abnormal elevation of intraocular pressure, called *glaucoma* (glaw-KŌ-ma), results in degeneration of the retina and blindness.

The second, larger cavity of the eyeball is the *vitreous chamber (posterior cavity).* It is located between the lens and retina and contains a soft, jellylike substance called the *vitreous body.* This substance contributes to intraocular pressure, helps to prevent the eyeball from collapsing, and holds the retina flush against the internal portions of the eyeball.

Label the parts of the eyeball in Figure 15.5.

3. Surface Anatomy

Refer to Figure 15.6 for a summary of several surface anatomy features of the eyeball and accessory structures of the eye.

4. Dissection of Vertebrate Eye (Beef or Sheep)

CAUTION! *Please reread Section D, ''Precautions Related to Dissection,'' at the beginning of the laboratory manual, on page xv, before you begin your dissection.*

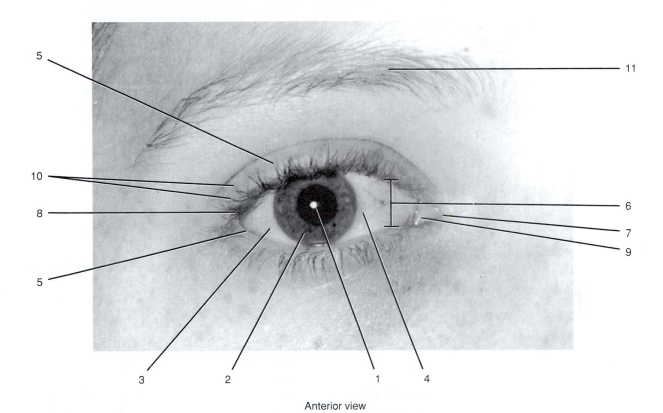

Anterior view

1. **Pupil**. Opening of center of iris of eyeball for light transmission.
2. **Iris**. Circular pigmented muscular membrane behind cornea.
3. **Sclera**. "White" of eye, a coat of fibrous tissue that covers entire eyeball except for cornea.
4. **Conjunctiva**. Membrane that covers exposed surface of eyeball and lines eyelids.
5. **Palpebrae (eyelids)**. Folds of skin and muscle lined by conjunctiva.
6. **Palpebral fissure**. Space between eyelids when they are open.
7. **Medial commissure**. Site of union of upper and lower eyelids near nose.
8. **Lateral commisure**. Site of union of upper and lower eyelids away from nose.
9. **Lacrimal caruncle**. Fleshy, yellowish projection of medial commissure that contains modified sweat and sebaceous glands.
10. **Eyelashes**. Hairs on margins of eyelids, usually arranged in two or three rows.
11. **Eyebrows**. Several rows of hair superior to upper eyelids.

FIGURE 15.6   Surface anatomy of the eyeball and accessory structures.

a. EXTERNAL EXAMINATION

PROCEDURE

1. Note any *fat* on the surface of the eyeball that protects the eyeball from shock in the orbit. Remove the fat.
2. Locate the *sclera,* the tough external white coat, and the *conjunctiva,* a delicate membrane that covers the anterior surface of the eyeball and is attached near the edge of the cornea. The *cornea* is the anterior, transparent portion of the sclera. It is probably opaque in your specimen due to the preservative.
3. Locate the *optic (II) nerve,* a solid, white cord of nerve fibers on the posterior surface of the eyeball.

4. If possible, identify six *extrinsic eye muscles* that appear as flat bands near the posterior part of the eyeball.

b. INTERNAL EXAMINATION

PROCEDURE

1. With a sharp scalpel, make an incision about 0.6 cm (¼ in.), lateral to the cornea (Figure 15.7).
2. Insert scissors into the incision and carefully and slowly cut all the way around the corneal region. The eyeball contains fluid, so take care that it does not squirt out when you make your first incision. Examine the inside of the anterior part of the eyeball.

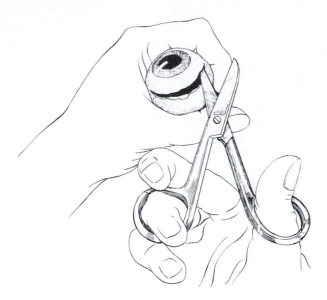

FIGURE 15.7   Procedure for dissecting a vertebrate eye.

3. The *lens* can be seen held in position by *suspensory ligaments,* which are delicate fibers. Around the outer margin of the lens, with a pleated appearance, is the black *ciliary body,* which also functions to hold the lens in place. Free the lens and notice how hard it is.
4. The *iris* can be seen just anterior to the lens and is also heavily pigmented or black.
5. The *pupil* is the circular opening in the center of the iris.
6. Examine the inside of the posterior part of the eyeball, identifying the thick *vitreous humor* that fills the space between the lens and retina.
7. The *retina* is the white inner coat beneath the choroid coat and is easily separated from it.
8. The *choroid coat* is a dark, iridescent-colored tissue that gets its iridescence from a special structure called the *tapetum lucidum.* The tapetum lucidum, which is not present in the human eye, functions to reflect some light back onto the retina.
9. Finally, identify the *blind spot,* the point at which the retina is attached to the back of the eyeball.

5. Testing for Visual Acuity

The acuteness of vision can be tested by means of a *Snellen chart.* It consists of letters of different sizes that are read at a distance normally designated at 20 ft. If the subject reads to the line that is marked "50," he/she is said to possess 20/50 vision in that eye, meaning that he/she is reading at 20 ft what a person who has normal vision can read at 50 ft. If he/she reads to the line marked "20," he/she has 20/20 vision in that eye. The normal eye can sufficiently refract light rays from an object 20 ft away to focus a clear object on the retina. Therefore, if you have 20/20 vision, your eyes are perfectly normal. The higher the bottom number, the larger the letter must be for you to see it clearly, and of course the worse, or weaker, your eyes are.

PROCEDURE

1. Have the subject stand 20 ft from the Snellen chart and cover the right eye with a 3-in.-by-5-in. card.
2. Instruct the subject to read slowly down the chart until he/she can no longer focus the letters.
3. Record the number of the last line (20/20, 20/30, or whichever) that can be successfully read.
4. Repeat this procedure covering the left eye.
5. Now the subject should read the chart using both eyes.
6. Record your results in Section C.5 of the LABORATORY REPORT RESULTS at the end of the exercise and change places.

6. Testing for Astigmatism

The eye, with normal ability to refract light, is referred to as an *emmetropic* (em'-e-TROP-ik) eye. It can refract light rays sufficiently well from an object 20 ft away to focus a clear object on the retina. If the lens is normal, objects as far away as the horizon and as close as about 20 ft will form images on the sensitive part of the retina. When objects are closer than 20 ft, however, the lens has to sharpen its focus by using the ciliary muscles. Many individuals, however, have abnormalities related to improper refraction. Among these are *myopia* (mī-Ō-pē-a) (nearsightedness), *hypermetropia* (hī'-per-mē-TRŌ-pē-a) (farsightedness), and *astigmatism* (a-STIG-ma-tizm) (irregularities in the surface of the lens or cornea).

Why do you think nearsightedness can be corrected with glasses containing biconcave lenses? _____

_____

_____

How would you correct farsightedness? _____

_____

_____

PROCEDURE

1. To determine the presence of astigmatism, remove any corrective lenses if you are wearing them and look at the center of the following astigmatism test chart, first with one eye, then the other:

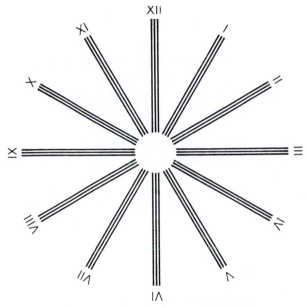

2. If all the radiating lines appear equally sharp and equally black, there is no astigmatism.
3. If some of the lines are blurred or less dark than others, astigmatism is present.
4. If you wear corrective lenses, try the test with them on.

7. Testing for the Blind Spot

PROCEDURE

1. Hold this page about 20 in. from your face with the cross directly in front of your right eye. You should be able to see the cross and the circle when you close your left eye.
2. Now, keeping the left eye closed, slowly bring the page closer to your face while fixing the right eye on the cross.

3. At a certain distance the circle will disappear from your field of vision because its image falls on the blind spot.

8. Image Formation

Formation of an image on the retina requires three basic processes, all concerned with focusing light rays. These are (1) refraction of light rays by the cornea and lens, (2) accommodation of the lens, and (3) constriction of the pupil.

When light rays traveling through a transparent medium (such as air) pass into a second transparent medium with a different density (such as water), the rays bend at the surface of the two media. This is called *refraction*. As light rays enter the eye, they are refracted at the anterior and posterior surfaces of the cornea. Both surfaces of the lens of the eye further refract the light rays so that they come into exact focus on the retina.

The lens of the eye has the unique ability to change the focusing power of the eye by becoming moderately curved at one moment and greatly curved the next. When the eye is focusing on a close object, the lens curves greatly in order to bend the rays toward the central fovea of the eye. This increase in the curvature of the lens is called *accommodation.*

a. TESTING FOR NEAR-POINT ACCOMMODATION

The following test determines your *near-point accommodation:*

PROCEDURE

1. Using any card that has a letter printed on it, close one eye and focus on the letter. (*Note:* Use a letter that is the size of typical newsprint.)
2. Measure the distance of the card from the eye using a ruler or a meter stick.
3. Now *slowly* bring the card as close to your open eye as possible. Stop when you no longer see a clear, detailed letter.
4. Measure and record this distance. This value is your near-point accommodation.
5. Repeat this procedure three times and then test your other eye.
6. Check Table 15.1 to see whether the near point for your eyes corresponds with that recorded for your age group.

b. TESTING FOR CONSTRICTION OF THE PUPIL

PROCEDURE

1. Place a 3-in.-by-5-in. card on the side of the nose so that a light shining on one side of

Table 15.1
Correlation of Age and Near-point Accommodation

Age	Inches	Centimeters
10	2.95	7.5
20	3.54	9.0
30	4.53	11.5
40	6.77	17.2
50	20.67	52.5
60	32.80	83.3

the face will not affect the eye on the other side.

2. Shine the light from a lamp or a flashlight on one eye, 6 in. away (approximately 15 cm), for about 5 sec. Note the change in the size of the pupil of this eye.
3. Remove the light, wait about 3 min, and repeat, but this time observe the pupil of the opposite eye.
4. Wait a few minutes, and repeat the test, observing the pupils of both eyes.

9. TESTING FOR CONVERGENCE

In humans, both eyes focus on only one set of objects—a characteristic called *single binocular vision.* The term *convergence* refers to a medial movement of the two eyeballs so that they are both directed toward the object being viewed. The nearer the object, the greater the degree of convergence necessary to maintain single binocular vision.

PROCEDURE

1. Hold a pencil or pen about 2 ft from your nose and focus on its point. Now, slowly bring the pencil toward your nose.
2. At some moment you should suddenly see two pencil points, or a blurring of the point.
3. Observe your partner's eyes when he/she does this test.

Images are actually focused upside down on the retina. They also undergo mirror reversal. That is, light reflected from the right side of an object hits the left side of the retina and vice versa. Reflected light from the top of the object crosses light from the bottom of the object and strikes the retina below the central fovea. Reflected light from the bottom of the object crosses

light from the top of the object and strikes the retina above the central fovea.

The reason we do not see a topsy-turvy world is that the brain learns early in life to coordinate visual images with the exact location of objects. The brain stores memories of reaching and touching objects and automatically turns visual images right-side up and right-side around.

10. Testing for Binocular Vision, Depth Perception, Diplopia, and Dominance

Humans are endowed with binocular vision. Each eye sees a different view, although there is a large degree of overlap. The cerebral cortex uses these discrepancies to produce *stereopsis,* or three-dimensional vision, an orientation of the object in space that allows us to perceive its distance from us *(depth perception).*

Even though there are slight differences in the views from each eye, we see a single image because the cerebral cortex integrates the images from each eye into a single perception. If this integration is not present, *diplopia* (dip-LŌ-pē-a), or double vision, results. We do not perceive an "average" or "mean" of both left and right views; rather the view from one of our eyes is *dominant*—i.e., it is the view we always perceive with both eyes open.

PROCEDURE

1. Place a book at arm's length on a table. Place your left hand at your side and touch the nearest corner of the book with your right index finger. Close the left eye and repeat; close the right eye and repeat. Note the accuracy of your attempts.

   How accurate was your attempt to touch a

   point with one eye closed? _____

   If the right eye was closed, did you have error

   to the left or the right? _____
2. Use a depth-perception tester with both eyes open and with each eye closed. Attempt to align the two arrows and measure monocular perception errors using the scale on the base.

   Length of depth perception error _____
3. Focus on a discrete object in the distance and press *very gently* on your left eyelid. Note how

   many objects you see: _____

4. Stab a pencil through a piece of paper and remove the scrap. Hold the paper at arm's length and focus on a small discrete object which just fills the hole. (You may want to use an "X" on the blackboard.) *Without moving the paper*, close your left eye and note whether the object is still present. Repeat with your right

eye closed. Note your dominant eye: _____

11. Testing for Afterimages

The rods and cones of the retina are photoreceptors—that is, they contain light-sensitive pigments that absorb light of different wavelengths. *Rhodopsin*, the light-sensitive pigment in rods consists of opsin and retinal. When light strikes a molecule of rhodopsin, the retinal changes from a curved to a straight shape and breaks away from the opsin. The energy released as a result of the exchange initiates the nerve impulse that causes us to perceive light.

In bright light, all the rhodopsin is decomposed, so that the rods are nonfunctional. Cones function in bright light, in much the same way as rods, but they absorb light of specific wavelengths (red, blue, or green). The particular color perceived depends on which combinations of cones are stimulated. If all three are stimulated, we see white; if none is stimulated, we see black.

If photoreceptors are stimulated by staring at a bright object for a long period of time, they will continue firing briefly after the stimulus has been removed *(positive afterimage)*. However, after prolonged firing, these photoreceptors become "bleached," or fatigued, so that they can no longer fire, resulting in a reversal of the image *(negative afterimage)*.

PROCEDURE

1. Stare at a light source, but not a bright light, for 30 sec. Close the eyes briefly and note the positive afterimage. Open the eyes and focus on a piece of white paper. Note the negative afterimage.
2. Draw a red cross on paper about 1 in. high, with lines ½ in. thick. Stare at the cross without moving your eyes for at least 30 sec. Close your eyes very briefly; open them and stare at a piece of white paper. Note the positive and negative afterimages. Record your observations in Section C.11 of the LABORATORY REPORT RESULTS at the end of the exercise.

12. Testing for Color Blindness

Color blindness is an inherited disorder in which certain cones are absent. It is a sex-linked disorder. About 0.5% of all females and about 8% of all males are affected.

PROCEDURE

1. View Ishihara plates in bright light and compare your responses to those in the plate book OR use Holmgren's test for matching colored threads.

2. Note the accuracy of your responses: _____

13. Visual Pathway

From the rods and cones, impulses are transmitted through bipolar cells to ganglion cells. The cell bodies of the ganglion cells lie in the retina and their axons leave the eye via the *optic (II) nerve*. The axons pass through the *optic chiasma* (kī-AZ-ma), a crossing point of the optic nerves. Fibers from the medial retina cross to the opposite side. Fibers from the lateral retina remain uncrossed. Upon passing through the optic chiasma, the fibers, now part of the *optic tract,* enter the brain and terminate in the lateral geniculate nucleus of the thalamus. Here the fibers synapse with the neurons whose axons pass to the visual centers located in the occipital lobes of the cerebral cortex. Label the visual pathway in Figure 15.8.

D. AUDITORY SENSATIONS AND EQUILIBRIUM

In addition to containing receptors for sound waves, the *ear* also contains receptors for equilibrium. The ear is subdivided into three principal regions: (1) external, or outer, ear, (2) middle ear, and (3) internal, or inner, ear.

1. Structure of Ear

a. EXTERNAL (OUTER) EAR

The *external (outer) ear* collects sound waves and directs them inward. Its structure consists of the auricle, external auditory canal, and tympanic membrane. The *auricle (pinna)* is a trumpet-shaped flap of elastic cartilage covered by thick

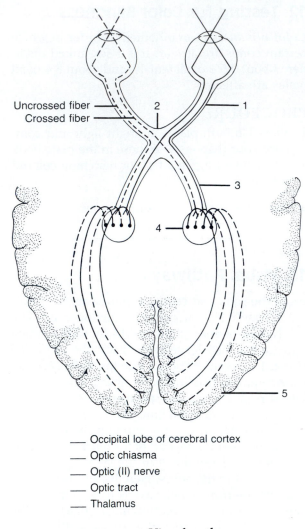

Uncrossed fiber
Crossed fiber

___ Occipital lobe of cerebral cortex
___ Optic chiasma
___ Optic (II) nerve
___ Optic tract
___ Thalamus

FIGURE 15.8 Visual pathway.

skin. The rim of the auricle is called the *helix,* and the inferior portion is referred to as the *lobule.* The auricle is attached to the head by ligaments and muscles. The *external auditory canal (meatus)* is a tube, about 2.5 cm (1 in.) in length, that leads from the auricle to the eardrum. The walls of the canal consist of bone lined with cartilage that is continuous with the cartilage of the auricle. Near the exterior opening, the canal contains a few hairs and specialized sebaceous glands called *ceruminous* (se-ROO-me-nus) *glands,* which secrete *cerumen* (earwax). The combination of hairs and cerumen prevents foreign objects from entering the ear. The *tympanic* (tim-PAN-ik) *membrane (eardrum)* is a thin, semitransparent partition of fibrous connective tissue located between the external auditory canal and the middle ear.

Examine a model or charts and label the parts of the external ear in Figure 15.9.

b. MIDDLE EAR

The *middle ear (tympanic cavity)* is a small, epithelium-lined, air-filled cavity hollowed out of the temporal bone. The area is separated from the external ear by the eardrum and from the internal ear by a very thin bony partition that contains two small membrane-covered openings, the oval window and the round window. The posterior wall of the middle ear communicates with the mastoid cells of the temporal bone through a chamber called the *tympanic antrum.*

The anterior wall of the middle ear contains an opening that leads into the *auditory (Eustachian) tube.* The auditory tube connects the middle ear with the nose and nasopharynx. The function of the tube is to equalize air pressure on both sides of the tympanic membrane. Any sudden pressure changes against the eardrum can be equalized by deliberately swallowing.

Extending across the middle ear are three exceedingly small bones called *auditory ossicles* (OS-si-kuls). These are known as the malleus, incus, and stapes. Based on their shape, they are commonly named the hammer, anvil, and stirrup, respectively. The "handle" of the *malleus* is attached to the internal surface of the tympanic membrane. Its head articulates with the base of the *incus,* the intermediate bone in the series, which articulates with the stapes. The base of the *stapes* fits into a small opening between the middle and inner ear called the *oval window.* Directly below the oval window is another opening, the *round window.* This opening, which separates the middle and inner ears, is enclosed by a membrane called the *secondary tympanic membrane.*

Examine a model or charts and label the parts of the middle ear in Figures 15.10 and 15.11.

c. INTERNAL (INNER) EAR

The *internal (inner) ear* is also known as the *labyrinth* (LAB-i-rinth). Structurally, this consists of two main divisions: (1) an outer bony labyrinth and (2) an inner membranous labyrinth that fits within the bony labyrinth. The *bony labyrinth* is a series of cavities within the petrous portion of the temporal bone that can be divided into three portions, named on the basis of shape: vestibule, cochlea, and semicircular canals. The bony labyrinth is lined with periosteum and contains a fluid called the *perilymph.* This fluid surrounds the *membranous labyrinth,* a series of sacs and tubes lying inside and having the same general form as the bony labyrinth. Epithelium lines the

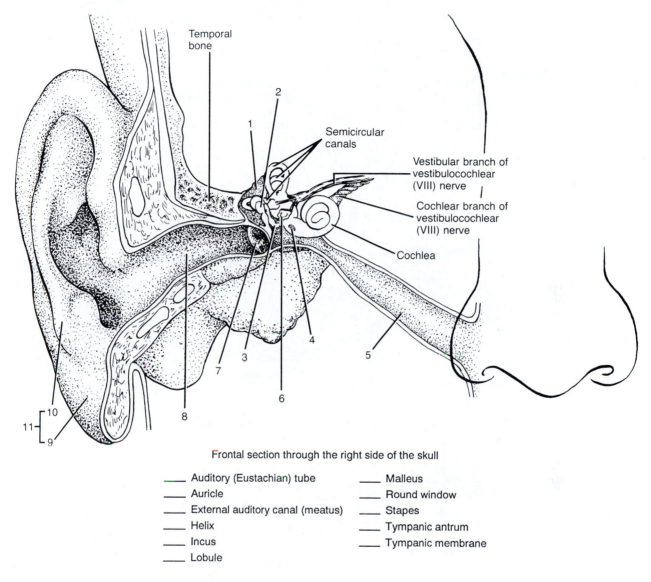

Frontal section through the right side of the skull

____ Auditory (Eustachian) tube   ____ Malleus
____ Auricle   ____ Round window
____ External auditory canal (meatus) ____ Stapes
____ Helix   ____ Tympanic antrum
____ Incus   ____ Tympanic membrane
____ Lobule

FIGURE 15.9 Principal subdivisions of the ear.

membranous labyrinth, which is filled with a fluid called the *endolymph.*

The *vestibule* is the oval, central portion of the bony labyrinth (see Figure 15.10). The membranous labyrinth within the vestibule consists of two sacs called the *utricle* (YOO-tri-kul) and *saccule* (SAK-yool). These sacs are connected to each other by a small duct.

Projecting superiorly and posteriorly from the vestibule are the three bony semicircular canals (see Figure 15.10). Each is arranged at approximately right angles to the other two. They are called the anterior, posterior, and lateral canals. One end of each canal enlarges into a swelling called the *ampulla* (am-POOL-la). Inside the bony semicircular canals lie portions of the mem-

branous labyrinth, the *semicircular ducts (membranous semicircular canals).* These structures communicate with the utricle of the vestibule. Label these structures in Figure 15.11.

Anterior to the vestibule is the *cochlea* (KOK-lē-a). Label it in Figure 15.11, which resembles a snail's shell. The cochlea consists of a bony spiral canal that makes about three turns around a central bony core called the *modiolus.* A cross section through the cochlea shows that the canal is divided by partitions into three separate channels resembling the letter "Y" lying on its side. The stem of the Y is a bony shelf that protrudes into the canal. The wings of the Y are composed of the vestibular and basilar membranes. The channel above the partition is called the *scala*

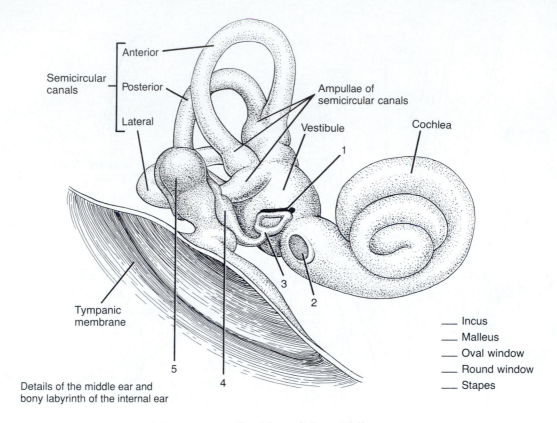

Semicircular canals
Anterior
Posterior
Lateral

Ampullae of semicircular canals

Vestibule

Cochlea

1

3

2

5

4

Tympanic membrane

Details of the middle ear and bony labyrinth of the internal ear

___ Incus
___ Malleus
___ Oval window
___ Round window
___ Stapes

FIGURE 15.10   Ossicles of the middle ear.

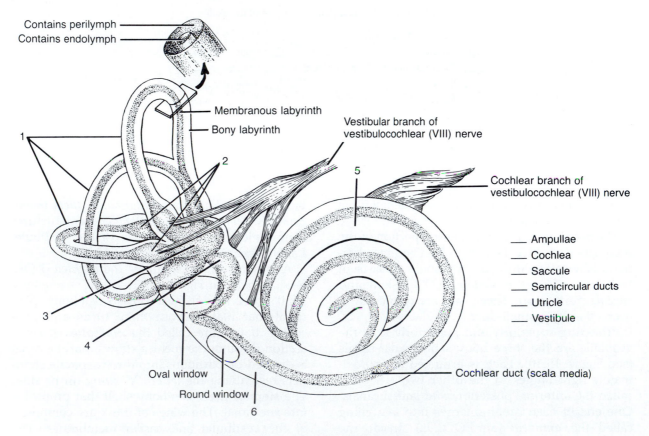

Contains perilymph
Contains endolymph

Membranous labyrinth
Bony labyrinth

Vestibular branch of vestibulocochlear (VIII) nerve

Cochlear branch of vestibulocochlear (VIII) nerve

1

2

5

3

4

6

Oval window

Round window

Cochlear duct (scala media)

___ Ampullae
___ Cochlea
___ Saccule
___ Semicircular ducts
___ Utricle
___ Vestibule

FIGURE 15.11   Details of the inner ear.

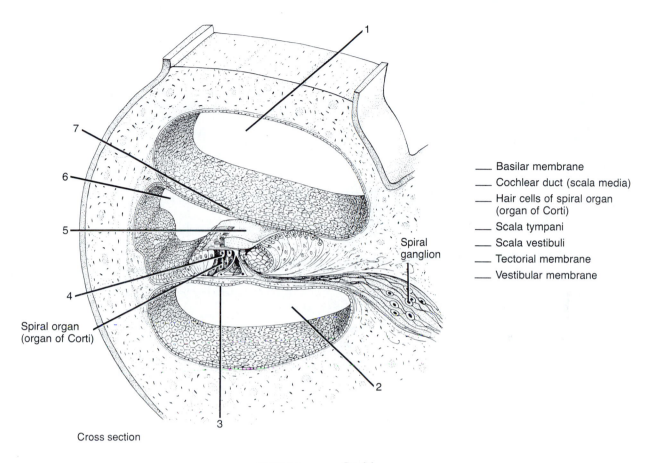

Cross section

_____ Basilar membrane
_____ Cochlear duct (scala media)
_____ Hair cells of spiral organ
        (organ of Corti)
_____ Scala tympani
_____ Scala vestibuli
_____ Tectorial membrane
_____ Vestibular membrane

Spiral
ganglion

Spiral organ
(organ of Corti)

FIGURE 15.12   Cochlea.

vestibuli. The channel below is known as the *scala tympani.* The cochlea adjoins the wall of the vestibule, into which the scala vestibuli opens. The scala tympani terminates at the round window. The perilymph of the vestibule is continuous with that of the scala vestibuli. The third channel (between the wings of the Y) is the membranous labyrinth, the *cochlear duct (scala media).* This duct is separated from the scala vestibuli by the *vestibular membrane* and from the scala tympani by the *basilar membrane.* Resting on the basilar membrane is the *spiral organ (organ of Corti),* the organ of hearing. Label these structures in Figure 15.12.

The spiral organ (organ of Corti) is a coiled sheet of epithelial cells on the inner surface of the basilar membrane. This structure is composed of supporting cells and hair cells, which are receptors for auditory sensations. The hair cells have long hairlike processes at their free ends that extend into the endolymph of the cochlear duct. The basal ends of the hair cells are in contact with fibers of the cochlear branch of the vestibulocochlear (VIII) nerve. Projecting over and in contact

with the hair cells of the spiral organ is the *tectorial membrane,* a very delicate and flexible gelatinous membrane. Label the tectorial membrane in Figure 15.12.

Obtain a prepared microscope slide of the spiral organ (organ of Corti) and examine is under high power. Now label Figure 15.13.

2. Surface Anatomy

Refer to Figure 15.14 for a summary of several surface anatomy features of the ear.

3. Tests for Auditory Acuity

Sound waves result from the alternate compression and decompression of air. The waves have both a frequency and an amplitude. The *frequency* is the distance between crests of a sound wave. It is measured in *Hertz (Hz),* or *cycles per second,* and is perceived as *pitch.* The higher the frequency of a sound, the higher its pitch. Different frequencies displace different areas of the basilar membrane. The *amplitude* of a sound

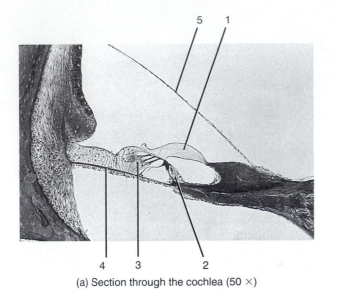

(a) Section through the cochlea (50 ×)

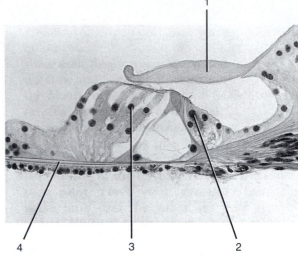

(b) Details of the spiral organ (organ of Corti) (250 ×)

___ Basilar membrane
___ Inner hair cell
___ Outer hair cell
___ Tectorial membrane
___ Vestibular membrane

FIGURE 15.13   Photomicrographs of the spiral organ (organ of Corti).

wave is perceived as *intensity (loudness)*. Intensity is measured in *decibels (db)*. Population norms have been calculated that measure the intensity of a sound that can just be heard—i.e., the *threshold* of sound. These levels are said to have an intensity of 0 db. Each 10 db reflects a tenfold increase in intensity; thus a sound is 10 times louder than threshold at 10 db, 100 times greater at 20 db, one million times greater at 60 db, and so on.

a. SOUND LOCALIZATION TEST

The cerebral cortex localizes the source of a sound by evaluating the time lag between the entry of sound into each ear. Without turning the head, it is impossible to distinguish between the origin of sounds that arise at 30° ahead of the right ear and 30° behind the right ear.

PROCEDURE

1. In a quiet room, have a subject occlude one ear with a cotton plug and close his/her eyes.
2. Move a ticking watch to various points 15 cm (6 in.) away from the ear (side, front, top, back) and ask the subject to point to the location of the sound.

3. Note in which positions the sound is best localized. In what regions was sound localization most accurate? _____

_____

b. WEBER'S TEST

Weber's test is diagnostic for both conduction and sensorineural deafness. *Conduction deafness* results from interference with sound waves reaching the inner ear (plugged external auditory canal, fusion of the ossicles, etc.); *sensorineural deafness* is caused by damage to the nerve pathway between the cochlea and the brain. This test should not be performed in a quiet room.

PROCEDURE

1. Strike a tuning fork with a mallet and place the tip of the handle on the median line of a subject's forehead.
2. Determine if the tone is equally loud in both ears. _____
3. Have the subject occlude the left ear with cotton and repeat the exercise; occlude the right ear and repeat.

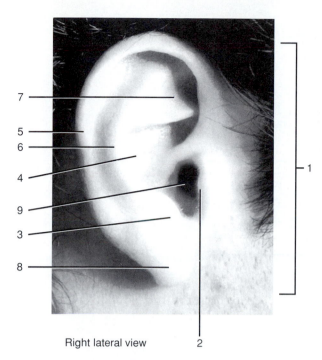

Right lateral view   2

1. **Auricle.** Portion of external ear not contained in head, also called pinna.
2. **Tragus.** Cartilaginous projection anterior to external opening of ear.
3. **Antitragus.** Cartilaginous projection opposite tragus.
4. **Concha.** Hollow of auricle.
5. **Helix.** Superior and posterior free margin of auricle.
6. **Antihelix.** Semicircular ridge posterior and superior to concha.
7. **Triangular fossa.** Depression in superior portion of anti-helix.
8. **Lobule.** Inferior portion of auricle devoid of cartilage.
9. **External auditory canal (meatus).** Canal extending from external ear to eardrum.

FIGURE 15.14   Surface anatomy of the ear.

Describe your results: _____

_____

If the sound is equally loud in both unoccluded ears, hearing is normal. If the sound is louder in the occluded ear, hearing is normal. This situation parallels conduction deafness, when the cochlea is not receiving environmental background noise but only the vibrations of the tuning fork. If there is sensorineural deafness, the sound is louder in the normal (or nonoccluded) ear, because sensorineural loss produces a deficit in transmission to the cerebral cortex.

4. Equilibrium Apparatus

The term *equilibrium* has two meanings. One kind of equilibrium, called *static equilibrium*, refers to the position of the body (mainly the head) relative to the ground (gravity). The second kind of equilibrium, called *dynamic equilibrium*, is the maintenance of the position of the body (mainly the head) in response to sudden movements (rotation, acceleration, and deceleration) or to a change in the rate or direction of movement. The receptor organs for equilibrium are the maculae in the saccule and utricle and the cristae in the semicircular ducts.

The *maculae* (MAK-yoo-lē) in the walls of the *utricle* and *saccule* are the receptors concerned mainly with static equilibrium. The maculae are small, flat, plaquelike regions that resemble the spiral organ (organ of Corti) microscopically. Maculae are located in planes perpendicular to each other and contain two kinds of cells: *hair (receptor) cells* and *supporting cells.* The hair cells project *stereocilia* (microvilli) and a *kinocilium* (conventional cilium). The supporting cells are scattered among the hair cells. Floating over the hair cells is a thick, gelatinous glycoprotein layer, the *otolithic membrane.* A layer of calium carbonate crystals, called *otoliths* (*oto* = ear, *lithos*=stone), extends over the entire surface of the otolithic membrane. When the head is tilted, the membrane slides over the hair cells in the direction determined by the tilt of the head. This sliding causes the membrane to pull on the stereocilia, thus initiating a nerve impulse that is conveyed via the vestibular branch of the vestibulocochlear (VIII) nerve to the brain (cerebellum). The cerebellum sends continuous nerve impulses to the motor areas of the cerebral cortex in response to input from the maculae in the utricle and saccule, causing the motor system to increase or decrease its nerve impulses to specific skeletal muscles to maintain static equilibrium.

Label the parts of the macula in Figure 15.15a.

Now consider the role of the cristae in the semicircular ducts in maintaining dynamic equilibrium. The three semicircular ducts are positioned at right angles to one another in three planes: frontal (anterior duct); sagittal (posterior duct); and lateral (lateral duct). This positioning permits correction of an imbalance in three planes. In the ampulla, the dilated portion of each duct, is a small elevation called the *crista.* Each crista is composed of a group of *hair (receptor) cells* and *supporting cells* covered by a

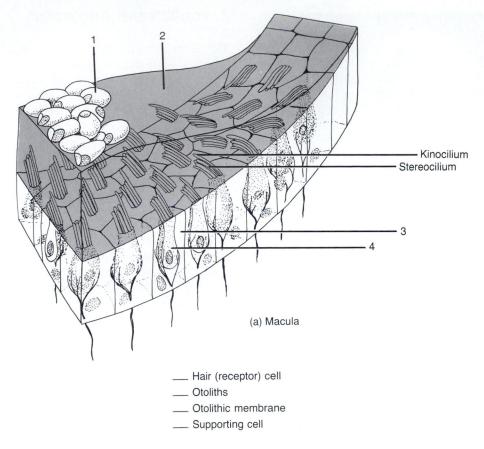

Kinocilium
Stereocilium

(a) Macula

___ Hair (receptor) cell
___ Otoliths
___ Otolithic membrane
___ Supporting cell

FIGURE 15.15   Equilibrium apparatus.

mass of gelatinous material called the *cupula.* When the head moves, endolymph in the semicircular ducts flows over the hairs and bends them as water in a stream bends the plant life growing at its bottom. Movement of the hairs stimulates sensory neurons, and nerve impulses pass over the vestibular branch of the vestibulocochlear (VIII) nerve. The nerve impulses follow the same pathway as those involved in static equilibrium and are eventually sent to the muscles that must contract to maintain body balance in the new position.

Label the parts of the crista in Figure 15.15b.

5. Tests for Equilibrium

You can test equilibrium by using a few simple procedures.

a. BALANCE TEST

This test is used to evaluate static equilibrium.

PROCEDURE

1. Instruct a subject to stand perfectly still with his/her arms at the sides and the eyes closed.
2. Observe any swaying movements. These are easier to detect if the subject stands in front of a blackboard with a light in front of him/her.
3. A mark is made at the edge of the shadow of the subject's shoulders and the movement of the shadow is then observed.

If the static equilibrium system is dysfunctional the subject will sway or fall, although other proprioreceptors may compensate for the defect.

b. BARANY TEST

This test evaluates the function of each of the semicircular canals. A subject is rotated, and when rotation is stopped, the momentum causes the endolymph in the lateral (horizontal) canal to continue spinning. The spinning bends the cupula in the direction of rotation. Other sensory input informs us that we are no longer rotating. *Nystagmus,* the reflex movement of the eyes

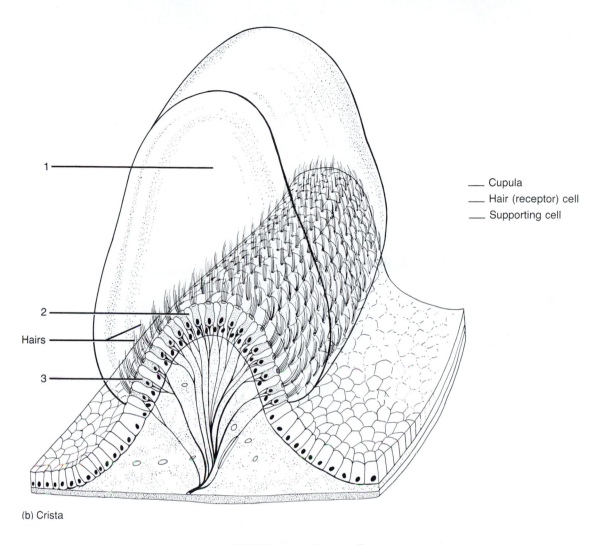

1

2

Hairs

3

—— Cupula
—— Hair (receptor) cell
—— Supporting cell

(b) Crista

FIGURE 15.15 *(Continued)*

rapidly and then slowly, attempts to compensate for the loss of balance by visual fixation on an object. When the head is tipped toward the shoulder, the anterior semicircular canals are stimulated, so nystagmus is vertical; when the head is tipped slightly forward, the lateral canals are stimulated by the rotation, so nystagmus occurs laterally; when the head is bent onto the chest, the posterior canal receptors are stimulated, so nystagmus is rotational.

Record your observations in Section D.5a of the LABORATORY REPORT RESULTS at the end of the exercise.

PROCEDURE

1. Use a person who is not subject to vertigo (dizziness).
2. The subject sits firmly anchored on a stool, legs up on the stool rung, head tilted onto one shoulder. Then the stool is *very carefully* revolved to the right about 10 times (about one turn every 2 sec) for a few seconds and suddenly stopped.
3. Observe the subject's eye movements.

CAUTION! *Watch the subject carefully and be prepared to provide support until the dizziness has passed.*

4. The subject will experience the sensation that the stool is still rotating, which means that the semicircular canals are functioning properly.
5. Repeat on a different subject with the head tipped slightly forward. Use still another subject with the chin touching the chest.
6. Record the direction of nystagmus in all three subjects in Section D.5.b of the LABORATORY REPORT RESULTS at the end of the exercise.

Special Senses

SECTION B. GUSTATORY SENSATIONS
2. Identifying Taste Zones

Areas of Tongue in Which Basic Tastes are Detected

	Sweet	Bitter	Salty	Sour
Tip of tongue				
Back of tongue				
Sides of tongue				

4. Taste and Smell

	Sensations when placed on dry tongue	Sensations while chewing (nostrils closed)	Sensations while chewing with nostrils opened
Carrot			
Onion			
Pototo			
Apple			

SECTION C. VISUAL SENSATIONS
5. Testing for Visual Acuity

Visual activity, left eye _____

Visual activity, right eye _____

Visual activity, both eyes _____

11. Testing for Afterimages

	Appearance of positive afterimage	Appearance of negative afterimage
Bright light		
Red cross		

SECTION D. AUDITORY SENSATIONS AND EQUILIBRIUM
5. Tests For Equilibrium

a. BALANCE TEST

Describe the response of the subject. _____

_____

_____

b. BARANY TEST

Subject 1 _____

Subject 2 _____

Subject 3 _____

Special Senses

STUDENT _____   DATE _____

LABORATORY SECTION _____   SCORE/GRADE _____

PART 1. Multiple Choice

_____ 1. The papillae located in an inverted V-shaped row at the posterior portion of the tongue are (a) circumvallate (b) filiform (c) fungiform (d) gustatory

_____ 2. The technical name for the "white of the eye" is the (a) cornea (b) conjunctiva (c) choroid (d) sclera

_____ 3. Which is *not* a component of the vascular tunic? (a) choroid (b) macula lutea (c) iris (d) ciliary body

_____ 4. The amount of light entering the eyeball is regulated by the (a) lens (b) iris (c) cornea (d) conjunctiva

_____ 5. Which region of the eye is concerned primarily with image formation? (a) retina (b) choroid (c) lens (d) ciliary body

_____ 6. The densest concentration of cones is found at the (a) blind spot (b) macula lutea (c) central fovea (d) optic disc

_____ 7. Which region of the eyeball contains the vitreous body? (a) anterior chamber (b) vitreous chamber (c) posterior chamber (d) conjunctiva

_____ 8. Among the structures found in the middle ear are the (a) vestibule (b) auditory ossicles (c) semicircular canals (d) external auditory canal

_____ 9. The receptors for dynamic equilibrium are the (a) saccules (b) utricles (c) cristae in semicircular ducts (d) spiral organs (organs of Corti)

_____ 10. The auditory ossicles are attached to the tympanic membrane, to each other, and to the (a) semicircular ducts (b) semicircular canals (c) oval window (d) labyrinth

_____ 11. Another name for the internal ear is the (a) labyrinth (b) fenestra (c) cochlea (d) vestibule

_____ 12. Orientation of the body relative to the ground is termed (a) the postural reflex (b) the tonal reflex (c) dynamic equilibrium (d) static equilibrium

_____ 13. Which sequence best describes the normal flow of tears from the eyes into the nose? (a) lacrimal canals, lacrimal sacs, nasolacrimal ducts (b) lacrimal sacs, lacrimal canals, nasolacrimal ducts (c) nasolacrimal ducts, lacrimal sacs, lacrimal canals (d) lacrimal sacs, nasolacrimal ducts, lacrimal canals

_____ 14. A patient whose lens has lost transparency is suffering from (a) glaucoma (b) conjunctivitis (c) cataract (d) trachoma

_____ 15. The portion of the eyeball that contains aqueous humor is the (a) anterior cavity (b) lens (c) posterior chamber (d) macula lutea

_____ 16. The organ of hearing located within the inner ear is the (a) vestibule (b) oval window (c) modiolus (d) spiral organ (organ of Corti)

_____ 17. The sense organs of static equilibrium are the (a) semicircular ducts (b) membranous labyrinths (c) maculae in the utricle and saccule (d) auricles

_____ 18. The membrane that is reflected from the eyelids onto the eyeball is the (a) retina (b) bulbar conjunctiva (c) sclera (d) choroid

_____ 19. Which region of the tongue reacts strongest to bitter tastes? (a) tip (b) center (c) back (d) sides

_____ 20. Which of the following values indicates the best visual acuity? (a) 20/30 (b) 20/40 (c) 20/50 (d) 20/60

_____ 21. Nearsightedness is referred to as (a) emmetropia (b) hypermetropia (c) eumetropia (d) myopia

PART 2. Completion

22. Structures that collectively produce and drain tears are referred to as the _____.

23. The three zones of the inner nervous layer of the retina (nervous tunic) are the photoreceptor layer, bipolar cell layer, and _____.

24. The small area of the retina where no image is formed is referred to as the

_____.

25. Abnormal elevation of intraocular pressure (IOP) is called _____.

26. In the visual pathway, nerve impulses pass from the optic chiasma to the _____ before passing to the thalamus.

27. The openings between the middle and inner ears are the oval and _____ windows.

28. The fluid within the bony labyrinth is called _____.

29. The neural pathway for olfaction includes olfactory receptors, olfactory bulbs,

_____, and cerebral cortex.

30. The posterior wall of the middle ear communicates with the mastoid air cells of the temporal bone

through the _____.

31. The thin, semitransparent partition of fibrous connective tissue that separates the external auditory

meatus from the inner ear is the _____.

32. The fluid in the membranous labyrinth is called _____.

33. The cochlear duct is separated from the scala vestibuli by the _____.

34. The gelatinous glycoprotein layer over the hair cells in the maculae is called the

_____.

35. _____ is blurred vision caused by an irregular curvature of the surface of the cornea or lens.

36. Image formation requires refraction, accommodation, constriction of the pupil, and

_____.

Endocrine System

You have learned how the nervous system controls the body through nerve impulses that are delivered over neurons. Now you will study the body's other great control system, the *endocrine system.* The endocrine organs and tissues affect bodily activities by releasing chemical messengers, called *hormones,* into the bloodstream (the term *hormone* means "to urge on"). The nervous and endocrine systems coordinate their activities as an interlocking supersystem. Certain parts of the nervous system stimulate or inhibit the release of hormones. The hormones, in turn, are quite capable of stimulating or inhibiting the flow of particular nerve impulses.

The body contains two different kinds of glands: exocrine and endocrine. *Exocrine glands* secrete their products into ducts. The ducts then carry the secretions into body cavities, into the lumens of various organs, or to the external surface of the body. Examples are sudoriferous (sweat), sebaceous (oil), mucous, and digestive glands. *Endocrine glands,* by contrast, secrete their products (hormones) into the extracellular space around the secretory cells. The secretion then passes through the membranes of cells lining blood vessels and into the blood. Because they have no ducts, endocrine glands are also called *ductless glands.*

A. ENDOCRINE GLANDS

The endocrine glands of the body are the pituitary (hypophysis), thyroid, parathyroids, adrenals (suprarenals), pineal (epiphysis cerebri), and thymus. In addition, several organs of the body contain endocrine tissue but are not endocrine glands exclusively. They include the pancreas, ovaries, testes, kidneys, stomach, small intestine, skin, heart, hypothalamus, and the placenta. The endocrine glands are organs that together form the *endocrine system.* Locate the endocrine glands on a torso, and, using your textbook or charts for reference, label Figure 16.1.

All hormones maintain homeostasis by changing the physiological activities of cells. A hormone may stimulate changes in the cells of an organ or in groups of organs. Or, the hormone may directly affect the activities of all the cells in the body. The cells that respond to the effects of a hormone are called *target cells.*

B. PITUITARY GLAND (HYPOPHYSIS)

The hormones of the *pituitary gland,* also called the *hypophysis* (hī-POF-i-sis), regulate so many body activities that the pituitary has been nicknamed the "master gland." This gland lies in the sella turcica of the sphenoid bone and is attached to the hypothalamus of the brain via a stalklike structure, the *infundibulum.* Not only is the hypothalamus an important regulatory center in the nervous system, it is also a crucial endocrine gland. Cells in the hypothalamus synthesize at least nine different hormones, and the pituitary gland secretes seven more. Together, they play important roles in regulation of virtually all

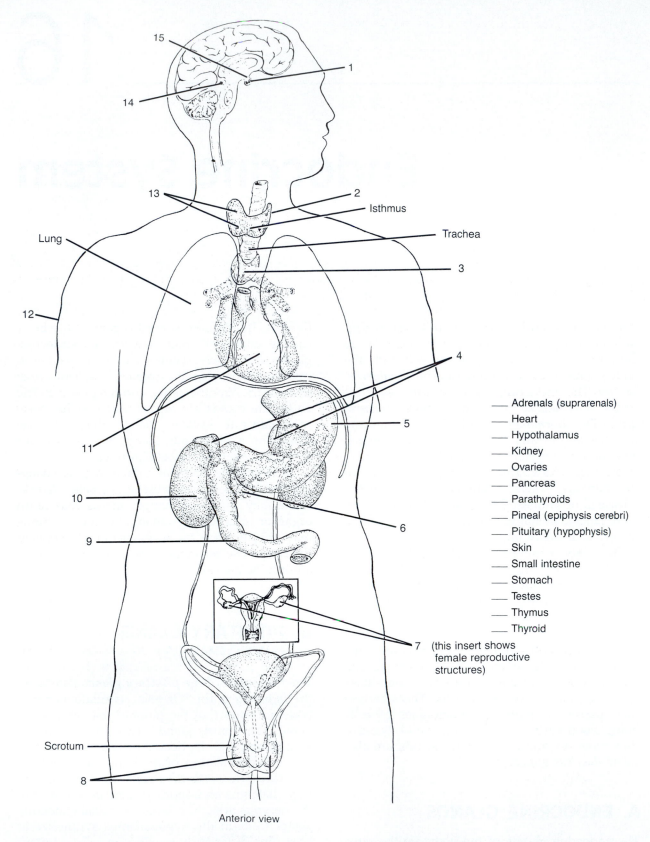

15

1

14

13

2

Isthmus

Lung

Trachea

3

12

4

5

11

10

9

6

7 (this insert shows female reproductive structures)

___ Adrenals (suprarenals)
___ Heart
___ Hypothalamus
___ Kidney
___ Ovaries
___ Pancreas
___ Parathyroids
___ Pineal (epiphysis cerebri)
___ Pituitary (hypophysis)
___ Skin
___ Small intestine
___ Stomach
___ Testes
___ Thymus
___ Thyroid

Scrotum

8

Anterior view

FIGURE 16.1 Location of the endocrine glands, organs containing endocrine tissue, and associated structures.

aspects of growth, development, metabolism, and homeostasis.

The pituitary gland is divided structurally and functionally into an *anterior pituitary*, also called the *adenohypophysis* (ad'-i-nō-hī-POF-i-sis) and a *posterior pituitary*, also called the *neurohypophysis* (noo'-rō-hī-POF-i-sis). The anterior pituitary contains many glandular epithelium cells and forms the glandular part of the pituitary. The posterior pituitary contains axon terminals of neurons whose cell bodies are in the hypothalamus. These terminals form the neural part of the pituitary. The hypothalamus is connected to the anterior and posterior pituitary by a series of blood vessels, the hypothalamic-hypophyseal portal system. Between the anterior and posterior pituitary is a small, relatively avascular rudimentary zone, the *pars intermedia,* whose role in humans is obscure.

1. Histology of Pituitary Gland

The anterior pituitary releases hormones that regulate a whole range of body activities, from growth to reproduction. However, the release of these hormones is stimulated by *releasing hormones* or inhibited by *inhibiting hormones* produced by neurosecretory cells in the hypothalamus of the brain. This is one interaction between the nervous system and the endocrine system. The hypothalamic hormones reach the anterior pituitary via a network of blood vessels.

When the anterior pituitary receives proper stimulation from the hypothalamus via releasing hormones, its glandular cells secrete any one of seven hormones. Recently, special staining techniques have established the division of glandular cells into five principal types:

1. *Somatotrophs*—Produce *Human growth hormone (hGH),* which controls general body growth and certain aspects of metabolism.
2. *Lactotrophs*—Synthesize *prolactin (PRL),* which initiates milk production by the mammary glands.
3. *Corticolipotrophs*—Synthesize *adrenocorticotropic* (ad-rē-nō-kor'-ti-kō-TRŌ-pik) *hormone (ACTH),* which stimulates the adrenal cortex to secrete its hormones and *melanocyte-stimulating hormone (MSH),* which is believed to be related to skin pigmentation.
4. *Thyrotrophs*—Manufacture the *thyroid-stimulating hormone (TSH),* which controls the thyroid gland.
5. *Gonadotrophs*—Produce *follicle-stimulating hormone (FSH),* which stimulates the production of eggs and sperm in the ovaries and testes, respectively, and *luteinizing hormone (LH),* which stimulates other sexual and reproductive activities.

Except for human growth hormone (hGH), melanocyte-stimulating hormone (MSH), and prolactin (PRL), all the secretions are referred to as *tropic hormones,* which means that their target organs are other endocrine glands. Follicle-stimulating hormone (FSH) and luteinizing hormone (LH) are also called *gonadotropic* (gō-nad-ō-TRŌ-pik) *hormones* because they regulate the functions of the gonads (ovaries and testes). The gonads are the endocrine glands that produce sex hormones.

Examine a prepared slide of the anterior pituitary and identify as many types of cells as possible with the aid of your textbook or a histology textbook.

The posterior pituitary is not really an endocrine gland. Instead of synthesizing hormones, it stores hormones synthesized by cells of the hypothalamus. The posterior pituitary consists of (1) cells called *pituicytes* (pi-TOO-i-sītz), which are similar in appearance to the neuroglia of the nervous system, and (2) axon terminations of secretory nerve cells of the hypothalamus. The cell bodies of the neurons, called *neurosecretory cells,* originate in nuclei in the hypothalamus. The fibers project from the hypothalamus, form the *supraopticohypophyseal (hypothalamic-hypophyseal) tract,* and terminate on blood capillaries in the posterior pituitary. The cell bodies of the neurosecretory cells produce the hormones *oxytocin* (ok'-sē-TŌsin), or *OT,* and *antidiuretic hormone (ADH).* These hormones are transported in the neuron fibers into the posterior pituitary and are stored in the axon terminals resting on the capillaries. When properly stimulated, the hypothalamus sends impulses over the neurons. The impulses cause release of hormones from the axon terminals into the blood.

Examine a prepared slide of the posterior pituitary under high power. Identify the pituicytes and axon terminations of neurosecretory cells with the aid of your textbook.

2. Pituitary Gland Hormones

Using your textbook as a reference, give the major functions for the hormones listed.

a. HORMONES SECRETED BY ANTERIOR PITUITARY

Human growth hormone (hGH). Also called so-matotropin and somatotropic hormone (STH).

_____

_____

_____

Thyroid-stimulating hormone (TSH). Also called thyrotropin. _____

_____

_____

_____

Adrenocorticotropic hormone (ACTH). _____

_____

_____

_____

Follicle-stimulating hormone (FSH).

In females: _____

_____

_____

In males: _____

_____

_____

Luteinizing hormone (LH)

In females: _____

_____

_____

In males: _____

_____

_____

Prolactin (PRL). Also called lactogenic hormone.

_____

_____

Melanocyte-stimulating hormone (MSH). _____

_____

_____

_____

_____

b. HORMONES STORED BY POSTERIOR PITUITARY

Oxytocin (OT). _____

_____

_____

_____

Antidiuretic hormone (ADH). _____

_____

_____

_____

C. THYROID GLAND

The *thyroid gland* is the endocrine gland located just below the larynx. The right and left *lateral lobes* lie on either side of the trachea and are connected by a mass of tissue called an *isthmus* (IS-mus) that lies in front of the trachea just below the cricoid cartilage. The *pyramidal lobe,* when present, extends upward from the isthmus.

1. Histology of Thyroid Gland

Histologically, the thyroid consists of spherical sacs called *thyroid follicles.* The walls of each follicle consist of epithelial cells that reach the surface of the lumen of the follicle *(follicular cells)* and epithelial cells that do not reach the lumen *(parafollicular,* or *C, cells).* The follicular cells manufacture the hormones *thyroxine* (thī-ROX-sēn) *(T$_4$,* or *tetraiodothyronine)* and *triiodo-thyronine* (trī-ī-ōd-ō-THĪ-rō-nēn) *(T$_3$).* Together these hormones are referred to as the *thyroid hormones.* Thyroxine is considered to be the principal hormone produced by follicular cells. The parafollicular cells produce the hormone *calcitonin* (kal-si-TŌ-nin) *(CT).* Each thyroid follicle is filled with a glycoprotein called *thyroglob-ulin (TBG)* and stored thyroid hormones, to-gether called *thyroid colloid.*

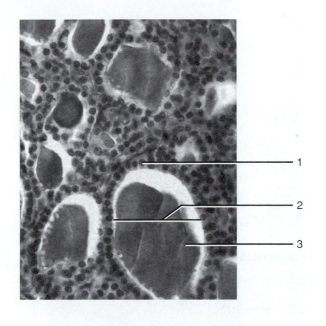

_____ Epithelium of follicle
_____ Thyroid colloid
_____ Thyroid follicle

FIGURE 16.2 Histology of the thyroid gland.

Examine a prepared slide of the thyroid gland under high power. Identify the thyroid follicles, epithelial cells forming the follicle, and thyroid colloid. Now label the photomicrograph in Figure 16.2.

2. Hormones of Thyroid Gland

Using your textbook as a reference, give the major functions of the hormones listed below.

Thyroxine (T₄) and *triiodothyronine (T₃).* _____

_____

_____

_____

Calcitonin (CT). _____

_____

_____

_____

D. PARATHYROID GLANDS

Attached to the posterior surfaces of the lateral lobes of the thyroid gland are small, round masses of tissue called the *parathyroid glands.* Typically, two parathyroids, superior and inferior, are attached to each lateral thyroid lobe.

1. Histology of Parathyroid Glands

Histologically, the parathyroids contain two kinds of epithelial cells. The first, a larger, more numerous cell called a *principal (chief) cell,* is believed to be the major synthesizer of *parathyroid hormone (PTH).* The second, a smaller cell called an *oxyphil cell,* may serve as a reserve cell for hormone synthesis.

Examine a prepared slide of the parathyroid glands under high power. Identify the principal and oxyphil cells. Now label the photomicrograph in Figure 16.3.

2. Hormone of Parathyroid Glands

Using your textbook as a reference, write the major functions of the hormone listed:
Parathyroid hormone (PTH). Also called *parathormone.* _____

_____

_____

_____

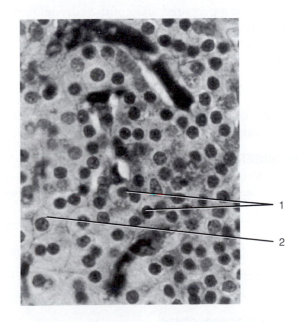

_____ Oxyphil cell
_____ Principal (chief) cells

FIGURE 16.3 Histology of the parathyroids.

E. ADRENAL (SUPRARENAL) GLANDS

The two *adrenal (suprarenal)* glands are superior to each kidney. Each is structurally and functionally differentiated into two regions: the outer *adrenal cortex,* which forms the bulk of the gland, and the inner *adrenal medulla.* Covering the gland are a thick layer of fatty tissue and an outer, thin, fibrous connective capsule.

1. Histology of Adrenal Cortex

Histologically, the adrenal cortex is subdivided into three zones, each of which has a different cellular arrangement and secretes different steroid hormones. The outer zone, directly underneath the connective tissue capsule, is called the *zona glomerulosa.* Its cells, arranged in arched loops or round balls, primarily secrete a group of hormones called *mineralocorticoids* (min'-er-al-ō-KŌR-ti-koyds). The middle zone, or *zona fasciculata,* is the widest of the three zones and consists of cells arranged in long, straight cords. The zona fasciculata secretes mainly *glucocorticoids* (gloo'-kō-KŌR-ti-koyds). The inner zone, the *zona reticularis,* contains cords of cells that branch freely. This zone synthesizes minute amounts of *gonadocorticoids* (gō-na-dō-KŌR-ti-koyds), or sex hormones, chiefly male hormones (androgens).

Examine a prepared slide of the adrenal cortex under high power. Identify the capsule, zona glomerulosa, zona fasciculata, and zona reticularis. Now label Figure 16.4.

2. Adrenal Cortex Hormones

Using your textbook as a reference, give the major functions of the hormones listed below.

A. MINERALOCORTICOIDS
Aldosterone (only one of three mineralocorticoids, but responsible for 95% of their activity).

_____

_____

_____

_____ .

B. GLUCOCORTICOIDS
Cortisol (hydrocortisone), corticosterone, and

cortisone. _____

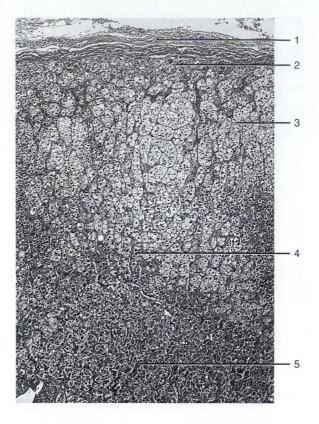

_____ Adrenal medulla   _____ Zona glomerulosa
_____ Capsule   _____ Zona reticularis
_____ Zona fasciculata

FIGURE 16.4 Histology of the adrenal (suprarenal) gland.

_____

_____

_____

C. GONADOCORTICOIDS
Estrogens (in female). See ''Estrogens'' under ''Hormones of Ovaries.''

Androgens (in male). See ''Testosterone'' under ''Hormones of Testes.''

3. Histology of Adrenal Medulla

The adrenal medulla consists of hormone-producing cells called *chromaffin* (krō-MAF-in) *cells.* These cells develop from the same source as the postganglionic cells of the sympathetic division of the nervous system. They are directly innervated by preganglionic cells of the sympathetic division of the autonomic nervous system and can be regarded as postganglionic cells that are specialized to secrete hormones. Secretion of hormones from the chromaffin cells is directly controlled by the sympathetic division of the autonomic nervous system; innervation by the preganglionic fibers allows the gland to respond extremely rapidly to a stimulus. The adrenal medulla secretes the hormones *epinephrine* and *norepinephrine (NE)*.

Examine a prepared slide of the adrenal medulla under high power. Identify the chromaffin cells. Now locate the cells in Figure 16.4.

4. Hormones of Adrenal Medulla

Using your textbook as a reference, give the major functions of the hormones listed below.

Epinephrine and norepinephrine (NE). _____

_____

_____

_____

F. PANCREAS

The *pancreas* is classified as both an endocrine and an exocrine gland. Thus, it is referred to as a *heterocrine gland.* We will discuss only its endocrine functions now. The pancreas is a flat-tened organ located posterior and slightly inferior to the stomach. The adult pancreas consists of a head, body, and tail.

1. Histology of Pancreas

The endocrine portion of the pancreas consists of clusters of cells called *pancreatic islets (islets of Langerhans).* They contain four kinds of cells: (1) *alpha cells,* which have more distinguishable plasma membranes, are usually peripheral in the islet and secrete the hormone *glucagon;* (2) *beta cells,* which generally lie deeper within the islet and secrete the hormone *insulin;* (3) *delta cells,* which secrete *growth hormone-inhibiting hormone factor (GHIH),* or *somatostatin;* and (4) *F-cells,* which secrete *pancreatic polypeptide.* The pancreatic islets are surrounded by blood capillaries and by cells called *acini* that form the exocrine part of the gland.

Examine a prepared slide of the pancreas under high power. Identify the alpha cells, beta cells, delta cells, and acini (clusters of cells that secrete digestive enzymes) around the pancreatic islets. Now label Figure 16.5.

2. Hormones of Pancreas

Using your textbook as a reference, give the major functions of the hormones listed.

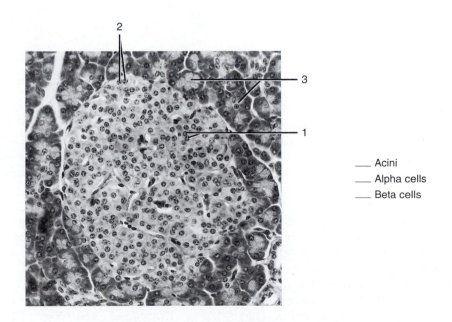

___ Acini
___ Alpha cells
___ Beta cells

FIGURE 16.5   Histology of the pancreas.

Glucagon. _____

_____

_____

Insulin. _____

_____

_____

Growth hormone-inhibiting hormone (GHIH).

_____

_____

_____

Pancreatic polypeptide. _____

_____

_____

G. TESTES

The *testes* (male gonads) are paired oval glands enclosed by the scrotum. They are covered by a dense layer of white fibrous tissue, the *tunica albuginea* (al'-byoo-JIN-ē-a), which extends inward and divides each testis into a series of internal compartments called *lobules.* Each lobule contains 1 to 3 tightly coiled tubules, the convoluted *seminiferous tubules,* which produce sperm by a process called *spermatogenesis* (sper'-ma-tō-JEN-e-sis).

1. Histology of Testes

A cross section through a seminiferous tubule reveals that it is packed with sperm cells in various stages of development. The most immature cells, called *spermatogonia* (sper'-ma-tō-GŌ-nē-a), are located near the basement membrane. Toward the lumen in the center of the tube, layers of progressively more mature cells can be observed. In order of advancing maturity, these are *primary spermatocytes* (SPER-ma-tō-sīts), *secondary spermatocytes,* and *spermatids.* By the time a *sperm cell (spermatozoon)* (sper'-ma-tō-ZŌ-on) has reached full maturity, it is in the lumen of the tubule and begins to be moved through a series of ducts. Between the developing sperm cells in the tubules are *sustentacular* (sus'-ten-TAK-yoo-lar) *cells (Sertoli cells).* They

produce secretions that supply nutrients to the spermatozoa and secrete the hormone *inhibin.* Between the seminiferous tubules are clusters of *interstitial endocrinocytes (interstitial cells of Leydig)* that secrete the primary male hormone *testosterone.*

Examine a prepared slide of the testes under high power. Identify the basement membrane, spermatogonia, spermatids, spermatozoa, sustentacular cells, interstitial endocrinocytes (interstitial cells of Leydig), and lumen of the seminiferous tubule. Now label Figure 16.6.

2. Hormones of Testes

Using your textbook as a reference, give the major functions of the hormones listed.

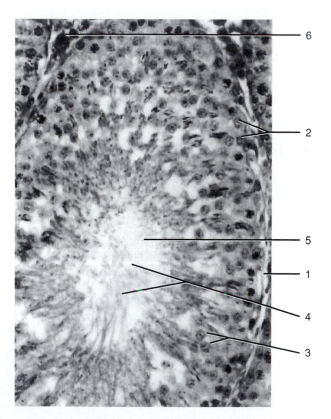

___ Basement membrane
___ Interstitial endocrinocytes
     (interstitial cells of Leydig)
___ Lumen of seminiferous tubule
___ Spermatids
___ Spermatogonia
___ Spermatozoa

FIGURE 16.6   Histology of these testes.

Testosterone _____

_____

_____

Inhibin _____

_____

_____

H. OVARIES

The *ovaries* (female gonads) are paired glands resembling unshelled almonds in size and shape. They are positioned in the superior pelvic cavity, one on each side of the uterus, and are maintained in position by a series of ligaments. They are (1) attached to the **broad ligament** of the uterus, which is itself part of the parietal peritoneum, by a fold of peritoneum called the **mesovarium;** (2) anchored to the uterus by the **ovarian ligament;** and (3) attached to the pelvic wall by the **suspensory ligament.** Each ovary also contains a **hilus,** which is the point of entrance for blood vessels and nerves.

1. Histology of Ovaries

Histologically, each ovary consists of the following parts:

1. *Germinal epithelium*—Layer of simple cuboidal epithelium that covers the free surface of the ovary.
2. *Tunica albuginea*—Capsule of dense irregular connective tissue immediately deep to the germinal epithelium.
3. *Stroma*—Region of connective tissue deep to the tunica albuginea. This tissue is composed of an outer, dense layer called the **cortex** and an inner, loose layer known as the **medulla.** The cortex contains ovarian follicles.
4. *Ovarian follicles*—Oocytes (immature ova) and their surrounding tissues in various stages of development. Primary follicles are smaller and secondary follicles are larger.
5. *Vesicular ovarian (Graafian) follicle*—Relatively large fluid-filled follicle containing an immature ovum and its surrounding tissues. The vesicular ovarian follicle is a stage ready for ovulation that secretes hormones called **estrogens.**

6. *Corpus luteum*—Mature vesicular ovarian follicle that has ruptured to expel a secondary oocyte (potential mature ovum), a process known as ovulation. It produces the hormones *progesterone (PROG)*, *estrogens*, *relaxin*, and *inhibin*.

Examine a prepared slide of the ovary under high power. Identify the germinal epithelial cells, tunica albuginea, cortex, medulla, primary follicle, secondary follicle, vesicular ovarian (Graafian) follicle, and corpus luteum. Label Figure 16.7.

2. Hormones of Ovaries

Using your textbook as a reference, give the major functions of the hormones listed below.

Estrogens. _____

_____

_____

_____

Progesterone (PROG). _____

_____

_____

_____

Relaxin (RLX). _____

_____

_____

_____

Inhibin. _____

_____

_____

_____

I. PINEAL GLAND (EPIPHYSIS CEREBRI)

The cone-shaped endocrine gland attached to the roof of the third ventricle is known as the *pineal* (PĪN-ē-al) *gland (epiphysis cerebri).*

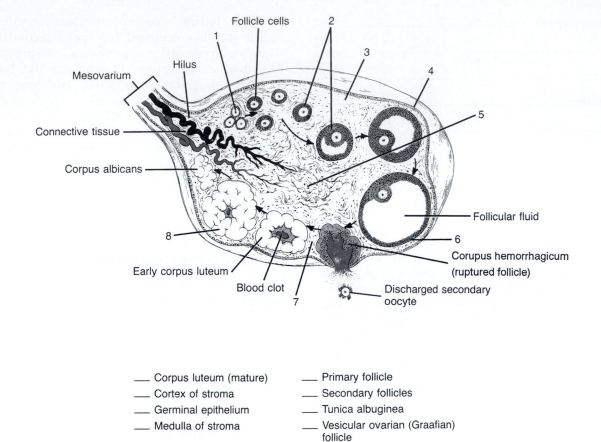

FIGURE 16.7   Histology of the ovary.

___ Corpus luteum (mature)   ___ Primary follicle
___ Cortex of stroma   ___ Secondary follicles
___ Germinal epithelium   ___ Tunica albuginea
___ Medulla of stroma   ___ Vesicular ovarian (Graafian)
    follicle

1. Histology of Pineal Gland

The gland is covered by a capsule formed by the pia mater and consists of masses of neuroglial cells and secretory cells called *pinealocytes* (pin-ē-AL-ō-sīts). Around the cells are scattered preganglionic sympathetic fibers. The pineal gland starts to calcify at about the time of puberty. Such calcium deposits are called *brain sand.* Contrary to a once widely held belief, no evidence suggests that the pineal atrophies with age and that the presence of brain sand indicates atrophy. Rather, brain sand may even denote increased secretory activity.

The physiology of the pineal gland is still somewhat obscure. The gland secretes *melatonin.*

2. Hormone of Pineal Gland

Using your textbook as a reference, give the major functions of the hormone listed.

Melatonin. _____

_____

_____

J. THYMUS GLAND

The *thymus gland* is a bilobed lymphatic gland located in the upper mediastinum posterior to the sternum and between the lungs. The gland is conspicuous in infants and reaches maximum size during puberty.

1. Histology of Thymus Gland

After puberty, thymic tissue, which consists primarily of *lymphocytes,* is replaced by fat. By the time a person reaches maturity, the gland has atrophied. Lymphoid tissue of the body consists primarily of two kinds of lymphocytes: B cells and T cells. Both are derived originally in the embryo from lymphocytic stem cells in bone marrow. Before migrating to their positions in lymphoid tissue, the descendants of the stem cells follow two distinct pathways. About half of them migrate to the thymus gland, where they are processed to become thymus-dependent lymphocytes, or *T cells.* The thymus gland confers on them the ability to destroy antigens (foreign microbes and substances). The remaining stem cells are processed in some as yet undetermined area of the body, possibly the fetal liver and spleen, and are known as *B cells.* Hormones produced by the thymus gland are *thymosin, thymic humoral factor (THF), thymic factor (TF),* and *thymopoietin.*

2. Hormones of Thymus Gland

Using your textbook as a reference, write the major functions of the hormones listed:
Thymosin, thymic humoral factor (THF), thymic

factor (TF), and thymopoietin. _____

_____

_____

K. OTHER ENDOCRINE TISSUES

Body tissues other than endocrine glands also secrete hormones. The gastrointestinal tract synthesizes several hormones that regulate digestion in the stomach and small intestine. Among these hormones are *gastrin, secretin, cholecystokinin (CCK),* and *gastric inhibitory peptide (GIP).*

The placenta produces *human chorionic gonadotropin (hCG), estrogens, progesterone (PROG), relaxin,* and *human chorionic somatomammotroph (hCS),* all of which are related to pregnancy.

When the kidneys (and liver, to a lesser extent) become hypoxic (subject to below-normal levels of oxygen), they release an enzyme called *renal erythropoietic factor.* This is secreted into the blood, where it acts on a plasma protein pro-

duced in the liver to form a hormone called *erythropoietin* (ē-rith'-rō-POY-ē-tin), which stimulates red bone marrow to produce more red blood cells and hemoglobin. This ultimately reverses the original stimulus (hypoxia).

Vitamin D, produced by the skin in the presence of sunlight, is converted to its active hormone, *calcitriol,* in the kidneys and liver.

The atria of the heart secrete a hormone called *atrial natriuretic peptide (ANP),* released in response to increased blood volume.

Using your textbook as a reference, write the major functions of the hormones listed:

Gastrin. _____

_____

Secretin. _____

_____

Cholecystokinin (CCK). _____

_____

Gastric inhibitory peptide (GIP). _____

_____

Human chorionic gonadotropin (hCG). _____

_____

Human chorionic somatomammotropin (hCS).

_____

_____

Erythropoietin. _____

_____

Calcitriol. _____

_____

Atrial natriuretic peptide (ANP). _____

_____

ANSWER THE LABORATORY REPORT QUESTIONS AT THE END OF THE EXERCISE.

Endocrine System

STUDENT _____   DATE _____

LABORATORY SECTION _____   SCORE/GRADE _____

PART 1. Multiple Choice

_____   1. Somatotrophs, gonadotrophs, and corticolipotrophs are associated with the (a) thyroid (b) anterior pituitary (c) parathyroids (d) adrenals

_____   2. The posterior pituitary is *not* an endocrine gland because it (a) has a rich blood supply (b) is not near the brain (c) does not make hormones (d) contains ducts

_____   3. Which hormone assumes a role in the development and discharge of a secondary oocyte? (a) hGH (b) TSH (c) LH (d) PRL

_____   4. The endocrine gland that is probably malfunctioning if a person has a high metabolic rate is the (a) thymus (b) posterior pituitary (c) anterior pituitary (d) thyroid

_____   5. The antagonistic hormones that regulate blood calcium level are (a) hGH-TSH (b) insulin-glucagon (c) aldosterone-cortisone (d) CT-PTH

_____   6. The endocrine gland that develops from the sympathetic nervous system is the (a) adrenal medulla (b) pancreas (c) thyroid (d) anterior pituitary

_____   7. The hormone that aids in sodium conservation and potassium excretion is (a) hydrocortisone (b) CT (c) ADH (d) aldosterone

_____   8. Which of the following hormones is sympathomimetic? (a) insulin (b) oxytocin (OT) (c) epinephrine (d) testosterone

_____   9. Which hormone lowers blood sugar level? (a) glucagon (b) melatonin (c) insulin (d) cortisone

_____   10. The hormone whose release is related to the diurnal dark-light cycle is (a) melatonin (b) insulin (c) thymosin (d) calcitonin

PART 2. Completion

11. The hypophysis is attached to the hypothalamus by a stalklike structure called the

_____ .

12. A hormone that acts on another endocrine gland and causes that gland to secrete its own hormones

is called a(n) _____ hormone.

13. _____ cells of the anterior pituitary synthesize ACTH.

14. The hormone that helps cause contraction of the smooth muscle of the pregnant uterus is

_____ .

15. Histologically, the spherical sacs that compose the thyroid gland are called thyroid

    _____.

16. The thyroid hormones associated with metabolism are triiodothyronine and

    _____.

17. Principal and oxyphil cells are associated with the _____ gland.

18. The zona glomerulosa of the adrenal cortex secretes a group of hormones called

    _____.

19. The hormones that promote normal metabolism, provide resistance to stress, and function as anti-

    inflammatories are _____.

20. The pancreatic hormone that raises blood sugar level is _____.

21. In spermatogenesis, the most immature cells near the basement membrane are called

    _____.

22. Cells within the testes that secrete testosterone are known as _____.

23. The ovaries are attached to the uterus by means of the _____ ligament.

24. The female hormones that help cause the development of secondary sex characteristics are called

    _____.

25. The endocrine gland that assumes a direct function in the proliferation and maturation of a

    lymphocyte is the _____.

26. Any hormone that regulates the functions of the gonads is classified as a(n)

    _____ hormone.

27. The hormone that is stored in the posterior pituitary that prevents excessive urine production is

    _____.

28. The region of the adrenal cortex that synthesizes androgens is the _____.

29. The hormone-producing cells of the adrenal medulla are called _____ cells.

30. Together, alpha cells, beta cells, delta cells, and F-cells constitute the _____.

31. Regulating hormones are produced by the _____ and reach the anterior pituitary
    by a network of blood vessels.

32. FSH and LH are produced by _____ of the anterior pituitary.

33. The structure in an ovary that produces progesterone (PROG), estrogens, and relaxin is the

    _____.

34. Calcium deposits in the pineal gland are referred to as _____.

35. A gland that is both an exocrine and endocrine gland is known as a _____ gland.

Blood

The blood, heart, and blood vessels constitute the *cardiovascular system.* In this exercise you will examine the characteristics of *blood,* a connective tissue also known as *vascular tissue.*

CAUTION! *Please reread Section A, "General Safety Precautions and Procedures," and Section B, "Precautions Related to Working with Blood, Blood Products, or Other Body Fluids," on pages xiii–xiv, at the beginning of the laboratory manual, before you begin any of the following experiments. Read the experiments before you perform them, to be sure that you understand all the procedures and safety precautions.*

When working with whole blood, wear tight-fitting surgical gloves and safety goggles. Avoid any kind of contact with an open sore, cut, or wound.

Work with your own blood only if blood is to be obtained by finger puncture. After you have completed your experiments, place all glassware in a fresh solution of household bleach or other comparable disinfectant, wash the laboratory tabletop with a fresh solution of household bleach or comparable disinfectant, and dispose of your gloves in the appropriate biohazard container provided by your instructor.

A. COMPONENTS OF BLOOD

Blood is a liquid connective tissue that is composed of two portions: (1) *plasma,* a liquid that contains dissolved substances, and (2) *formed elements,* cells and cell fragments suspended in the plasma. In clinical practice, the following is the most common classification of the formed elements of the blood:

Erythrocytes (e-RITH-rō-sīts), or *red blood cells*
Leukocytes (LOO-kō-sīts), or *white blood cells*
Granular leukocytes (granulocytes)
  Neutrophils
  Eosinophils
  Basophils
Agranular leukocytes (agranulocytes)
  Lymphocytes (T cells and B cells)
  Monocytes
Thrombocytes (THROM-bō-sīts), or *platelets*

The origin and subsequent development of these formed elements can be seen in Figure 17.1. Blood cells are formed by a process called *hemopoiesis* (hē-mō-poy-Ē-sis) or *hematopoiesis* (hē-ma-tō-poy-Ē-sis); the process by which erythrocytes are formed is called *erythropoiesis* (e-rith'-rō-poy-Ē-sis). The immature cells that are eventually capable of developing into mature blood cells are called *pluripotent hematopoietic stem cells,* or *hemocytoblasts* (hē-mō-SĪ-tō-blasts) (see Figure 17.1). Mature blood cells are constantly being replaced, so special cells called *reticuloendothelial cells* (fixed macrophages that line liver sinusoids) have the responsibility of clearing away the dead, disintegrating cell bodies so that small blood vessels are not clogged.

The shapes of the nuclei, staining characteristics, and color of cytoplasmic granules are all useful in the differentiation and identification of the various white blood cells. Red blood cells are biconcave discs without nuclei and can be identified easily.

321

B. PLASMA

When the formed elements are removed from blood, a straw-colored liquid called *plasma* is left. This liquid consists of about 91.5% water and about 8.5% solutes. Among the solutes are proteins (albumins, globulins, and fibrinogen), nonprotein nitrogen (NPN) substances (urea, uric acid, and creatine), foods (amino acids, glucose, fatty acids, glycerides, and glycerol), regulatory substances (enzymes and hormones), gases (oxygen and carbon dioxide), and electrolytes (Na^+, K^+, Ca^{2+}, Mg^{2+}, Cl^-, HCO_3^-, SO_4^{2-}, and HPO_4^{3-}).

1. Physical Characteristics

CAUTION! *Please reread Section A, "General Safety Precautions and Procedures," and Section B, "Precautions Related to Working with Blood, Blood Products, or Other Body Fluids" on pages xiii–xiv, at the beginning of the laboratory manual before you begin any of the following experiments. Read the experiments before you perform them, to be sure that you understand all the procedures and safety precautions.*

When working with whole blood, wear tight-fitting surgical gloves and safety goggles. Avoid any kind of contact with an open sore, cut, or wound.

PROCEDURE

Note: Obtain the plasma or whole blood for this procedure from a clinical lab-oratory where it has been tested and certified as noninfectious* or from a mammal (other than a human).

1. Obtain about 5 ml of plasma as follows.
2. Centrifuge the whole blood obtained from a clinical laboratory where it has been tested and certified as noninfectious or from a mammal (other than a human). After centrifugation, the bottom of the test tube will contain an anticoagulant (usually EDTA). Right above this will be a layer of red blood cells, followed by a layer of leukocytes and platelets (buffy coat). The top layer is plasma. Visual inspection of these layers will give an indication of the relative proportions of the various components of blood.
3. Remove 5 ml of plasma using a disposable sterile Pasteur pipette attached to a bulb and place the plasma in a disposable test tube.

*Whole blood that has been tested for syphilis; hepatitis A, B, and C; and HIV is available from Carolina Biological Supply Company.

Note its color. Test the pH of the plasma by dipping the end of a length of pH paper into the plasma and comparing the paper color with the color scale on the pH container. Record the pH in Section B.1 of the LABORATORY REPORT RESULTS at the end of the exercise.
4. Hold the test tube up to a source of natural light and note the color and transparency of the plasma. Record the color and transparency in Section B.1 of the LABORATORY REPORT RESULTS at the end of the exercise.
5. *Dispose of the pipettes and test tubes in the appropriate biohazard container, provided by your instructor. Dispose of the pH paper and plasma as per your instructor's directions.*

2. Chemical Constituents

CAUTION! *Please reread Section A, "General Safety Precautions and Procedures," and Section B, "Precautions Related to Working with Blood, Blood Products, or Other Body Fluids," on pages xiii–xiv, at the beginning of the laboratory manual, before you begin any of the following experiments. Read the experiments before you perform them, to be sure that you understand all the procedures and safety precautions.*

When working with whole blood, wear tight-fitting surgical gloves and safety goggles. Avoid any kind of contact with an open sore, cut, or wound.

PROCEDURE

Note: Obtain the plasma or whole blood for this procedure from a clinical laboratory where it has been tested and certified as noninfectious or from a mammal (other than a human).

1. Using a medicine dropper, add 10 ml of distilled water to 5 ml of plasma in a Pyrex test tube.

CAUTION! *Place the test tube in a test tube rack because it will become too hot to handle.*

2. With forceps, add one clinitest tablet.

CAUTION! *The concentrated sodium hydroxide in the tablet generates enough heat to make the liquid in the test tube boil. Make sure that the mouth of the test tube is pointed away from you and anyone else in the area.*

3. The color of the solution is graded as follows:

Color	Results
Blue	Negative
Greenish-yellow	1 + (0.5 g/100 ml)
Olive green	2 + (1 g/100 ml)
Orange-yellow	3 + (1.5 g/100 ml)
Brick red (with precipitate)	4 + (more than 2 g/100ml)

4. Fifteen seconds after the boiling has stopped, shake the test tube *gently* and evaluate the color according to the test table above.

CAUTION! *Make sure that the mouth of the test tube is pointed away from you and others.*

A green, yellow, orange, or red color indicates the presence of glucose. Record your results in Section B.2 of the LABORATORY REPORT RESULTS at the end of the exercise.

5. Test the contents of the filter paper for the presence of protein. If protein is present, it will be coagulated by the acetic acid and filtered out of the plasma. Place the contents of the filter paper in a test tube and add 3 ml of water and 3 ml of Biuret reagent. A violet or purple color indicates the presence of protein. Record your results in Section B.2 of the LABORATORY REPORT RESULTS at the end of the exercise.

C. ERYTHROCYTES

Erythrocytes (red blood cells, RBCs) are biconcave in appearance, have no nucleus, and can neither reproduce nor carry on extensive metabolic activities (Figure 17.1). The interior of the cell contains a red pigment called *hemoglobin,* which is responsible for the red color of blood. The heme portion of hemoglobin combines with oxygen and, to a lesser extent, carbon dioxide and transports them through the blood vessels. An average red blood cell has a life span of about 120 days. A healthy male has about 5.4 million red blood cells per cubic millimeter (mm^3) or per deciliter (dl) of blood, a healthy female, about 4.8 million. Erythropoiesis and red cell destruction normally proceed at the same pace. A diagnostic test that informs the physician about the rate of erythropoiesis is the *reticulocyte* (re-TIK-yoo-lō-sīt) *count. Reticulocytes* (see Figure 17.1) are precursor cells of mature red blood cells. Nor-

mally, a reticulocyte count is 0.5% to 1.5%. The reticulocyte count is an important diagnostic tool because it is a relatively accurate predictor of the status of red cell production in bone marrow. Normally, bone marrow replaces about 1% of the adult red blood cells each day. A decreased reticulocyte count (reticulocytopenia) is seen in aplastic anemia and in conditions in which the bone marrow is not producing red blood cells. An increase in reticulocytes (reticulocytosis) is found in acute and chronic blood loss and certain kinds of anemias such as iron-deficiency anemia.

NOTE: Although many blood tests in this exercise are sufficient for laboratory demonstration, they are not necessarily clinically accurate.

D. RED BLOOD CELL TESTS

CAUTION! *Please reread Section A, ''General Safety Precautions and Procedures,'' and Section B, ''Precautions Related to Working with Blood, Blood Products, or Other Body Fluids,'' on pages xiii–xiv, at the beginning of the laboratory manual before you begin any of the following experiments. Read the experiments before you perform them, to be sure that you understand all the procedures and safety precautions.*

When working with whole blood, wear tight-fitting surgical gloves and safety goggles. Avoid any kind of contact with an open sore, cut, or wound.

1. Finger Puncture Procedure*

CAUTION! *SOURCE OF BLOOD*

At your discretion and that of your instructor and within the policies of your institution regarding laboratory safety procedures, blood for various tests

---

*If the finger puncture procedure is not performed, blood can be obtained from a clinical laboratory where it has been tested and certified as noninfectious or from a mammal (other than a human).

Whole blood that has been tested for syphilis; hepatitis A, B, and C; and HIV is available from Carolina Biological Supply Company. This blood can be used for blood grouping and typing, blood glucose, hematocrit, hemoglobin, red blood cell count, white blood cell count, differential white blood cell count, platelet count, sedimentation rate, osmotic fragility studies, and various plasma chemistries. Carolina Biological Supply Company also provides aseptic red blood cells that can be used for blood grouping and typing, blood glucose, hematocrit, hemoglobin, red blood cell count, and osmotic fragility studies. These blood cells are suspended in a modified alsevere's solution, to which several antibiotics have been added. The cells have been tested for syphilis; hepatitis A, B, and C; and HIV.

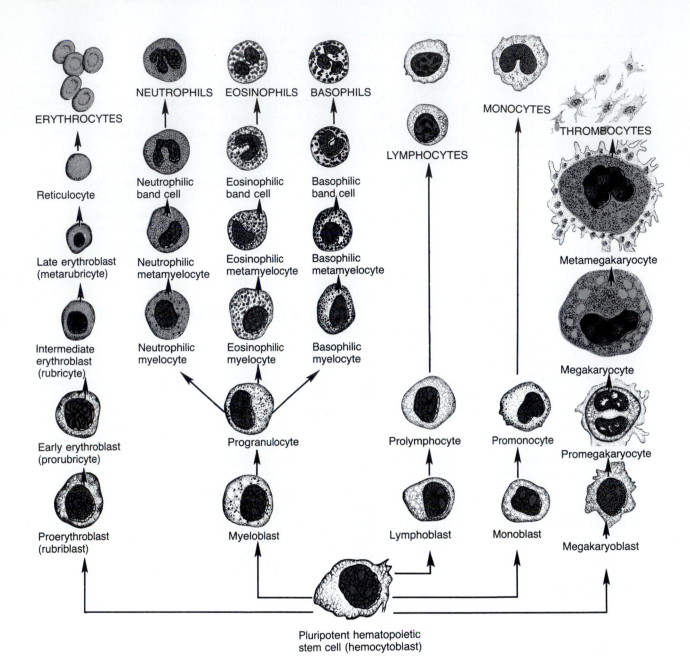

ERYTHROCYTES

Reticulocyte

Late erythroblast
(metarubricyte)

Intermediate
erythroblast
(rubricyte)

Early erythroblast
(prorubricyte)

Proerythroblast
(rubriblast)

NEUTROPHILS   EOSINOPHILS   BASOPHILS

Neutrophilic
band cell

Eosinophilic
band cell

Basophilic
band cell

Neutrophilic
metamyelocyte

Eosinophilic
metamyelocyte

Basophilic
metamyelocyte

Neutrophilic
myelocyte

Eosinophilic
myelocyte

Basophilic
myelocyte

Progranulocyte

Myeloblast

LYMPHOCYTES

Prolymphocyte

Lymphoblast

MONOCYTES

Promonocyte

Monoblast

THROMBOCYTES

Metamegakaryocyte

Megakaryocyte

Promegakaryocyte

Megakaryoblast

Pluripotent hematopoietic
stem cell (hemocytoblast)

FIGURE 17.1   Origin, development, and structure of blood cells.

*may be obtained by finger puncture. All students do
not need to provide blood samples for all the tests.
Each student should provide blood for only one or
two tests from a single puncture. The results from
the various tests can be recorded on the blackboard
and shared by the entire class.*

PROCEDURE

1. Before you begin: To enhance your chances
   for a successful finger puncture the first time,
   warm your finger (if it is cold), shake your

finger several times to dry the alcohol and
force more blood into it, and gently milk
blood toward the puncture site.

2. Obtain 70% alcohol, a sterile lancet, and
   cotton.

3. *Thoroughly* cleanse the end of the third or
   fourth finger with alcohol.

4. Open the lancet packet at the blunt end; *do
   not touch the sharp end.*

5. *Take a sample of YOUR OWN BLOOD ONLY as
   follows.* Remove the rubber glove from the

hand from which you will take blood. Holding your finger firmly, take the lancet and puncture the finger, using a quick, jabbing motion. The puncture is made toward the side of the finger rather than in the full fleshy part. (Do not begin the motion from a great distance away from the finger. The closer you are and the quicker the motion, the less the discomfort that results.)

6. If the puncture is made properly, the blood should flow freely from the wound. If not, squeeze the finger at its base, moving toward the tip.

7. If an additional puncture is required, discard the first lancet and *use a new sterile one.*

CAUTION! *Make sure that used lancets are discarded immediately in the appropriate biohazard container provided by your instructor, to prevent subsequent injury and to prevent the passing of diseases such as AIDS and hepatitis from one person to another.*

8. Do not use the first drop, which usually coagulates quickly.

9. Finally, place a piece of cotton between the finger and the thumb and press them together firmly.

10. *Use an adhesive bandage to cover the puncture site.*

2. Filling of Hemocytometer (Counting Chamber)

The procedure for filling the hemocytometer follows. Please read it carefully *before* doing Procedure 3, Red Blood Cell Count.

PROCEDURE

Note: The blood used for this procedure can be obtained by finger puncture or from a clinical laboratory where it has been tested and certified as noninfectious or from a mammal (other than a human).

1. Obtain a hemocytometer and cover slip (Figure 17.2).
2. Clean the hemocytometer thoroughly and carefully with alcohol.
3. Place the cover slip on the hemocytometer.
4. Fill an Unopette Reservoir System® pipette

with the proper dilution of blood. Place the tip of the pipette on the polished surface of the hemocytometer next to the edge of the cover slip. Although we recommend this system, your instructor may wish to use an alternate procedure. *If you use a different system for pipetting and diluting, use rubber suction bulbs or pipette pumps. Do not pipette by mouth.*

5. Deposit a small drop of diluted blood by squeezing the sides of the reservoir, but do not leave the tip of the pipette in contact with the hemocytometer for more than an instant because this will cause the chamber to overfill. The diluted blood must not overflow the moat; overfilling the moat results in an inaccurate cell count. A properly filled hemocytometer has a blood specimen only within the space between the cover glass and counting area.

3. Red Blood Cell Count

The purpose of a red blood cell count is to determine the number of circulating red blood cells per cubic millimeter (mm 3) or per deciliter (dl) of blood. Red blood cells carry oxygen to all tissues; thus, a drastic reduction in the red cell count will cause immediate reduction in available oxygen.

A decrease in red blood cells can result from a variety of conditions, including impaired cell production, increased cell destruction, and acute blood loss. When the red cell count is increased above normal limits, the condition is called *polycythemia* (pol'-ē-sī-THĒ-mē-a).

PROCEDURE

WARD'S Natural Science Establishment, Inc., has made available WARD'S *Simulated Blood.* It is a solution that contains microcomponents that simulate erythrocytes, leukocytes, and thrombocytes. The microcomponents are similar in relative proportion to those found in human blood and can be observed under a microscope without staining.

1. Obtain a Simulated Blood Activity kit.
2. Follow the instructions to perform the erythrocyte count.
3. Complete the portion of the Data Table related to erythrocytes.
4. Answer the questions related to erythrocytes.

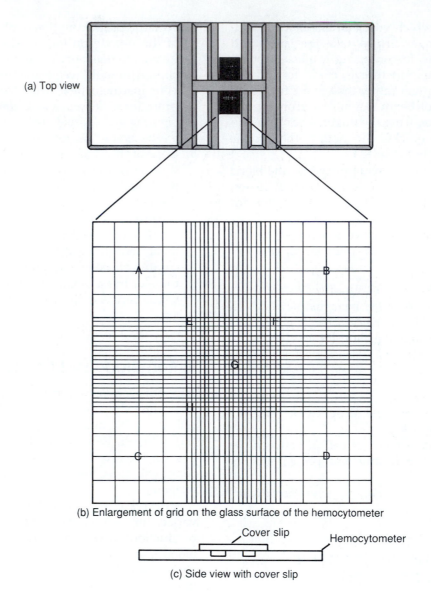

(a) Top view

(b) Enlargement of grid on the glass surface of the hemocytometer

Cover slip

Hemocytometer

(c) Side view with cover slip

FIGURE 17.2 Various parts of a hemocytometer (counting chamber). In (b), the areas used for red cell counts are labeled E, F, G, H, and I; those for white cell counts are labeled A, B, C, and D.

ALTERNATE PROCEDURE

Note: The blood used for this procedure can be obtained by finger puncture or from a clinical laboratory where it has been tested and certified as noninfectious or from a mammal (other than a human).

1. Obtain a Unopette Reservoir System® for red blood cell determination (the color of the reservoir's bottom surface is red). Identify the reservoir chamber, diluent fluid inside the chamber, pipette, and protective shield for the pipette (Figure 17.3). The diluent fluid

contains isotonic saline and sodium azide, an antibacterial agent.

CAUTION! *Do not ingest this fluid; it is for in vitro diagnostic use only.*

2. Hold the reservoir on a flat surface in one hand and grasp the pipette assembly in the other hand.
3. Push the tip of the pipette shield firmly through the diaphragm in the neck of the reservoir (Figure 17.4a).

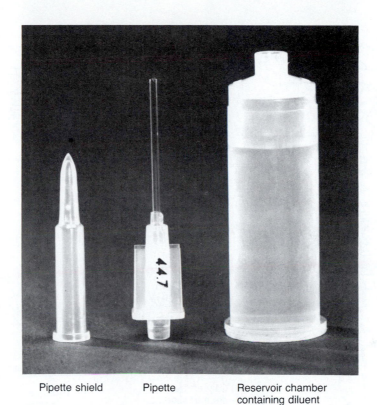

| Pipette shield | Pipette | Reservoir chamber containing diluent fluid |

FIGURE 17.3 Photograph of the Unopette Reservoir System®.

4. Pull out the assembly unit and, with a twist, remove the protective shield from the pipette assembly.

5. Hold the pipette *almost* horizontally and touch the tip of it to a forming drop of blood that has been placed on a microscope slide (Figure 17.4b). The pipette will fill by capillary action and, when the blood reaches the end of the capillary bore in the neck of the pipette, the filling action will stop.

6. Carefully wipe any excess blood from the tip of the pipette, using a Kimwipe or similar wiping tissue, *making certain that no sample is removed from the capillary bore.*

7. Squeeze the reservoir slightly to expel a small amount of air.

CAUTION! *Do not expel any liquid. If the reservoir is squeezed too hard, the specimen may be expelled through the overflow chamber, resulting in contamination of the fingers. Reagent contains sodium azide, which is extremely toxic and yields explosive products in metal sinks or pipes. Azide compounds should be diluted with running water before being discarded and disposed of as directed by your instructor.*

While maintaining pressure on the reservoir, cover the opening of the overflow chamber of the pipette with your index finger and push the pipette securely into the reservoir neck (Figure 17.4c).

8. Release the pressure on the reservoir and remove your finger from the pipette opening. This will draw the blood into the diluent fluid.

9. Mix the contents of the reservoir chamber by squeezing the reservoir gently several times. It is important to *squeeze gently* so that the diluent fluid is not forced out of the chamber. In addition to squeezing the reservoir chamber, invert it *gently* several times (Figure 17.4d).

10. Remove the pipette from the reservoir chamber, reverse its position, and replace it on the reservoir chamber (Figure 17.4e). This con-

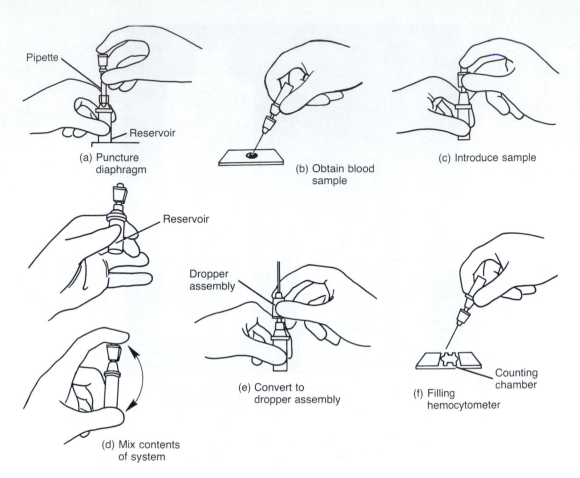

Pipette

Reservoir

(a) Puncture
diaphragm

(b) Obtain blood
sample

(c) Introduce sample

Reservoir

Dropper
assembly

(e) Convert to
dropper assembly

(f) Filling
hemocytometer

Counting
chamber

(d) Mix contents
of system

FIGURE 17.4   Preparation of the Unopette Resevior System®. (Modified from *RBC Determination for Manual Methods* and *WBC Determination for Manual Methods*, Becton-Dickinson, Division of Becton, Dickinson and Company.)

verts the apparatus into a dropper assembly (Figure 17.4f).

11. Squeeze a few drops out of the reservoir chamber into a disposal container or wipe with Kimwipe or similar wiping tissue. You are now ready to fill the hemocytometer.

12. Follow the procedure outlined in Number 2, filling the hemocytometer.

13. Place the hemocytometer on the microscope stage and allow the red blood cells to settle in the hemocytometer (approximately 1 to 2 min).

14. When counting red blood cells with the hemocytometer, the loaded (filled) chamber should be first located using the low power objective of the microscope. Upon finding that, switch to the high-power objective via parfocal procedure. Focus the field using the fine adjustment of the microscope. Each red cell counted is equal to 10,000, so the degree

of error in manual blood cell counting is quite high, generally in the range of 15% to 20%.

15. Using the high-power lens, count the cells in each of the five squares (E, F, G, H, I) as shown in Figure 17.5. It is suggested that you use a hand counter. As you count, move up and down in a systematic manner. *Note:* To avoid overcounting cells at the boundaries, count the cells that touch the lines on the left and top sides of the hemocytometer, but not the ones that touch the boundary lines on the right and bottom.

16. Multiply the total number of red blood cells counted in the five squares by 10,000 to obtain the number of red blood cells in 1 mm^3 of blood. Record your value in Section D.1 of the LABORATORY REPORT RESULTS at the end of the exercise.

17. The hemocytometer and cover slip should be saved, not discarded.

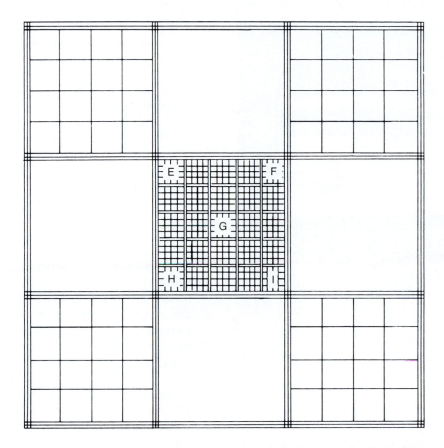

FIGURE 17.5 Hemocytometer (counting chamber). Areas E, F, G, H, and I are all counted in the red blood cell count. The large square containing the smaller squares E, F, G, H, and I is seen in microscopic field under 10× magnification. To count the smaller squares (E, F, G, H, and I), the 45× lens is used; therefore, E, F, G, H, and I each encompass the entire microscopic field.

CAUTION! *Place the hemocytometer and cover slip in a container of fresh bleach solution and dispose of all other materials as directed by your instructor.*

4. Red Blood Cell Volume (Hematocrit)

The percentage of blood volume occupied by the red blood cells is called the *hematocrit (Hct)* or *packed cell volume (PCV).* It is an important component of a complete blood count and is a standard test for almost everyone having a physical examination or for diagnostic purposes. When a tube of blood is centrifuged, the erythrocytes pack into the bottom part of the tube with the plasma on top. The white blood cells and platelets are found in a thin area, the buffy layer, above the red blood cells.

The Readacrit® (Figure 17.6) incorporates a built-in hematocrit scale and tube-holding compartments which, when used with special precalibrated capillary tubes, permits direct reading of the hematocrit value by measuring the length of the packed red cell column. Readacrit present the final hematocrit value without requiring computation by the operator. If the Readacrit or tube reader is not used for direct reading, measure the total length of blood volume, divide this quantity into length of packed red cells, and multiply by 100 for percentage.

The calculation follows:

$$\frac{\text{length of packed red cells}}{\text{total length of blood volume}} \times 100 =$$

% volume of whole blood occupied by red cells (hematocrit)

FIGURE 17.6   Centrifuge used for spinning blood to determine hematocrit.

In males, the normal range is between 40% and 54%, with an average of 47%. In females the normal range is between 38% and 47%, with an average of 42%. Anemic blood may have a hematocrit of 15%; polycythemic blood may have a hematocrit of 65%.

The following procedure for testing red blood cell volume is a micromethod requiring only a drop of blood:

PROCEDURE

Note: The blood used for this procedure can be obtained by finger puncture or from a clinical laboratory where it has been tested and certified as noninfectious or from a mammal (other than a human).

1. Place a drop of blood on a microscope slide.
2. Place the unmarked (clear) end of a disposable capillary tube into the drop of blood. Hold the tube slightly below the level of the blood and do not move the tip from the blood or an air bubble will enter the column.

CAUTION! *If an air bubble is present, immediately dispose of the capillary tube in the appropriate biohazard container provided by your instructor and begin again.*

3. Allow blood to flow about two-thirds of the way into the tube. The tube will fill easily if you hold the open end level with or below the blood source.
4. Place your finger over the marked (red) end of the tube and keep it in that position as you seal the blood end of the tube with Seal-ease® clay or other similar sealing material.
5. According to the directions of your laboratory instructor, place the tube into the centrifuge, making sure that the sealed end is against the rubber ring at the circumference of the centrifuge rotor. Have your laboratory instructor make sure that the tubes are properly balanced. Write the number there that indicates the location of your tube. _____
6. When all the tubes from your laboratory section are in place, secure the inside cover of the centrifuge and close the outside cover. Centrifuge on high speed for 4 min.
7. Remove your tube and determine the hematocrit value by reading the length of packed red cells directly on the centrifuge scale, or by placing the tube in the tube reader (Figure 17.7) and following the instructions on the reader, or by calculation using the equation mentioned previously (page 329). When using the tube reader, place the base of the Seal-ease clay on the base line and move the tube until the top of the plasma is at the line marked 100.

FIGURE 17.7   Microhematocrit tube reader.

CAUTION! *Immediately dispose of your capillary tube in the appropriate biohazard container provided by your instructor.*

Record your results in Section D.2 of the LABORATORY REPORT RESULTS at the end of the exercise.

What effect would dehydration have on hematocrit (Hct)? _____

_____

5. Hemoglobin Determination

It is possible to have anemia even if your red blood cell count is normal. For example, the red blood cells present may be deficient in hemoglobin or they may be smaller than normal. Thus, the amount of hemoglobin per unit volume of blood, and not necessarily the number of red blood cells, is the determining factor for anemia. The hemoglobin content of blood is expressed in grams (g) per 100 ml of blood and the normal ranges vary with the technique used. An average range is 12 to 15 g/100 ml in females and 13 to 16 g/100 ml in males.

One procedure that can be used to determine hemoglobin uses an apparatus called a *hemoglobinometer* (see Figure 17.8a). This instrument compares the absorption of light by hemoglobin in a blood sample of known depth to that of a standardized glass plate.[1]

PROCEDURE

Note: The blood used for this procedure can be obtained by finger puncture or from a clinical laboratory where it has been tested and certified as noninfectious or from a mammal (other than a human).

1. Obtain a hemoglobinometer and examine its parts (Figure 17.8a). Note the four scales on the side of the instrument. The top scale gives hemoglobin content in g/100 ml of blood; the other three scales give comparative readings, that is, the amount of hemoglobin expressed as a percent of 15.6, 14.5, or 13.8 g/100 ml of blood.

_____

[1]If Unopette tests are available, follow the procedure outlined in Unopette test No. 5857, "Cyanmethemoglobin Determination for Manual Methods," for hemoglobin estimation.

2. Remove the two pieces of glass from the blood chamber assembly (Figure 17.8b). Clean them with alcohol and wipe them with lens paper. The piece of glass that contains the moat will receive the blood sample; the other piece of glass serves as a cover glass.
3. Replace the glass pieces halfway onto the clip with the moat plate on the bottom.
4. Obtain a sample of blood.
5. Apply the blood to the moat and completely cover the raised surface of the plate with blood.
6. Hemolyze the blood by agitating it with the tip of a hemolysis applicator for about 30 to 45 sec. Hemolysis is complete when the appearance of the blood changes from cloudy to transparent. This change reflects the lysis of the red blood cell membranes and the liberation of hemoglobin.
7. Push the specimen chamber into the clip and push the clip into the hemocytometer.
8. Holding the hemocytometer in your left hand, look through the eyepiece while depressing the light switch button. You will see a split field.
9. Move the slide button back and forth with the right index finger until the two sides of the field match in color and shading.
10. Read the value indicated on the scale marked 15.6. Determine the grams of hemoglobin per 100 ml of the blood sample by reading the number above the index mark on the hemocytometer. Record your results in Section D.3 of the LABORATORY REPORT RESULTS at the end of the exercise.

CAUTION! *Remove the glass pieces from the chamber and wash them in a fresh bleach solution. Dry the glass pieces with lens paper.*

11. What type of anemia would reveal a normal or slightly below normal hematocrit, yet a significantly reduced hemoglobin level?

_____

ALTERNATE PROCEDURE

The *Tallquist measurement* of hemoglobin is an old and somewhat inaccurate technique, although it is quick and inexpensive. The hemoglobinometer gives more accurate (±5%) results.

Note: The blood used for this procedure may be obtained by finger puncture or from a clinical

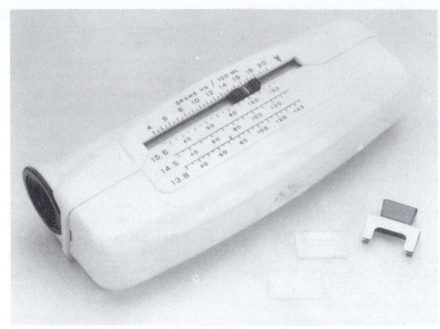

(a) Cambridge Instruments Hemoglobin-Meter

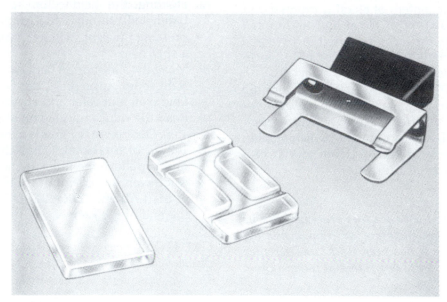

(b) Specimen chamber

FIGURE 17.8   Hemoglobinometer.

laboratory where it has been tested and certified as noninfectious or from a mammal (other than a human).

1. Obtain a sample of blood.
2. Place one drop of blood on the Tallquist® paper.
3. As soon as the blood no longer appears shiny, match its color with the scale provided.

Record your observations in Section D.3 of the LABORATORY REPORT RESULTS at the end of the exercise.

6. Oxyhemoglobin Saturation

The ability of blood to carry oxygen is determined by proper respiratory system function, hematocrit (Hct), and the ability of hemoglobin to bind reversibly with oxygen within the alveoli of the lungs and release oxygen at the tissue level. Hemoglobin's interaction with oxygen is measured by an *oxygen-hemoglobin dissociation curve.* Such a curve demonstrates that the number of available oxygen-binding sites on hemoglobin that actually bind to oxygen is proportional to the partial pressure of oxygen (pO_2). Arterial blood usually has a pO_2 of 95, causing 97% of the available hemoglobin to be saturated with oxygen. Venous blood, however, has a pO_2 of 40, so only 75% is saturated with oxygen; 25% is given off to tissues. The determination of percent saturation for hemoglobin is a very sensitive test of pulmonary function. However, such a test needs to be interpreted carefully, as an individual with a normally functioning pulmonary system can exhibit abnormally low oxyhemoglobin saturation due to a variety of possible causes, such as *methemoglobinemia* (transformation of normal oxyhemoglobin due to the reduction of normal Fe^{2+} to Fe^{3+} within the hemoglobin molecule) or *carbon monoxide (CO) poisoning.* In order to determine the percent hemoglobin saturation in an unknown sample of blood, the absorption spectrum of the blood sample would need to be compared to that obtained from a pure sample of *oxyhemoglobin (HbO₂)*, *carboxyhemoglobin (Hb-CO)*, and *reduced hemoglobin (Hb)* (carboxyhe-

moglobin is included due to the normal presence of carbon monoxide in today's polluted atmosphere). This is shown in Figure 17.9. The spectrum obtained with the unknown sample of blood would be some combination of these three, because it would contain a certain percentage of each form of hemoglobin. Such a test is possible because all three of these types of hemoglobin compounds are different colors and would therefore absorb different portions of the light spectrum. The percentage of each hemoglobin form can be determined through a complex evaluation of the relative contribution of each type of hemoglobin to the resulting spectrum for the unknown sample.

7. Spectrum of Oxyhemoglobin and Reduced Hemoglobin

The following procedure details a simplified, although less accurate method for determining the absorption spectra for fully oxygenated and deoxygenated hemoglobin. Exposing blood to air saturates it with oxygen, while sodium dithionate reduces it by removing oxygen.

PROCEDURE

Note: The blood used for this procedure can be obtained by finger puncture or from a clinical

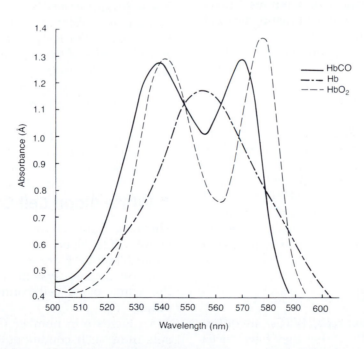

FIGURE 17.9 Absorption spectra of carboxyhemoglobin (HbCO), reduced hemoglobin (Hb), and oxyhemoglobin (Hb₂).

laboratory where it has been tested and certified as noninfectious or from a mammal (other than a human).

1. Turn on the spectrophotometer and let it warm up for several minutes.
2. After the instrument has warmed up, set the wavelength at 500 nm and adjust the meter needle to 0% transmittance (absorbance at infinity) by using the zero control knob.
3. Place a cuvette containing distilled water into the cuvette holder and adjust the needle to 100% by using the light control knob.
4. Place 8 ml of distilled water in a clean test tube.
5. Obtain several drops of blood.
6. Mix the distilled water with the blood by placing a stopper in the mouth of the test tube and then inverting the test tube, thereby adding the blood to the distilled water.
7. Transfer half of the test tube contents (4 ml) into a second clean test tube.
8. Add 0.2 ml of 1% sodium dithionite solution to the second test tube and mix thoroughly. (*Note:* In order for the procedure to work properly, the sodium dithionite solution must be fresh and made up just prior to use, and a spectral determination must be completed within 5 min of adding sodium dithionite to the second tube.)
9. Record the absorbances of solutions 1 (oxyhemoglobin) and 2 (reduced hemoglobin) at 500 nm wavelength.
10. Standardize the spectrophotometer at 510 nm utilizing the cuvette containing only distilled water as outlined in steps 2 and 3.
11. Determine the absorbances for solutions 1 and 2 at 510 nm.
12. Repeat procedures 2, 3, and 9 at each of the following wavelengths: 520, 530, 540, 560, 570, 580, 590, 600 nm.
13. Record and graph your results in Section D.4 of the LABORATORY REPORT RESULTS at the end of the exercise. Be sure to compare and give physiological reasons for the observed differences in the two spectra.

E. LEUKOCYTES

Leukocytes (white blood cells, WBCs) are different from red blood cells in that they have nuclei and do not contain hemoglobin (see Figure 17.1). They are less numerous than red blood cells, ranging from 5,000 to 10,000 cells per cubic millimeter (mm^3) or per deciliter (dl) of blood. The ratio, therefore, of red blood cells to white blood cells is about 700:1.

As Figure 17.1 shows, leukocytes can be differentiated by their appearance. They are divided into two major groups, granular leukocytes and agranular leukocytes. *Granular leukocytes,* which are formed from red bone marrow, have large granules in the cytoplasm that can be seen under a microscope and possess lobed nuclei. The three types of granular leukocytes are *neutrophils, eosinophils,* and *basophils. Agranular leukocytes,* which are also formed from red bone marrow, contain small granules that cannot be seen under a light microscope and usually have spherical nuclei. The two types of agranular leukocytes are *lymphocytes* and *monocytes.*

Leukocytes as a group function in phagocytosis, producing antibodies, and combatting allergies. The life span of a leukocyte usually ranges from a few hours to a few months. Some lymphocytes, called T and B memory cells, can live throughout a person's life once they are formed.

F. WHITE BLOOD CELL TESTS

CAUTION! *Please reread Section A, "General Safety Precautions and Procedures," and Section B, "Precautions Related to Working with Blood, Blood Products, or Other Body Fluids," on pages xiii–xiv at the beginning of the laboratory manual before, you begin any of the following experiments. Read the experiments before you perform them, to be sure that you understand all the procedures and safety precautions.*

When working with whole blood, wear tight-fitting surgical gloves and safety goggles. Avoid any kind of contact with an open sore, cut, or wound.

1. White Blood Cell Count

This procedure determines the number of circulating white blood cells in the body. Because white blood cells are a vital part of the body's immune defense system, any abnormalities in the white blood cell count must be carefully noted.

An increase in number *(leukocytosis)* can result from such conditions as bacterial or viral infection, metabolic disorders, chemical and drug poisoning, and acute hemorrhage. A decrease in number *(leukopenia)* may result from typhoid infection, measles, infectious hepatitis, tuberculosis, or cirrhosis of the liver.

PROCEDURE

WARD'S Natural Science Establishment, Inc., has made available WARD'S *Simulated Blood.* It is a solution that contains microcomponents that simulate erythrocytes, leukocytes, and thrombocytes. The microcomponents are similar in relative proportion to those found in human blood and can be observed under a microscope without staining.

1. Obtain a Simulated Blood Activity kit.
2. Follow the instructions to perform the leukocyte count.
3. Complete the portion of the Data Table related to count leukocytes.
4. Answer the questions related to leukocytes.

ALTERNATE PROCEDURE

Note: The blood used for this procedure can be obtained by finger puncture or from a clinical laboratory where it has been tested and certified as noninfectious or from a mammal (other than a human).

Whole blood that has been tested for syphilis; hepatitis A, B, and C; and HIV is available from Carolina Biological Supply Company. This blood can be used for blood grouping and typing, blood glucose, hematocrit, hemoglobin, red blood cell count, white blood cell count, differential white blood cell count, platelet count, sedimentation rate, osmotic fragility studies, and various plasma chemistries.

1. Follow steps 1 through 14 of D.3, "Red Blood Cell Count, Alternate Procedure," but use the Unopette Reservoir System® for white blood cell determination (the color of the reservoir's bottom surface is white).
2. Using the low-power objective (10X), count the cells in each of the four corner squares (A, B, C, D, in Figure 17.10) of the hemocytometer. The direction to follow when counting white blood cells is indicated in square A in Figure 17.10. *Note:* To avoid overcounting of cells at the boundaries, the cells that touch the lines on the left and top sides of the hemocytometer should be counted, but not the ones that touch the boundary lines on the right and bottom sides. It is suggested that you use a hand counter.

3. Multiply the results by 50 to obtain the amount of circulating white blood cells per mm^3 of blood and record your results in Section F.5 of the LABORATORY REPORT RESULTS at the end of the exercise.
4. The factor of 50 is the dilution factor, or 2.5 × 20 = 50. The volume correction factor of 2.5 is arrived at in this manner: Each of the corner areas (A, B, C, and D) is exactly 1 mm^3 by 0.1 mm deep (Figure 17.10). Therefore, the volume of each of these corner areas is 0.1 mm^3. Because four of them are counted, the total volume of diluted blood examined is 0.4 mm^3. However, because we want to know the number of cells in 1 mm^3 instead of 0.4 mm^3, we must multiply our count by 2.5 (0.4 × 2.5 = 1.0).

CAUTION! *Place the hemocytometer and cover slip in a container of fresh bleach solution and dispose of all other materials as directed by your instructor.*

5. The hemocytometer should be saved, not discarded.

2. Differential White Blood Cell Count

The purpose of a differential white blood cell count is to determine the relative *percentages* of each of the five normally circulating types of white blood cells in a total count of 100 white blood cells. A normal differential white blood cell count might appear as follows:

Type of leukocyte	Normal
Neutrophils	60–70%
Eosinophils	2–4%
Basophils	0.5–1%
Lymphocytes	20–25%
Monocytes	3–8%

Significant elevations of different types of white blood cells usually indicate specific pathological conditions. For example, a high neutrophil count may indicate acute appendicitis or infection. Lymphocytes predominate in antigen-antibody reactions, specific leukemias, and in

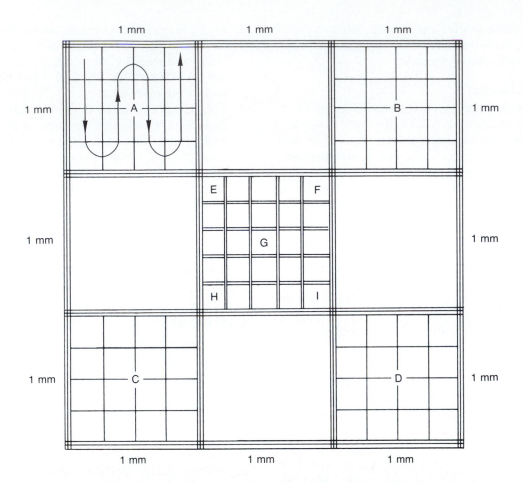

FIGURE 17.10 Hemocytometer (counting chamber) for a white blood cell count (as seen through scanning lens). A, B, C, and D are the areas counted in a white blood cell count (when viewed through 10× lens, 1 sq mm is seen in a microscopic field). Areas A, B, C, and D each equal 1 sq mm; therefore, a total of 4 sq mm is counted in a white cell count.

infectious mononucleosis. An increase in eosinophils can be seen in allergic reactions and parasitic infections. An elevated percentage of monocytes can result from chronic infections. An increase in basophils is rare and denotes a specific type of leukemia and allergic reactions.

The procedure for making a differential white blood cell count follows:

Type of leukocyte	Number observed	Percentage
Nuetrophils		
Eosinophils		
Basophils		
Lymphocytes		
Monocytes		

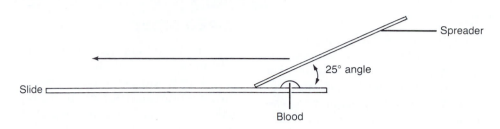

FIGURE 17.11 Procedure to prepare a blood smear.

PROCEDURE

NOTE: The blood used for this procedure can be obtained by finger puncture or from a clinical laboratory where it has been tested and certified as noninfectious or from a mammal (other than a human).

1. Obtain a drop of blood.
2. Use a second slide as a spreader (Figure 17.11).
3. Draw the spreader toward the drop of blood (in this direction →) until it touches the drop. The blood should fan out to the edges of the spreader slide.
4. Keeping the spreader at a 25° angle, press the edge of the spreader firmly against the slide, and push the spreader rapidly over the entire length of the slide (in this direction ←). The drop of blood will thin out toward the end of the slide.
5. Let the smear dry.
6. To stain the slide, follow this procedure:
   a. Place the slide on a staining rack.
   b. Cover the entire slide with Wright's stain.
   c. Let the stain stand for 1 min.
   d. Add 25 drops of buffer solution and mix it completely with Wright's stain.
   e. Let the mixture stand for 8 min.
   f. Wash the mixture off completely with distilled water.
7. Let the slide dry completely before counting.
8. To proceed with counting the cells, use an area of the slide where the blood is thinnest (one cell thick). This is called the *feathering edge*. Count cells under an oil-immersion lens. Count a total of 100 white cells.
9. The actual counting can be done by moving the slide either up and down or from side to side as follows:

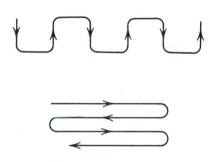

10. Record each white blood cell you observe by making a mark in the chart on page 336 until you have recorded 100 cells.

ALTERNATE PROCEDURE

1. Obtain a prepared slide of stained blood.
2. Repeat steps 9 and 10 in the preceding procedure.

G. THROMBOCYTES

Thrombocytes (platelets) are formed from fragments of the cytoplasm of megakaryocytes (see Figure 17.1). The fragments become enclosed in pieces of cell membrane from the megakaryocytes and develop into platelets. Platelets are very small, disc-shaped cell fragments without nuclei. Between 250,000 and 400,000 are found in each cubic millimeter or deciliter of blood. They function to prevent fluid loss by starting a chain of reactions that result in blood clotting. They have a short life span, probably only one week, because they are expended in clotting and are just too simple to carry on extensive metabolic activity.

PROCEDURE

WARD'S Natural Science Establishment, Inc., has made available WARD'S *Simulated Blood.* It is a solution that contains microcomponents that simulate erythrocytes, leukocytes, and thrombocytes. The microcomponents are similar in relative proportion to those found in human blood and can be oberved under a microscope without staining.

1. Obtain a Simulated Blood Activity kit.
2. Follow the instructions to perform the thrombocyte count.
3. Complete the portion of the Data Table related to thrombocytes.
4. Answer the questions related to thrombocytes.

Examine a prepared slide of a stained smear of blood cells using the oil-immersion objective. With the aid of your textbook, identify red blood cells, neutrophils, basophils, eosinophils, lymphocytes, monocytes, and thrombocytes. Using colored crayons or pencil crayons, draw and label all of the different blood cells in Section F.6 of the LABORATORY REPORT RESULTS at the end of the exercise. Draw and color the granules and the nuclear shapes very accutately, because both are used in identifying the various types of cells.

H. BLOOD GROUPING (TYPING)

The plasma membranes of red blood cells contain genetically determined antigens called *agglutinogens* (ag'-loo-TIN-ō-jens). The plasma of blood contains genetically determined antibodies called *agglutinins* (a-GLOO-ti-nins). The antibodies cause the *agglutination (clumping)* of the red blood cells carrying the corresponding antigen.

These proteins, agglutinogens and agglutinins, are responsible for the two major classifications of blood groups: the ABO group and the Rh system. In addition to the ABO group and the Rh system, other human blood groups include MNSs, P, Lutheran, Kell, Lewis, Duffy, Kidd, Diego, and Sutter. Fortunately, these different antigenic factors do not exhibit extreme degrees of antigenicity and, therefore, usually cause very weak transfusion reactions or no reaction at all.

1. ABO Group

This major blood grouping is based on two agglutinogens symbolized as *A* and *B* (Figure 17.12). Individuals whose red blood cells produce only agglutinogen *A* have blood type A. Individuals who produce only agglutinogen *B* have blood type B. If both *A* and *B* agglutinogens are produced, the result is type AB, whereas the absence of both *A* and *B* agglutinogens results in the so-called type O.

The agglutinins in blood plasma are antibody *a* (anti-A), which attacks agglutinogen *A*, and antibody *b* (anti-B), which attacks agglutinogen *B*. The agglutinogens and agglutinins formed by each of the four blood types and the various agglutination reactions that occur when whole blood samples are mixed with serum are shown in Figure 17.12. You do not have agglutinins that attack the agglutinogens of your own erythrocytes. For example, a type A person has agglutinogen *A* but not agglutinin *a* (anti-A).

Agglutinogens and agglutinins are critically important in blood transfusions. In an incompatible blood transfusion, the donated erythrocytes are attacked by the recipient agglutinins, causing the blood cells to agglutinate. Agglutinated cells become lodged in small capillaries throughout the body and, over a period of hours, the cells swell, rupture, and release hemoglobin into the blood. Such a reaction, as it relates to erythrocytes, is called *hemolysis* (*lysis* = dissolve). The degree of agglutination depends on the titer (strength or amount) of agglutinin in the blood. Agglutinated cells can block blood vessels and can lead to kidney or brain damage and death; and the liberated hemoglobin may also cause kidney damage.

Because cells of type O blood contain neither of the two antigens (*A* or *B*), moderate amounts of this blood can be transfused into a recipient of any ABO blood type without *immediate* agglutination. For this reason, type O blood is referred to as the *universal donor*. However, transfusing large amounts of type O blood into a recipient of type A, B, or AB can cause *delayed* agglutination of the recipient's erythrocytes, because the transfused antibodies (type O blood contains *a* and *b* antibodies) are not sufficiently diluted to prevent the reaction.

People with type AB blood are sometimes called *universal recipients* because their blood contains no antibodies to agglutinate donor erythrocytes. *Small quantities* of blood from all

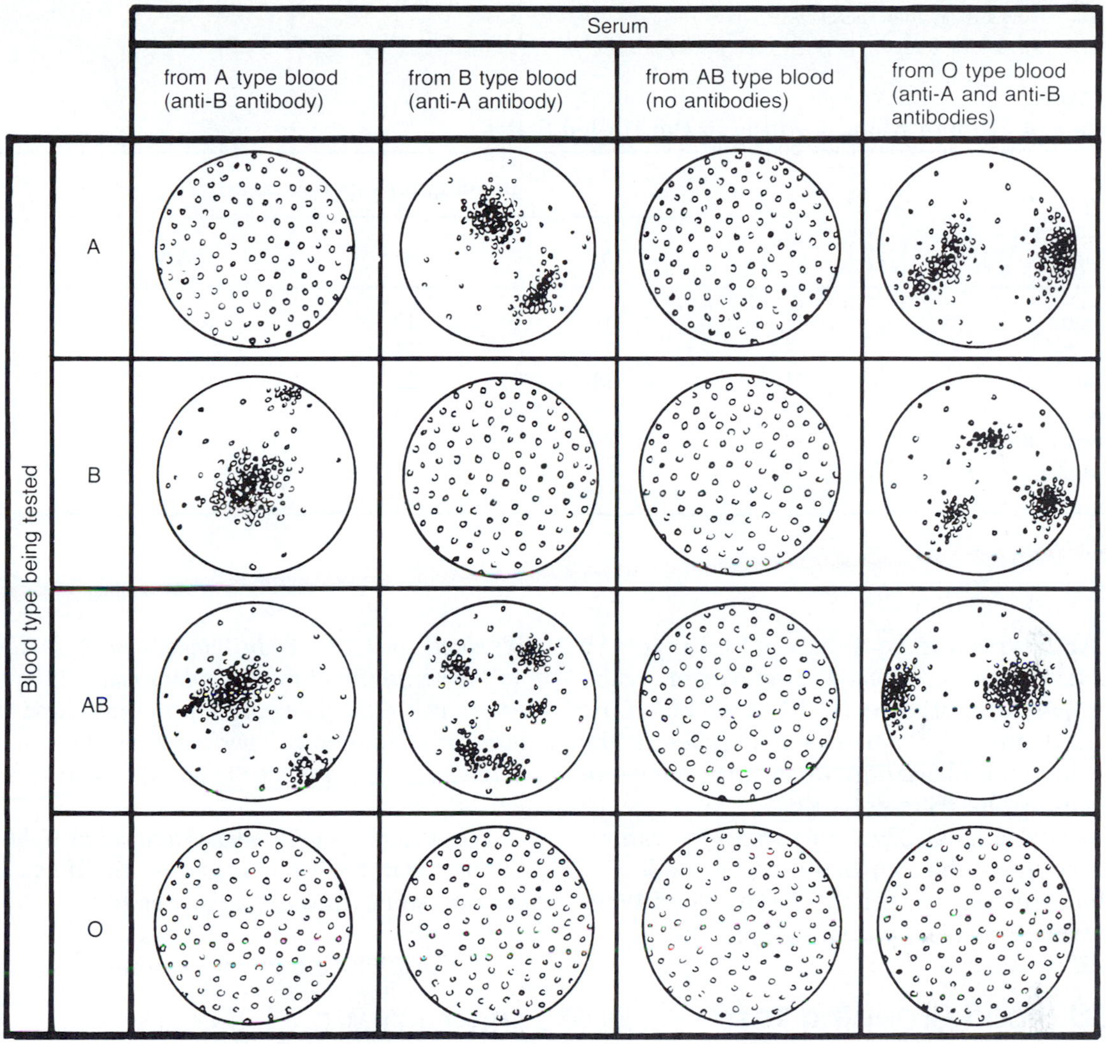

FIGURE 17.12 Origin, development, and structure of blood cells.

TABLE 17.1
Summary of ABO System Interactions

Blood type	A	B	AB	O
Agglutinogen (antigen) on RBCs	*A*	*B*	*A* and *B*	Neither *A* nor *B*
Agglutinin (antibody) in plasma	*b*	*a*	Neither *a* nor *b*	*a* and *b*
Compatible donor blood types	A, O	B, O	A, B, AB, O	O
Incompatible donor blood types	B, AB	A, AB	—	A, B, AB
Genotype (genetic make-up)	*AO* and *AA*	*BO* and *BB*	*AB*	*OO*

TABLE 17.2
Incidence of Human Blood Groups in the United States

	Blood groups (percentages)				
	O	A	B	AB	Rh+
Caucasions	45	41	10	4	85
Blacks	48	27	21	4	88
Japanese	31	38	22	9	~100
Chinese	36	28	23	13	~100
American Indians	23	76	0	1	~100
Hawaiians	37	61	1.5	0.5	~100

other ABO types can be transfused into type AB blood without adverse reaction. However, again, if *large quantities* of type A, B, or O blood are transfused into type AB, the antibodies in the donor blood might accumulate in sufficient quantity to clump the recipient's RBCs.

Table 17.1 on page 339 summarizes the various interreactions of the four blood types. Table 17.2 lists the incidence of these different blood types in the United States, comparing some races.

2. ABO Blood Grouping Test*

CAUTION! *Please reread Section A, "General Safety Precautions and Procedures," and Section B, "Precautions Related to Working with blood, Blood*

Products, or Other Body Fluids," on pages xiii–xiv, at the beginning of the laboratory manual, before you begin any of the following experiments. Read the experiments before you perform them, to be sure that you understand all the procedures and safety precautions.

When working with whole blood, wear tight-fitting surgical gloves and safety goggles. Avoid any kind of contact with open sore, cut or wound.

The procedure for ABO sampling follows:

PROCEDURE
1. Using a wax pencil, divide a glass slide in half and label the left side A and the right side B.
2. On the left side, place one large drop of anti-A serum; on the right side, place one large drop of anti-B serum.
3. Next to the drops of antisera, place one drop of blood, being careful not to mix the samples.
4. Using a mixing stick or a toothpick, mix the blood on the left side with the anti-A serum, and the, *using a different stick or different toothpick*, mix the blood on the right side with the anti-B.

CAUTION! *Immediately dispose of the stick or toothpick in the appropriate biohazard container provided by your instructor.*

6. *Gently* tilt the slide back and forth and observe it for 1 min.
7. Record your results in Section H.1 of the LABORATORY REPORT RESULTS at the end of the exercise, using "+" for clumping (agglutination) and "−" for no clumping.

If the finger puncture procedure is not performed, blood can be obtained from a clinical laboratory where it has been tested and certified as noninfectious or from a mammal (other than a human).

Whole blood that has been tested for syphilis; hepatitis A, B, and C; and HIV is available from Carolina Biological Supply Company. This blood can be used for blood grouping and typing, blood glucose, hematocrit, hemoglobin, red blood cell count, white blood cell count, differential white blood cell count, platelet count, sedimentation rate, osmotic fragility studies, and various plasma chemistries. Carolina Biological Supply Company also provides aseptic red blood cells that can be used for blood grouping and typing, blood glucose, hematocrit, hemoglobin, red blood cell count, and osmotic fragility studies. These blood cells are suspended in a modified alsever's solution to which several antibiotics have been added. The cells have been tested for syphilis; hepatitis A, B, and C; and HIV.

Also available for this exercise is WARD'S Simulated ABO and Rh Blood Typing Activity.

CAUTION! *Dispose of your slide in the appropriate biohazard container provided by your instructor.*

Identify your own ABO system blood type, based on your observations, and record it in Section H.2 of the LABORATORY REPORT RESULTS at the end of the exercise.

If your instructor wishes, you can also summarize the results obtained by your class in the table provided in Section H.4.

3. Rh System

The Rh factor is the other major classification of blood grouping. This group was designated Rh because the blood of the rhesus monkey was used in the first research and development. As is the ABO grouping, this classificaiton is based on agglutinogens that lie on the surfaces of erythrocytes. The designatin Rh^+ (positive) is given to those that have the agglutinogen, and Rh^- (negative) is for those that lack the agglutinogen. The estimation is that 85% of whites and 88% of blacks in the United States are Rh^+, whereas 15% of whites and 12% of blacks are Rh^- (see Table 17.2.)

The Rh factor is extremely important in pregnancy and childbirth. Under normal circumstances, human plasma does not contain anti-Rh antibodies. If however, a woman who is Rh^- becomes pregnant with an Rh^+ child, her blood may produce antibodies that will react with the blood of a subsequent child.[2] The first child is unaffected because the mother's body has not yet produced these antibodies. This is a serious reaction and hemolysis may occur in the fetal blood.

The hemolysis produced by this fetal-maternal incompatibility is called *hemolytic disease of newborn (HDN)* or *erythroblastosis fetalis,* and could be fatal for the newborn. A drug called Rho-GAM, given to RH^- mothers immediately after delivery or abortion, prevents the production of antibodies by them, so that the fetus of the next pregnancy is protected.

---

[2]Be sure to note the important difference in the production of antibodies in the ABO and Rh systems. An Rh^- person can *produce antibodies in response to the stimulus of invading Rh antigen;* by contrast, any antibodies of the ABO system that exist in a person's blood *occur naturally and are present regardless of whether ABO antigens are introduced.*

4. Rh Blood Grouping Test*

CAUTION! *Please reread Section A, ''General Safety Precautions and Procedures,'' and Section B, ''Precautions Related to Working with Blood, Blood Products, or Other Body Fluids,'' on pages xiii–xiv, at the beginning of the laboratory manual, before you begin any of the following experiments. Read the experiments before you perform them, to be sure that you understand all the procedures and safety precautions.*

When working with whole blood, wear tight-fitting surgical gloves and safety goggles. Avoid any kind of contact with an open sore, cut, or wound.

NOTE: Remember that this procedure is only sufficient for laboratory demonstration. It is not clinically accurate because the anti-Rh (anti-D) serum deteriorates rapidly at room temperature and the anti-Rh agglutinins are far less potent in their agglutinizing capability than the anti-A or anti-B antibodies. This test is less accurate than the ABO determination.

The procedure for Rh grouping follows:

PROCEDURE

1. Place one large drop of anti-Rh (anti-D) serum on a glass slide.[3]
2. Add one drop of blood and mix, using a mixing stick or toothpick.

If the finger puncture procedure is not performed, blood can be obtained from a clinical laboratory where it has been tested and certified as noninfectious or from a mammal (other than a human).

Whole blood that has been tested for syphilis; hepatitis A, B, and C; and HIV is available from Carolina Biological Supply Company. This blood can be used for blood grouping and typing, blood glucose, hematocrit, hemoglobin, red blood cell count, white blood cell count, differential white blood cell count, platelet count, sedimentation rate, osmotic fragility studies, and various plasma chemistries. Carolina Biological Supply Company also provides aseptic red blood cells that can be used for blood grouping and typing, blood glucose, hematocrit, hemoglobin, red blood cell count, and osmotic fragility studies. These blood cells are suspended in a modified alsevere's solution to which several antibiotics have been added. The cells have been tested for syphilis; hepatitis A, B, and C; and HIV.

Also available for this exercise is WARD'S Simulated ABO and Rh Blood Typing Activity.

[3]The Rh antigen is more specifically termed the D antigen after the Fisher-Race nomenclature, which is based on genetic concepts, or theories of inheritance.

CAUTION! *Immediately dispose of the lancet and stick or toothpick in the appropriate biohazard container provided by your instructor.*

3. Place the slide on a preheated warming box and gently rock the box back and forth for 2 min. (Unlike ABO typing, Rh typing is better done on a heated warming box.)
4. Record your results, using "+" for clumping and "−" for no clumping. Record whether you are Rh^+ or Rh^- in Section H.5 of the LABORATORY REPORT RESULTS at the end of the exercise. If your instructor wishes, you can also summarize your class's results in the table provided in Section H.6 of the LABORATORY REPORT RESULTS.

CAUTION! *Dispose of your slide in the appropriate biohazard container provided by your instructor.*

ANSWER THE LABORATORY REPORT QUESTIONS AT THE END OF THE EXERCISE.

Blood

STUDENT _____   DATE _____

LABORATORY SECTION _____   SCORE/GRADE _____

SECTION B. PLASMA

1. *Physical Characteristics*

pH _____

Color _____

Transparency (clear, translucent, opaque) _____
2. *Chemical Constituents*

Is glucose present? _____

Is protein present? _____

SECTION D. RED BLOOD CELL TESTS

1. Red blood cell count results: _____ RBCs per mm^3.

2. Red blood cell volume (hematocrit) results: _____%.

3. Hemoglobin determination results: hemoglobinometer, _____ g per 100 ml.

Tallqvist® paper, _____ g per 100 ml.

4. Complete the following table:

	Wavelength (nm)										
	500	510	520	530	540	550	560	570	580	590	600
Solution 1											
Solution 2											

Graph your results here:

SECTION F. WHITE BLOOD CELL TESTS

5. White blood cell count results: _____ WBCs per mm^3.
6. Drawings of various blood cells.

SECTION H. BLOOD GROUPING (TYPING)

1. In determining your ABO blood grouping, did you observe clumping when your blood was mixed with

   _____ anti-A serum only

   _____ anti-B serum only

   _____ both anti-A and anti-B serums

   _____ neither anti-A nor anti-B serum

2. Based on your observations, what is your ABO grouping?

   _____ A   _____ B   _____ AB   _____ O

3. Based on your observations, briefly explain why you identify your ABO blood grouping as you do.

_____

_____

_____

4. Record the results of the ABO blood grouping test done by your class.

Type	Anti-A (present or absent)	Anti-B (present or absent)	Number of individuals	Class percentage
A				
B				
AB				
O				

5. Based on your observations, are you Rh^+ or Rh^-?

_____ Rh^+ _____ Rh^-

6. Record the results of the Rh test done by your class.

Type	Number of individuals	Class percentage
Rh^+		
Rh^-		

Blood

STUDENT _____   DATE _____

LABORATORY SECTION _____   SCORE/GRADE _____

PART **1. Multiple Choice**

_____ **1.** The process by which all blood cells are formed is called (a) hemocytoblastosis (b) erythropoiesis (c) hemaopoiesis (d) leukocytosis

_____ **2.** An inability of body cells to receive adequate amounts of oxygen may indicate a malfunction of (a) neutrophils (b) leukocytes (c) lymphocytes (d) erythrocytes

_____ **3.** Special cells of the body that have the responsibility of clearing away dead, disintegrating bodies of red and white blood cells are called (a) agranular leukocytes (b) reticuloendo-thelial cells (c) erythrocytes (d) thrombocytes

_____ **4.** The name of the test procedure that informs the physician about the rate of erythropoiesis is called the (a) reticulocyte count (b) sedimentation rate (c) hemoglobin count (d) differential white blood cell count

_____ **5.** The normal red blood cell count per cubic millimeter in a male is about (a) 5.4 million (b) 7 million (c) 4 million (d) more than 9 million

_____ **6.** The normal number of leukocytes per cubic millimeter is (a) 5,000 to 10,000 (b) 8,000 to 12,000 (c) 2,000 to 4000 (d) more than 15,000

_____ **7.** Under the microscope, red blood cells appear as (a) circular discs with centrally located nuclei (b) circular discs with lobed nuclei (c) oval discs with many nuclei (d) biconcave discs without nuclei

_____ **8.** An increase in the number of white blood cells is called (a) leukopenia (b) hematocrit (c) polycythemia (d) leukocytosis

_____ **9.** Thrombocytes are formed from a special large cell that breaks up into small fragments. This cell is called a(n) (a) eosinophil (b) hemocytoblast (c) megakaryocyte (d) platelet

_____ **10.** The blood type with the highest incidence in Caucasians in the United States is (a) A (b) O (c) AB (d) B

PART **2. Completion**

11. Another name for red blood cells is _____.

12. Blood gets its red color from the presence of _____.

13. The life span of a red blood cell is approximately _____.

14. A good method for routine testing for anemia is _____.

15. The normal sedimentation rate value for adults is _____.

16. The normal ratio of red blood cells to white blood cells is about _____.

17. The granular leukocytes are formed from _____ tissue.

18. The number of thrombocytes per cubic millimeter found normally in blood is _____.

19. The function of thrombocytes is to prevent blood loss by starting a chain of reactions resulting in

_____.

20. In blood groupings (typing), the antigens are also called _____.

21. The hemolysis produced by fetal-maternal incompatability of blood cells is called _____.

22. The part of the cell where agglutinogens are located is _____.

PART 3. Matching

_____	**23.** Hemocytoblast	**A.** Response to tissue destruction by invading bacteria
_____	**24.** Polycythemia	
_____	**25.** A high neutrophil count	**B.** An increase in the normal red blood cell count
_____	**26.** A high monocyte count	**C.** Leukemia and infectious mononucleosis
_____	**27.** Leukopenia	**D.** Immature cells that develop into mature blood cells
_____	**28.** A high lymphocyte count	
_____	**29.** Plasma	**E.** Chronic infections
_____	**30.** A high eosinophil count	**F.** Liquid portion of blood without the formed elements and clotting substances
_____	**31.** Serum	**G.** An allergic reaction
		H. A decrease in the normal white blood cell count
		I. Liquid portion of blood without the formed elements

18

Heart

The **heart** is a hollow, muscular organ that pumps blood through miles and miles of blood vessels. This organ is located in the mediastinum, between the lungs. Two thirds of its mass lies to the left of the body's midline. Its pointed end, the **apex**, projects downward to the left, and its broad end, the **base**, projects upward to the right. The main parts of the heart and associated structures to be discussed here are the pericardium, wall, chambers, great vessels, and valves.

A. PERICARDIUM

A loose-fitting serous membrane called the **pericardium (pericardial sac)** encloses the heart (Figure 18.1). The membrane is composed of two layers, the fibrous pericardium and the serous pericardium. The **fibrous pericardium** forming the outer layer is tough, inelastic fibrous connective tissue that adheres to the parietal pleura and anchors the heart in the mediastinum. The inner layer, the **serous pericardium,** is a thinner, delicate membrane that is a double-layered structure. The outer **parietal layer** of the serous pericardium is directly beneath the fibrous pericardium. The inner **visceral layer** of the serous pericardium is also called the **epicardium.** Between the parietal and visceral layers of the serous pericardium is a potential space, the **pericardial cavity,** that contains pericardial fluid and functions to prevent friction between the layers as the heart beats. Identify these structures using a specimen, model, or chart of a heart and label Figure 18.1.

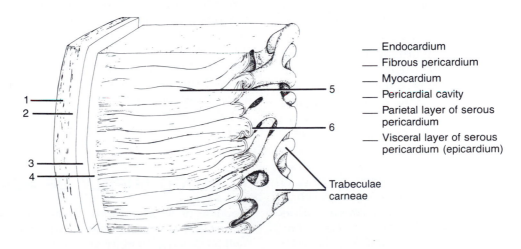

_____ Endocardium
_____ Fibrous pericardium
_____ Myocardium
_____ Pericardial cavity
_____ Parietal layer of serous pericardium
_____ Visceral layer of serous pericardium (epicardium)

Trabeculae carneae

FIGURE 18.1   Structure of the pericardium and heart wall.

B. HEART WALL

Three layers of tissue compose the heart: the visceral layer of the serous pericardium (external layer), the myocardium (middle layer), and the endocardium (inner layer). The *visceral layer of the serous pericardium (epicardium)* is the thin, transparent outer layer of the heart wall. The *myocardium,* which is composed of cardiac muscle tissue, forms the bulk of the heart and is responsible for contraction. The *endocardium* is a thin layer of endothelium and areolar connective tissue that lines the inside of the myocardium and covers the heart valves and the tendons that hold them open. Label the layers of the heart in Figure 18.1.

C. CHAMBERS OF HEART

The interior of the heart is divided into four cavities, called *chambers,* that receive circulating blood. The two superior chambers are known as *right* and *left atria* (atrium = entry hall) and are separated internally by a partition called the *interatrial septum* (septum = partition). A prominent feature of this septum is an oval depression, the *fossa ovalis,* which corresponds to the site of the *foramen ovale,* an opening in the interatrial septum of the fetal heart that helps blood bypass the nonfunctioning lungs. Each atrium has an appendage called an *auricle* (OR-i-kul), so named because its shape resembles a dog's ear. The auricles increase the surface area of the atria. The lining of the atria is smooth, except for the anterior walls and linings of the auricles, which contain projecting muscle bundles called *pectinate* (PEK-ti-nāt) *muscles.*

The two inferior and larger chambers, called the *right* and *left ventricles* (ventricle = little belly), are separated internally by a partition called the *interventricular septum.* The irregular surface of ridges and folds of the myocardium in the ventricles is known as the *trabeculae carneae* (tra-BEK-yoo-lē KAR-nē-ē). Externally, a groove known as the *coronary sulcus* (SUL-kus) separates the atria from the ventricles. The groove encircles the heart and houses the coronary sinus (a large cardiac vein) and the circumflex branch of the left coronary artery. The *anterior interventricular sulcus* and *posterior interventricular sulcus* separate the right and left ventricles externally. They also contain coronary blood vessels and a variable amount of fat.

Label the chambers and associated structures of the heart in Figures 18.2 and 18.3.

D. GREAT VESSELS AND VALVES OF HEART

The right atrium receives deoxygenated blood from every area of the body except the lungs. The blood enters through three veins: the *superior vena cava (SVC)* bringing blood from the upper body, the *inferior vena cava (IVC)* bringing blood from the lower body, and the *coronary sinus* bringing blood from most of the veins supplying the heart wall.

Deoxygenated blood is passed from the right atrium into the right ventricle through the atrioventricular valve called the *tricuspid valve,* which consists of three cusps (flaps). The right ventricle then pumps the blood through the *pulmonary semilunar valve* into the *pulmonary trunk.* The pulmonary trunk divides into a *right* and *left pulmonary artery,* each of which carries blood to the lungs, where the blood releases its carbon dioxide and takes on oxygen. The oxygenated blood returns to the heart via four *pulmonary veins* that empty the blood into the left atrium. The blood is then passed into the left ventricle through another atrioventricular valve, called the *bicuspid (mitral) valve,* which consists of two cusps. The cusps of the tricuspid and bicuspid valves are connected to cords called *chordae tendineae* (KOR-dē TEN-din-ē-ē), which in turn attach to projections in the ventricular walls called *papillary muscles.* The left ventricle pumps oxygenated blood through the *aortic semilunar valve* into the *ascending aorta.* From this vessel, blood is passed into the *coronary arteries, arch of the aorta, thoracic aorta,* and *abdominal aorta.* These blood vessels transport the blood to all areas of the body except the lungs. The function of the heart valves is to permit the blood to flow in only one direction.

Label the great vessels and valves of the heart in Figures 18.2 and 18.3a and b.

E. BLOOD SUPPLY OF HEART

Because of its importance in myocardial infarction (heart attack), the blood supply of the heart will be described briefly at this point. The arterial supply of the heart is provided by the right and

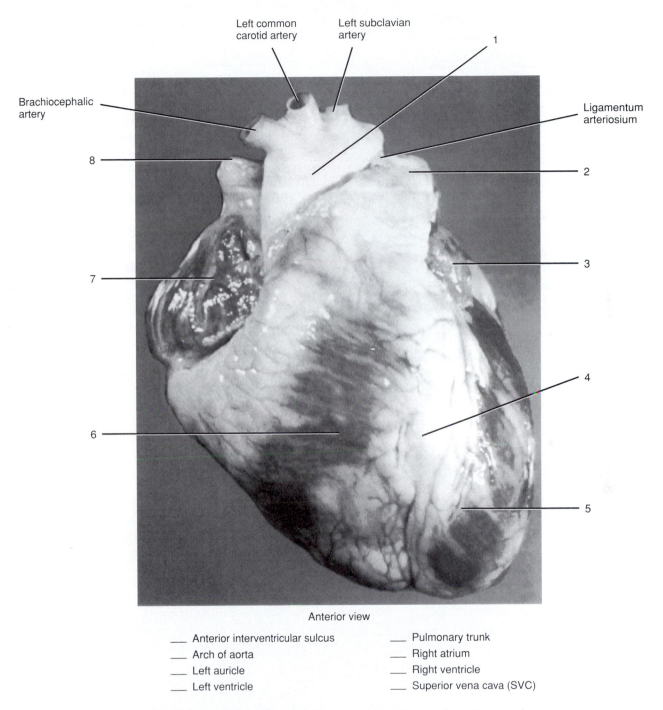

Anterior view

___ Anterior interventricular sulcus	___ Pulmonary trunk
___ Arch of aorta	___ Right atrium
___ Left auricle	___ Right ventricle
___ Left ventricle	___ Superior vena cava (SVC)

FIGURE 18.2 Photograph of the external surface of the human heart.

left coronary arteries. The ***right coronary artery*** originates as a branch of the ascending aorta, descends in the coronary sulcus, and gives off a ***marginal branch*** that supplies the right ventricle, right atrium, and interatrial septum. The right coronary artery continues around the posterior surface of the heart in the posterior interventric-

ular sulcus. This portion of the artery is known as the ***posterior interventricular branch*** and supplies the right and left ventricles and interventricular septum. The ***left coronary artery*** also originates as a branch of the ascending aorta. Between the pulmonary trunk and left auricle, the left coronary artery divides into two branches: ante-

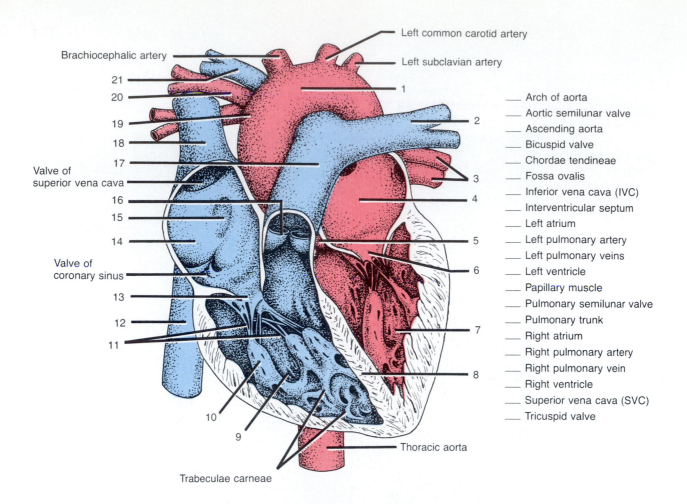

Brachiocephalic artery

Left common carotid artery

Left subclavian artery

21
20
19
18
17

Valve of
superior vena cava

16
15
14

Valve of
coronary sinus

13
12
11

10
9

Trabeculae carneae

1
2
3
4
5
6
7
8

Thoracic aorta

___ Arch of aorta
___ Aortic semilunar valve
___ Ascending aorta
___ Bicuspid valve
___ Chordae tendineae
___ Fossa ovalis
___ Inferior vena cava (IVC)
___ Interventricular septum
___ Left atrium
___ Left pulmonary artery
___ Left pulmonary veins
___ Left ventricle
___ Papillary muscle
___ Pulmonary semilunar valve
___ Pulmonary trunk
___ Right atrium
___ Right pulmonary artery
___ Right pulmonary vein
___ Right ventricle
___ Superior vena cava (SVC)
___ Tricuspid valve

(a) Diagram of internal structure in anterior view

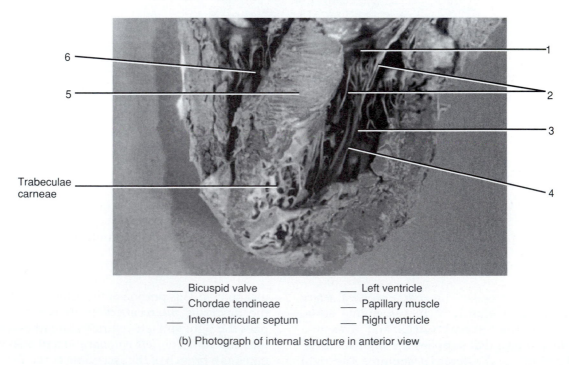

6
5

Trabeculae
carneae

1
2
3
4

___ Bicuspid valve   ___ Left ventricle
___ Chordae tendineae   ___ Papillary muscle
___ Interventricular septum ___ Right ventricle

(b) Photograph of internal structure in anterior view

FIGURE 18.3   Structure of the human heart. Red-colored vessels carry
oxygenated blood; blue-colored vessels carry deoxygenated blood.

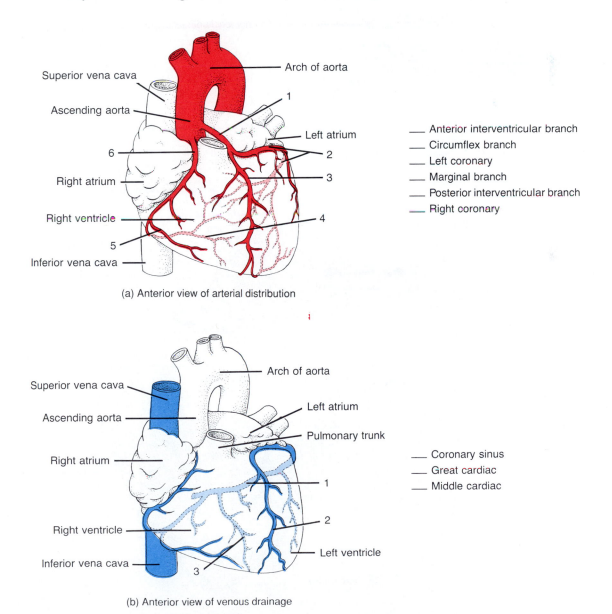

rior interventricular and circumflex. The ***anterior interventricular branch*** passes in the anterior interventricular sulcus and supplies the right and left ventricles and interventricular septum. The ***circumflex branch*** circles toward the posterior surface of the heart in the coronary sulcus and distributes blood to the left ventricle, left atrium, and interventricular septum.

Label these arteries in Figure 18.4a.

Most blood from the heart drains into the ***coronary sinus,*** a venous channel in the posterior portion of the coronary sulcus between the left atrium and left ventricle. The principal tributaries of the coronary sinus are the ***great cardiac vein,***

which drains the anterior aspect of the heart, and the ***middle cardiac vein,*** which drains the posterior aspect of the heart.

Label these veins in Figure 18.4b.

F. DISSECTION OF SHEEP HEART

The anatomy of the sheep heart closely resembles that of the human heart. Use Figures 18.5 and 18.6 as references for this dissection. In addition, models of human hearts can also be used as references.

Superior vena cava

Ascending aorta

6

Right atrium

Right ventricle

5

Inferior vena cava

Arch of aorta

1

Left atrium

2

3

4

(a) Anterior view of arterial distribution

___ Anterior interventricular branch
___ Circumflex branch
___ Left coronary
___ Marginal branch
___ Posterior interventricular branch
___ Right coronary

Superior vena cava

Ascending aorta

Right atrium

Right ventricle

Inferior vena cava

3

Arch of aorta

Left atrium

Pulmonary trunk

1

2

Left ventricle

___ Coronary sinus
___ Great cardiac
___ Middle cardiac

(b) Anterior view of venous drainage

FIGURE 18.4   Coronary (cardiac) circulation.

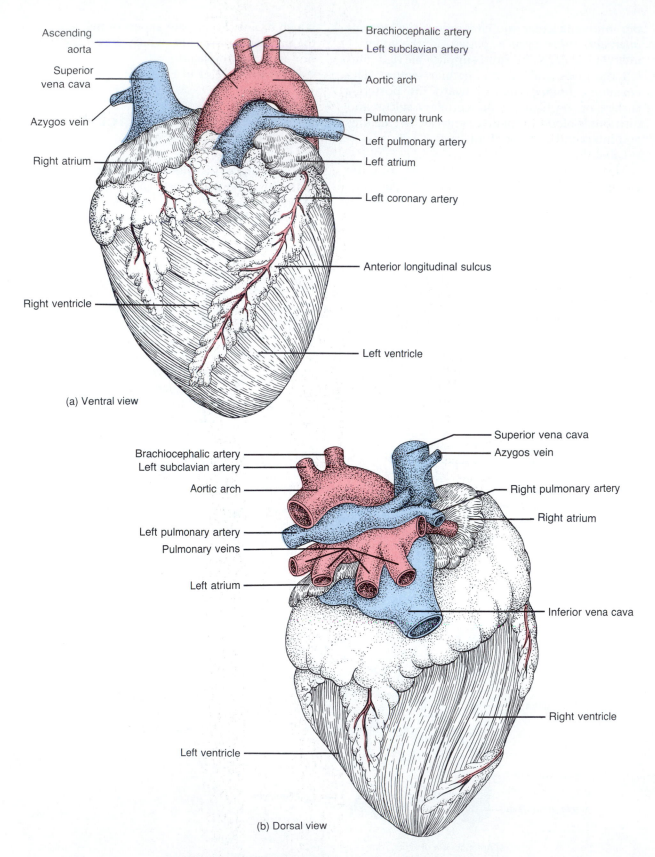

Ascending aorta

Superior vena cava

Azygos vein

Right atrium

Right ventricle

Brachiocephalic artery

Left subclavian artery

Aortic arch

Pulmonary trunk

Left pulmonary artery

Left atrium

Left coronary artery

Anterior longitudinal sulcus

Left ventricle

(a) Ventral view

Brachiocephalic artery

Left subclavian artery

Aortic arch

Left pulmonary artery

Pulmonary veins

Left atrium

Left ventricle

Superior vena cava

Azygos vein

Right pulmonary artery

Right atrium

Inferior vena cava

Right ventricle

(b) Dorsal view

FIGURE 18.5 External structure of a cat or sheep heart.

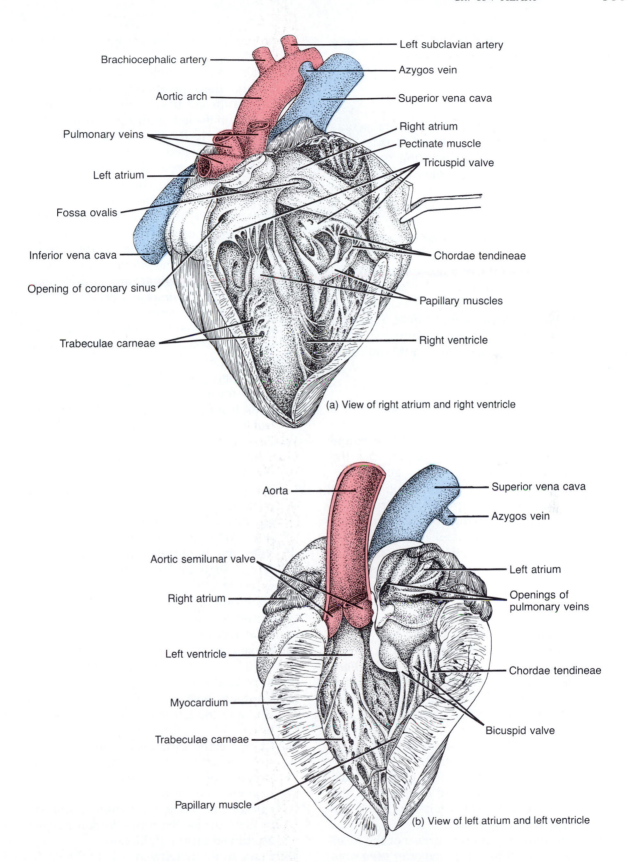

Brachiocephalic artery

Left subclavian artery

Aortic arch

Azygos vein

Superior vena cava

Pulmonary veins

Right atrium

Pectinate muscle

Tricuspid valve

Left atrium

Fossa ovalis

Inferior vena cava

Chordae tendineae

Opening of coronary sinus

Papillary muscles

Trabeculae carneae

Right ventricle

(a) View of right atrium and right ventricle

Aorta

Superior vena cava

Azygos vein

Aortic semilunar valve

Right atrium

Left atrium

Openings of pulmonary veins

Left ventricle

Myocardium

Chordae tendineae

Trabeculae carneae

Bicuspid valve

Papillary muscle

(b) View of left atrium and left ventricle

FIGURE 18.6   Internal structure of a cat or sheep heart.

CAUTION! *Please reread Section D, "Precautions Related to Dissection," at the beginning of the laboratory manual, on page xv, before you begin your dissection.*

First examine the *pericardium,* a fibroserous membrane that encloses the heart. It may have already been removed in preparing the sheep heart for preservation. The *myocardium* is the middle layer and constitutes the main muscle portion of the heart. The *endocardium* (the third layer) is the inner lining of the heart. Use the figures to determine which is the ventral surface of the heart and then identify the *pulmonary trunk* emerging from the anterior ventral surface, near the midline, and medial to the *left auricle.* A longitudinal depression on the ventral surface, called the *anterior longitudinal sulcus,* separates the right ventricle from the left ventricle. Locate the *coronary blood vessels* lying in this sulcus.

PROCEDURE

1. Remove any fat or pulmonary tissue that is present.
2. In cutting the sheep heart open to examine the chambers, valves, and vessels, use the anterior longitudinal sulcus as a guide.
3. Carefully make a shallow incision through the ventral wall of the pulmonary trunk and the right ventricle, trying not to cut the dorsal surface of either structure.
4. The incision is best made *less than an inch to the right of, and parallel to,* the previously mentioned anterior longitudinal sulcus.
5. If necessary, the incision can be continued to where the pulmonary trunk branches into a *right pulmonary artery,* which goes to the right lung, and a *left pulmonary artery,* which goes to the left lung. The *pulmonary semilunar valve* of the pulmonary artery can be clearly seen upon opening it. In any of these internal dissections of the heart, any coagulated blood or latex should be immediately removed so that all important structures can be located and identified.
6. Keeping the cut still parallel to the sulcus, extend the incision around and through the dorsal ventricular wall until you reach the *interventricular septum.*
7. Now examine the dorsal surface of the heart and locate the thin-walled *superior vena cava* directly above the *right auricle.* This vein proceeds posteriorly straight into the right atrium.
8. Make a second longitudinal cut, this time through the superior vena cava (dorsal wall).
9. Extend the cut posteriorly through the right atrium on the left of the right auricle. Proceed posteriorly to the dorsal right ventricle wall and join your first incision.
10. The entire internal right side of the heart should now be clearly seen when carefully spread apart. The interior of the superior vena cava, right atrium, and right ventricle will now be examined. Start with the right auricle and locate the *pectinate muscle,* the large opening of the *inferior vena cava* on the left side of the right atrium, and the opening of the *coronary sinus* just below the opening of the inferior vena cava. By using a dull probe and gentle pressure, most of the vessels can be traced to the dorsal surface of the heart.
11. Now find the wall that separates the two atria, the *interatrial septum.* Also find the *fossa ovalis,* an oval-shaped depression ventral to the entrance of the inferior vena cava.
12. Examine the *tricuspid valve* between the right atrium and the right ventricle to locate the three cusps, as its name indicates. From the cusps of the valve itself, and tracing posteriorly, identify the *chordae tendineae,* which hold the valve in place. Still tracing posteriorly, the chordae are seen to originate from the *papillary muscles,* which themselves originate from the wall of the right ventricle.
13. Look carefully again at the dorsal surface of the left atrium and locate as many *pulmonary veins* (normally, four) as possible.
14. Make your third longitudinal cut through the most lateral of the pulmonary veins that you have located.
15. Continue posteriorly through the left atrial wall and the left ventricle to the *apex* of the heart.
16. Compare the difference in the thickness of the wall between the right and left ventricles. Explain your answer.
17. Examine the *bicuspid (mitral) valve,* again counting the cusps. Determine if the left side of the heart has basically the same structures as studied on the right side.
18. Probe from the left ventricle to the *aorta* as it emerges from the heart, examining the *aortic*

semilunar valve. Find the openings of the right and left main coronary arteries.

19. Now locate the *brachiocephalic artery,* which is one of the first branches from the arch of the aorta. This artery continues branching and terminates by supplying the arms and head, as its name indicates.

20. Connecting the aorta with the pulmonary artery is the remnant of the *ductus arteriosus,* called the *ligamentum arteriosum.* It may not be present in your sheep heart.

ANSWER THE LABORATORY REPORT QUESTIONS AT THE END OF THE EXERCISE.

Heart

STUDENT _____   DATE _____

LABORATORY SECTION _____   SCORE/GRADE _____

PART 1. Multiple Choice

_____ 1. Which of the following veins drains the blood from most of the vessels supplying the heart wall? (a) vasa vasorum (b) superior vena cava (c) coronary sinus (d) inferior vena cava

_____ 2. The atrioventricular valve on the same side of the heart as the origin of the aorta is the (a) aortic semilunar (b) tricuspid (c) bicuspid (d) pulmonary semilunar

_____ 3. Which valve does the blood go through just before entering the pulmonary trunk on the way to the lungs? (a) tricuspid (b) pulmonary semilunar (c) aortic semilunar (d) bicuspid

_____ 4. The pointed end of the heart that projects downward and to the left is the (a) costal surface (b) base (c) apex (d) coronary sulcus

_____ 5. Which of these structures is more internal? (a) fibrous pericardium (b) visceral layer of serous pericardium (c) parietal layer of serous pericardium (d) myocardium

_____ 6. The musculature of the heart is referred to as the (a) endocardium (b) myocardium (c) epicardium (d) pericardium

_____ 7. The depression in the interatrial septum corresponding to the foramen ovale of fetal circulation is the (a) interventricular sulcus (b) pectinate muscle (c) chordae tendineae (d) fossa ovalis

PART 2. COMPLETION

8. Malfunction of the _____ valve would interfere with the flow of blood from the right atrium to the right ventricle.

9. Deoxygenated blood is sent to the lungs through the _____.

10. The loose-fitting serous membrane that encloses the heart is called the _____.

11. The two inferior chambers of the heart are separated by the _____.

12. The earlike flap of tissue on each atrium is called a(n) _____.

13. The large vein that drains blood from superior parts of the body and empties into the right atrium

    is the _____.

14. The cusps of atrioventricular valves are prevented from inverting by the presence of cords

    called _____, which are attached to papillary muscle.

15. A groove on the surface of the heart that houses blood vessels and a variable amount of fat is called

    a(n) _____.

16. The branch of the left coronary artery that distributes blood to the left atrium and left ventricles is

    the _____.

PART 3. Special Exercise

Draw a model of the heart and carefully label the four chambers, the four valves in their proper places, and the major blood vessels entering and exiting the heart.

Blood Vessels

Blood vessels are networks of tubes that carry blood throughout the body. Blood vessels are called arteries, arterioles, capillaries, venules, or veins. In this exercise you will study the histology of blood vessels and identify the principal arteries and veins of the human cardiovascular system.

A. ARTERIES AND ARTERIOLES

Arteries are blood vessels that carry blood *away* from the heart to body tissues. Arteries are constructed of three coats of tissue, called *tunics,* around a hollow core, called a *lumen,* through which blood flows (Figure 19.1). The inner coat is called the *tunica interna* and consists of a lining of endothelium in contact with the blood and a layer of elastic tissue called the internal elastic membrane. The middle coat, or *tunica media,* is usually the thickest layer and consists of elastic fibers and smooth muscle fibers. This tunic is responsible for two major properties of arteries: *elasticity* and *contractility.* The outer coat, or *tunica externa,* is composed principally of elastic and collagen fibers. An external elastic membrane may separate the tunica externa from the tunica media.

Obtain a prepared slide of a cross section of an artery and identify the tunics, using Figure 19.1 as a guide.

As arteries approach various tissues of the body, they become smaller and are known as *arterioles.* Arterioles play a key role in regulating blood flow from arteries into capillaries. When arterioles enter a tissue, they branch into countless microscopic blood vessels called capillaries.

B. CAPILLARIES

Capillaries are microscopic blood vessels that connect arterioles and venules. Their function is to permit the exchange of nutrients and wastes between blood and body tissues. This function is related to the fact that capillaries consist of only a single layer of endothelium.

C. VENULES AND VEINS

When several capillaries unite, they form small veins called *venules.* They collect blood from capillaries and drain it into veins.

Veins are composed of the same three tunics as arteries, but there are variations in their relative thicknesses. The tunica interna of veins is extremely thin compared to that of their accompanying arteries. In addition, the tunica media of veins is much thinner than that of accompanying arteries, and the tunica externa is thicker in veins (Figure 19.1). Functionally, veins return blood from tissues *to* the heart.

Obtain a prepared slide of a cross section of an artery and its accompanying vein and compare them, using Figure 19.1 as a guide.

361

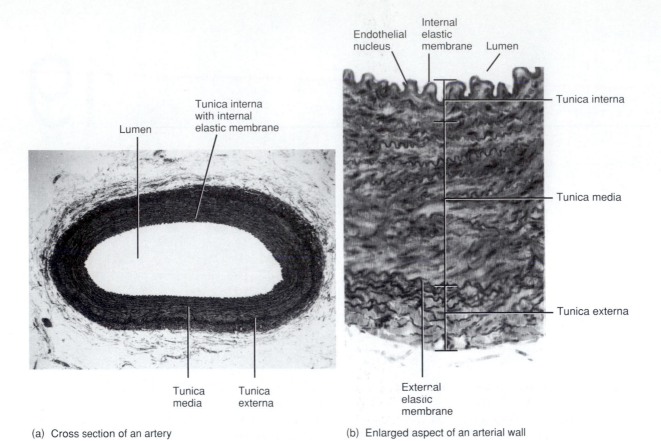

Lumen

Tunica interna
with internal
elastic membrane

Tunica
media

Tunica
externa

(a) Cross section of an artery

Endothelial
nucleus

Internal
elastic
membrane

Lumen

Tunica interna

Tunica media

Tunica externa

External
elastic
membrane

(b) Enlarged aspect of an arterial wall

Adipose
tissue

Vein

Blood cells

Artery

(c) Comparison of structure of an artery
and its accompanying vein

Lumen   Tunica externa   Tunica media   Tunica interna

(d) Scanning electron micrograph of an artery
at a magnification of 5170x

FIGURE 19.1 Histology of blood vessels. (d) from *Tissues and Organs: A Text-Atlas of Scanning Electron Microscopy* by Richard G. Kessel and Randy H. Kardon. W. H. Freeman and Company. Copyright © 1979.)

D. CIRCULATORY ROUTES

The two basic postnatal (after birth) circulatory routes are systemic and pulmonary circulation (Figure 19.2). Some other circulatory routes, which are all subdivisions of systemic circulation, include hepatic portal circulation, coronary (cardiac) circulation, fetal circulation, and the cerebral arterial circle (circle of Willis). The latter is found at the base of the brain (see Table 19.2).

1. Systemic Circulation

The largest route is the *systemic circulation* (see Figures 19.3 through 19.12 and Tables 19.1 through 19.11). This route includes the flow of blood from the left ventricle to all parts of the body. The function of the systemic circulation is to carry oxygen and nutrients to all body tissues (except the air sacs of the lungs) and to remove carbon dioxide and other wastes from the tissues.

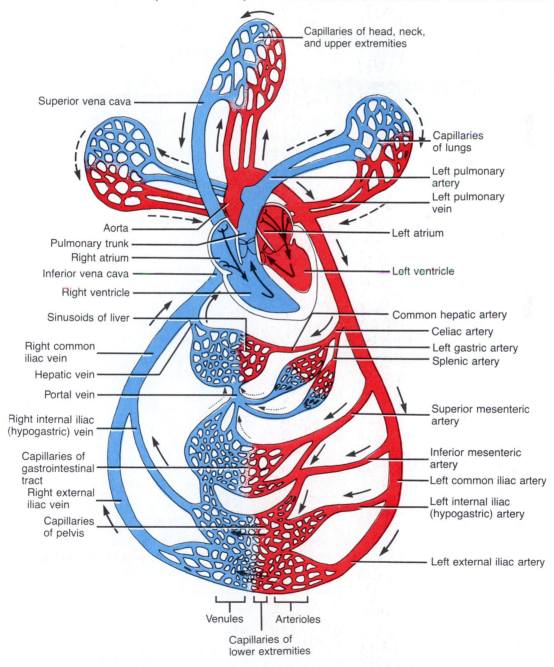

FIGURE 19.2 Circulatory routes. Systemic circulation is indicated by solid arrows; pulmonary circulation by broken arrows; and hepatic portal circulation by dotted arrows.

TABLE 19.1
Aorta and Its Branches (Figure 19.3)

OVERVIEW: The **aorta** (ā-OR-ta) is the largest artery of the body, about 2 to 3 cm (0.8 to 1.2 in.) in diameter. It begins at the left ventricle and contains a valve at its origin, called the aortic semilunar valve (see Figure 18.3a), which prevents backflow of blood into the left ventricle during its diastole (relaxation). The principal divisions of the aorta are the ascending aorta, arch of the aorta, thoracic aorta, and abdominal aorta.

Division of aorta	Arterial branch	Region supplied
Ascending aorta	Right and left coronary	Heart
Arch of aorta	Brachiocephalic trunk { Right common carotid Right subclavian	Right side of head and neck Right upper extremity
	Left common carotid	Left side of head and neck
	Left subclavian	Left upper extremity
Thoracic aorta	Intercostals	Intercostal and chest mucles, pleurae
	Superior phrenics	Posterior and superior surfaces of diaphragm
	Bronchials	Bronchi of lungs
	Esophageals	Esophagus
Abdominal aorta	Inferior phrenics	Inferior surface of diaphragm
	Celiac trunk { Common hepatic Left gastric Splenic	Liver Stomach and esophagus Spleen, pancreas, stomach
	Superior mesenteric	Small intestine, cecum, ascending and transverse colons, pancreas
	Adrenals (suprarenals)	Adrenal (suprarenal) glands
	Renals	Kidneys
	Gonadals { Testiculars or Ovarians	Testes Ovaries
	Inferior mesenteric	Transverse, descending, sigmoid colons; rectum
	Common iliacs { External iliacs Internal iliacs (hypogastrics)	Lower extremities Uterus, prostate gland, muscles of buttocks, urinary bladder

Label Figure 19.3.

All systemic arteries branch from the *aorta.* As the aorta emerges from the left ventricle, it passes upward and deep to the pulmonary trunk. At this point, it is called the *ascending aorta.* The ascending aorta gives off two coronary branches to the heart muscle. Then it turns to the left, forming the *arch of the aorta* before descending to the level of the disc between the fourth and fifth thoracic vertebrae as the *descending aorta.* The descending aorta lies close to the vertebral bodies, passes through the diaphragm, and divides at the level of the fourth lumbar vertebra into two *common iliac arteries,* which carry blood to the lower extremities. The section of the descending aorta between the arch of aorta and the diaphragm is referred to as the *thoracic aorta.* The section between the diaphragm and the common iliac arteries is termed the *abdominal aorta.* Each section of the aorta gives off arteries that continue to branch into distributing arteries leading to organs and finally into the arterioles and capillaries that service the tissues of that organ.

Deoxygenated blood is returned to the heart through the systemic veins. All the veins of the systemic circulation flow into either the *superior* or *inferior vena cava.* They, in turn, empty into the right atrium.

Refer to Tables 19.1 through 19.11 and Figures 19.3 through 19.12.

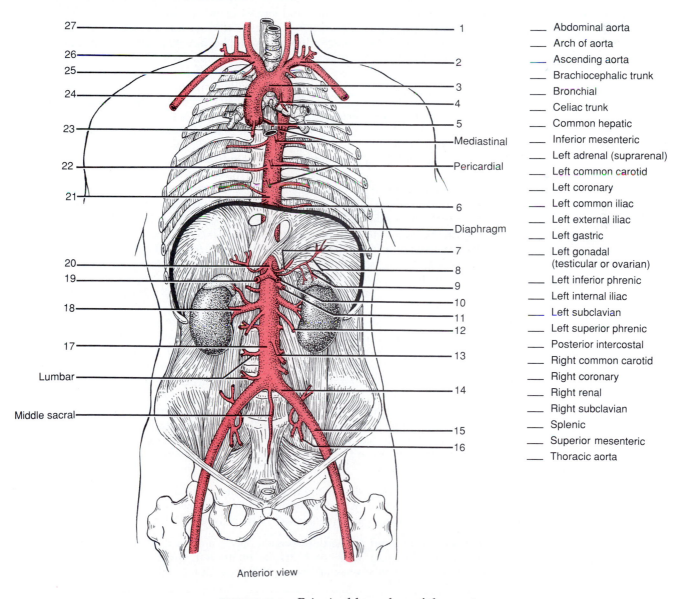

Abdominal aorta
Arch of aorta
Ascending aorta
Brachiocephalic trunk
Bronchial
Celiac trunk
Common hepatic
Inferior mesenteric
Left adrenal (suprarenal)
Left common carotid
Left coronary
Left common iliac
Left external iliac
Left gastric
Left gonadal (testicular or ovarian)
Left inferior phrenic
Left internal iliac
Left subclavian
Left superior phrenic
Posterior intercostal
Right common carotid
Right coronary
Right renal
Right subclavian
Splenic
Superior mesenteric
Thoracic aorta

Mediastinal
Pericardial
Diaphragm
Lumbar
Middle sacral

Anterior view

FIGURE 19.3 Principal branches of the aorta.

TABLE 19.2
Arch of Aorta (Figures 19.4, 19.5, and 19.6)

OVERVIEW: The **arch of the aorta** is about 4.5 cm (1.8 in.) in length and is the continuation of the ascending aorta that emerges from the pericardium behind the sternum at the level of the sternal angle. Initially, the arch is directed upward, backward and to the left, and then downward on the left side of the body of the fourth thoracic vertebra Actually, the arch is directed not only from right to left, but from anterior to posterior as well. The arch of the aorta terminates at the level of the disc between the fourth and fifth thoracic vertebrae, where it becomes the thoracic aorta. The thymus gland lies in front of the arch of the aorta; the trachea lies behind it.

Three major arteries branch from the arch of the aorta. In order of their origination, they are the brachiocephalic trunk, left common carotid artery, and left subclavian artery.

Branch	Description and region supplied
Brachiocephalic	The *brachiocephalic trunk* is the first and largest branch off the arch of the aorta. It divides to form a right subclavian artery and right common carotid artery. The *right subclavian artery* extends from the brachiocephalic to the first rib, passes into the armpit (axilla), and supplies the arm, forearm, and hand. This artery is a good example of the same vessel having different names as it passes through different regions. Continuation of the right subclavian into the axilla is called the *axillary artery.* From here, it continues into the arm as a *brachial artery.* At the bend of the elbow, the brachial artery divides into medial *ulnar* and lateral *radial arteries.* These vessels pass down to the palm, one on each side of the forearm. In the palm, branches of two arteries anastomose to form two palmar arches—*superficial palmar arch* and *deep palmar arch.* From these arches arise *digital arteries,* which supply the fingers and thumb (Figure 19.4).
	Before passing into the axilla, the right subclavian gives off a major branch to the brain called the *vertebral artery.* The right vertebral artery passes through the foramina of the transverse processes of the cervical vertebrae and enters the skull through the foramen magnum to reach the undersurface of the brain. Here it unites with the left vertebral artery to form the *basilar artery* (Figures 19.5 and 19.6).
	The *right common carotid artery* passes upward in the neck. At the upper level of the larynx, it divides into the *right external* and *right internal carotid arteries.* The external carotid supplies the right side of the thyroid gland, tongue, face, ear, scalp, and dura mater. The internal carotid supplies the brain, right eye, and right sides of the forehead and nose (Figure 19.5).
	Inside the cranium, anastomoses of the left and right internal carotids along with the basilar artery form a somewhat hexagonal arrangement of blood vessels at the base of the brain near the sella turcica called the *cerebral arterial circle (circle of Willis).* From this anastomosis arise arteries supplying the brain. Essentially, the cerebral arterial circle is formed by the union of the *anterior cerebral arteries* (branches of the internal carotids) and *posterior cerebral arteries* (branches of basilar artery). Posterior cerebral arteries are connected with internal carotids by *posterior communicating arteries.* Anterior cerebral arteries are connected by *anterior communicating arteries.* The *internal carotid arteries* are also considered part of the cerebral arterial circle. The cerebral arterial circle equalizes blood pressure to the brain and provides alternate routes for blood to the brain, should the arteries become damaged.
Left common carotid	The *left common carotid* is the second branch off the arch of the aorta. Corresponding to the right common carotid, it divides into basically the same branches with the same names—except that arteries are now labeled "left" instead of "right."
Left subclavian	The *left subclavian artery* is the third branch off the arch of the aorta. It distributes blood to the left vertebral artery and vessels of the left upper extremity. Arteries branching from the left subclavian are named like those of the right subclavian.

Label figures 19.4, 19.5, and 19.6.

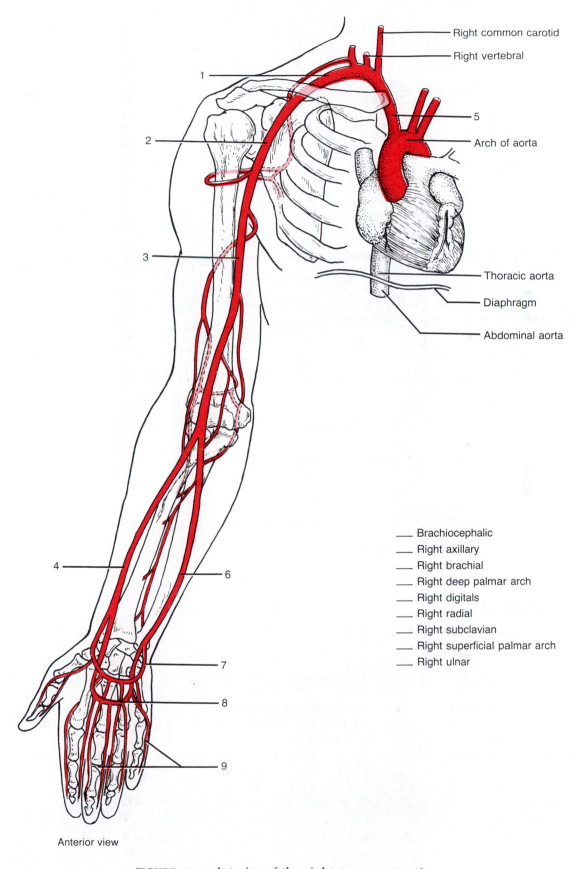

Right common carotid

Right vertebral

5

Arch of aorta

Thoracic aorta

Diaphragm

Abdominal aorta

___ Brachiocephalic
___ Right axillary
___ Right brachial
___ Right deep palmar arch
___ Right digitals
___ Right radial
___ Right subclavian
___ Right superficial palmar arch
___ Right ulnar

Anterior view

FIGURE 19.4   Arteries of the right upper extremity.

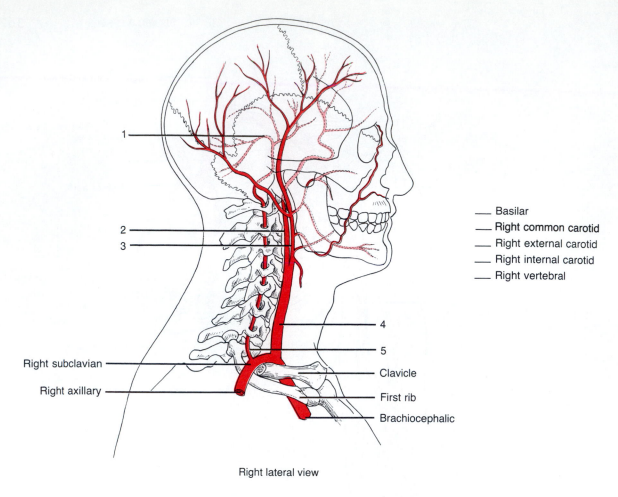

1 ——

2 ——
3 ——

— Basilar
— Right common carotid
— Right external carotid
— Right internal carotid
— Right vertebral

4

5

Right subclavian

Right axillary

Clavicle

First rib

Brachiocephalic

Right lateral view

FIGURE 19.5   Arteries of the neck and head.

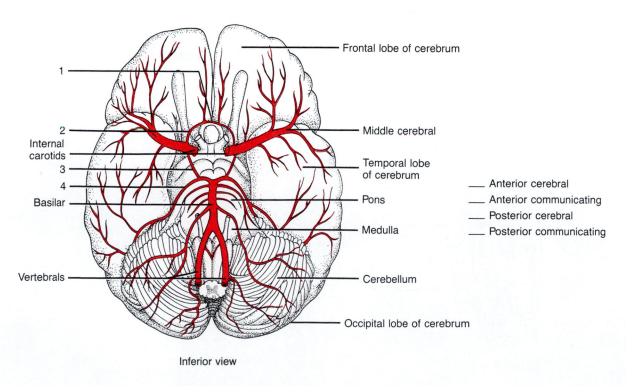

Frontal lobe of cerebrum

1

2

Internal
carotids

3

4

Basilar

Middle cerebral

Temporal lobe
of cerebrum

Pons

Medulla

Vertebrals

Cerebellum

— Anterior cerebral
— Anterior communicating
— Posterior cerebral
— Posterior communicating

Occipital lobe of cerebrum

Inferior view

FIGURE 19.6   Arteries of the base of the brain.

Write the names of the missing arteries in the following scheme of circulation. Be sure to indicate "left" or "right," where applicable.

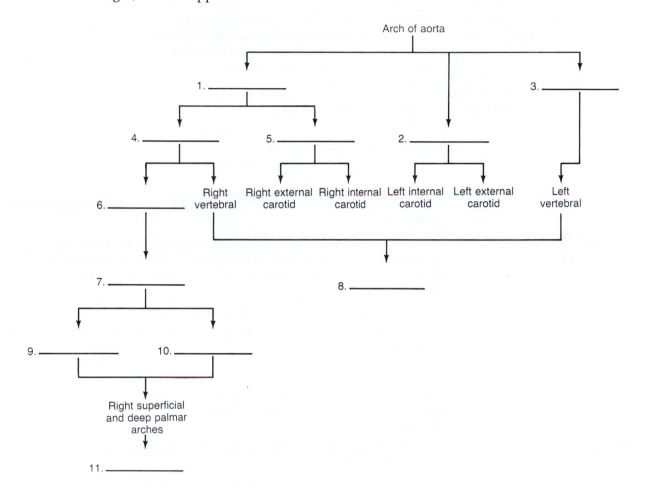

TABLE 19.3
Thoracic Aorta (see Figure 19.3)

OVERVIEW: The **thoracic aorta** is about 20 cm (8 in.) long and is a continuation of the arch of the aorta. It begins at the level of the disc between the fourth and fifth thoracic vertebrae, where it lies to the left of the vertebral column. As it descends, it moves closer to the midline and terminates at an opening in the diaphragm (aortic hiatus) in front of the vertebral column at the level of the intervertebral disc between the twelfth thoracic and first lumbar vertebrae.

Along its course, the thoracic aorta sends off numerous small arteries to viscera **(visceral branches)** and body wall structures **(parietal branches).**

Branch	Description and region supplied
VISCERAL	
Pericardial	Several minute *pericardial arteries* supply blood to the posterior aspect of the pericardium.
Bronchial	One *right* and two *left bronchial arteries* supply the bronchial tubes, visceral pleurae, bronchial lymph nodes, and esophagus. (Whereas the right bronchial artery arises from the third posterior intercostal artery, the two left bronchial arteries arise from the thoracic aorta.)
Esophageal	Four or five *esophageal arteries* supply the esophagus.
Mediastinal	Numerous small *mediastinal arteries* supply blood to structures in the posterior mediastinum.
PARIETAL	
Posterior intercostal	Nine pairs of *posterior intercostal arteries* supply (1) the intercostal, pectoral, and abdominal muscles; (2) overlying subcutaneous tissue and skin; (3) mammary glands; and (4) vertebral canal and its contents.
Subcostal	The *left* and *right subcostal arteries* have a distribution similar to that of the posterior intercostals.
Superior phrenic	Small *superior phrenic arteries* supply the superior surface of the diaphragm.

Label Figure 19.3.

TABLE 19.4
Abdominal Aorta (see Figure 19.7)

OVERVIEW: The **abdominal aorta** is the continuation of the thoracic aorta. It begins at the aortic hiatus in the diaphragm and ends at about the level of the fourth lumbar vertebra, where it divides into right and left common iliac arteries. The abdominal aorta lies in front of the vertebral column.

As with the thoracic aorta, the abdominal aorta gives off **visceral** and **parietal** branches. The unpaired visceral branches arise from the anterior surface of the aorta and include the celiac, superior mesenteric, and inferior mesenteric arteries. The paired visceral branches arise from the lateral surfaces of the aorta and include the adrenal (suprarenal), renal, and gonadal arteries. The paired parietal branches arise from the posterolateral surfaces of the aorta and include the inferior phrenic and lumbar arteries. The unpaired parietal artery is the middle sacral.

Branch	Description and region supplied
VISCERAL	
Celiac	The *celiac artery (trunk)* is the first visceral aortic branch below the diaphragm. The artery has three branches: (1) *common hepatic artery,* (2) *left gastric artery,* and (3) *splenic artery.*
	The common hepatic artery has three main branches: (1) *hepatic artery proper,* a continuation of the common hepatic artery, which supplies the liver and gallbladder; (2) *right gastric artery,* which supplies the stomach and duodenum; and (3) *gastroduodenal artery,* which supplies the stomach, duodenum, and pancreas.
	The left gastric artery supplies the stomach and its *eophageal branch* supplies the esophagus.
	The splenic artery supplies the spleen and has three main branches: (1) *pancreatic arteries,* which supply the pancreas; (2) *left gastroepiploic artery,* which supplies the stomach; and (3) *short gastric arteries,* which supply the stomach.
Superior mesenteric	The *superior mesenteric artery* has several principal branches: (1) *inferior pancreaticoduodenal artery,* which supplies the pancreas and duodenum; (2) *jejunal* and *ileal arteries,* which supply the jejunum and ileum, respectively; (3) *ileocolic artery,* which supplies the ileum and ascending colon; (4) *right colic artery,* which supplies the ascending colon; and (5) *middle colic artery,* which supplies the transverse colon.
Adrenals (suprarenals)	The right and left *adrenal (suprarenal) arteries* supply blood to the adrenal (suprarenal) glands. The glands are also supplied by branches of the renal and inferior phrenic arteries.
Renals	The right and left *renal arteries* carry blood to the kidneys and adrenal (suprarenal) glands.
Gonadals (testiculars or ovarians)	The right and left *testicular arteries* extend into the scrotum and terminate in the testes; the right and left *ovarian arteries* are distributed to the ovaries.
Inferior mesenteric	The principal branches of the *inferior mesenteric artey* are the (1) *left colic artery,* which supplies the transverse and descending colons; (2) *sigmoid arteries,* which supply the descending and sigmoid colons; and (3) *superior rectal artery,* which supplies the rectum.
PARIETAL	
Inferior phrenics	*Inferior phrenic arteries* are distributed to the undersurface of the diaphragm and the adrenal (suprarenal) glands.
Lumbars	*Lumbar arteries* supply the spinal cord and its meninges and muscles and the skin of the lumbar region of the back.
Middle sacral	The *middle sacral artery* supplies the sacrum, coccyx, and rectum.

Label Figure 19.7.

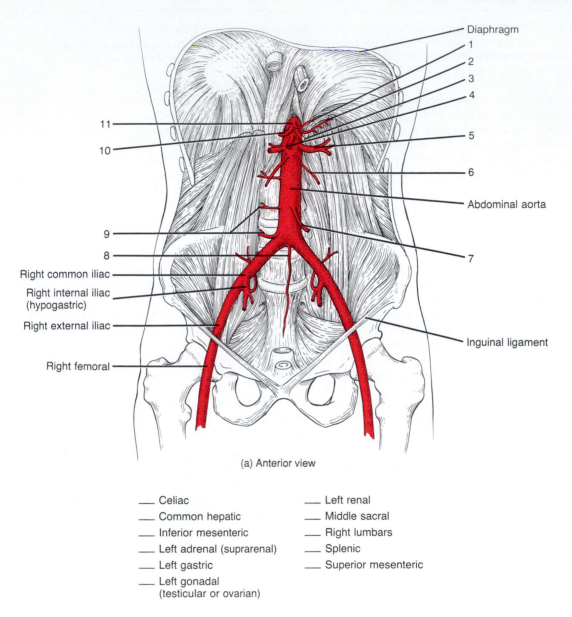

(a) Anterior view

___ Celiac

___ Common hepatic

___ Inferior mesenteric

___ Left adrenal (suprarenal)

___ Left gastric

___ Left gonadal
    (testicular or ovarian)

___ Left renal

___ Middle sacral

___ Right lumbars

___ Splenic

___ Superior mesenteric

FIGURE 19.7 Abdominal arteries. (a) Abdominal aorta and its principal branches.

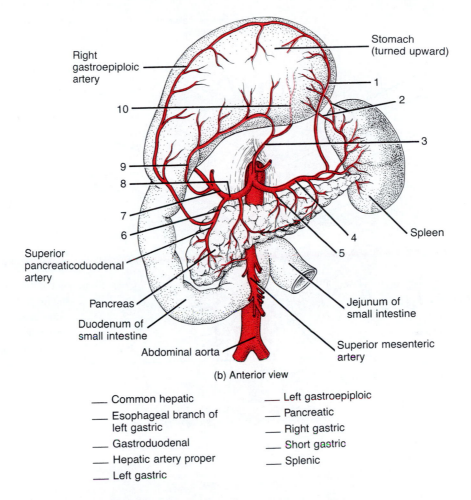

Right gastroepiploic artery

Stomach (turned upward)

10

9

8

7

6

Superior pancreaticoduodenal artery

Pancreas

Duodenum of small intestine

Abdominal aorta

1

2

3

4

5

Spleen

Jejunum of small intestine

Superior mesenteric artery

(b) Anterior view

___ Common hepatic

___ Esophageal branch of left gastric

___ Gastroduodenal

___ Hepatic artery proper

___ Left gastric

___ Left gastroepiploic

___ Pancreatic

___ Right gastric

___ Short gastric

___ Splenic

FIGURE 19.7 *(Continued)*   Abdominal arteries. (b) Branches of the common hepatic, left gastric, and splenic arteries.

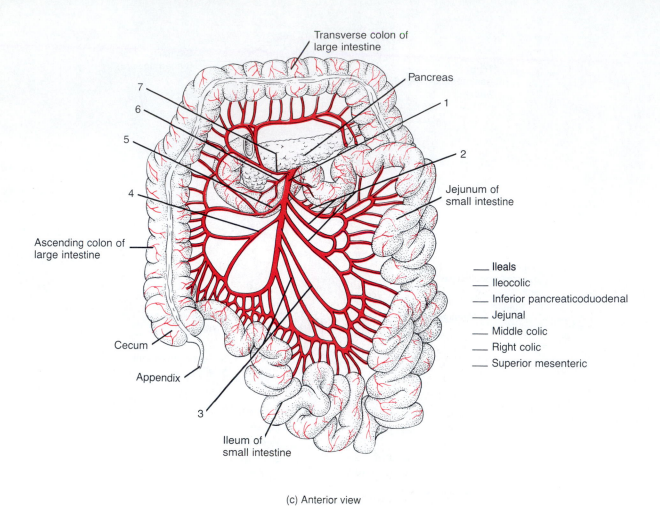

Transverse colon of
large intestine

Pancreas

7

6

1

5

2

4

Jejunum of
small intestine

Ascending colon of
large intestine

___ Ileals

___ Ileocolic

___ Inferior pancreaticoduodenal

___ Jejunal

___ Middle colic

___ Right colic

___ Superior mesenteric

Cecum

Appendix

3

Ileum of
small intestine

(c) Anterior view

FIGURE 19.7 *(Continued)*   Abdominal arteries. (c) Branches of the superior
mesenteric artery.

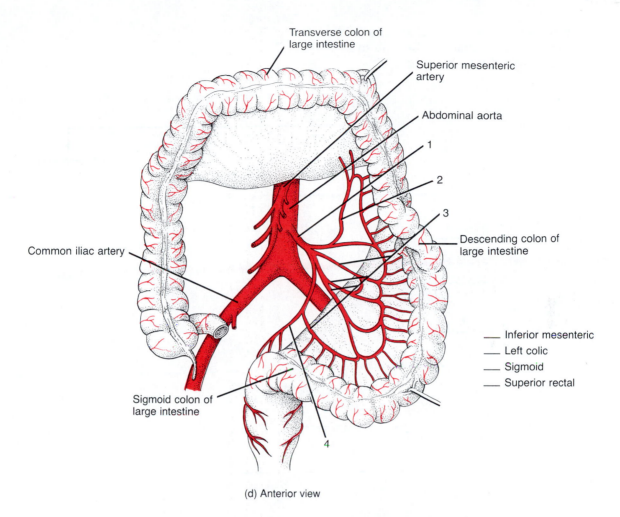

Transverse colon of
large intestine

Superior mesenteric
artery

Abdominal aorta

1

2

3

Descending colon of
large intestine

Common iliac artery

Sigmoid colon of
large intestine

4

___ Inferior mesenteric
___ Left colic
___ Sigmoid
___ Superior rectal

(d) Anterior view

FIGURE 19.7 *(Continued)*   Abdominal arteries. (d) Branches of the inferior
mesenteric artery.

TABLE 19.5
Arteries of Pelvis and Lower Extremities (Figure 19.8)

OVERVIEW: The **internal iliac arteries** enter the pelvic cavity in front of the sacroiliac joint and supply most of the blood to the pelvic viscera and wall. The external
iliacs travel along the brim of the lesser (true) pelvis. Behind the midportion of the
inguinal ligament, each external iliac artery enters the thigh, where its name
changes to the femoral artery.

Branch	Description and region supplied
Common iliacs	At about the level of the fourth lumbar vertebra, the abdominal aorta divides into the right and left *common iliac arteries.* Each passes downward about 5 cm (2 in.) and gives rise to two branches: the internal iliac and external iliac.
Internal iliacs	*Internal iliac (hypogastric) arteries* form branches that supply the psoas major, quadratus lumborum, medial side of each thigh, urinary bladder, rectum, prostate gland, uterus, ductus (vas) deferens, and vagina.
External iliacs	*External iliac arteries* diverge through the greater (false) pelvis, enter the thighs, and become right and left *femoral arteries.* Both femorals send branches back up to the genitals and the wall of abdomen. Other branches run to the muscles of the thigh. The femoral continues down the medial and posterior side of the thigh at the back of the knee joint, where it becomes the *popliteal artery.* Between the knee and ankle, the popliteal runs down the back of the leg and is called the *posterior tibial artery.* Below the knee, the *peroneal artery* branches off the posterior tibial to supply the structures on the medial side of the fibula and calcaneus. In the calf, *anterior tibial artery* branches off the popliteal and runs along the front of the leg. At the ankle, it becomes the *dorsalis pedis artery.* At the ankle, the posterior tibial divides into the *medial* and *lateral plantar arteries.*

Label Figure 19.8.

Write the names of the missing arteries in the following scheme of circulation. Be sure to indicate "left" or "right," where applicable.

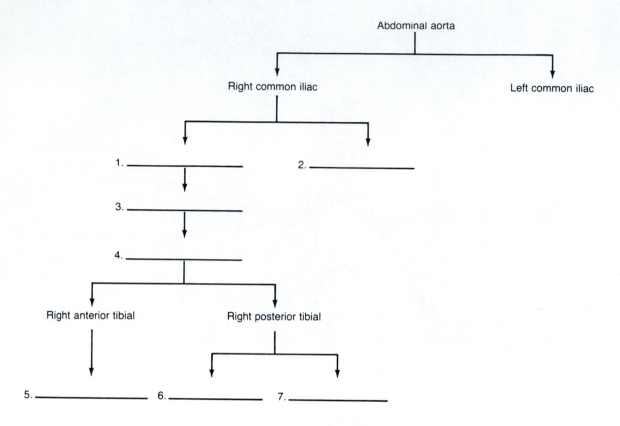

Abdominal aorta

Right common iliac   Left common iliac

1. _____   2. _____

3. _____

4. _____

Right anterior tibial   Right posterior tibial

5. _____ 6. _____ 7. _____

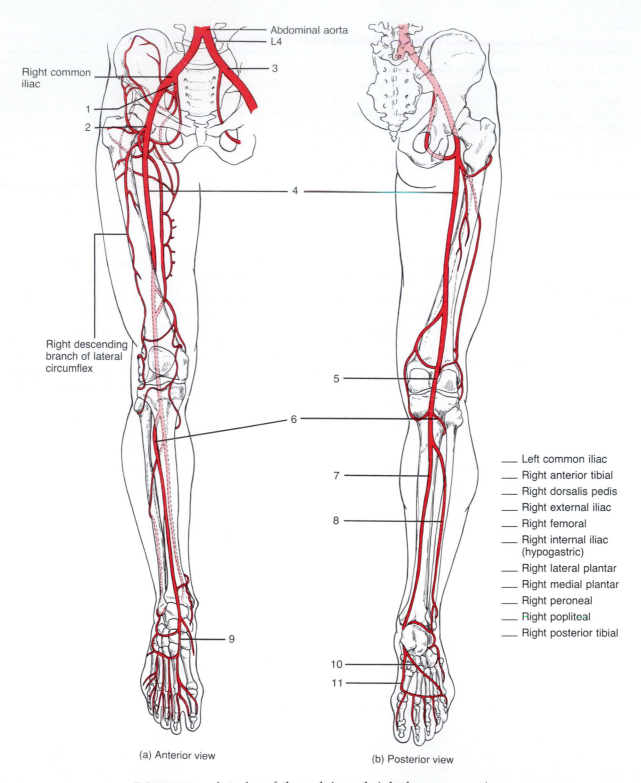

Abdominal aorta
L4
3

Right common
iliac

1
2

Right descending
branch of lateral
circumflex

4

5

6

7

8

9

10
11

___ Left common iliac
___ Right anterior tibial
___ Right dorsalis pedis
___ Right external iliac
___ Right femoral
___ Right internal iliac
    (hypogastric)
___ Right lateral plantar
___ Right medial plantar
___ Right peroneal
___ Right popliteal
___ Right posterior tibial

(a) Anterior view

(b) Posterior view

FIGURE 19.8   Arteries of the pelvis and right lower extremity.

TABLE 19.6
Veins of Systemic Circulation (See Figure 19.10)

OVERVIEW: Deoxygenated blood returns to the right atrium from three veins: **coro-nary sinus, superior vena cava,** and **inferior vena cava.** The coronary sinus receives blood from the cardiac veins; the superior vena cava receives blood from veins superior to the diaphragm, except the lungs. This includes the head, neck, upper extremities, and thoracic wall. The inferior vena cava receives blood from veins inferior to the diaphragm. This includes the lower extremities, most of the abdominal wall, and abdominal viscera.

Vein	Description and region drained
Coronary sinus	**Coronary sinus** receives almost all venous blood from the myocardium. It is located in the coronary sulcus (see Figure 18.4b) and opens into the right atrium between the orifice of the inferior vena cava and the tricuspid valve.
Superior vena cava (SVC)	**SVC** is about 7.5 cm (3 in.) long and empties its blood into the upper part of the right atrium. It begins posterior to the right first costal cartilage by the union of the right and left brachiocephalic veins and ends at the level of the right third costal cartilage, where it enters the right atrium.
Inferior vena cava (IVC)	**IVC** is the largest vein in the body, about 2.5 cm (1.4 in.) in diameter. It begins anterior to the fifth lumbar vertebra by the union of the common iliac veins, ascends behind the peritoneum to the right of the midline, pieces the costal tendon of the diaphragm at the level of the eighth thoracic vertebra, and enters the lower part of the right atrium.

Write the names of the missing veins in the following scheme of circulation. Be sure to indicate "left" or "right," where applicable.

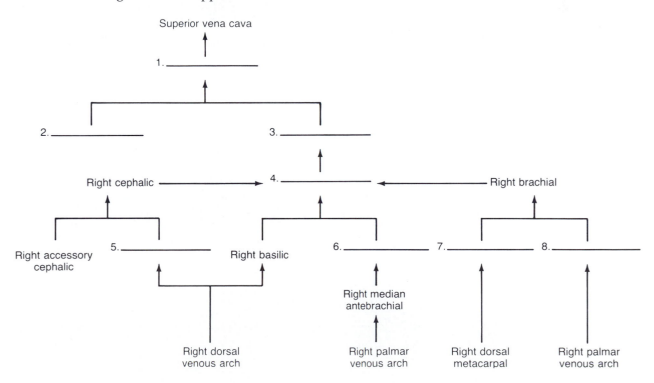

TABLE 19.7
Veins of Head and Neck (Figure 19.9)

OVERVIEW: The majority of blood draining from the head is pushed into three pairs of veins: **Internal jugular, external jugular,** and **vertebral.** Within the cranium, all veins lead to the external jugular veins.

Vein	Description and region drained
Internal jugulars	Right and left *internal jugular veins* receive blood from face and neck. They arise as the continuation of the *sigmoid sinuses* at the base of the skull. Intracranial vascular sinuses are located between the layers of the dura mater and receive blood from the brain. Other sinuses that drain into the internal jugular include the *superior sagittal sinus, inferior sagittal sinus, straight sinus,* and *transverse (lateral) sinuses.* Internal jugulars descend on either side of the neck and pass behind the clavicles, where they join with the right and left subclavian veins. Unions of internal jugulars and subclavians form right and left brachiocephalic veins. From here blood flows into the superior vena cava.
External jugulars	Right and left *external jugular veins* run down the neck along the outside of the internal jugulars. They drain blood from the parotid (salivary) glands, facial muscles, scalp, and other superficial structures into subclavian veins.
Vertebrals	Right and left *vertebral veins* descend through transverse foramina of cervical vertebrae and enter the brachiocephalic veins. They drain the deep structures of the neck, such as vertebrae and muscles.

Label Figure 19.9.

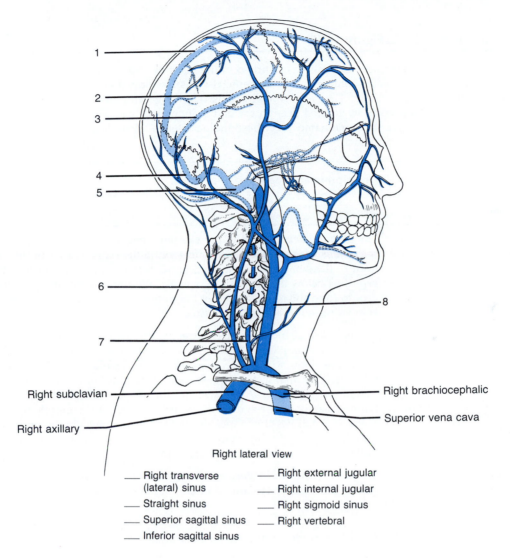

1

2
3

4
5

6

7

8

Right subclavian

Right brachiocephalic

Right axillary

Superior vena cava

Right lateral view

___ Right transverse   ___ Right external jugular
    (lateral) sinus   ___ Right internal jugular
___ Straight sinus   ___ Right sigmoid sinus
___ Superior sagittal sinus ___ Right vertebral
___ Inferior sagittal sinus

FIGURE 19.9 Veins of the head and neck.

TABLE 19.8
Veins of Upper Extremities (Figure 19.10)

OVERVIEW: Blood from each upper extremity is returned to the heart by superficial and deep veins. Both sets of veins contain valves. **Superficial veins** are located just below the skin and are often visible. They anastomose extensively with each other and with deep veins. **Deep veins** are located deep in the body. They usually accompany arteries, and many have the same names as corresponding arteries. Most deep veins are paired vessels.

Vein	Description and region drained
SUPERFICIAL	
Cephalics	The *cephalic vein* of each upper extremity begins in the medial part of the ***dorsal venous arch*** and winds upward around the radial border of the forearm. In front of the elbow, it is connected to the basilic vein by the ***median cubital vein.*** Just below the elbow, the cephalic vein unites with the ***accessory cephalic vein*** to form the cephalic vein of the upper extremity. Ultimately, the cephalic vein empties into the axillary vein.
Basilics	The *basilic vein* of each upper extremity originates in the ulnar part of the ***dorsal venous arch.*** It extends along the posterior surface of the ulna to the point below the elbow where it receives the ***median cubital vein.*** If a vein must be punctured for an injection, transfusion, or removal of blood sample, median cubitals are preferred. After receiving the median cubital vein, the basilic continues ascending on the medial side until it reaches the middle of the arm. There it penetrates the tissues deeply and runs alongside the brachial artery until it joins the brachial vein. As the basilic and brachial veins merge in the axillary area, they form the axillary vein.
Median antebrachials	*Median antebrachial veins* drain the ***palmar venous arch,*** ascend on the ulnar side of the anterior forearm, and end in the median cubital veins.
DEEP	
Radials	*Radial veins* receive the dorsal metacarpal veins.
Ulnars	*Ulnar veins* receive tributaries from the ***palmar venous arch.*** Radial and ulnar veins unite in the bend of the elbow to form brachial veins.
Brachials	Located on either side of the brachial artery, *brachial veins* join into axillary veins.
Axillaries	*Axillary veins* are a continuation of brachials and basilics. Axillaries end at the first rib, where they become subclavians.
Subclavians	*Right* and *left subclavian veins* unite with internal jugulars to form ***brachiocephalic veins.*** The thoracic duct of the lymphatic system delivers lymph into the left subclavian vein at the junction with the internal jugular. The right lymphatic duct delivers lymph into the right subclavian vein at the corresponding junction.

Label Figure 19.10.

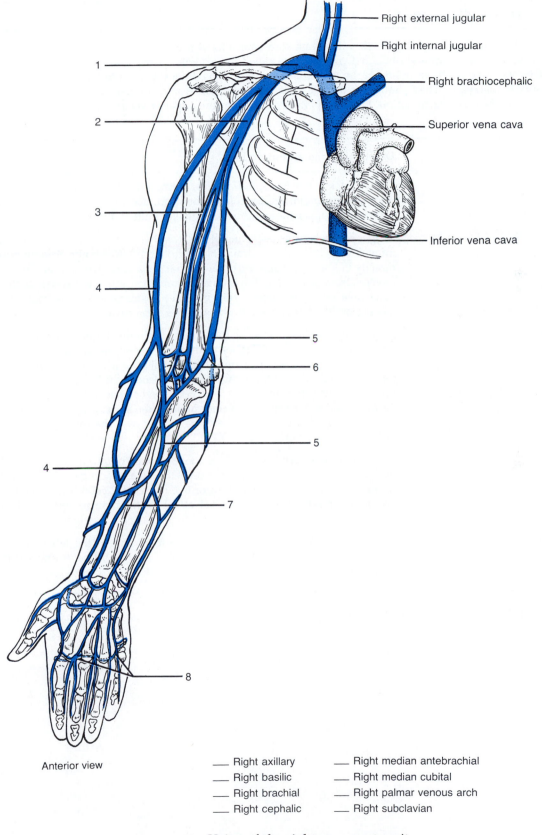

Right external jugular

Right internal jugular

Right brachiocephalic

Superior vena cava

Inferior vena cava

Anterior view

___ Right axillary

___ Right basilic

___ Right brachial

___ Right cephalic

___ Right median antebrachial

___ Right median cubital

___ Right palmar venous arch

___ Right subclavian

FIGURE 19.10 Veins of the right upper extremity.

TABLE 19.9
Veins of Thorax (Figure 19.11)

OVERVIEW: Although brachiocephalic veins drain some portions of the thorax, most thoracic structures are drained by a network of veins called the **azygos system.** This is a network of veins on each side of the vertebral column: azygos, hemiazygos, and accessory hemiazygos. They show considerable variation in origin, course, tributaries, anastomoses, and termination. Ultimately, they empty into the superior vena cava.

Vein	Description and region drained
Brachiocephalic	Right and left *brachiocephalic veins,* formed by the union of subclavians and internal jugulars, drain blood from the head, neck, upper extremities, mammary glands, and upper thorax. Brachiocephalics unite to form the *superior vena cava.*
Azygos veins	*Azygos veins,* besides collecting blood from the thorax, serve as a bypass for the inferior vena cava, which drains blood from the lower body. Several small veins directly link azygos veins with the inferior vena cava. Large veins that drain the lower extremities and abdomen may drain blood into the azygos. If the inferior vena cava or hepatic portal vein becomes obstructed, azygos veins can return blood from the lower body to the superior vena cava.
Azygos	The *azygos vein* lies in front of the vertebral column, slightly right of the midline. The vein begins as a continuation of the right ascending lumbar vein and connects with the inferior vena cava, right common iliac, and lumbar veins. The azygos receives blood from (1) the right *intercostal veins,* which drain the chest muscles: (2) the hemiazygos and accessory hemiazygos veins; (3) several *esophageal, mediastinal,* and *pericardial veins;* and (4) the right *bronchial vein.* The vein ascends to the fourth thoracic vertebra, arches over the right lung, and empties into the superior vena cava.
Hemiazygos	The *hemiazygos vein* is in front of the vertebral column and slightly left of the midline, beginning as the continuation of the left ascending lumbar vein. The vein receives blood from the lower four or five *intercostal veins* and some *esophageal* and *mediastinal veins.* At the level of the ninth thoracic vertebra, it joins the azygos vein.
Accessory hemiazygos	The *accessory hemiazygos vein* is also in front and to the left of the vertebral column. It receives blood from three or four *intercostal veins* and left *bronchial vein* and joins azygos at the level of the eighth thoracic vertebra.

Label Figure 19.11, using Figure 19.3 for reference, if necessary.

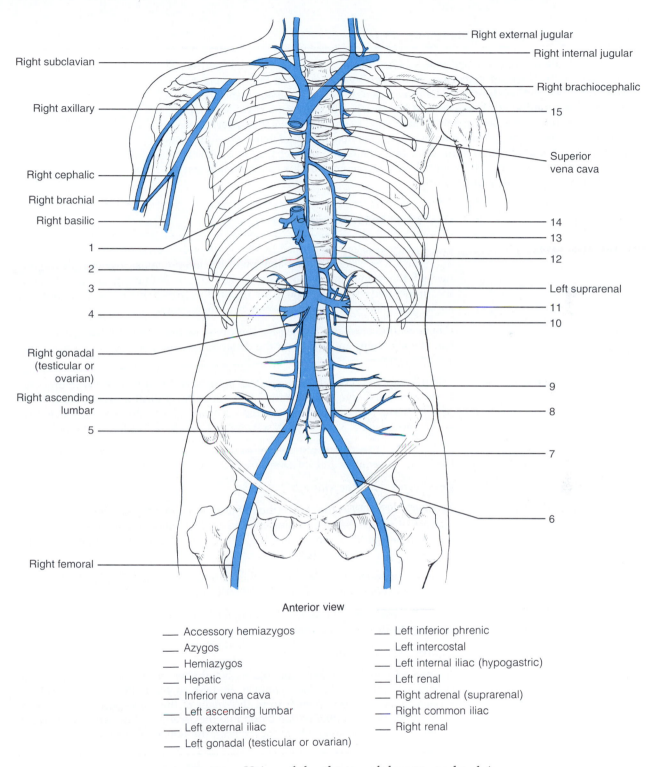

Right external jugular

Right internal jugular

Right subclavian

Right brachiocephalic

Right axillary

15

Superior
vena cava

Right cephalic

Right brachial

Right basilic

14

13

1

12

2

3

Left suprarenal

4

11

10

Right gonadal
(testicular or
ovarian)

Right ascending
lumbar

9

8

5

7

6

Right femoral

Anterior view

___ Accessory hemiazygos   ___ Left inferior phrenic
___ Azygos   ___ Left intercostal
___ Hemiazygos   ___ Left internal iliac (hypogastric)
___ Hepatic   ___ Left renal
___ Inferior vena cava   ___ Right adrenal (suprarenal)
___ Left ascending lumbar   ___ Right common iliac
___ Left external iliac   ___ Right renal
___ Left gonadal (testicular or ovarian)

FIGURE 19.11 Veins of the thorax, abdomen, and pelvis.

TABLE 19.10
Veins of Abdomen and Pelvis (Figure 19.11)

OVERVIEW: Blood from the abdominopelvic viscera and abdominal wall returns to the heart via the **inferior vena cava.** Many small veins enter the inferior vena cava. Most carry return flow from parietal branches of the abdominal aorta and their names correspond to the names of the arteries. The inferior vena cava does not receive veins from the gastrointestinal tract, spleen, pancreas, or gallbladder. These organs pass their blood into a common vein, the hepatic portal vein, which delivers the blood to the liver. From here, the blood drains into the hepatic veins, which enter the inferior vena cava. This special flow of venous blood is called **hepatic portal circulation,** which is described shortly.

Vein	Description and region drained
Inferior vena cava	The *inferior vena cava* is formed by the union of the two common iliac veins that drain the lower extremities and abdomen. The inferior vena cava extends upward through the abdomen and thorax to the right atrium. Numerous small veins enter the inferior vena cava. Most carry the return flow from branches of the abdominal aorta; their names correspond to names of the arteries.
Common iliacs	*Common iliac veins* are formed by the union of the internal (hypogastric) and external iliac veins and represent the distal continuation of the inferior vena cava at its bifurcation.
Internal iliacs (hypogastrics)	Tributaries of *internal iliac (hypogastric) veins* basically correspond with branches of the external iliac arteries. Internal iliacs drain the gluteal muscles, medial side of the thigh, urinary bladder, rectum, prostate gland, ductus (vas) deferens, uterus, and vagina.
External iliacs	*External iliac veins* are continuations of the femoral veins and receive blood from the lower extremities and inferior part of the anterior abdominal wall.
Renals	*Renal veins* drain the kidneys.
Gonadals (testiculars or ovarians)	*Testicular veins* drain the testes (the left testicular vein empties into the left renal vein and the right testicular drains into the inferior vena cava); *ovarian veins* drain the ovaries (the left ovarian vein empties into the left renal vein and the right ovarian drains into the inferior vena cava).
Adrenals (suprarenals)	*Adrenal (suprarenal) veins* drain the adrenal (suprarenal) glands (the left suprarenal vein empties into the left renal vein).
Inferior phrenics	*Inferior phrenic veins* drain the diaphragm (the left inferior phrenic vein sends a tributary to the left renal vein).
Hepatics	*Hepatic veins* drain the liver.
Lumbars	A series of parallel *lumbar veins* drains blood from both sides of the posterior abdominal wall. The lumbars connect at right angles with the right and left *ascending lumbar veins,* which form the origin of the corresponding azygos or hemiazygos vein. Lumbars drain blood into the ascending lumbars and then run to the inferior vena cava, where they release the remainder of the flow.

Label Figure 19.11.

TABLE 19.11
Veins of Lower Extermities (Figure 19.12)

OVERVIEW: Blood from each lower extremity is drained by **superficial** and **deep veins.** The superficial veins often anastomose with each other and with deep veins along their length. Deep veins, for the most part, have the same names as their accompanying arteries. Both superficial and deep veins have valves.

Vein	Description and region drained
SUPERFICIAL	
Great saphenous	The *great saphenous vein,* the longest vein in the body, begins at the medial end of the *dorsal venous arch* of the foot. The vein passes in front of the medial malleolus and then upward along the medial aspect of the leg and thigh. It receives tributaries from superficial tissues, connects with deep veins, and empties into the femoral vein in the groin.
Small saphenous	The *small saphenous vein* begins at the lateral end of the dorsal venous arch of the foot. The vein passes behind the lateral malleolus and ascends under the skin of the back of the leg. It receives blood from the foot and posterior portion of the leg and empties into the popliteal vein behind the knee.
DEEP	
Posterior tibial	The *posterior tibial vein* is formed by the union of the *medial* and *lateral plantar veins* behind the medial malleolus. The vein ascends deep in the muscles at the back of the leg, receives blood from the *peroneal vein,* and unites with the anterior tibial vein just below the knee. These veins begin in the *plantar arch* under the bones of the foot.
Anterior tibial	The *anterior tibial vein* is the upward continuation of the *dorsalis pedis veins* in the foot. The veins runs between the fibia and fibula and unite with the posterior tibial to form the popliteal vein.
Popliteal	The *popliteal vein,* just behind the knee, receives blood from the anterior and posterior tibials and small saphenous vein.
Femoral	The *femoral vein* is the upward continuation of the popliteal just above the knee. Femorals run up the posterior of the thighs and drain the deep stuctures of the thighs. After receiving the great saphenous veins in the groin, they continue as the right and left external iliac veins.

Label Figure 19.12.

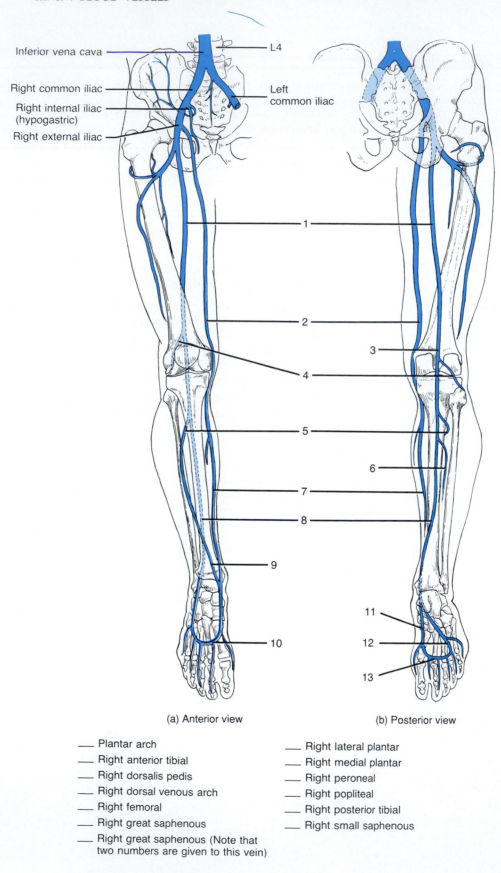

Inferior vena cava

L4

Right common iliac

Left common iliac

Right internal iliac (hypogastric)

Right external iliac

1

2

3

4

5

6

7

8

9

11

12

13

10

(a) Anterior view

(b) Posterior view

____ Plantar arch

____ Right anterior tibial

____ Right dorsalis pedis

____ Right dorsal venous arch

____ Right femoral

____ Right great saphenous

____ Right great saphenous (Note that two numbers are given to this vein)

____ Right lateral plantar

____ Right medial plantar

____ Right peroneal

____ Right popliteal

____ Right posterior tibial

____ Right small saphenous

FIGURE 19.12   Veins of the pelvis and right lower extremity.

Write the names of the missing veins in the following scheme of circulation. Be sure to indicate "left" or "right," where applicable.

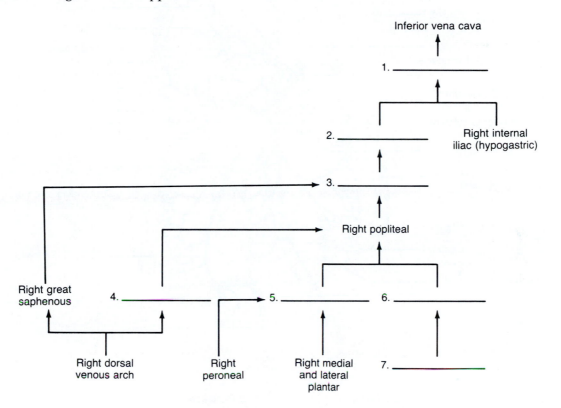

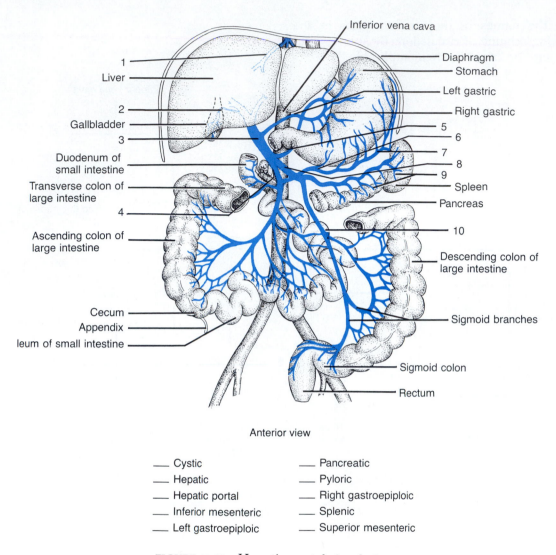

Anterior view

___ Cystic	___ Pancreatic
___ Hepatic	___ Pyloric
___ Hepatic portal	___ Right gastroepiploic
___ Inferior mesenteric	___ Splenic
___ Left gastroepiploic	___ Superior mesenteric

FIGURE 19.13   Hepatic portal circulation.

2. Hepatic Portal Circulation

Hepatic portal circulation refers to the flow of venous blood from the organs of the gastrointestinal tract to the liver before it is returned to the heart (Figure 19.13). The blood vessels contained in this circulatory route include the *hepatic portal vein, superior mesenteric vein, splenic vein, gastric vein, pyloric vein, gastroepiploic vein, pancreatic veins, inferior mesenteric veins,* and *cystic vein.* Ultimately, blood leaves the liver through the hepatic veins, which enter the inferior vena cava.

Using your textbook, charts, or models for reference, label Figure 19.13.

3. Pulmonary Circulation

Pulmonary circulation refers to the flow of deoxygenated blood from the right ventricle to the air

sacs of the lungs and the return of oxygenated blood from the air sacs of the lungs to the left atrium (Figure 19.14). The *right* and *left pulmonary arteries* are the only postnatal (after birth) arteries that carry deoxygenated blood, and the *pulmonary veins* are the only postnatal veins that carry oxygenated blood.

Study a chart or model of the pulmonary circulation, trace the path of blood through it, and label Figure 19.14.

4. Fetal Circulation

The developing fetus has nonfunctional lungs and a nonfunctional gastrointestinal tract. This difference from an adult's circulatory system is called *fetal circulation* (Figure 19.15). The lungs of the fetus are not active in respiration because

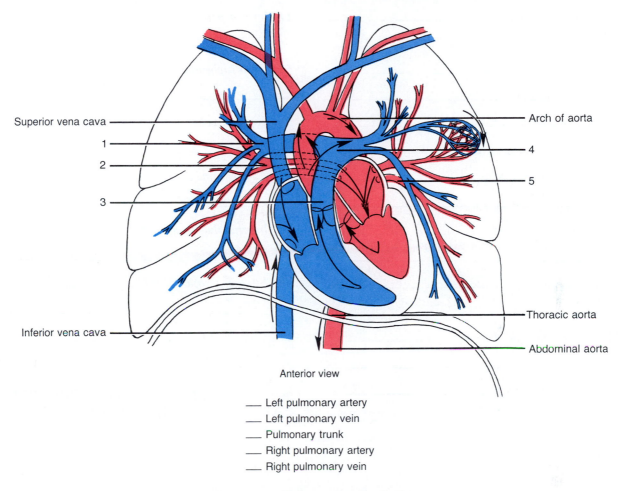

Superior vena cava

1

2

3

Arch of aorta

4

5

Inferior vena cava

Thoracic aorta

Abdominal aorta

Anterior view

___ Left pulmonary artery
___ Left pulmonary vein
___ Pulmonary trunk
___ Right pulmonary artery
___ Right pulmonary vein

FIGURE 19.14   Pulmonary circulation.

the fetus derives its oxygen and nutrients and eliminates its carbon dioxide and wastes through the maternal blood. An opening, the *foramen ovale* (fo-RĀ-men ō-VAL-ē), is found between the right and left atria. Most blood leaving the right atrium passes through this opening into the left atrium rather than into the right ventricle. Blood that does enter the right ventricle is pumped into the pulmonary trunk, but little of this blood reaches the lungs. Most is conveyed from the pulmonary trunk to the aorta via a small vessel, the *ductus arteriosus* (DUK-tus ar-tē-rē-Ō-sus), which closes shortly after birth.

A highly specialized structure called the *placenta* (pla-SEN-ta) accomplishes the exchange of substances between maternal and fetal circulation. Other special blood vessels include two *umbilical arteries* and an *umbilical vein,* which extend through the *umbilical cord,* and the *ductus venosus* (DUK-tus ve-NŌ-sus), which bypasses the fetal liver.

Label Figure 19.15 on page 393.

E. BLOOD VESSEL EXERCISE

For each vessel listed, indicate the region supplied (if an artery) or the region drained (if a vein):

1. *Coronary artery.* _____

_____

2. *Internal iliac veins.* _____

_____

3. *Lumbar arteries.* _____

_____

4. *Renal artery.* _____

_____

5. *Left gastric artery.* _____

_____

6. *External jugular vein.* _____

_____

7. *Left subclavian artery.* _____

_____

8. *Axillary vein.* _____

_____

9. *Brachiocephalic veins.* _____

_____

10. *Transverse sinuses.* _____

_____

11. *Hepatic artery.* _____

_____

12. *Inferior mesenteric artery.* _____

_____

13. *Suprarenal artery.* _____

_____

14. *Inferior phrenic artery.* _____

_____

15. *Great saphenous vein.* _____

_____

16. *Popliteal vein.* _____

_____

17. *Azygos vein.* _____

_____

18. *Internal iliac (hypogastric) artery.* _____

_____

19. *Internal carotid artery.* _____

_____

20. *Cephalic vein.* _____

ANSWER THE LABORATORY REPORT QUESTIONS AT THE END OF THE EXERCISE.

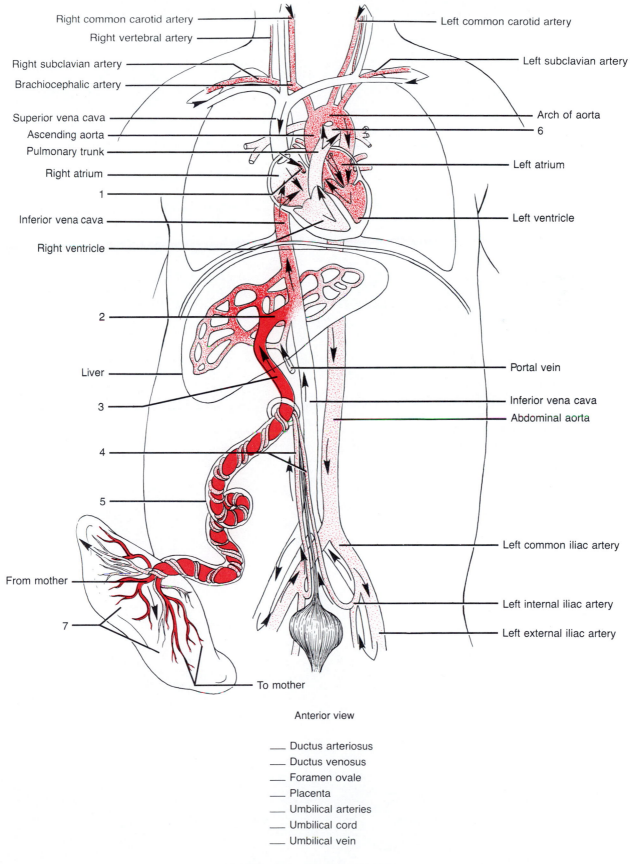

Right common carotid artery

Right vertebral artery

Right subclavian artery

Brachiocephalic artery

Superior vena cava

Ascending aorta

Pulmonary trunk

Right atrium

1

Inferior vena cava

Right ventricle

2

Liver

3

4

5

From mother

7

To mother

Left common carotid artery

Left subclavian artery

Arch of aorta

6

Left atrium

Left ventricle

Portal vein

Inferior vena cava

Abdominal aorta

Left common iliac artery

Left internal iliac artery

Left external iliac artery

Anterior view

___ Ductus arteriosus
___ Ductus venosus
___ Foramen ovale
___ Placenta
___ Umbilical arteries
___ Umbilical cord
___ Umbilical vein

FIGURE 19.15   Fetal circulation.

Blood Vessels

STUDENT _____   DATE _____

LABORATORY SECTION _____   SCORE/GRADE _____

PART 1. Multiple Choice

_____ 1. The largest of the circulatory routes is (a) systemic (b) pulmonary (c) coronary (d) hepatic portal

_____ 2. All arteries of systemic circulation branch from the (a) superior vena cava (b) aorta (c) pulmonary artery (d) coronary artery

_____ 3. The arterial system that supplies the brain with blood is the (a) hepatic portal system (b) pulmonary system (c) cerebral arterial circle (circle of Willis) (d) carotid system

_____ 4. An obstruction in the inferior vena cava would hamper the return of blood from the (a) head and neck (b) upper extremities (c) thorax (d) abdomen and pelvis

_____ 5. Which statement best describes arteries? (a) all carry oxygenated blood to the heart (b) all contain valves to prevent the backflow of blood (c) all carry blood away from the heart (d) only large arteries are lined with endothelium

_____ 6. Which statement is *not* true of veins? (a) they have less elastic tissue and smooth muscle than arteries (b) they contain more fibrous tissue than arteries (c) most veins in the extremities have valves (d) they always carry deoxygenated blood

_____ 7. A thrombus in the first branch of the arch of the aorta would affect the flow of blood to the (a) left side of the head and neck (b) myocardium of the heart (c) right side of the head and neck and right upper extremity (d) left upper extremity

_____ 8. If a vein must be punctured for an injection, transfusion, or removal of a blood sample, the likely site would be the (a) median cubital (b) subclavian (c) hemiazygous (d) anterior tibial

_____ 9. In hepatic portal circulation, blood is eventually returned to the inferior vena cava through the (a) superior mesenteric vein (b) hepatic portal vein (c) hepatic artery (d) hepatic veins

_____ 10. Which of the following are involved in pulmonary circulation? (a) superior vena cava, right atrium, and left ventricle (b) inferior vena cava, right atrium, and left ventricle (c) right ventricle, pulmonary artery, and left atrium (d) left ventricle, aorta, and inferior vena cava

_____ 11. If a thrombus in the left common iliac vein dislodged, into which arteriole system would it first find its way? (a) brain (b) kidneys (c) lungs (d) left arm

_____ 12. In fetal circulation, the blood containing the highest amount of oxygen is found in the (a) umbilical arteries (b) ductus venosus (c) aorta (d) umbilical vein

_____ 13. The greatest amount of elastic tissue found in the arteries is located in which coat? (a) tunica interna (b) tunica media (c) tunica externa (d) tunica adventitia

_____ 14. Which coat of an artery contains endothelium? (a) tunica interna (b) tunica media (c) tunica externa (d) tunica adventitia

_____ 15. Permitting the exchange of nutrients and gases between the blood and tissue cells is the primary function of (a) capillaries (b) arteries (c) veins (d) arterioles

_____ 16. The circulatory route that runs from the gastrointestinal tract to the liver is called (a) coronary circulation (b) pulmonary circulation (c) hepatic portal circulation (d) cerebral circulation

_____ 17. Which of the following statements about systemic circulation is *not* correct? (a) its purpose is to carry oxygen and nutrients to body tissues and to remove carbon dioxide (b) all systemic arteries branch from the aorta (c) it involves the flow of blood from the left ventricle to all parts of the body except the lungs (d) it involves the flow of blood from the body to the left atrium

_____ 18. The opening in the septum between the right and left atria of a fetus is called the (a) foramen ovale (b) ductus venosus (c) foramen rotundum (d) foramen spinosum

_____ 19. The branch of the umbilical vein in the fetus that connects with the inferior vena cava, bypassing the liver, is the (a) foramen ovale (b) ductus venosus (c) ductus arteriosus (d) patent ductus

_____ 20. Which of the vessels does *not* belong with the others? (a) brachiocephalic artery (b) left common carotid artery (c) celiac artery (d) left subclavian artery

PART 2. Matching

_____ 21. Aortic branch that supplies the head and associated structures

_____ 22. Artery that distributes blood to the small intestine and part of the large intestine

_____ 23. Vessel into which veins of the head and neck, upper extremities, and thorax enter

_____ 24. Vessel into which veins of the abdomen, pelvis, and lower extremities enter

_____ 25. Vein that drains the head and associated structures

_____ 26. Longest vein in the body

_____ 27. Vein just behind the knee

_____ 28. Artery that supplies a major part of the large intestine and rectum

_____ 29. First branch off of the arch of the aorta

_____ 30. Arteries supplying the heart

A. Inferior vena cava

B. Superior vena cava

C. Superior mesenteric

D. Common carotid

E. Jugular

F. Brachiocephalic

G. Coronary

H. Inferior mesenteric

I. Popliteal

J. Great saphenous

20

Cardiovascular Physiology

A. CARDIAC CONDUCTION SYSTEM AND ELECTROCARDIOGRAM (ECG OR EKG)

The heart is innervated by the autonomic nervous system (ANS), which modulates, but does not initiate, the cardiac cycle. The heart can continue to contract if separated from the ANS. This is possible because the heart has an *intrinsic pacemaker* termed the *sinoatrial (SA) node.* The SA node is called an intrinsic pacemaker because of its ability to rhythmically, spontaneously depolarize, thereby reaching threshold value and initiating an action potential. This ability to spontaneously depolarize is due to the sinoatrial node's permeability characteristics to sodium, potassium, and calcium ions. The SA node is connected to a series of specialized muscle cells that comprise the *cardiac conduction system.* This system distributes the electrical impulses that stimulate the cardiac muscle fibers to contract.

The SA node is located in the right atrial wall, inferior to the opening of the superior vena cava. Once the SA node spontaneously depolarizes and reaches threshold value, the impulse is conducted through the remainder of the heart via the cardiac conduction system. From the SA node, the action potential travels throughout the right atrium via a wavelike conduction process. In addition, the impulse is conducted to the left atrium and the *atrioventricular (AV) node,* which is located near the inferior portion of the intera-

trial septum, via the *anterior interatrial myocardial band (Bachmann's bundle).* The band is composed of *anterior, posterior,* and *middle pathways,* which are three groups of specialized cardiac muscle cells. As the action potential spreads to the atrial musculature, it initiates atrial contraction. When the action potential reaches the AV node, it too depolarizes, but in a manner that significantly slows the conduction velocity of the cardiac action potential. From the atrioventricular node, a tract of conducting fibers called the *atrioventricular (AV) bundle (bundle of His)* extends to the top of the interventricular septum and continues down both sides of the septum as the *right* and *left bundle branches.* The AV bundle distributes the electrical impulses over the medial surfaces of the ventricles via the *conduction myofibers (Purkinje fibers)* that emerge from the bundle branches and pass into the cells of the ventricular myocardium. Because of the very large diameter of these fibers, the conduction velocity of the action potential reaches its highest speed within the conduction myofibers, thereby significantly increasing the speed with which the electrical activity is spread throughout the ventricles. As the electrical impulses reach the ventricular fibers, they are depolarized, thereby initiating ventricular contraction.

Label the components of the cardiac conduction system shown in Figure 20.1

The electrocardiogram (ECG) provides a record of the electrical events happening within the heart. The ECG is obtained by placing recording electrodes on the surface of the body. These recording electrodes detect the changing poten-

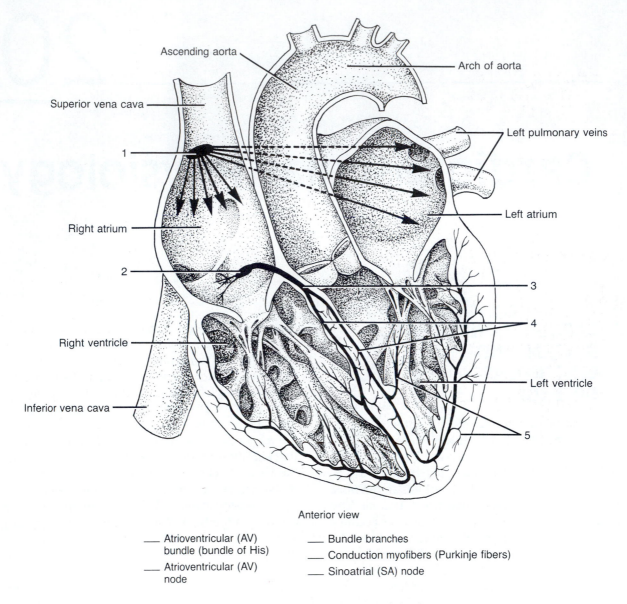

Anterior view

___ Atrioventricular (AV)   ___ Bundle branches
    bundle (bundle of His)   ___ Conduction myofibers (Purkinje fibers)
___ Atrioventricular (AV)   ___ Sinoatrial (SA) node
    node

FIGURE 20.1   Conduction system of the heart.

tial differences on the body's surfaces, and a time-dependence plot of these differences is called an *electrocardiogram (ECG).* An *electrocardiograph* is an instrument used to record these changes. As the electrical impulses are transmitted throughout the cardiac conduction system and myocardial cells, a different electrical impulse is generated. These impulses are transmitted from the electrodes to a recording needle that graphs the impulses as a series of up-and-down (vertical) waves called *deflection waves* (See Figure 20.2).

1. Electrocardiographic Recordings

Electrocardiograms are usually recorded by *indirect leads* (recording electrodes placed some distance from the heart on the skin of the subject, rather than directly on the heart). Because of the differences in the size and shape of any one individual's heart and body dimensions, the electrocardiogram varies from individual to individual. In addition, within any one individual the ECG pattern varies with the placement of the leads.

The ECG recordings that will be obtained in the laboratory utilize the ***three standard limb leads.*** A **lead** is defined as two electrodes working in pairs. ***Electrodes*** are sensing devices of sensitive metal plates or small rods that can detect electrophysiological phenomena such as changes in the electrical potential of skin or nerves (see Exercise 9). For convenience, the three standard limb leads are typically attached to the right and left wrists and the left leg. Lead I records the potential difference between the left and right arms (LA and RA, respectively). Lead II records the potential difference between the right arm (RA) and left leg (LL). Lead III, therefore, would record the potential differences between the left arm (LA) and the left leg (LG).

A typical lead II ECG record (Figure 20.2) is composed of an orderly series of deflections and intervals. The first deflection, or wave, is termed the ***P wave*** and represents atrial depolarization. After the P wave, the ECG returns to baseline level because the electrodes are no longer recording any potential differences. This time period is termed the ***P-Q interval.*** During this time period, however, the electrical activity of the heart is not quiescent, as the action potential is being propagated through the AV node, the AV bundle branches, and the conduction myofibers. This deflection of the ECG is termed the ***QRS complex*** and represents ventricular depolarization and atrial repolarization. (No separate deflection for atrial repolarization is found because of the simultaneous nature of atrial depolarization and ventricular depolarization. In addition, atrial repolarization is masked because of the significantly smaller amount of atrial muscle mass as compared to the ventricular muscle mass.) In the time period between the QRS complex and the repolarization of the ventricle (during the T wave described below) the ECG again returns to, or very closely to, its baseline level. This time period is termed the ***S-T segment.*** During the S-T segment of the ECG, all of the ventricular muscle is depolarized. This segment of the ECG represents the plateau phase of the ventricular action potential. The final wave is termed the ***T wave***, and represents ventricular repolarization.

In reading and interpreting an electrocardiogram, you must note the **lead** type of configuration from which the ECG was recorded, the size

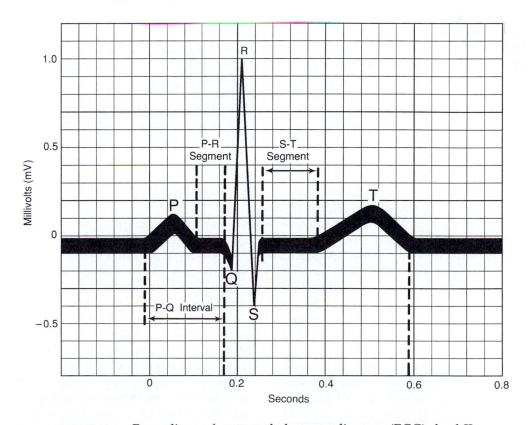

FIGURE 20.2 Recordings of a normal electrocardiogram (ECG), lead II.

of the deflection waves, and the lengths of the P-Q interval and S-T segment. An attempt will not be made to interpret an electrocardiogram, but only to become familiar with a normal ECG. The ECG is valuable in diagnosing abnormal cardiac rhythms and conducting patterns, detecting the presence of fetal life, determining the presence of more than one fetus, and following the course of recovery from a heart attack.

2. Recording an ECG

In the following procedure, a physiograph-type polygraph is used, but an electrocardiograph or oscilloscope can also be used (Figure 20.3a).

In recording the ECG, we will utilize only the standard three limb leads. Usually, only two electrodes will be in actual use at any one time. The third electrode will be automatically switched off by the *lead-selector switch* of the polygraph.

For most general purposes, three leads are used. However, electrocardiologists use additional leads, including several chest-wall electrodes (Figure 20.3b). These leads are positioned around the chest and encircle the heart so that there are six intersecting lines on a horizontal plane through the atrioventricular node.

The procedure for recording electrocardiograms using the physiograph follows:

PROCEDURE

1. Either you or your laboratory partner should lie on a table or cot, roll down any long stockings or socks to the ankles, and *remove all metal jewelry.*
2. Swab the skin with alcohol where the electrodes will be applied. Electrode cream, jelly, or saline paste is applied to the skin only where the electrode will make contact, such as to the inner forearms just above the wrists and to the inside of the legs just above the ankles.
3. The cream is then applied on the concave surface of the electrode plates, which are then fastened securely to the limb area surfaces with rubber straps or adhesive tape (rings). Adjust the straps to fit snugly, *but not so tight as to impede blood flow.*
4. Firmly connect the proper end of the electrode cables to the electrode plates, using the standard limb leads previously described, and connect the other end to a cardiac preamplifier or to a self-contained electrocardiograph.

5. After you apply the electrodes and connect the instrument to the subject, ascertain the following before actually recording:
   a. Recording power switch is on.
   b. Paper is sufficient for the entire recording, and the pen is centered properly on the paper with ink flowing freely.
   c. Preamplifier sensitivity has been set at 10 mm per millivolt.
   d. Subject is lying quietly and is relaxed.
   Note: Usually five or six ECG complex recordings from each lead are sufficient. All electrocardiographers have chosen a standard paper speed of 25 mm per second.
6. Turn the instrument on and record for 30 sec from each lead. Obtain directions from your laboratory instructor on the procedure to be followed when switching from lead to lead. When the recording leads are switched, note the change by marking it on the electrocardiogram. The subject must remain very still and relax completely during the recording period.
7. When the recording is finished, turn off the reading instrument and disconnect the leads from the subject and remove the electrodes, cleaning off the excess cream or paste. The skin can be washed with water to remove the electrode cream.
8. Identify and letter the P, QRS, and T waves, and compare the waves with those shown in Figure 20.2.
9. Calculate the duration of the waves and the P-Q interval. Attach this recording to Section A of the LABORATORY REPORT RESULTS at the end of the exercise.

B. CARDIAC CYCLE

In a normal *cardiac cycle* (heartbeat) the contraction (*systole*) (SIS-tō-lē) and relaxation (*diastole*) (dī-AS-tō-lē) of the four chambers of the heart are synchronized in such a way as to facilitate the proper filling and emptying of the artria and ventricles. Normally, the cardiac cycle is subdivided into four phases, or parts, according to events occurring within the ventricles. The different phases of the cardiac cycle (Figure 20.4a) are (1) *ventricular ejection,* which is subdivided into fast ejection and slow ejection subphases; (2) *isovolumetric ventricular relaxation;* (3) *ventricular filling,* which is also subdivided into fast-filling and slow-filling subphases; and (4) *isovol-*

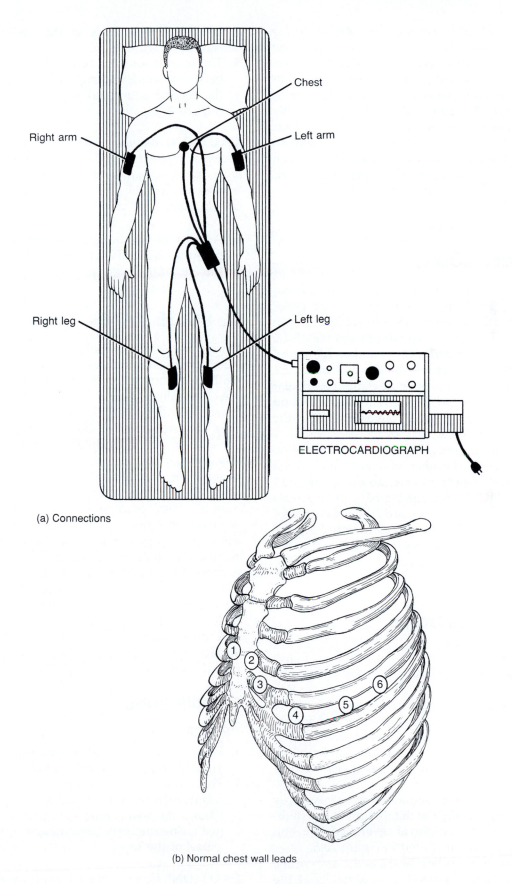

(a) Connections

ELECTROCARDIOGRAPH

(b) Normal chest wall leads

FIGURE 20.3 Electrocardiograph.

umetric ventricular contraction. Atrial filling occurs continuously during the cardiac cycle (except during atrial systole), while *atrial ejection* occurs simultaneously with ventricular filling. In a complete cardiac cycle, the atria are in systole for 0.1 sec and in diastole for 0.7 sec. By contrast, the ventricles are in systole for 0.3 sec and in diastole for 0.5 sec. If we can assume an average heartbeat of 75 cardiac cycles each minute (or a heartbeat of 75 per min), each cardiac cycle requires about 0.8 sec.

C. HEART SOUNDS

The beating of a human heart consists of two well-defined sounds during each cardiac cycle. These two sounds can be easily heard by using an ordinary *stethoscope.* Listening to sounds of the body is called *auscultation* (aws-kul-TĂ-shun). The first sound is created by the closure of the atrioventricular valves soon after the ventricular systole begins (Figure 20.4e). The second sound is created as the semilunar valves close near the end of the ventricular systole.

Heart sounds provide valuable information about the valves. Peculiar sounds may be called *murmurs.* Some murmurs are caused by the noise made by a little blood flowing back in an atrium because of improper closure of an atrioventricular valve. Murmurs do not always indicate that the valves are not functioning properly, and many have no clinical significance.

1. Use of Stethoscope

PROCEDURE

1. The stethoscope should be used in a quiet room.
2. The earpieces of the stethoscope *should be cleaned with alcohol* just before use and should also be pointed slightly forward when placed in the ears. They will be more comfortable in this position, and it will be easier to hear through them.
3. Listen to the heart sounds of your laboratory partner by placing the diaphragm of the stethoscope next to the skin at several positions on the chest wall illustrated in Figure 20.5.
4. The first sound is best heard at the apex of the heart, which is located approximately at the

fifth intercostal space at the midline of the clavicles (Figure 20.5).
5. The second sound is best heard over the area between the second right costal cartilage and the second intercostal space on the left side (Figure 20.5).

CAUTION! *Assuming that your partner has no known or apparent cardiac or other health problems and is capable of such an activity, ask her/him to run in place for about 25 steps. Listen to the heart sounds again.*

6. Answer all questions pertaining to Section C of the LABORATORY REPORT RESULTS at the end of the exercise.

D. PULSE RATE

During ventricular systole, a wave of pressure, called a *pulse,* is produced in the arteries due to ventricular contraction. Pulse rate and heart rate are essentially the same. Pulse can be felt readily where an artery is near the surface of the skin and over the surface of a bone. Pulse rates vary considerably in individuals because of time of day, temperature, emotions, stress, and other factors. The normal adult pulse rate of the heart at rest is within a range of 72 to 80 beats per minute. With practice you can learn to take accurate pulse rates by counting the beats per 15 sec and multiplying by 4 to obtain beats per minute. The term *tachycardia* (tak'-e-KAR-dē-a) is applied to a rapid heart rate or pulse rate (over 100 beats per min). *Bradycardia* (brād-e-KAR-dē-a) indicates a slow heart rate or pulse rate (below 50 beats per min).

Although the pulse can be detected in most surface arteries, the pulse rate is usually determined on the *radial artery* of the wrist.

1. Radial Pulse

PROCEDURE

1. Using your index and middle finger, palpate your laboratory partner's radial artery.
2. The thumb should never be used because it has its own prominent pulse.
3. Palpate the area behind your partner's thumb, just inside the bony prominence on the lateral aspect of the wrist.

CAUTION! *Do not apply too much pressure.*

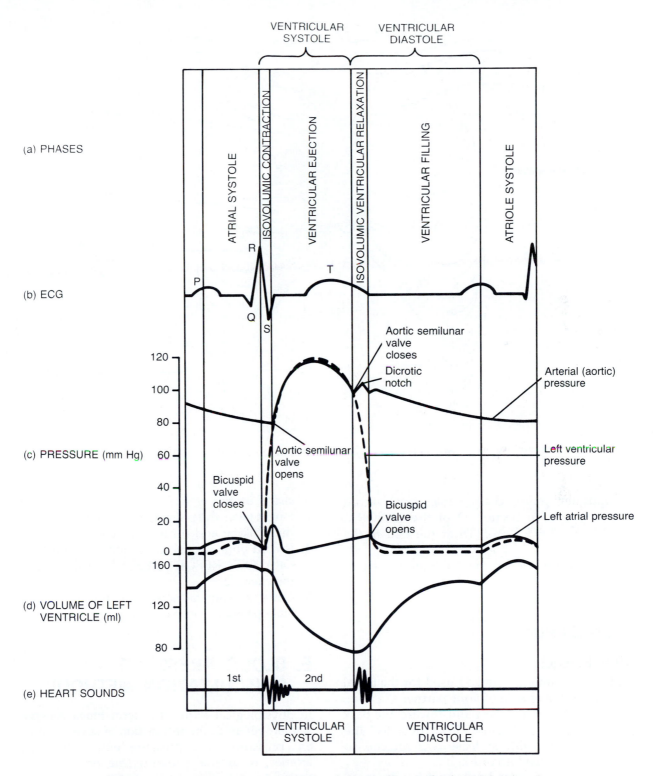

FIGURE 20.4   Cardiac cycle. (a) Phases of the cardiac cycle (b) ECG related to the cardiac cycle. (c) Left atrial, left ventricular, and arterial (aortic) pressure changes along with the opening and closing of the valves during cardiac cycle. (d) Left ventricular volume during the cardiac cycle. (e) Heart sounds related to the cardiac cycle.

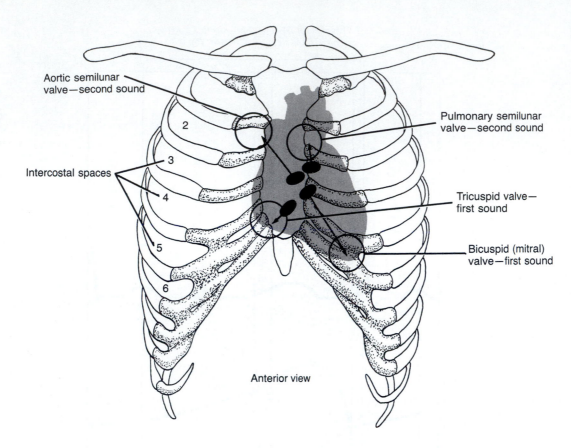

Aortic semilunar
valve—second sound

Pulmonary semilunar
valve—second sound

Intercostal spaces

2

3

4

5

6

Tricuspid valve—
first sound

Bicuspid (mitral)
valve—first sound

Anterior view

FIGURE 20.5   Surface areas where heart valve sounds are best heard.

4. Count the pulse, change roles, and record your results in Section D of the LABORATORY REPORT RESULTS at the end of the exercise.
5. Calculate the *median pulse rate* of the entire class.

2. Carotid Pulse

PROCEDURE

1. Place the same fingers you used for the radial pulse on either side of your partner's larynx.
2. *Gently* press downward and toward the back until you feel the pulse. You must feel the pulse clearly with at least two fingers, so adjust your hand accordingly.
3. The radial and carotid pulse can be compared under the following conditions: (a) sitting quietly, (b) standing quietly, (c) right after walking 60 steps, (d) right after running in place 60 steps, *assuming that your partner has no known or apparent cardiac or other health problems*

and is capable of such an activity. Notice how long it takes the pulse to return to normal after the walking and running exercises.
4. Record your results in Section D of the LABORATORY REPORT RESULTS at the end of the exercise.
5. Compare your radial and carotid pulse in the table provided on page 410.

E. BLOOD PRESSURE (AUSCULTATION METHOD)

In physiological terms, the term *blood pressure* actually refers to the interaction of several different pressures (the pressure only within the arteries, or arterial pressure; the pressure only within the veins, or venous pressure; the pressure within the pulmonary system, or pulmonary pressure; and the pressure within all of the other vascular beds, termed systemic pressure). Clinically, however, the term *blood pressure* only refers to the pressure within the large arteries.

Arterial pressure can be measured either directly, by the insertion of a needle or catheter directly into the artery in such a way that the needle is pointing "upstream," or against the flow of blood within the vessel, or indirectly, by an instrument called a *sphygmomanometer* (sfig'-mō-ma-NOM-e-ter) (*sphygmo* = pulse). Regardless of which method is utilized, blood pressure is recorded in millimeters of mercury (mm Hg) and is normally taken in the brachial artery.

A sphygmomanometer consists of an inflatable rubber cuff attached by a rubber tube to a compressible hand pump or bulb. Another tube attaches to a cuff and to a mercury column marked off in millimeters or an anaeroid gauge that measures the pressure in millimeters of mercury (mm Hg). The highest pressure in the artery, occurring during ventricular systole, is termed *systolic blood pressure.* The lowest pressure, occurring during ventricular diastole, is termed *diastolic blood pressure.* Blood pressures are usually expressed as a ratio of systolic to diastolic. The average blood pressure of a young adult is about 120 mm Hg systolic and 80 mm Hg diastolic, abbreviated as 120/80. The difference between systolic and diastolic pressure is called *pulse pressure.* Pulse pressure can be used clinically to indicate several physiological and pathological parameters, and usually averages 40 mm Hg in a healthy individual. The ratio of systolic pressure to diastolic pressure to pulse pressure can also be utilized clinically as a diagnostic tool and is usually 3:2:1.

This exercise employs the indirect *ausculatory method,* in which the sounds of blood flow are heard with a stethoscope. Blood flow in an artery is impeded by increasing pressure within a sphygmomanometer. When the cuff of the sphygmomanometer applies sufficient pressure to completely occlude blood flow, no sounds can be heard distal to the cuff because no blood can flow through the artery. When cuff pressure drops below the maximal (systolic) pressure in the artery, blood is heard passing through the vessel. When cuff pressure drops below the lowest (diastolic) pressure in the vessel, the sound becomes muffled and usually disappears. The sounds heard through the stethoscope via this procedure are termed *Korotkoff* (kō-ROT-kof) *sounds.*

Indirect blood pressure can be taken in any artery that can be occluded easily. The brachial artery has the advantage of being at approximately the same level as the heart, so brachial pressure closely reflects aortic pressure.

The procedure for determining blood pressure using the sphygmomanometer is as follows:

PROCEDURE

1. Either you or your laboratory partner should be comfortably seated, at ease, with your arm bared, slightly flexed, abducted, and perfectly relaxed. You may, for convenience, rest your forearm on a table in a supine position.

2. Wrap the deflated cuff of the sphygmomanometer around the arm with the lower edge about 2.54 cm (1 in.) above the antecubital space. Close the valve on the neck of the rubber bulb.

3. Clean the earpieces of the stethoscope with alcohol before using it. Using the diaphragm of the stethoscope, find the pulse in the brachial artery just above the bend of the elbow, on the inner margin of the biceps brachii muscle.

4. Inflate the cuff by squeezing the bulb until the air pressure within it just exceeds 170 mm Hg. At this point, the wall of the brachial artery is compressed tightly, and no blood should be able to flow through.

5. Place the diaphragm of the stethoscope firmly over the brachial artery and while watching the pressure gauge, slowly turn the valve, releasing air from the cuff. Listen carefully for Korotokoff sounds as you watch the pressure fall. The first loud, rapping sound you hear will be the *systolic pressure.*

6. Continue listening as the pressure falls. The pressure recorded on the mercury column when the sounds become faint or disappear is the *diastolic pressure* reading. It measures the force of blood in arteries during ventricular relaxation and specifically reflects the peripheral resistance of the arteries.

7. Repeat this procedure for both readings two or three times, to see if you get consistent results. *Allow a few minutes between readings.* Record all results in Section E of the LABORATORY REPORT RESULTS at the end of the exercise.

8. Have a partner stand and record his/her blood pressure several times for each arm. Record all results in the table provided in Section E of the LABORATORY REPORT RESULTS at the end of the exercise.

9. *Now, assuming that your partner has no known or apparent cardiac or other health problems and is*

capable of such an activity, have him/her do some exercise, such as running in place 40 or 50 steps, and measure the blood pressure again *immediately after the completion of the exercise*. Record the pulse pressure in the table provided in Section E of the LABORATORY REPORT RESULTS at the end of the exercise.

F. CONTROL OF PERIPHERAL BLOOD FLOW

Peripheral blood flow is controlled by a variety of factors, one of which is the *autonomic nervous system (ANS)*. Such control is accomplished by an interaction between the *sympathetic* and *parasympathetic* branches of the ANS. A peripheral vascular bed is composed of *arterioles, metarterioles, capillaries,* and *venules*. Metarterioles contain *precapillary smooth muscle sphincters* that assist in limiting blood flow to capillary beds by controlled contraction and relaxation. The sympathetic nervous system innervates most, if not all, components of a peripheral vascular bed, with the exception of the capillaries. Sympathetic outflow to the peripheral vasculature is controlled by many factors, some of which are arterial blood pressure (as sensed by the baroreceptors) and pCO_2, pO_2, and pH (as sensed by central and peripheral chemoreceptors). With increased sympathetic nervous system output to the vascular smooth muscle of a particular vascular bed, contraction will occur and vessel diameter will decrease, thereby increasing resistance and decreasing peripheral blood flow through that particular capillary bed.

Peripheral blood flow is also controlled by a series of local factors that alter smooth muscle contraction within the vessels. Changes in pH, pCO_2, and pO_2, endothelium-derived relaxing factor (EDRF), endothelium-derived contracting factor (EDCF), histamine, bradykinin, or prostaglandin concentrations will cause either contraction or relaxation of smooth muscle within peripheral vascular beds, thereby altering capillary bed blood flow. Various factors—such as inflammation, sympathetic or parasympathetic nervous system activity, or exercise—will alter the local concentration of one or more of these substances and, therefore, alter smooth muscle contraction.

G. PERIPHERAL BLOOD FLOW IN THE FROG

PROCEDURE

1. Strap a frog* to a frog board in the prone position by wrapping a piece of moist cloth around the body, head, and legs and securing them with several rubber bands. Be sure not to secure the animal too tightly, thereby causing internal injury.
2. Pin the webbing of one foot over the hole in the frog board. Secure the frog board to the stage of the microscope with rubber bands in such a way that the hole in the board is located directly over the hole in the microscope stage.
3. Moisten the restraining cloth and foot webbing with water throughout the experiment.

1. Control Observations

PROCEDURE

1. Using an adequate amount of light, observe the blood flow through the peripheral vasculature of the frog's foot with either the 10× or 20× objective.
2. Identify an arteriole by its characteristic pulsating, rapid flow of blood.
3. A venule can be characterized by a slower, more constant blood flow; a capillary can be recognized by its small diameter and slow blood flow. Red blood cells usually pass through a capillary in single file, thereby aiding in the vessel's identification.
4. Attempt to locate the junction of an arteriole and capillary (metarterioles cannot be distinguished readily in such a preparation) in the vascular bed. Although the precapillary sphincter probably will not be visible, you should be able to observe the irregular blood flow from the arteriole into the capillary due to the activity of the precapillary sphincter.
5. Draw the peripheral vascular bed visible in your microscope and label the various vessels in the space provided in Section G.1 of the LABORATORY REPORT RESULTS at the end of the exercise.

*Your instructor may wish to substitute a goldfish (tail) instead of a frog (foot).

2. Effect of Histamine

PROCEDURE

1. Dry the webbing of the foot by gently blotting it with paper towel.
2. *While observing the peripheral blood flow through the microscope,* use a medicine dropper to add several drops of 1:10,000 histamine solution and note the changes, if any, in peripheral blood flow.
3. Record your observations in Section G.2 of the LABORATORY REPORT RESULTS at the end of the excercise.

3. Effect of Antihistamines

PROCEDURE

1. After rinsing the foot several times with plain water, gently blot it dry. Add several drops of antihistamine solution to the webbing of the foot *while observing the peripheral blood flow through the microscope.*

2. *Without rinsing the foot,* use a medicine dropper and add several drops of 1:10,000 histamine solution. Note the changes, if any, in peripheral blood flow.
3. Record and explain your observations in Section G.3 of the LABORATORY REPORT RESULTS at the end of the exercise.

4. Effect of Epinephrine

PROCEDURE

1. After rinsing the foot several times with plain water, gently blot it dry. Use a medicine dropper to add several drops of 1:1000 epinephrine solution to the webbing of the foot *while observing the peripheral blood flow through the microscope.*
2. Record and explain your observations in Section G.4 of the LABORATORY REPORT RESULTS at the end of the exercise.

ANSWER THE LABORATORY REPORT QUESTIONS AT THE END OF THE EXERCISE.

Cardiovascular Physiology

STUDENT _____   DATE _____

LABORATORY SECTION _____   SCORE/GRADE _____

SECTION A. CARDIAC CONDUCTION SYSTEM AND ELECTROCARDIOGRAM (ECG OR EKG)

Attach examples of the electrocardiogram strips you obtained.

LEAD I

Electrocardiogram Strip

LEAD II

Electrocardiogram Strip

LEAD III

Electrocardiogram Strip

SECTION C. HEART SOUNDS

1. Which heart sound is the loudest? _____

2. Did you hear a third sound? _____

3. Where does the first sound originate? _____

4. Where does the second sound originate? _____

5. After you exercised, how did the heart sounds differ from before? _____

6. Did they differ in rate and intensity? _____

7. Did the first or second increase in loudness? _____

SECTION D. PULSE RATE

1. Radial pulse rate count results: _____ pulses per minute

2. Have all radial pulse rates put on the blackboard, arranging them from the highest to the lowest. The median pulse rate is found exactly halfway down from the top.

What is the median radial pulse rate of the class? _____

What was the highest rate? _____

What was the lowest rate? _____

3. Compare your radial and carotid pulse rates by filling in the following table.

	Radial pulse rate	Carotid pulse rate
Sitting quietly		
Standing quietly		
After walking		
After running in place		

SECTION E. BLOOD PRESSURE (AUSCULTATION METHOD)

Record your systolic and diastolic blood pressures in the following table.

	Systolic pressure		Diastolic pressure		Pulse pressure
	Left arm	Right arm	Left arm	Right arm	
Sitting					
Standing					
After running					

SECTION G. PERIPHERAL BLOOD FLOW IN THE FROG

1. Draw a diagram indicating the vascular bed observed during your control observations.

2. Indicate whether blood flow increased, decreased, or remained unchanged following the addition of the histamine. Explain the physiological mechanism for your observations. _____

_____

3. a. Indicate how blood flow changed, if at all, following the addition of the antihistamine solution to the webbing of the foot. Explain the physiological mechanism for your observation. _____

_____

b. Indicate how blood flow changed following the addition of the histamine solution while the antihistamine solution was still in place and indicate a possible physiological mechanism for antihistamine. _____

_____

4. Indicate how blood flow changed, if at all, following the addition of the epinephrine solution to the webbing of the frog's foot. Provide a physiological mechanism for your observation. _____

_____

Also indicate how the autonomic nervous system could give rise to a physiological condition opposite to that observed in your experiment. _____

_____

Cardiovascular Physiology

STUDENT _____   DATE _____

LABORATORY SECTION _____   SCORE/GRADE _____

PART 1. Multiple Choice

_____ 1. When the semilunar valves are open during a cardiac cycle, which of the following occur?
I—atrioventricular valves are closed
II—ventricles are in systole
III—ventricles are in diastole
IV—blood enters the aorta
V—blood enters the pulmonary trunk
VI—atrial contraction
(a) I, II, IV, and V (b) I, II, and VI (c) II, IV, and V (d) I, III, IV, and VI

_____ 2. Pressure within the ventricles will be the lowest during which phase of the cardiac cycle?
(a) ventricle filling (b) ventricle ejection (c) isovolumetric ventricular contraction
(d) isovolumetric ventricular relaxation

_____ 3. Pressure within the aorta will be the lowest during which phase of the cardiac cycle?
(a) ventricle filling (b) ventricle ejection (c) isovolumetric ventricular contraction
(d) isovolumetric ventricular relaxation

_____ 4. Which of the following statements is *not true?* (a) Pulse pressure is the difference between systolic and diastolic blood pressures. (b) Both systolic and diastolic blood pressures can be obtained via the pulse method. (c) Systolic blood pressure is recorded at the first loud, rapping sound you hear when you are measuring blood pressure via the ausculatory method. (d) Diastolic blood pressure is obtained when the sound heard through the stethoscope when you are measuring blood pressure via the ausculatory method becomes muffled and usually disappears.

_____ 5. The two distinct heart sounds, described phonetically as lubb and dupp, represent (a) contraction of the ventricles and relaxation of the atria (b) contraction of the atria and relaxation of the ventricles (c) closing of the atrioventricular and semilunar valves (d) surging of blood into the pulmonary artery and aorta.

PART 2. Completion

6. Systole and diastole of both atria plus systole and diastole of both ventricles are called

_____.

7. Blood flow through the heart is controlled by the speed of the cardiac cycle, venous return to the

heart, opening and closing of the valves, and _____.

8. Heart sounds provide valuable information about the _____.

9. Abnormal or peculiar heart sounds are called _____.

10. Although the pulse can be detected in most surface arteries, pulse rate is usually determined on the

    _____.

11. The heart has an intrinsic regulating system called the cardiac _____ system.

12. Electrical impulses accompanying the cardiac cycle are recorded by the _____.

13. The typical ECG produces three clearly recognizable waves. The first wave, which indicates

    depolarization of the atria, is called the _____.

14. Various up-and-down impulses produced by an ECG are called _____.

15. The instrument normally used to measure blood pressure is called a(n) _____.

16. The artery that is normally used to evaluate blood pressure is the _____.

17. Rapping or thumping sounds heard clinically when blood pressure is being taken are called

    _____ sounds.

18. The difference between systolic and diastolic pressure is called _____.

19. An average blood pressure value for an adult is _____.

20. An average pulse pressure is _____.

21

Lymphatic System

The *lymphatic* (lim-FAT-ik) *system* is composed of a pale yellow fluid called lymph, vessels that transport lymph called lymphatic vessels (lymphatics), and a number of structures and organs that contain lymphatic (lymphoid) tissue. Lymphatic tissue is a specialized form of reticular connective tissue that contains large numbers of lymphocytes and is found in lymph nodes, tonsils, the thymus gland, spleen, and bone marrow (see Figure 21.3a).

A. LYMPHATIC VESSELS

Lymphatic vessels originate as *lymph capillaries,* microscopic vessels in spaces between cells. They are found in most parts of the body; they are absent in avascular tissue, the CNS, splenic pulp, and bone marrow. They are slightly larger and more permeable in only one direction than blood capillaries. Lymph capillaries also differ from blood capillaries in that they end blindly; blood capillaries have an arterial and a venous end. In addition, lymph capillaries are structurally adapted to ensure the return of proteins to the cardiovascular system when they leak out of blood capillaries.

Just as blood capillaries converge to form venules and veins, lymph capillaries unite to form larger and larger lymph vessels called *lymphatic vessels* (Figure 21.1). Lymphatic vessels resemble veins in structure but have thinner walls and more valves and contain lymph nodes at various intervals along their length (Figure 21.1). Ultimately, lymphatic vessels deliver lymph into two main channels: the thoracic duct

and the right lymphatic duct. These will be described shortly.

Lymphangiography (lim-fan'-jē-OG-ra-fē) is the X-ray examination of lymphatic vessels and lymph organs after they are filled with a radiopaque substance. Such an X ray is called a *lymphangiogram* (lim-FAN-jē-ō-gram). Lymphangiograms are useful in detecting edema and carcinomas and in locating lymph nodes for surgical and radiotherapeutic treatment.

B. LYMPHATIC TISSUE
1. Lymph Nodes

The oval or bean-shaped structures located along the length of lymphatic vessels are called *lymph nodes.* A lymph node contains a slight depression on one side called a *hilus* (HĪ-lus), where blood vessels and efferent lymphatic vessels leave the node. Each node is covered by a *capsule* of dense connective tissue that extends into the node. The capsular extensions are called *trabeculae* (tra-BEK-yoo-lē). Internal to the capsule is a supporting network of reticular fibers and reticular cells (fibroblasts and macrophages). The capsule, trabeculae, and reticular fibers and cells constitute the stroma (framework) of a lymph node. The interior of a lymph node is specialized into two regions: cortex and medulla. The outer *cortex* contains densely packed lymphocytes arranged in masses called *lymphatic nodules.* The nodules often contain lighter-staining central areas, the *germinal centers,* where lymphocytes are produced. The inner region of a lymph node is called

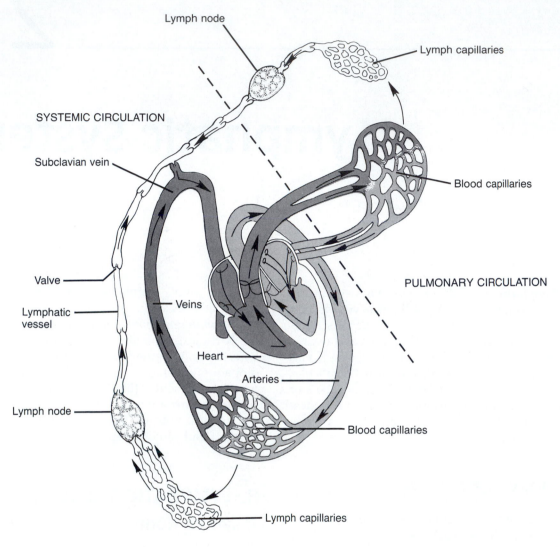

FIGURE 21.1 Relationship of the lymphatic system to the cardiovascular system.

the *medulla.* In the medulla, the lymphocytes are arranged in strands called *medullary cords.* These cords also contain macrophages and plasma cells and constitute the parenchyma (functional portion) of the lymph node.

The circulation of lymph through a node involves afferent (to convey toward a center) lymphatic vessels, sinuses in the node, and efferent (to convey away from a center) lymphatic vessels. *Afferent lymphatic vessels* enter the convex surface of the node at several points. They contain valves that open toward the node so that the lymph is directed *inward.* Once inside the node, the lymph enters the sinuses, which are a series of irregular channels. Lymph from the afferent lymphatic vessels enters the *cortical*

sinuses just inside the capsule. From here it circulates to the *medullary sinuses* between the medullary cords. From these sinuses the lymph usually circulates into one or two *efferent lymphatic vessels,* located at the hilus of the lymph node. Efferent lymphatic vessels are wider than afferent vessels and contain valves that open away from the node to convey lymph *out* of the node.

Lymph passing from tissue spaces through lymphatic vessels on its way back to the cardiovascular system is filtered through lymph nodes. As lymph passes through the nodes it is filtered of foreign substances. These substances are trapped by the reticular fibers within the node. Then, macrophages destroy the foreign sub-

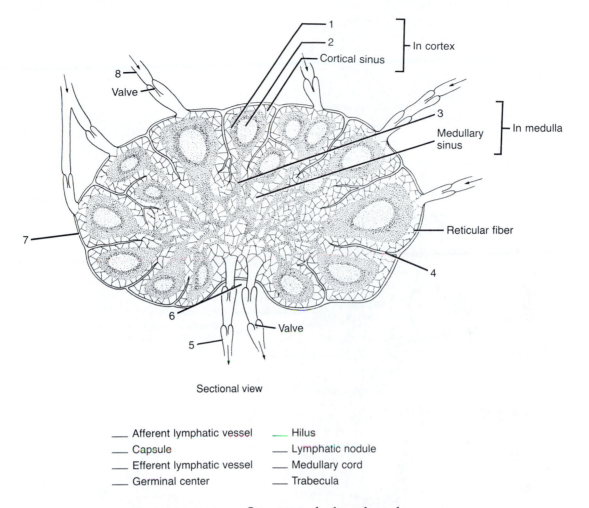

1 _____
2 _____
Cortical sinus

In cortex

8 _____
Valve

3 _____
Medullary
sinus

In medulla

Reticular fiber

7 _____

4 _____

6 _____

Valve

5 _____

Sectional view

___ Afferent lymphatic vessel   ___ Hilus
___ Capsule   ___ Lymphatic nodule
___ Efferent lymphatic vessel   ___ Medullary cord
___ Germinal center   ___ Trabecula

FIGURE 21.2   Structure of a lymph node.

stances by phagocytosis, T cells may destroy them by releasing various products, and/or B cells may develop into plasma cells that produce antibodies that destroy them. Lymph nodes also produce lymphocytes, some of which can circulate to other parts of the body.

Label the parts of a lymph node in Figure 21.2 and the various groups of lymph nodes in Figure 21.3.

2. Tonsils

Tonsils are multiple aggregations of large lymphatic nodules embedded in a mucous membrane. The tonsils are arranged in a ring at the junction of the oral cavity and pharynx. The tonsils are situated strategically to protect against the invasion of foreign substances (Figure 21.3a). The *pharyngeal* (fa-RIN-jē-al) *tonsil,* or *adenoid,* is embedded in the posterior wall of the naso-

pharynx. The paired *palatine* (PAL-a-tīn) *tonsils* are situated in the tonsillar fossae between the pharyngopalatine and palatoglossal arches. They are the ones commonly removed by a tonsillectomy. The *lingual* (LIN-gwal) *tonsil* is located at the base of the tongue and may also have to be removed by a tonsillectomy. Functionally, the tonsils produce lymphocytes and antibodies.

3. Spleen

The oval *spleen* is the largest mass of lymphatic tissue in the body. It is situated in the left hypochondriac region between the fundus of the stomach and diaphragm (Figure 21.3a).

The splenic artery and vein and the efferent lymphatics pass through the hilus. Because the spleen has no afferent lymphatic vessels or lymph sinuses, it does not filter lymph. One key splenic function related to immunity is the pro-

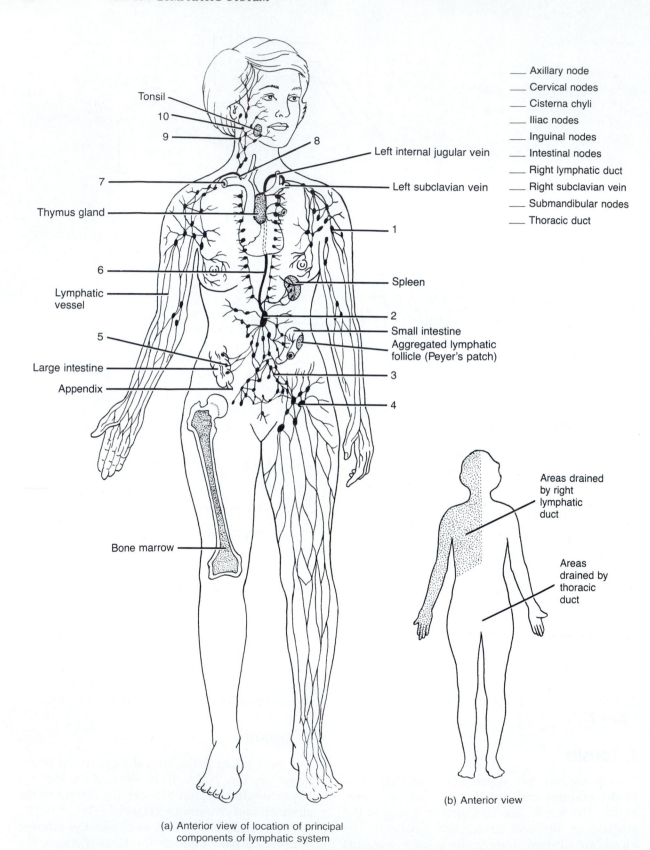

Tonsil

10

9

8

7

Thymus gland

6

Lymphatic vessel

5

Large intestine

Appendix

Bone marrow

Left internal jugular vein

Left subclavian vein

1

Spleen

2

Small intestine
Aggregated lymphatic follicle (Peyer's patch)

3

4

___ Axillary node

___ Cervical nodes

___ Cisterna chyli

___ Iliac nodes

___ Inguinal nodes

___ Intestinal nodes

___ Right lymphatic duct

___ Right subclavian vein

___ Submandibular nodes

___ Thoracic duct

Areas drained by right lymphatic duct

Areas drained by thoracic duct

(b) Anterior view

(a) Anterior view of location of principal components of lymphatic system

FIGURE 21.3 Lymphatic system.

duction of B cells, which develop into antibody-producing plasma cells. The spleen also phagocytizes bacteria and worn-out and damaged red blood cells and platelets. In addition, the spleen stores and releases blood in case of demand, such as during hemorrhage. Sympathetic impulses cause the smooth muscle of the capsule of the spleen to contract. During early fetal development, the spleen participates in blood cell formation.

4. Thymus Gland

Usually, a bilobed lymphatic organ, the *thymus gland* is located in the anterior mediastinum, posterior to the sternum and between the lungs (Figure 21.3a). Its role in immunity is to synthesize hormones that help produce T cells that destroy invading microbes, including the AIDS (acquired immune deficiency syndrome) virus, directly or indirectly, by producing various substances.

C. LYMPH CIRCULATION

When plasma is filtered by blood capillaries, it passes into the interstitial spaces; it is then known as interstitial fluid. When this fluid passes from interstitial spaces into lymph capillaries, it is called *lymph* (*lympha* = clear water). Lymph from lymph capillaries flows into lymphatic vessels that run toward lymph nodes. At the nodes, afferent vessels penetrate the capsules at numerous points, and the lymph passes through the sinuses of the nodes. Efferent vessels from the nodes unite to form *lymph trunks.*

The principal trunks pass their lymph into two main channels, the thoracic duct and the right lymphatic duct. The *thoracic (left lymphatic) duct* begins as a dilation in front of the second lumbar vertebra called the *cisterna chyli* (sis-TER-na KĪ-lē). The thoracic duct is the main collecting duct of the lymphatic system and receives lymph from the left side of the head, neck, and chest, the left upper extremity, and the entire body below the ribs (Figure 21.3b).

The *right lymphatic duct* drains lymph from the upper right side of the body (Figure 21.3b). Ultimately, the thoracic duct empties all of its lymph into the junction of the left internal jugular vein and left subclavian vein, and the right lymphatic duct empties all of its lymph into the junction of the right internal jugular vein and right subclavian vein. Thus, lymph is drained back into the blood and the cycle repeats itself continuously.

Edema, an excessive accumulation of interstitial fluid in tissue spaces, can be caused by an obstruction, such as an infected node or a blockage of vessels, in the pathway between the lymphatic capillaries and the subclavian veins. Another cause is excessive lymph formation and increased permeability of blood capillary walls. A rise in capillary blood pressure, in which interstitial fluid is formed faster than it is passed into lymphatics, also may result in edema.

ANSWER THE LABORATORY REPORT QUESTIONS AT THE END OF THE EXERCISE.

Lymphatic System

STUDENT _____   DATE _____

LABORATORY SECTION _____   SCORE/GRADE _____

PART 1. Completion

1. Small masses of lymphatic tissue located along the length of the lymphatic vessels are called

   _____.

2. Lymphatic vessels have thinner walls than veins but resemble veins in that they also have

   _____.

3. All lymphatic vessels converge, get larger, and eventually merge into two main channels, the

   thoracic duct and the _____.

4. Cells in the lymph nodes that carry on phagocytosis are _____.

5. Lymph is conveyed out of lymph nodes in _____ vessels.

6. B cells in lymph nodes produce certain cells that are responsible for the production of antibodies.

   These cells are called _____.

7. The X-ray examination of lymphatic vessels and lymph organs after they are filled with a radiopaque

   substance is called _____.

8. This X-ray examination is useful in detecting edema and _____.

9. The largest mass of lymphatic tissue is the _____.

10. The main collecting duct of the lymphatic system is the _____ duct.

22

Respiratory System

Cells need a continuous supply of oxygen (O_2) for various metabolic reactions that release energy from nutrient molecules and produce ATP for cellular use. As a result of these reactions, cells also release quantities of carbon dioxide (CO_2). Because an excessive amount of carbon dioxide produces acid conditions that are poisonous to cells, the excess CO_2 must be eliminated quickly and efficiently. The two systems that supply O_2 and eliminate CO_2 are the cardiovascular system and the respiratory system. The *respiratory system* consists of the nose, pharynx, larynx, trachea, bronchi, and lungs (Figure 22.1). The cardiovascular system transports the gases in the blood between the lungs and the cells. The term *upper respiratory system* refers to the nose, pharynx (throat), and associated structures. The *lower respiratory system* refers to the larynx (voice box), trachea (windpipe), bronchi, and lungs.

The overall exchange of gases between the atmosphere, blood, and cells is called *respiration.* Three basic processes are involved. The first process, *pulmonary ventilation,* or breathing, is the inspiration (inflow) and expiration (outflow) of air between the atmosphere and the lungs. The second and third processes involve the exchange of gases within the body. *External respiration* is the exchange of gases between the lungs and blood. *Internal respiration* is the exchange of gases between the blood and the cells.

Using your textbook, charts, or models for reference, label Figure 22.1.

A. ORGANS OF THE RESPIRATORY SYSTEM

1. Nose

The *nose* has an external portion and an internal portion found inside the skull. The inside of both the external and internal portions of the nose is called the *nasal cavity.* It is divided into right and left sides by a vertical partition called the *nasal septum.* The undersurface of the external nose contains two openings called the *external nares* (NA-rēz), or *nostrils.* The anterior portion of the nasal cavity, just inside the nostrils, is called the *vestibule,* which is lined with coarse hairs. The internal nose is formed by the ethmoid, maxillae, inferior conchae, and palatine bones; it communicates with the paranasal sinuses. Sometimes the palatine and maxillary bones fail to fuse completely during embryonic life and the condition called *cleft palate* results. The nasal cavity communicates with the throat through two openings, the *internal nares (choanae).* The nasal cavity is separated from the *oral cavity* below it by the palate, which is composed of the anterior, bony, *hard palate* and the posterior, muscular *soft palate.* Three shelves formed by projections of the *superior, middle,* and *inferior nasal conchae* (KON-kē), or *turbinates,* extend out of the lateral wall of each nasal cavity. The conchae almost reach the nasal septum and subdivide the nasal cavity into a series of groovelike passageways called the *superior, middle,* and *inferior meatuses.*

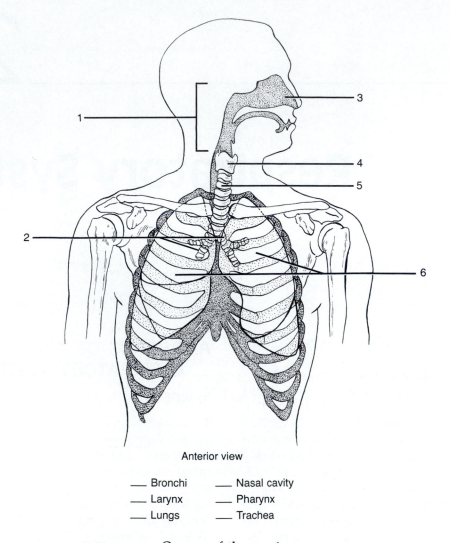

Anterior view

___ Bronchi   ___ Nasal cavity

___ Larynx   ___ Pharynx

___ Lungs   ___ Trachea

FIGURE 22.1   Organs of the respiratory system.

The interior structures of the nose are specialized for three functions: (1) they warm, moisten, and filter incoming air; (2) they receive olfactory stimuli; and (3) large hollow resonating chambers modify speech sounds. In addition, mucous membranes trap dust particles and, with the help of cilia, move unwanted particles to the throat for elimination.

Label the hard palate, inferior meatus, inferior nasal concha, internal naris, middle meatus, middle nasal concha, nasal cavity, oral cavity, soft palate, superior meatus, superior nasal concha, and vestibule in Figure 22.2.

The surface anatomy of the nose is shown in Figure 22.3.

2. Pharynx

The *pharynx* (FAR-inks) (throat) is a somewhat funnel-shaped tube about 13 cm (5 in.) long that starts at the internal nares and extends to the level of the cricoid cartilage. Lying in back of the nasal and oral cavities and just in front of the cervical vertebrae, the pharynx is a passageway for air and food and a resonating chamber for speech sounds.

The pharynx is composed of a superior portion, called the *nasopharynx,* a middle portion, the *oropharynx,* and an inferior portion, the *laryngopharynx* (la-rin'-gō-FAR-inks), or *hypopharynx.* The nasopharynx consists of *pseudo-*

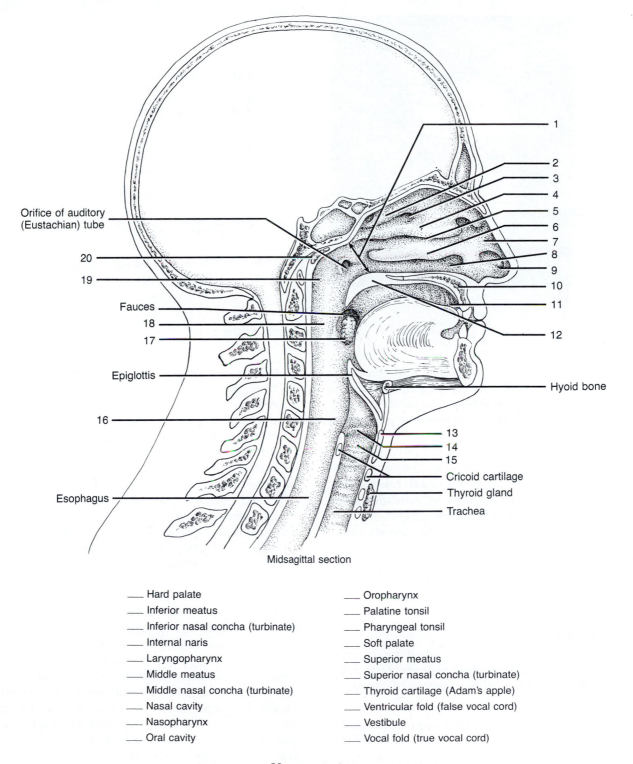

Orifice of auditory
(Eustachian) tube

20

19

Fauces

18

17

Epiglottis

16

Esophagus

1
2
3
4
5
6
7
8
9
10
11
12

Hyoid bone

13
14
15

Cricoid cartilage
Thyroid gland
Trachea

Midsagittal section

____ Hard palate

____ Inferior meatus

____ Inferior nasal concha (turbinate)

____ Internal naris

____ Laryngopharynx

____ Middle meatus

____ Middle nasal concha (turbinate)

____ Nasal cavity

____ Nasopharynx

____ Oral cavity

____ Oropharynx

____ Palatine tonsil

____ Pharyngeal tonsil

____ Soft palate

____ Superior meatus

____ Superior nasal concha (turbinate)

____ Thyroid cartilage (Adam's apple)

____ Ventricular fold (false vocal cord)

____ Vestibule

____ Vocal fold (true vocal cord)

FIGURE 22.2   Upper respiratory system.

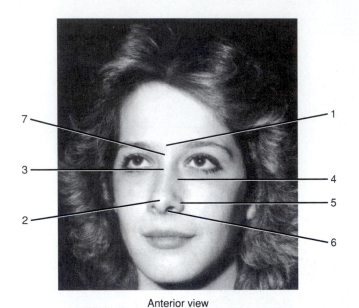

Anterior view

1. **Root.** Superior attachment of nose at forehead located between eyes.
2. **Apex.** Tip of nose.
3. **Dorsum nasi.** Rounded anterior border connecting root and apex; in profile, may be straight, convex, concave, or wavy.
4. **Nasofacial angle.** Point at which side of nose blends with tissues of face.
5. **Ala.** Convex flared portion of inferior lateral surface.
6. **External nares.** External openings into nose (nostrils).
7. **Bridge.** Superior portion of dorsum nasi, superficial to nasal bones.

FIGURE 22.3   Surface anatomy of the nose.

stratified columnar epithelium and has four openings in its wall: two *internal nares* plus two openings into the *auditory (Eustachian) tubes.* The nasopharynx also contains the *pharyngeal tonsil (adenoid).* The oropharynx is lined by *stratified squamous epithelium* and receives one opening: the *fauces* (FAW-sēz). The oropharynx contains the *palatine* and *lingual tonsils.* The laryngopharynx is also lined by *stratified squamous epithelium* and becomes continuous with the esophagus posteriorly and the larynx anteriorly.

Label the laryngopharynx, nasopharynx, oropharynx, palatine tonsil, and pharyngeal tonsil in Figure 22.2.

3. LARYNX

The *larynx* (voice box) is a short passageway connecting the laryngopharynx with the trachea. Its wall is composed of nine pieces of cartilage.

a. *Thyroid cartilage (Adam's apple)*—Large anterior piece that gives the larynx its triangular shape.
b. *Epiglottis* (*epi* = above, *glotta* = tongue) —Leaf-shaped cartilage on top of the larynx that closes off the larynx so that foods and liquids are routed into the esophagus and kept out of the respiratory system.
c. *Cricoid* (KRĪ-koyd) *cartilage*—Ring of cartilage forming the inferior portion of the larynx

that is attached to the first ring of tracheal cartilage.
d. *Arytenoid* (ar-i-TĒ-noyd) *cartilages*—Paired, pyramid-shaped cartilages at the superior border of the cricoid cartilage that attach the vocal folds to the intrinsic pharyngeal muscles.
e. *Corniculate* (kor-NIK-yoo-lāt) *cartilages*—Paired, horn-shaped cartilages at the apex of the arytenoid cartilages.
f. *Cuneiform* (kyoo-NĒ-i-form) *cartilages*—Paired, club-shaped cartilages anterior to the corniculate cartilages.

With the aid of your textbook, label the laryngeal cartilages shown in Figure 22.4. Also label the thyroid cartilage in Figure 22.2.

The mucous membrane of the larynx is arranged into two pairs of folds, an upper pair called the *ventricular folds (false vocal cords)* and a lower pair called the *vocal folds (true vocal cords).* The space between the vocal folds when they are apart is called the *rima glottidis.* Together, the vocal folds and rima glottidis are referred to as the *glottis.* Movement of the vocal folds produces sounds; variations in pitch result from (1) varying degrees of tension and (2) varying lengths in males and females.

With the aid of your textbook, label the ventricular folds, vocal folds, and rima glottidis in Figure 22.5. Also label the ventricular and vocal folds in Figure 22.2.

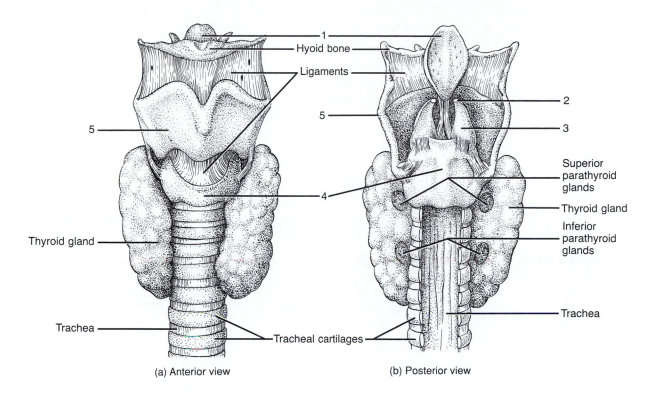

1
Hyoid bone
Ligaments
5
2
3
5
Superior parathyroid glands
4
Thyroid gland
Inferior parathyroid glands
Thyroid gland
Trachea
Trachea
Tracheal cartilages

(a) Anterior view   (b) Posterior view

___ Arytenoid cartilage
___ Corniculate cartilage
___ Cricoid cartilage
___ Epiglottis
___ Thyroid cartilage (Adam's apple)

FIGURE 22.4 Larynx.

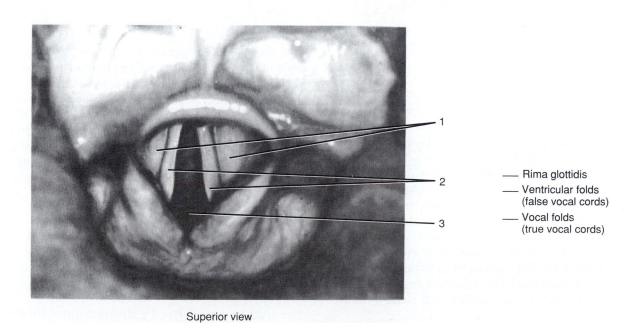

1
2
3

___ Rima glottidis
___ Ventricular folds (false vocal cords)
___ Vocal folds (true vocal cords)

Superior view

FIGURE 22.5 Photograph of the larynx.

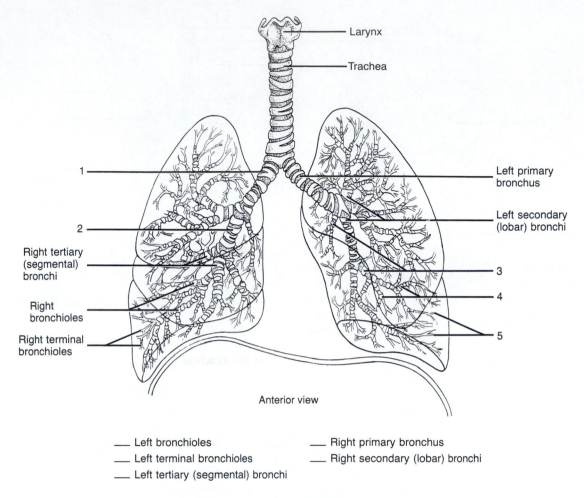

— Larynx

— Trachea

Left primary bronchus

Left secondary (lobar) bronchi

1

2

Right tertiary (segmental) bronchi

Right bronchioles

Right terminal bronchioles

3

4

5

Anterior view

__ Left bronchioles   __ Right primary bronchus
__ Left terminal bronchioles   __ Right secondary (lobar) bronchi
__ Left tertiary (segmental) bronchi

FIGURE 22.6   Air passageways of the lungs. The bronchial tree is shown in relationship to the lungs.

4. Trachea

The *trachea* (TRĀ-kē-a) (windpipe) is a tubular air passageway about 12 cm (4½ in.) in length and 2.5 cm (1 in.) in diameter. It lies in front of the esophagus and, at its inferior end (T5), divides into right and left primary bronchi (Figure 22.6). The epithelium of the trachea consists of *pseudostratified columnar epithelium.* This epithelium consists of ciliated columnar cells, goblet cells, and basal cells. The epithelium offers the same protection against dust as the membrane lining the larynx.

Obtain a prepared slide of pseudostratified columnar epithelium from the trachea and, with the aid of your textbook, label the ciliated columnar cells, cilia, goblet cells, and basal cells in Figure 22.7.

The trachea consists of smooth muscle, elastic connective tissue, and incomplete *rings of carti-*

lage shaped like a series of letter Cs. The open ends of the Cs are held together by the *trachealis muscle.* The cartilage provides a rigid support so that the tracheal wall does not collapse inward and obstruct the air passageway. Because the open parts of the Cs face the esophagus, the latter can expand into the trachea during swallowing. If the trachea should become obstructed, a *tracheostomy* (trā-kē-OS-tō-mē) may be performed. Another method of opening the air passageway is called *intubation,* in which a tube is passed into the mouth and down through the larynx and the trachea.

5. Bronchi

The trachea terminates by dividing into a *right primary bronchus* (BRON-kus), going to the right lung, and a *left primary bronchus,* going to the

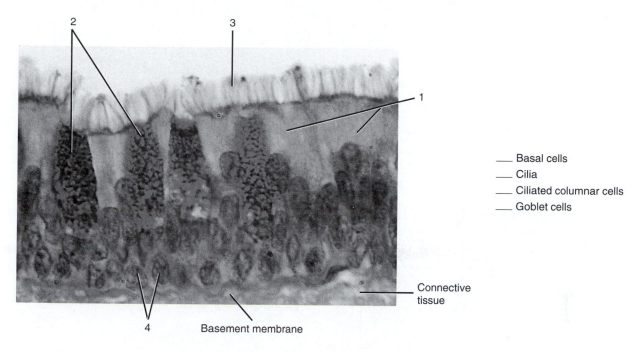

Basal cells
Cilia
Ciliated columnar cells
Goblet cells

Connective tissue

Basement membrane

FIGURE 22.7 Histology of the trachea.

left lung. They continue dividing in the lungs into smaller bronchi, the *secondary (lobar) bronchi* (BRON-kē), one for each lobe of the lung. These bronchi, in turn, continue dividing into still smaller bronchi called *tertiary (segmental) bronchi,* which divide into *bronchioles.* The next division is into even smaller tubes called *terminal bronchioles.* This entire branching structure of the trachea is commonly referred to as the *bronchial tree.*

Label Figure 22.6.

Bronchography (brong-KOG-ra-fē) is a technique for examining the bronchial tree. With this procedure, an intratracheal catheter is passed transorally or transnasally through the rima glottidis into the trachea. Then an opaque iodinated medium is introduced, by means of gravity, into the trachea and distributed through the bronchial branches. X-rays of the chest in various positions are taken and the developed film, called a *bronchogram* (BRONG-kō-gram), provides a picture of the bronchial tree.

6. Lungs

The *lungs* (*lunge* = light, because the lungs float) are paired, cone-shaped organs lying in the thoracic cavity (see Figure 22.1). The *pleural membrane* encloses and protects each lung. Whereas the *parietal pleura* lines the wall of the thoracic cavity, the *visceral pleura* covers the lungs; the potential space between parietal and visceral pleurae, the *pleural cavity,* contains a lubricating fluid to reduce friction as the lungs expand and recoil.

Major surface features of the lungs include

a. *Base*—Broad inferior portion resting on the diaphragm.
b. *Apex*—Narrow superior portion just above the clavicles.
c. *Costal surface*—Surface lying against the ribs.
d. *Mediastinal surface*—Medial surface.
e. *Hilus*—Region in mediastinal surface through which the bronchial tubes, blood vessels, lymphatic vessels, and nerves enter and exit the lung.
f. *Cardiac notch*—Medial concavity in the left lung in which the heart lies.

Each lung is divided into *lobes* by one or more *fissures.* The right lung has three lobes, *superior, middle,* and *inferior;* the left lung has two lobes, *superior* and *inferior.* The *horizontal fissure* separates the superior lobe from the middle lobe in the right lung; an *oblique fissure* separates the middle lobe from the inferior lobe in the right lung and the superior lobe from the inferior lobe in the left lung.

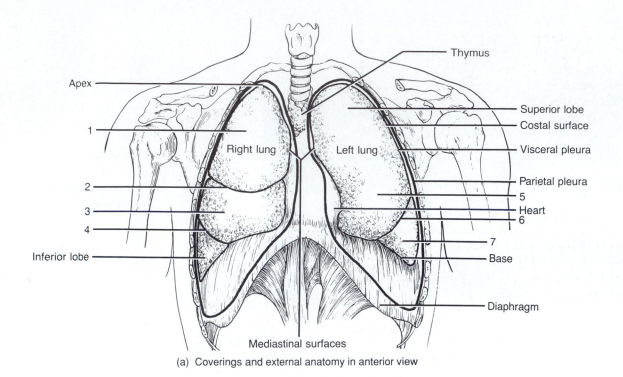

Thymus

Apex

1

Right lung   Left lung

2

3

4

Inferior lobe

Superior lobe

Costal surface

Visceral pleura

Parietal pleura

5

Heart

6

7

Base

Diaphragm

Mediastinal surfaces

(a) Coverings and external anatomy in anterior view

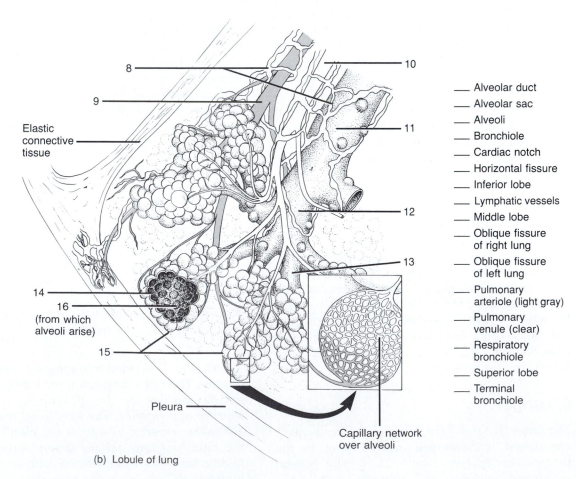

8

9

Elastic
connective
tissue

10

11

12

13

14

16
(from which
alveoli arise)

15

Pleura

Capillary network
over alveoli

(b) Lobule of lung

—— Alveolar duct

—— Alveolar sac

—— Alveoli

—— Bronchiole

—— Cardiac notch

—— Horizontal fissure

—— Inferior lobe

—— Lymphatic vessels

—— Middle lobe

—— Oblique fissure
of right lung

—— Oblique fissure
of left lung

—— Pulmonary
arteriole (light gray)

—— Pulmonary
venule (clear)

—— Respiratory
bronchiole

—— Superior lobe

—— Terminal
bronchiole

FIGURE 22.8   Lungs.

Using your textbook as a reference, label Figure 22.8a.

Each lobe of a lung is divided into many small compartments called *lobules.* Each lobule is wrapped in elastic connective tissue and contains a lymphatic vessel, arteriole, venule, and branch from a terminal bronchiole. Terminal bronchioles divide into *respiratory bronchioles,* which, in turn, divide into several *alveolar* (al-VĒ-ō-lar) *ducts.* Around the circumference of alveolar ducts are numerous alveoli and alveolar sacs. *Alveoli* (al-VĒ-ō-lī) are cup-shaped outpouchings lined by epithelium and supported by a thin elastic membrane. *Alveolar sacs* are two or more alveoli that share a common opening. Over the alveoli, an arteriole and venule disperse into a network of capillaries. Gas is exchanged between the lungs and blood by diffusion across the alveolar and the capillary walls.

Using your textbook as a reference, label Figure 22.8b.

The alveolar wall (Figure 22.9) consists of

a. *Type I alveolar (squamous pulmonary epithelial) cells*—Large cells that form a continuous lining of the alveolar wall, except for occasional type II alveolar (septal) cells.

b. *Type II alveolar (septal) cells*—Cuboidal cells dispersed among type I alveolar cells that secrete a phospholipid substance called *surfactant* (sur-FAK-tant), a surface tension-lowering agent.

c. *Alveolar macrophages (dust cells)*—Phagocytic cells that remove dust particles and other debris from the alveolar spaces.

Obtain a slide of normal lung tissue and examine it under high power. Using your textbook as a reference, see if you can identify a terminal bronchiole, respiratory bronchiole, alveolar duct, alveolar sac, and alveoli.

If available, examine several pathological slides of lung tissue, such as slides that show emphysema and lung cancer. Compare your observations to the normal lung tissue.

The exchange of respiratory gases between the lungs and blood takes place by diffusion across the alveolar and capillary walls. This membrane, through which the respiratory gases move, is collectively known as the *alveolar-capillary (respiratory) membrane* (Figure 22.10). It consists of:

1. A layer of type I alveolar (squamous pulmonary epithelial) cells with type II alveolar

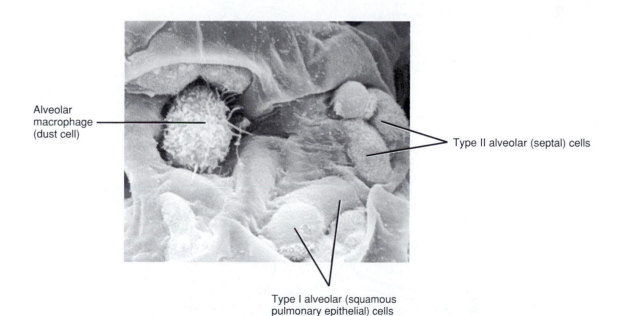

Alveolar macrophage (dust cell)

Type II alveolar (septal) cells

Type I alveolar (squamous pulmonary epithelial) cells

FIGURE 22.9   Scanning electron micrograph of the alevolar wall at a magnification of 3420×. (From *Tissues and Organs: A Text-Atlas of Scanning Electron Microscopy* by Richard G. Kessel and Randy H. Kardon. W. H. Freeman and Company. Copyright © 1979.)

(septal) cells and free alveolar macrophages (dust cells) that constitute the alveolar (epithelial) wall.

2. An epithelial basement membrane underneath the alveolar wall.

3. A capillary basement membrane that is often fused to the epithelial basement membrane.

4. The endothelial cells of the capillary.

Label the components of the alveolar-capillary (respiratory) membrane in Figure 22.10.

B. DISSECTION OF SHEEP PLUCK

CAUTION! *Please reread Section D, "Precautions Related to Dissection," at the beginning of the laboratory manual, on page xv, before you begin your dissection.*

PROCEDURE

1. Preserved sheep pluck may or may not be available for dissection.

2. Pluck consists mainly of a sheep *trachea, bronchi, lungs, heart,* and *great vessels* and a small portion of the *diaphragm.* It is a good demonstration because it is large and shows the close anatomical correlation between these structures and the systems to which they belong, namely, the respiratory and the cardiovascular systems.

3. The heart and its great blood vessels have been described in detail in Exercise 18.

4. Pluck can also be used to examine in great detail the trachea and its relationship to the distribution of the bronchi until they branch into each lung. In addition, this specimen is sufficiently large that the bronchial tree can be exposed by careful dissection.

5. This dissection is done by starting at the primary and secondary bronchi and slowly

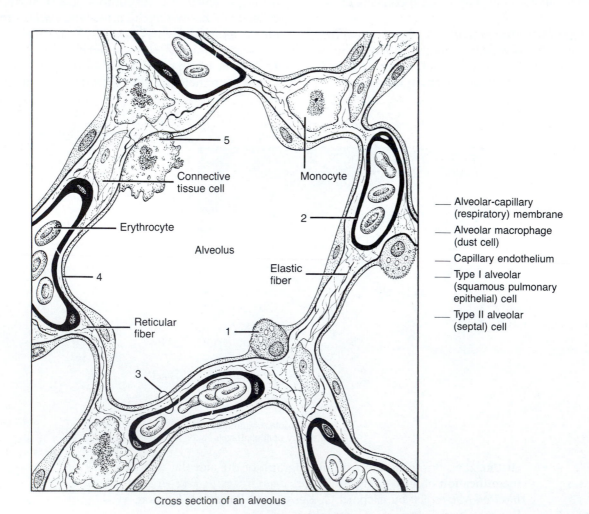

Cross section of an alveolus

Labels in figure: 5, Connective tissue cell, Monocyte, Erythrocyte, Alveolus, 2, Elastic fiber, 4, Reticular fiber, 1, 3

___ Alveolar-capillary (respiratory) membrane
___ Alveolar macrophage (dust cell)
___ Capillary endothelium
___ Type I alveolar (squamous pulmonary epithelial) cell
___ Type II alveolar (septal) cell

FIGURE 22.10 Alveolar-capillary (respiration) membrane.

and carefully removing lung tissue as the trees form even smaller branches into lungs.

6. These specimens should be used primarily by the instructor for demonstration, but you can help expose the bronchial tree.

C. LABORATORY TESTS ON RESPIRATION

1. Observation of Living Lung Tissue

CAUTION! *Please reread Section D, "Precautions Related to Dissection," at the beginning of the laboratory manual, on page xv, before you begin your dissection.*

PROCEDURE

1. *Note: It is strongly suggested that either your instructor or a laboratory assistant perform the procedure for pithing a frog. The procedure requires some practice and may prove difficult for inexperienced individuals.*
2. Pin the freshly pithed frog to the dissecting tray so the ventral surface is upward.
3. As illustrated in Figure 2.5, make a midsagittal cut through the skin from the abdomen to the level of the forelimbs. Make two lateral cuts to create flaps of skin that can be reflected and pinned to the tray.
4. *Taking care not to cut too deeply,* carefully make an incision parallel to the midline over either the right or left lung. Observe the inflated lung.
5. With a probe, gently lift the lung out through the incision and keep the lung moist with Ringer's solution that has been maintained at room temperature.
6. Place the tray under the dissecting microscope and locate the capillary network that surrounds the alveoli. Observe the blood flow through the capillaries of the lungs.

Describe your observation. _____

_____

_____

2. Mechanics of Pulmonary Ventilation (Breathing)

Pulmonary ventilation (breathing) is the process by which gases are exchanged between the atmosphere and the alveoli. Air moves throughout the respiratory system as a result of pressure gradients (differences) between the external environment and the respiratory system. Oxygen and carbon dioxide then move between the respiratory alveoli and the pulmonary capillaries of the cardiovascular system, the peripheral capillaries of the cardiovascular system, and the tissues of the body, as a result of diffusion gradients. We breathe in (inhale) when the pressure inside the thoracic cavity and lungs is less than the air pressure in the atmosphere; similarly, we breathe out (exhale) when the pressure inside the lungs and thoracic cavity is greater than the pressure in the atmosphere.

a. INSPIRATION

Breathing in is called *inspiration (inhalation).* When the thoracic cavity is at rest, and no air movement is occurring, the pressure inside the lungs equals the pressure of the atmosphere, which is approximately 760 mm Hg, or 1 atmosphere (atm), at sea level. For air to flow into the lungs, something must happen to reduce the pressure within the lungs to a value lower than the pressure of the atmosphere. This condition is achieved by increasing the volume of the thoracic cavity.

In order for inspiration to occur, the thoracic cavity must be expanded. This increases lung volume and thus decreases pressure in the lungs. The first step toward increasing lung volume involves contraction of the principal inspiratory muscles: the diaphragm and external intercostals.

Contraction of the diaphragm causes it to flatten. This increases the vertical length of the thoracic cavity and accounts for more than two-thirds of the air that enters the lungs during inspiration. At the same time that the diaphragm contracts, the external intercostals contract. As a result, the ribs are pulled upward and outward, increasing the anterior-posterior diameter of the thoracic cavity. This movement of the ribs by the external intercostals is much like the movement of a bucket handle when a bucket is placed on its side and the handle is moved from a vertical to a more horizontal position.

During normal breathing, the pressure between the two pleural layers, called *pleural (intrathoracic) pressure,* is always subatmospheric. (It may become temporarily positive only during modified respiratory movements such as coughing or straining during defecation.) When the respiratory system is at rest, no air movement in or out of the respiratory tree is occurring, and pleural pressure is approximately 4 mm Hg lower than atmospheric pressure (approximately

756 mm Hg). The overall increase in the size of the thoracic cavity causes pleural pressure to fall to approximately 754 mm Hg. The parietal and visceral pleurae are normally strongly attached to each other due to surface tension created by their moist adjoining surfaces. Therefore, as the walls of the thoracic cavity expand, the parietal pleura lining the cavity is pulled in all directions, and the visceral pleura is pulled along with it. Consequently, the size of the lungs increases, thereby decreasing the pressure within the lungs, called *alveolar (intrapulmonic) pressure,* and causing a pressure gradient between the alveoli of the lungs and e external environment. Because the respiratory ee is open to the external environment via t oral and nasal cavities, air moves down the p   sure gradient from the atmosphere into the pu onary alveoli. Air continues to move into t lungs until alveolar pressure equals atmos eric pressure.

b. EXPIRATIO

Breathing out,   lled *expiration (exhalation),* is also achieved b pressure gradient, but in this case the gradient   reversed, so that the pressure in the lungs is g ter than the pressure of the atmosphere. Norr   l quiet expiration, unlike inspiration, is a pass e process, in that it does not involve the active   traction of muscles. When inspiration is comp ted, the diaphragm and external intercostal   uscles relax. As the diaphragm returns to   domelike position, the vertical length of the horacic cavity decreases, returning to its original dimension. Similarly, when the external intercostals relax, the ribs move posteriorly and inferiorly, decreasing the anterior-posterior diameter of the thoracic cavity. These movements return pleural pressure to its normal resting value of 756 mm Hg, or 4 mm Hg below atmospheric pressure.

As pleural pressure returns to its preinspiration level, the walls of the lungs are no longer pulled outward by the parietal pleura. The elastic recoil of the connective tissue within the lungs and respiratory tree allows the lungs to return to their resting shape and volume, thereby causing alveolar pressure to become slightly greater than atmospheric pressure (or approximately 763 mm Hg). Now, air again moves down its pressure gradient, moving from the alveoli through the respiratory tree and out the oral and nasal openings.

In order to demonstrate pulmonary ventilation, a model lung (a bell-jar demonstrator) will

be used. This apparatus is basically an artificial thoracic cavity that mimics the organs of the respiratory system and allows the pressure/ratios to be manipulated. A diagram of such a device is shown in Figure 22.11. Some of the models will not have a stopcock (rima glottidis); some will not have a tube opening into the pleural space.

PROCEDURE

1. Label the diagram in Figure 22.11 by writing the name of the anatomical structure that corresponds to the following parts of the apparatus: rima glottidis, trachea, primary bronchus, lungs, alveolar pressure, pleural space, and diaphragm.

2. Pull on the rubber membrane (diaphragm) to simulate inspiration. What changes occur in pleural pressure? _____

_____

_____

In alveolar pressure? _____

_____

3. Push the rubber membrane (diaphragm) upward to simulate expiration. What changes occur in pleural pressure? _____

_____

In alveolar pressure? _____

_____

Under what normal processes would such changes in pleural and alveolar pressures be observed? _____

_____

4. Open the clamp on the tube leading to the intrapleural space. This simulates pneumothorax (air in the pleural cavity). Try to cause inspiration and expiration.

Explain your results. _____

_____

5. If present, close the stopcock (rima glottidis) and simulate expiration. This procedure mimics the Valsalva maneuver (forced expiration against a closed rima glottidis as during periods of straining). What changes occur in pleural and alveolar pressure? _____

_____

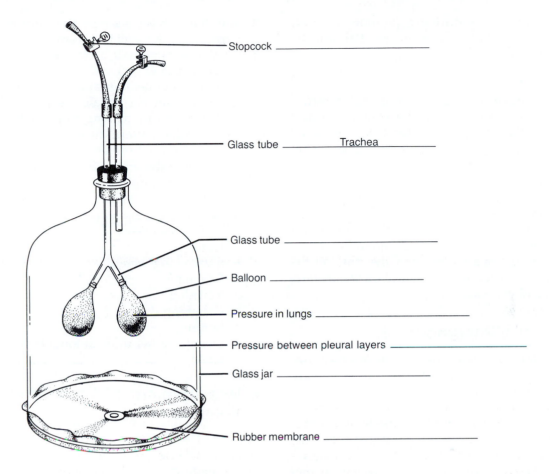

Stopcock _____

Glass tube _____ Trachea _____

Glass tube _____

Balloon _____

Pressure in lungs _____

Pressure between pleural layers _____

Glass jar _____

Rubber membrane _____

FIGURE 22.11   Model of a lung.

3. Measurement of Chest and Abdomen in Respiration

When the diaphragm contracts, the dome shape of this muscle flattens; the flattening pushes against the abdominal viscera and increases thoracic volume, which, in turn, decreases alveolar pressure. The volume changes in the abdomen and chest can easily be measured.

PROCEDURE

1. Place a tape measure at the level of the fifth rib (about armpit level) and determine the size of the chest immediately after a
   a. normal inspiration
   b. normal expiration
   c. maximal inspiration
   d. maximal expiration
2. Repeat Step 1 with the tape measure at the level of the waist to determine abdominal size.

3. Record your results in Section C.1 of the LABORATORY REPORT RESULTS at the end of the exercise.
   Which measurement increased during inspiration? _____

   Which measurement increased during expiration? _____

4. Respiratory Sounds

Air flowing through the respiratory tree creates characteristic sounds that can be detected through the use of a stethoscope. Normal breathing sounds include a soft, breezy sound caused by air filling the lungs during inspiration. As the air exits the lungs during expiration, a short, low-pitched sound may be heard. Perform the following exercises.

PROCEDURE

1. *Clean the earplugs of a stethoscope with alcohol.*
2. Place the diaphragm of the stethoscope just below the larynx and listen for bronchial sounds during both inspiration and expiration.
3. Move the stethoscope slowly downward, toward the bronchial tubes, until the sounds are no longer heard.
4. Place the stethoscope under the scapula, under the clavicle, and over different intercostal spaces (the spaces between the ribs) on the chest and listen for any sound during inspiration and expiration.

5. Use of a Pneumograph

The *pneumograph* is an instrument that measures variations in breathing patterns caused by various physical or chemical factors. The chest pneumograph, which is attached to a polygraph recorder via electrical leads, consists of a rubber bellows that fits around the chest just below the rib cage. As the subject breathes, chest movements cause changes in the air pressure within the pneumograph that are transmitted to the recorder. Normal inspiration and expiration can thus be recorded, and the effects of a wide range of physical and chemical factors on these movements can be studied.

PROCEDURE

Perform the following exercises and record your values in Section C.2 of the LABORATORY REPORT RESULTS at the end of the exercise. Label and save all recordings and attach them to Section C.2 of the LABORATORY REPORT RESULTS. In addition, label the inspiratory and expiratory phases of each recording, determine their duration, and then calculate the respiratory rate.

1. Place a respiratory pneumograph around the chest at the level of the sixth rib and attach it at the back. Connect the electrical lead to the recorder and adjust the instrument's centering and sensitivity so that the needle deflects as the subject breathes. If it does not, adjust the pneumograph bellows up or down on the chest or loosen the degree of the tightness around the subject's chest.

2. Seat the subject so that the pneumograph recording is not visible to him/her.
3. Set the polygraph at a slow speed (approximately 1 cm/per sec) and record normal, quiet breathing (eupnea) for approximately 30 sec.
4. Have the subject inhale deeply and hold his/her breath for as long as possible. Record the breathing pattern during the breath holding and 30 to 60 sec after the resumption of breathing at the end of breath holding.
5. Record the respiratory movements of a subject in Section C.2 of the LABORATORY REPORT RESULTS at the end of the exercise during the following activities:
   a. Reading
   b. Swallowing water
   c. Laughing
   d. Yawning
   e. Coughing
   f. Sniffing
   g. Doing a rather difficult long-division calculation

6. Measurement of Respiratory Volumes

The word *respiration* refers to one complete respiratory cycle, including one inspiration and one expiration. A normal adult has 14 to 18 respirations in a minute, during which the lungs exchange specific volumes of air with the atmosphere.

As the following respiratory volumes and capacities are discussed, keep in mind that the values given vary with age, height, sex, and physiological state. The volume of air expired under normal, quiet inspiration is approximately 500 ml. The volume of air expired under normal, quiet breathing conditions is equal to that of inspiration, and this volume of air is called *tidal volume* (Figure 22.12). Only approximately 350 ml of this tidal volume reaches the alveoli. The other 150 ml of air is called *dead air volume* because it remains in the spaces not designed for air exchange *(anatomical dead space)* (nose, pharynx, larynx, trachea, and bronchi), as well as those areas designed for air exchange but not currently being utilized by the lungs *(physiological dead space)*.

If we take a very deep breath, we can inspire much more than the 500 ml tidal volume taken in during normal, quiet respiration. The additional inhaled air, called the *inspiratory reserve volume,* averages 3100 ml above the tidal volume of quiet respiration. Thus, the *inspiratory capacity* of our

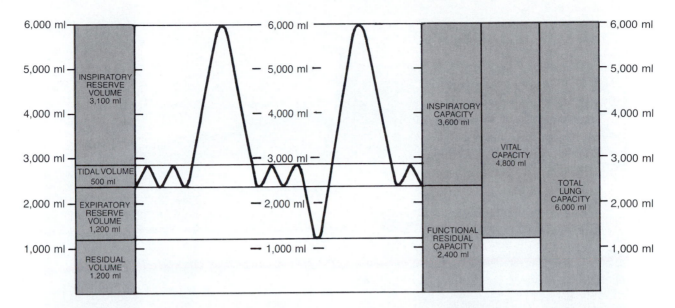

FIGURE 22.12   Spirogram of lung volumes and capacities.

lungs, or the volume of air that can be inhaled from a normal, resting chest position is approximately 3600 ml (500 ml of tidal volume + 3100 ml of inspiratory reserve volume). If we inspire *normally* and then expire *as forcibly as possible,* a normal individual can exhale the 500 ml of air taken in during a normal, quiet tidal volume, as well as an additional 1200 ml of air, which is termed ***expiratory reserve volume.*** This total amount of air (tidal volume + expiratory reserve volume) is called ***expiratory reserve capacity.*** Even after a forceful expiration some air remains in the lungs, which is termed ***residual volume.*** This volume of air ensures gas exchange between the lungs and the cardiovascular system during brief time periods between respiratory cycles, or during extended periods of no respiratory activity (apnea).

   Vital capacity of the lungs is equal to the volume of air that can be forcibly exhaled following a maximal inspiratory effort (inspiratory reserve volume + tidal volume + expiratory reserve capacity). ***Functional residual capacity*** is the sum of residual volume plus expiratory reserve volume. ***Total lung capacity*** is the sum of vital capacity plus residual volume.

a. USE OF VITAL CAPACITY APPARATUS

Using the vital capacity apparatus (Figure 22.13), determine your own particular vital capacity value.

   The procedure for determining forced vital capacity follows:

PROCEDURE

1. Place a *sterile* air shield over the mouthpiece.
2. Push the black pointer to the red pointer at zero.
3. Stand, holding the apparatus horizontally at eye level, and completely fill your lungs with air.
4. Apply your mouth to the air shield and blow air into the apparatus as forcefully as possible, completely emptying your lungs.
5. Read the vital capacity, in liters, at the upper edge of the black pointer.
6. Compare the value with the printed tables on the apparatus, using your height in centimeters as a guide.
7. Discard the air shield.

8. Record your vital capacity: _____

b. USE OF HANDHELD RESPIROMETER

A ***respirometer (spirometer)*** is an instrument used to measure volumes of air exchanged in breathing. Of the several different respirometers available, we will make use of two: a hand-held respirometer and the Collins respirometer. One type of hand-held respirometer is the Pulmometer (Figure 22.14). It measures and provides a direct digital display of certain respiratory volumes and capacities.

PROCEDURE

1. Set the needle or the digital indicator to zero.
2. Place a *clean* mouthpiece in the spirometer tube and hold the tube in your hand while

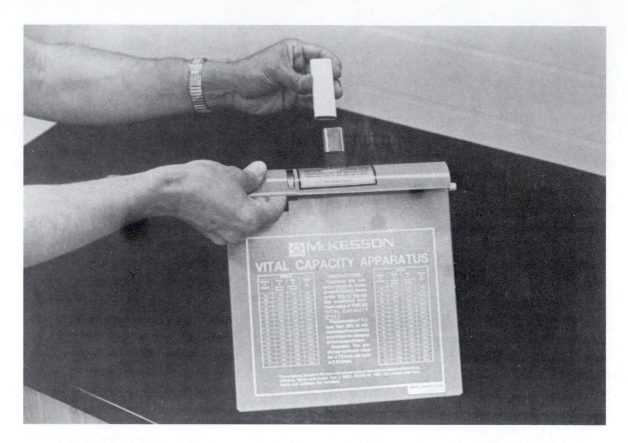

FIGURE 22.13   Placing a disposable air shield on a vital capacity apparatus.

you breathe normally for a few respirations. (*NOTE:* Inhale through the nose and exhale through the mouth.)

3. Place the mouthpiece in your mouth and inhale; exhale three normal breaths through the mouthpiece.
4. Divide the total volume of air expired by 3 to determine your average tidal volume. Record this value in Section C.3 of the LABORATORY REPORT RESULTS on page 447.
5. Return the scale indicator to zero.
6. Breathe normally for a few respirations. Following a normal expiration of tidal volume, forcibly exhale as much air as possible into the mouthpiece.
7. Repeat Step 6 twice. Divide the total amount of air expired by 3 to determine your average expiratory reserve volume. Record this value in Section C.3 of the LABORATORY REPORT RESULTS at the end of the exercise.
8. To determine vital capacity, breathe deeply a few times and inhale as much air as possible. Exhale as fully and as steadily as possible into the spirometer tube.
9. Repeat Step 8 twice. Divide the total volume

of expired air by 3 to determine your average vital capacity. Record this value in Section C.3 of the LABORATORY REPORT RESULTS at the end of the exercise.

10. Calculate your inspiratory reserve volume by substituting the known volumes in this equation: vital capacity = inspiratory reserve volume + tidal volume + expiratory reserve volume. Record this volume in Section C.3 of the LABORATORY REPORT RESULTS at the end of the exercise.
11. Compare your vital capacity with the normal values shown in Tables 22.1 and 22.2 on pages 440 and 441.
12. Sit quietly and count your respirations per minute. Calculate your minute volume of respirations (MVR) by multiplying the breaths per minute by your tidal volume.

$$\frac{\# \text{ breaths}}{\text{minute}} \times \frac{\# \text{ ml}}{\text{breath}} = \frac{\# \text{ ml}}{\text{minute}}$$

Record this value in Section C.3 of the LABORATORY REPORT RESULTS at the end of the exercise.

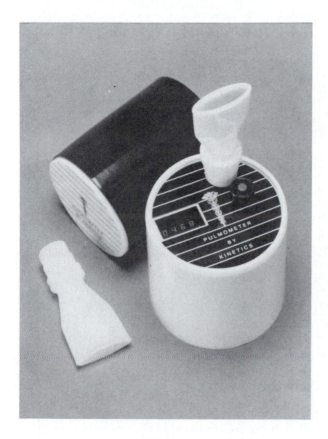

FIGURE 22.14   Hand-held respirometer.

The air in the body is at a different temperature than the air contained in the respirometer; body air is also saturated with water vapor. To make the correction for these differences, the temperature of the respirometer measured. Use Table 22.3 on page 442 to obtain a conversion factor [*BTPS* factor: *body temperature, pressure (atmospheric), saturated* (with water vapor)]. All measured volumes are then multiplied by this factor.

PROCEDURE

1. *When using the Collins respirometer, discard the disposable mouthpiece after use by each subject. Then detach the hose and rinse it with 70% alcohol.*

CAUTION! *This procedure must be repeated with every student using the equipment.*

2. Students should work in pairs, with one student operating the respirometer while the other is tested.
3. Before starting the recording, a little practice may be necessary to learn to inhale and exhale only through your mouth and into the hose. A nose clip can be used to prevent leakage from the nose.
4. Make sure that the respirometer is functioning. Your instructor may have already prepared the instrument for use.
   a. The leveling screws must be raised or lowered so that the bell does not rub on the metal body.
   b. Paper must be fed from a roll onto the kymograph *or* taped onto the kymograph in single sheets.
   c. The soda-lime that absorbs carbon dioxide must be pink (fresh) and not purple (saturated with CO_2).
   d. Pen reservoirs must be full and primed.
5. Raise and lower the drum several times to flush the respirometer of stale air. Position the ventilometer pen (the upper pen, usually with black ink) so that it will begin writing in the center of the spirogram paper.
6. Set the free-breathing valve so that the opening can be seen through the side of the valve.
7. Place a *sterile* or *disposable mouthpiece in your mouth* and close your nose with a clamp or your fingers.
8. Breathe several times to become accustomed to using this instrument.
9. Have your lab partner close the free-breathing valve (turn it completely in the opposite

C. USE OF COLLINS RESPIROMETER

The Collins respirometer, shown in Figure 22.15 on page 442 is a closed system that allows measurement of the volumes of air inhaled and exhaled. The instrument consists of a weighted drum, containing air, inverted over a chamber of water. The air-filled chamber is connected to the subject's mouth by a tube. When the subject inspires, air is removed from the chamber, causing the drum to sink and producing an upward deflection. This deflection is recorded by the stylus on the graph paper on the kymograph (rotating drum). When the subject expires, air is added, causing the drum to rise and producing a downward deflection. These deflections are recorded as a *spirogram* (see Figure 22.12). The horizontal *(x)* axis of a spirogram is graduated in millimeters and records elapsed time; the vertical *(y)* axis is graduated in milliliters (ml) and records air volumes. Spirometric studies measure lung capacities and rates and depths of ventilation for diagnostic purposes. Spirometry is indicated for individuals with labored breathing and is used to diagnose respiratory disorders such as bronchial asthma and emphysema.

TABLE 22.1
Predicted Vital Capacities for Females

Height in Centimeters and Inches

Age	cm 152 / in. 59.8	154 / 60.6	156 / 61.4	158 / 62.2	160 / 63.0	162 / 63.7	164 / 64.6	166 / 65.4	168 / 66.1	170 / 66.9	172 / 67.7	174 / 68.5	176 / 69.3	178 / 70.1	180 / 70.9	182 / 71.7	184 / 72.4	186 / 73.2	188 / 74.0
16	3070	3110	3150	3190	3230	3270	3310	3350	3390	3430	3470	3510	3550	3590	3630	3670	3715	3755	3800
17	3055	3095	3135	3175	3215	3255	3295	3335	3375	3415	3455	3495	3535	3575	3615	3655	3695	3740	3780
18	3040	3080	3120	3160	3200	3240	3280	3320	3360	3400	3440	3480	3520	3560	3600	3640	3680	3720	3760
20	3010	3050	3090	3130	3170	3210	3250	3290	3330	3370	3410	3450	3490	3525	3565	3605	3645	3695	3720
22	2980	3020	3060	3095	3135	3175	3215	3255	3290	3330	3370	3410	3450	3490	3530	3570	3610	3650	3685
24	2950	2985	3025	3065	3100	3140	3180	3200	3260	3300	3335	3375	3415	3455	3490	3530	3570	3610	3650
26	2920	2960	3000	3035	3070	3110	3150	3190	3230	3265	3300	3340	3380	3420	3455	3495	3530	3570	3610
28	2890	2930	2965	3000	3040	3070	3115	3155	3190	3230	3270	3305	3345	3380	3420	3460	3495	3535	3570
30	2860	2895	2935	2970	3010	3045	3085	3120	3160	3195	3235	3270	3310	3345	3385	3420	3460	3495	3535
32	2825	2865	2900	2940	2975	3015	3050	3090	3125	3160	3200	3235	3275	3310	3350	3385	3425	3460	3495
34	2795	2835	2870	2910	2945	2980	3020	3055	3090	3130	3165	3200	3240	3275	3310	3350	3385	3425	3460
36	2765	2805	2840	2875	2910	2950	2985	3020	3060	3095	3130	3165	3205	3240	3275	3310	3350	3385	3420
38	2735	2770	2810	2845	2880	2915	2950	2990	3025	3060	3095	3130	3170	3205	3240	3275	3310	3350	3385
40	2705	2740	2775	2810	2850	2885	2920	2955	2990	3025	3060	3095	3135	3170	3205	3240	3275	3310	3345
42	2675	2710	2745	2780	2815	2850	2885	2920	2955	2990	3025	3060	3100	3135	3170	3205	3240	3275	3310
44	2645	2680	2715	2750	2785	2820	2855	2890	2925	2960	2995	3030	3060	3095	3130	3165	3200	3235	3270
46	2615	2650	2685	2715	2750	2785	2820	2855	2890	2925	2960	2995	3030	3060	3095	3130	3165	3200	3235
48	2585	2620	2650	2685	2715	2750	2785	2820	2855	2890	2925	2960	2995	3030	3060	3095	3130	3160	3195
50	2555	2590	2625	2655	2690	2720	2755	2785	2820	2855	2890	2925	2955	2990	3025	3060	3090	3125	3155
52	2525	2555	2590	2625	2655	2690	2720	2755	2790	2820	2855	2890	2925	2955	2990	3020	3055	3090	3125
54	2495	2530	2560	2590	2625	2655	2690	2720	2755	2790	2820	2855	2885	2920	2950	2985	3020	3050	3085
56	2460	2495	2525	2560	2590	2625	2655	2690	2720	2755	2790	2820	2855	2885	2920	2950	2980	3015	3045
58	2430	2460	2495	2525	2560	2590	2625	2655	2690	2720	2750	2785	2815	2850	2880	2920	2945	2975	3010
60	2400	2430	2460	2495	2525	2560	2590	2625	2655	2685	2720	2750	2780	2810	2845	2875	2915	2940	2970
62	2370	2405	2435	2465	2495	2525	2560	2590	2620	2655	2685	2715	2745	2775	2810	2840	2870	2900	2935
64	2340	2370	2400	2430	2465	2495	2525	2555	2585	2620	2650	2680	2710	2740	2770	2805	2835	2865	2895
66	2310	2340	2370	2400	2430	2460	2495	2525	2555	2585	2615	2645	2675	2705	2735	2765	2800	2825	2860
68	2280	2310	2340	2370	2400	2430	2460	2490	2520	2550	2580	2610	2640	2670	2700	2730	2760	2795	2820
70	2250	2280	2310	2340	2370	2400	2425	2455	2485	2515	2545	2575	2605	2635	2665	2695	2725	2755	2780
72	2220	2250	2280	2310	2335	2365	2395	2425	2455	2480	2510	2540	2570	2600	2630	2660	2685	2715	2745
74	2190	2220	2245	2275	2305	2335	2360	2390	2420	2450	2475	2505	2535	2565	2590	2620	2650	2680	2710

Source: E. A. Gaensler and G. W. Wright, *Archives of Environmental Health* **12**:146–189 (February 1966).

TABLE 22.2
Predicted Vital Capacities for Males

Height in Centimeters and Inches

Age (cm / in.)	152 / 59.8	154 / 60.6	156 / 61.4	158 / 62.2	160 / 63.0	162 / 63.7	164 / 64.6	166 / 65.4	168 / 66.1	170 / 66.9	172 / 67.7	174 / 68.5	176 / 69.3	178 / 70.1	180 / 70.9	182 / 71.7	184 / 72.4	186 / 73.2	188 / 74.0
16	3920	3975	4025	4075	4130	4180	4230	4285	4335	4385	4440	4490	4540	4590	4645	4695	4745	4800	4850
18	3890	3940	3995	4045	4095	4145	4200	4250	4300	4350	4405	4455	4505	4555	4610	4660	4710	4760	4815
20	3860	3910	3960	4015	4065	4115	4165	4215	4265	4320	4370	4420	4470	4520	4570	4625	4675	4725	4775
22	3830	3880	3930	3980	4030	4080	4135	4185	4235	4285	4335	4385	4435	4485	4535	4585	4635	4685	4735
24	3785	3835	3885	3935	3985	4035	4085	4135	4185	4235	4285	4330	4380	4430	4480	4530	4580	4630	4680
26	3755	3805	3855	3905	3955	4000	4050	4100	4150	4200	4250	4300	4350	4395	4445	4495	4545	4595	4645
28	3725	3775	3820	3870	3920	3970	4020	4070	4115	4165	4215	4265	4310	4360	4410	4460	4510	4555	4605
30	3695	3740	3790	3840	3890	3935	3985	4035	4080	4130	4180	4230	4275	4325	4375	4425	4470	4520	4570
32	3665	3710	3760	3810	3855	3905	3950	4000	4050	4095	4145	4195	4240	4290	4340	4385	4435	4485	4530
34	3620	3665	3715	3760	3810	3855	3905	3950	4000	4045	4095	4140	4190	4225	4285	4330	4380	4425	4475
36	3585	3635	3680	3730	3775	3825	3870	3920	3965	4010	4060	4105	4155	4200	4250	4295	4340	4390	4435
38	3555	3605	3650	3695	3745	3790	3840	3885	3930	3980	4025	4070	4120	4165	4210	4260	4305	4350	4400
40	3525	3575	3620	3665	3710	3760	3805	3850	3900	3945	3990	4035	4085	4130	4175	4220	4270	4315	4360
42	3495	3450	3590	3635	3680	3725	3770	3820	3865	3910	3955	4000	4050	4095	4140	4185	4230	4280	4325
44	3450	3495	3540	3585	3630	3675	3725	3770	3815	3860	3905	3950	3995	4040	4085	4130	4175	4220	4270
46	3420	3465	3510	3555	3600	3645	3690	3735	3780	3825	3870	3915	3960	4005	4050	4095	4140	4185	4230
48	3390	3435	3480	3525	3570	3615	3655	3700	3745	3790	3835	3880	3925	3970	4015	4060	4105	4150	4190
50	3345	3390	3430	3475	3520	3565	3610	3650	3695	3740	3785	3830	3870	3915	3960	4005	4050	4090	4135
52	3315	3353	3400	3445	3490	3530	3575	3620	2660	3705	3750	3795	3835	3880	3925	3970	4010	4055	4100
54	3285	3325	3370	3415	3455	3500	3540	3585	3630	3670	3715	3760	3800	3845	3890	3930	3975	4020	4060
56	3255	3295	3340	3380	3425	3465	3510	3550	3595	3640	3680	3725	3765	3810	3850	3895	3940	3980	4025
58	3210	3250	3290	3335	3375	3420	3460	3500	3545	3585	3630	3670	3715	3755	3800	3840	3880	3925	3965
60	3175	3220	3260	3300	3345	3385	3430	3470	3500	3555	3595	3635	3680	3720	3760	3805	3845	3885	3930
62	3150	3190	3230	3270	3310	3350	3390	3440	3480	3520	3560	3600	3640	3680	3730	3770	3810	3850	3890
64	3120	3160	3200	3240	3280	3320	3360	3400	3440	3490	3530	3570	3610	3650	3690	3730	3770	3810	3850
66	3070	3110	3150	3190	3230	3270	3310	3350	3390	3430	3470	3510	3550	3600	3640	3680	3720	3760	3800
68	3040	3080	3120	3160	3200	3240	3280	3320	3360	3400	3440	3480	3520	3560	3600	3640	3680	3720	3760
70	3010	3050	3090	3130	3170	3210	3250	3290	3330	3370	3410	3450	3480	3520	3560	3600	3640	3680	3720
72	2980	3020	3060	3100	3140	3180	3210	3250	3290	3330	3370	3410	3450	3490	3530	3570	3610	3650	3680
74	2930	2970	3010	3050	3090	3130	3170	3200	3240	3280	3320	3360	3400	3440	3470	3510	3550	3590	3630

Source: E. A. Gaensler and G. W. Wright, *Archives of Environmental Health* **12:**146–189 (February 1966).

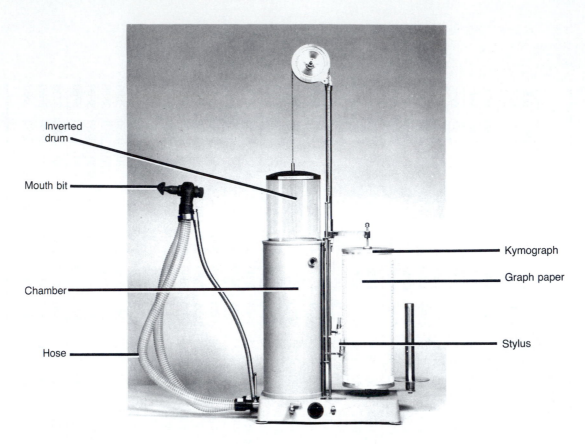

FIGURE 22.15 Collins respirometer. This type is commonly used in college biology laboratories.

Table 22.3
Temperature Variation Conversion Factors

Temperature °C (°F)	Conversion factor
20 (68.0)	1.102
21 (69.8)	1.096
22 (71.6)	1.091
23 (73.4)	1.085
24 (75.2)	1.080
25 (77.0)	1.075
26 (78.8)	1.068
27 (80.6)	1.063
28 (82.4)	1.057
29 (84.2)	1.051
30 (86.0)	1.045
31 (87.8)	1.039
32 (89.6)	1.032
33 (91.4)	1.026
34 (93.2)	1.020
35 (95.0)	1.014
36 (96.8)	1.007
37 (98.6)	1.000

direction so that no opening is seen) and turn on the respirometer to the slow speed. At this setting, the kymograph moves 32 mm (the distance between two vertical lines) each minute.

10. Perform the following exercises and record your values in Section C.4 of the LABORATORY REPORT RESULTS at the end of the exercise. In each case, * indicates the placement of the mouth on the mouthpiece.

11. Inspire normally, then * exhale normally three times. This volume is your tidal volume. Repeat twice more and average and record the values.

12. Expire normally, then * exhale as much air as possible three times, recording this volume. This value is your expiratory reserve volume. Repeat twice more and record the values.

13. After taking a deep breath, * exhale as much air as possible three times. This volume is your vital capacity. Repeat twice more and record the values. Compare your vital capac-

ity to the normal value shown in Tables 22.1 and 22.2 on pages 440 and 441.

14. Because your vital capacity consists of tidal volume, inspiratory reserve volume, and expiratory reserve volume, and because you have already measured tidal volume in Step 11 and expiratory reserve volume in Step 12, you can calculate your inspiratory reserve volume by subtracting tidal volume and expiratory reserve volume from vital capacity. Inspiratory reserve volume = (vital capacity) − (tidal volume + expiratory reserve volume). Record this value.

In some cases, an individual with a pulmonary disorder has a nearly normal vital capacity. If the rate of expiration is timed, however, the extent of the pulmonary disorder becomes apparent. In order to do this, an individual expels air into a Collins respirometer as fast as possible and the expired volume is measured per unit of time. Such a test is called *forced expiratory volume (FEV_T)*. The T indicates that the volume of air is timed. FEV_1 is the volume of air forcefully expired in 1 sec; FEV_2 is the volume expired in 2 sec, and so on. A normal individual should be able to expel 83% the total capacity during the first second, 94% in 2 sec, and 97% in 3 sec. For individuals and disorders such as emphysema and asthma, the percentage can be considerably lower, depending on the extent of the problem.

FEV_1 is determined according to the following procedure.

PROCEDURE

1. Apply a noseclip to prevent leakage of air through the nose.
2. Turn on the kymograph.
3. Before placing the mouthpiece in your mouth, inhale as deeply as possible.
4. Expel all the air you can into the mouthpiece.
5. Turn off the kymograph.
6. Draw a vertical line on the spirogram at the starting point of exhalation. Mark this "A."
7. Using the Collins VC timed interval ruler, draw a vertical line to the left of A and label it line "B." The time between lines A and B is 1 sec.
8. The FEV_1 is the point where the spirogram tracing crosses line B.
9. Record your value here: _____
10. Now read your vital capacity from the spirogram and record the value here: _____

11. In order to adjust for differences in temperature in the respirometer, use Table 22.3 as a guide. Determine the temperature in the respirometer, find the appropriate conversion factor, and multiply the conversion factor by your vital capacity.
12. If, for example, the temperature of the respirometer is 75.2°F (24°C), the conversion factor is 1.080. And, if your vital capacity is 5600 ml, then

$$1.080 \times 5600 \text{ ml} = 6048 \text{ ml.}$$

13. Use the same conversion factor and multiply it by your FEV_1. If your FEV_1 is 4000 ml, then

$$1.080 \times 4000 \text{ ml} = 4320 \text{ ml.}$$

14. To calculate FEV_1, divide 4320 by 6048:

$$FEV_1 = \frac{4320}{6048} = 71\%.$$

15. Repeat the procedure three times and record your FEV_1 in Section D.4 of the LABORATORY REPORT RESULTS at the end of the exercise.

Compare your results using the Collins respirometer with those using the hand-held respirometer.

7. Chemical Regulation of Respiration

The size of the thorax is affected by the action of the respiratory muscles. These muscles contract and relax as a result of nerve impulses transmitted to them from centers in the brain. Although the exact mechanisms involved in the rhythmic inspiratory and expiratory activity of the respiratory system are unknown, current theories discuss three respiratory centers found within and around the reticular formation of the brain stem. These respiratory centers are composed of widely dispersed groups of neurons that are functionally divided into three areas: (1) the medullary rhythmicity area, contained within the medulla, that appears to control the basic rhythm of respiration; (2) the pneumotaxic (noo-mō-TAK-sik) area in the pons that facilitates expiration; and (3) the apneustic (ap-NOO-stik) center in the pons that inhibits expiration.

Although the basic rhythm of respiration is set and coordinated by the respiratory center, the rhythm can be modified in response to the demands of the body by neural input to the center. Among the factors that can alter the rate of respiration are the pH of the blood and the levels of CO_2 and O_2 within the blood.

Under normal circumstances, the *partial pressure of carbon dioxide within the arteries* (arterial pCO_2) is 40 mm Hg. If there is even a slight increase in arterial pCO_2—a condition known as **hypercapnia**—the chemosensitive area in the medulla and peripheral chemoreceptors in the carotid and aortic bodies are stimulated. Stimulation of the chemosensitive area by peripheral chemoreceptors causes the inspiratory area to become highly active, and the rate of respiration increases. This increased rate allows the body to expel more CO_2 until the arterial pCO_2 returns to 40 mm Hg. If arterial pCO_2 is lower than 40 mm Hg **(hypocapnia)**, the chemosensitive area and peripheral chemoreceptors are stimulated to a lesser extent, and fewer stimulatory impulses are sent to the inspiratory center. Consequently, respiratory rate decreases, and pCO_2 returns to 40 mm Hg. A decreased respiratory rate that raises arterial pCO_2 to levels significantly above 40 mm Hg is termed **hypoventilation,** while an increased respiratory rate that reduces arterial pCO_2 to levels significantly below 40 mm Hg is termed **hyperventilation.**

The oxygen chemoreceptors found within the medulla and the aortic and carotid bodies are sensitive only to large decreases in the pO_2 because hemoglobin remains about 85% or more saturated at pO_2 values all the way down to 50 mm Hg. If arterial pO_2 falls from a normal of 105 mm Hg to approximately 50 mm Hg, the oxygen chemoreceptors become stimulated and send impulses to the inspiratory center and respirations increase. But if the pO_2 falls much below 50 mm Hg, the cells of the inspiratory area will suffer oxygen starvation and will not respond well to any chemical receptors. They would therefore send fewer impulses to the inspiratory muscles, and the respiration rate would decrease or breathing would cease altogether.

We can measure the effect of various factors on respiration by measuring (1) the CO_2 content of exhaled air or (2) the rate of respiration.

In order to measure CO_2 content, a gas is bubbled through lime water and the time is measured until the liquid becomes cloudy. Lime water is a solution of calcium hydroxide

$[Ca(OH)_2]$. The cloudiness occurs because carbon dioxide combines with calcium hydroxide to form calcium carbonate, a white precipitate, in the following reaction.

$$CO_2 + CaOH_2 \rightarrow CaCO_3 + H_2O$$

Carbon dioxide   Calcium hydroxide   Calcium carbonate   Water

PROCEDURE

1. Label three test tubes *N, H,* and *E* (Normal, Hyperventilation, Exercise). Fill each about half full of calcium hydroxide $[Ca(OH)_2]$ solution.
2. Start the stopwatch and *exhale* normal expirations through the straw into Tube N. Note the time elapsed until the solution becomes turbid.
3. Hyperventilate for about 30 sec.

CAUTION! *Stop hyperventilation at the first sign of dizziness or lightheadedness.*
Repeat Step 2, *exhaling* normally into Tube H.

4. After a 3-min rest, exercise briskly for 2 min.
5. Repeat step 2, *exhaling* into Tube E.
6. Record all time in Section C.5 of the LABORATORY REPORT RESULTS at the end of the exercise.
   In which situation was the CO_2 content of

   exhaled air the greatest? _____

   Why? _____

   _____

D. LABORATORY TESTS COMBINING RESPIRATORY AND CARDIOVASCULAR INTERACTIONS

1. *Experimental setup*—Review earlier explanations on recording the following:
   a. Respiratory movements with a pneumograph (Section C.5 in Exercise 22).
   b. Blood pressure with a sphygmomanometer and stethoscope (Section E in Exercise 20).
   c. Radial pulse (Section D.1 in Exercise 20).

CAUTION! *Assuming that your partner has no known or apparent cardiac or other health problems* *and is capable of such an activity, ask her/him to perform the following activities after attaching the various pieces of apparatus in order to record respiratory movements and blood pressure.*

2. *Experimental procedure*—Some form of regulated exercise is necessary for this experiment. Choose one of the following forms of exercise to have your subject participate in:
   a. *Riding exercise cycle* If this form is chosen, set the resistance to be felt while riding the cycle, but *not so high* that the subject cannot complete the 4-min exercise period without difficulty.
   b. *Harvard Step Test* In this form of exercise, the subject is to step up onto a 20-in. platform (a chair will substitute quite well) with one foot at a time. The subject is to bring *both* feet up onto the platform before stepping back down to the floor. The subject is also to remain erect at all times, and to do 30 complete cycles (up onto the platform and back down) per minute.
3. *Experimental protocol*—Record respiratory movements, blood pressure, and pulse during each of the procedures listed. Record your data in the table provided in Section D of the LABORATORY REPORT RESULTS at the end of the exercise. All data should be recorded *simultaneously*, thereby requiring participation of all members of the experimental group.
   a. *Basal readings* Have the subject sit erect and quiet for 3 min. Obtain readings for

   Respiratory rate and depth
   Systolic and diastolic blood pressure
   Pulse rate

When the basal readings have been recorded, obtain additional readings after (1) sitting quietly on the exercise cycle for 3 min, if this form of exercise is to be utilized, or (2) standing quietly for 3 min in front of the platform that is to be utilized for the Harvard Step Test.
   b. *Readings after 1 min of exercise* After the subject has exercised for 1 min, obtain additional readings.
   c. *Readings after 2 and 3 min of exercise* Again, obtain additional readings after the subject has completed 2 and 3 min of exercise.
   d. *Readings upon completion of exercise* When the subject has completed 4 min of exercise, obtain readings for respiratory rate, respiratory depth, heart rate, and systolic and diastolic blood pressure immediately upon completion *while the subject remains seated on the exercise cycle or stands erect on the floor,* depending upon the type of exercise utilized.
   e. *Readings 1, 2, 3, and 5 min after completion of exercise* With the subject *still sitting on the exercise cycle or still standing erect on the floor,* obtain additional readings at the preceding time intervals after completion of the exercise period.

ANSWER THE LABORATORY REPORT QUESTIONS AT THE END OF THE EXERCISE.

Respiratory System

STUDENT _____   DATE _____

LABORATORY SECTION _____   SCORE/GRADE _____

SECTION C. LABORATORY TEST ON RESPIRATION

1. Measurement of Chest and Abdomen in Respiration

Record the results of your chest measurements, in inches, in the following table:

Size after a	Chest	Abdomen
Normal inspiration		
Normal expiration		
Maximal inspiration		
Maximal expiration		

2. Use of Pneumograph

Attach a sample of any one of the following activities:

Reading
Swallowing water
Laughing
Yawning
Coughing
Sniffing

3. Use of Handheld Respirometer

Tidal volume _____

Expiratory reserve volume _____

Vital capacity _____

Inspiratory reserve volume _____

Minute volume of respiration (MVR) _____

4. Use of Collins Respirometer

Record the results of your exercises using the Collins respirometer in the following table:

	Tidal volume (1)	Expiratory reserve (2)	Vital capacity (3)	Inspiratory reserve (4)	FEV$_1$ (5)
First time	ml	ml	ml	ml	ml
Second time	ml	ml	ml	ml	ml
Third time	ml	ml	ml	ml	ml
Your average	ml	ml	ml	ml	ml
Normal value	ml	ml	ml	ml	ml

5. Chemical Regulation of Respiration

	Time until turbidity
Tidal respirations	
After hyperventilation	
After exercise	

SECTION D. COMBINED RESPIRATORY AND CARDIOVASCULAR INTERACTIONS

Record the results of your exercises in the following table:

	Respiratory rate and depth	Systolic and Diastolic pressure	Pulse rate
Basal readings while sitting erect and quiet for 3 min			
Basal readings while standing oncycle or in front of platform for 3 min			
Readings after 1 min of exercise			
Readings after 2 min of exercise			
Readings after 3 min of exercise			
Readings after 4 min of exercise			
Readings after 1 min following completion of exercise			
Readings after 2 min following completion of exercise			
Readings after 3 min following completion of exercise			
Readings after 5 min following completion of exercise			

Respiratory System

STUDENT _____   DATE _____

LABORATORY SECTION _____   SCORE/GRADE _____

PART 1. Multiple Choice

_____ 1. The overall exchange of gases between the atmosphere, blood, and cells is called (a) inspiration (b) respiration (c) expiration (d) none of these

_____ 2. The internal nose is formed by the ethmoid bone, maxillae, inferior conchae, and the (a) hyoid bone (b) nasal bone (c) palatine bone (d) none of these

_____ 3. The portion of the pharynx that contains the pharyngeal tonsils is the (a) oropharynx (b) laryngopharynx (c) nasopharynx (d) pharyngeal orifice

_____ 4. The Adam's apple is a common term for the (a) thyroid cartilage (b) cricoid cartilage (c) epiglottis (d) none of these

_____ 5. The C-shaped rings of cartilage of the trachea not only prevent the trachea from collapsing, but also aid in the process of (a) lubrication (b) removing foreign particles (c) gas exchange (d) swallowing

_____ 6. Of the following structures, the smallest in diameter is the (a) left primary bronchus (b) bronchioles (c) secondary bronchi (d) alveolar ducts

_____ 7. The structures of the lung that actually contain the alveoli are the (a) respiratory bronchioles (b) fissures (c) lobules (d) terminal bronchioles

_____ 8. From superficial to deep, the structure(s) that you would encounter first among the following is (are) the (a) bronchi (b) parietal pleura (c) pleural cavity (d) secondary bronchi

PART 2. Completion

9. An advantage of nasal breathing is that the air is warmed, moistened, and

_____.

10. Improper fusion of the palatine and maxillary bones results in a condition called

_____.

11. The protective lid of cartilage that prevents food from entering the trachea is the

_____.

12. After removal of the _____, an individual would be unable to speak.

13. The upper respiratory tract is able to trap and remove dust because of its lining of

_____.

14. The passage of a tube into the mouth and down through the larynx and trachea to bypass an obstruction is called _____.

15. An X ray of the bronchial tree after the administration of an opaque iodinated media is called a

    _____.

16. The sequence of respiratory tubes from largest to smallest is the trachea, primary bronchi, secondary bronchi, bronchioles, _____, respiratory bronchioles, and alveolar ducts.

17. Both the external and internal nose are divided internally by a vertical partition called the

    _____.

18. The undersurface of the external nose contains two openings called the nostrils or

    _____.

19. The functions of the pharynx are to serve as a passageway for air and food and to provide a resonating chamber for _____.

20. An inflammation of the membrane that encloses and protects the lungs is called

    _____.

21. The anterior portion of the nasal cavity just inside the nostrils is called the

    _____.

22. Groovelike passageways in the nasal cavity formed by the conchae are called

    _____.

23. The portion of the pharynx that contains the palatine and lingual tonsils is the

    _____.

24. Each lobe of a lung is subdivided into compartments called _____.

25. A(n) _____ is an outpouching lined by squamous epithelium and supported by a thin elastic membrane.

26. The _____ cartilage attaches the larynx to the trachea.

27. The portion of a lung that rests on the diaphragm is the _____.

28. Phagocytic cells in the alveolar wall are called _____.

29. The surface of a lung lying against the ribs is called the _____ surface.

30. The _____ is a structure in the medial surface of a lung through which bronchi, blood vessels, lymphatic vessels, and nerves pass.

PART 3. Matching

_____ 31 Tidal volume   A. 1200 ml of air

_____ 32. Inspiratory reserve volume   B. 3600 ml of air

_____ 33. Inspiratory capacity   C. 4800 ml of air

_____ 34. Expiratory reserve volume   D. 500 ml of air

_____ 35. Vital capacity   E. 3100 ml of air

_____ 36. Total lung capacity   F. 6000 ml of air

23

Digestive System

Digestion occurs basically as two events—mechanical digestion and chemical digestion. *Mechanical digestion* consists of various movements of the gastrointestinal tract that help chemical digestion. These movements include the physical breakdown of food by the teeth and complete churning and mixing of this food with enzymes by the smooth muscles of the stomach and small intestine. *Chemical digestion* consists of a series of catabolic (hydrolysis) reactions that break down the large nutrient molecules we eat—such as carbohydrates, lipids, and proteins—into much smaller molecules that can be absorbed and used by body cells.

A. GENERAL ORGANIZATION OF DIGESTIVE SYSTEM

Digestive organs are usually divided into two main groups. The first is the *gastrointestinal (GI) tract,* or *alimentary canal,* a continuous tube running from the mouth to the anus, and measuring about 9 m (30 ft) in length in a cadaver. This tract is composed of the mouth, pharynx, esophagus, stomach, small intestine, and large intestine. The small intestine has three regions: duodenum, jejunum, and ileum. The large intestine has four regions: cecum, colon, rectum, and anal canal. The colon is divided into ascending colon, transverse colon, descending colon, and sigmoid colon.

The second group of organs composing the digestive system consists of the *accessory struc-*

tures such as the teeth, tongue, salivary glands, liver, gallbladder, and pancreas (see Figure 23.1).

Using your textbook, charts, or models for reference, label Figure 23.1.

The wall of the gastrointestinal tract, especially from the esophagus to the anal canal, has the same basic arrangement of tissues. The four layers (tunics) of the tract, from the inside to the outside, are the *mucosa, submucosa, muscularis,* and *serosa,* or *adventitia* (see Figures 23.7 and 23.9).

The outermost layer of the gastrointestinal tract is a serous membrane composed of connective tissue and epithelium and called the *peritoneum* (per′-i-tō-NĒ-um; *peri* = around, *tonos* = tension) or *serosa.* The *parietal peritoneum* lines the wall of the abdominal cavity, and the *visceral peritoneum* covers some of the organs. The potential space between the parietal and visceral portions of the peritoneum is called the *peritoneal cavity.* Unlike the two other serous membranes of the body, the pericardium and the pleura, which smoothly cover the heart and lungs, the peritoneum contains large folds that weave in between the viscera. The important extensions of the peritoneum are the *mesentery* (MEZ-en-ter′-ē; *meso* = middle, *enteron* = intestine) *mesocolon, falciform* (FAL-si-form) *ligament, lesser omentum* (ō-MENT-um), and *greater omentum.*

Inflammation of the peritoneum, called *peritonitis,* is a serious condition because the peritoneal membranes are continuous with one another, enabling the infection to spread to all the organs in the cavity.

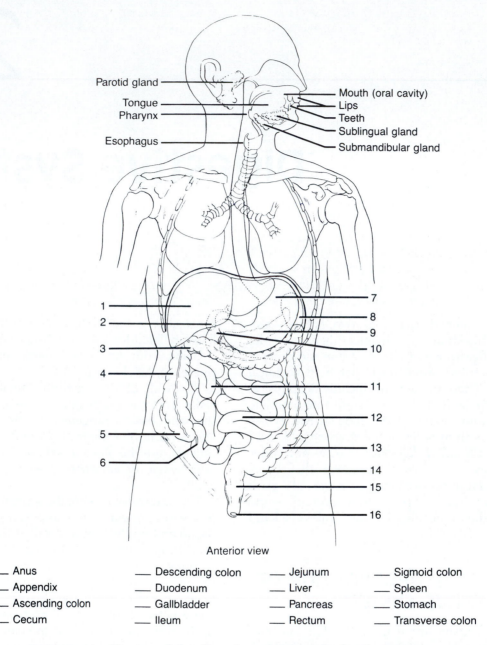

Parotid gland ——————
Tongue ——————
Pharynx ——————
Esophagus ——————

—————— Mouth (oral cavity)
—————— Lips
—————— Teeth
—————— Sublingual gland
—————— Submandibular gland

1 ——————
2 ——————
3 ——————
4 ——————
5 ——————
6 ——————

—————— 7
—————— 8
—————— 9
—————— 10
—————— 11
—————— 12
—————— 13
—————— 14
—————— 15
—————— 16

Anterior view

___ Anus   ___ Descending colon   ___ Jejunum   ___ Sigmoid colon
___ Appendix   ___ Duodenum   ___ Liver   ___ Spleen
___ Ascending colon ___ Gallbladder   ___ Pancreas   ___ Stomach
___ Cecum   ___ Ileum   ___ Rectum   ___ Transverse colon

FIGURE 23.1   Organs of the digestive system and related structures.

B. ORGANS OF DIGESTIVE SYSTEM

1. Mouth (Oral Cavity)

The *mouth,* also called the *oral* or *buccal* (BUK-al) *cavity,* is formed by the cheeks, hard and soft palates, and tongue. The *hard palate* forms the anterior portion of the roof of the mouth and the *soft palate* forms the posterior portion. The *tongue* forms the floor of the oral cavity and is composed of skeletal muscle covered by mucous membrane. Partial digestion of carbohydrates is the only chemical digestion that occurs in the mouth.

a. *Cheeks*—Lateral walls of the oral cavity. Muscular structures covered by skin and lined by stratified squamous epithelium; anterior portions terminate in the *superior* and *inferior labia* (lips).

b. *Vermilion* (ver-MIL-yon)—Transition zone of the lips, where outer skin and inner mucous membranes meet.

c. *Labial frenulum* (LĀ-bē-al FREN-yoo-lum)—Midline fold of the mucous membrane that attaches the inner surface of each lip to its corresponding gum.

d. *Vestibule*—Space bounded externally by the cheeks and lips and internally by the gums and teeth.

e. *Oral cavity proper*—Space extending from the vestibule to the *fauces* (FAW-sēz; *fauces* = passages), the opening of the oral cavity proper into pharynx. The area is enclosed by the dental arches.

f. *Hard palate*—Formed by maxillae and palatine bones and covered by mucous membrane.

g. *Soft palate*—Arch-shaped muscular partition between the oropharynx and nasopharynx. Hanging from the free border of the soft palate is a muscular projection, the *uvula* (YOU-vyoo-la).

h. *Palatoglossal arch (anterior pillar)*—Muscular fold that extends inferiorly, laterally, and anteriorly to the side of the base of the tongue.

i. *Palatopharyngeal* (PAL-a-tō-fa-rin'-jē-al) *arch (posterior pillar)*—Muscular fold that extends inferiorly, laterally, and posteriorly to the side of the pharynx. *Palatine tonsils* are between the arches and the *lingual tonsil* is at the base of the tongue.

j. *Tongue*—Movable, muscular organ on the floor of the mouth. *Extrinsic muscles* originate outside the tongue, insert into it, and move the tongue from side to side and in and out to maneuver food for chewing and swallowing; *intrinsic muscles* originate and insert within the tongue and alter the shape and size of the tongue for speech and swallowing.

k. *Lingual frenulum*—Midline fold of the mucous membrane on the undersurface of the tongue that helps restrict its movement posteriorly.

l. *Papillae* (pa-PIL-ē)—Projections of lamina propria on the surface of the tongue covered with epithelium; *filiform papillae* are conical projections in parallel rows over the anterior two-thirds of the tongue; *fungiform papillae* are mushroomlike elevations distributed among filiform papillae and more numerous near the tip of the tongue (appear as red dots and most contain taste buds); *circumvallate papillae* are arranged in the form of an inverted V on the posterior surface of the tongue (all contain taste buds).

Using a mirror and tongue depressor, examine your mouth or your partner's and locate as many of the structures (a) through (l) as you can. Label Figure 23.2.

2. Salivary Glands

Most saliva is secreted by the *salivary glands,* which lie outside the mouth and pour their contents into ducts that empty into the oral cavity. The carbohydrate-digesting enzyme in saliva is salivary amylase. The three pairs of salivary glands are the *parotid glands* (in front of and under the ears), *submandibular glands* (deep to the base of the tongue in the posterior part of the floor of the mouth), and *sublingual glands* (superior to the submandibular glands).

Label Figure 23.3 on page 457.

The parotid glands are compound tubuloacinar glands consisting of all serous acini, whereas the submandibulars (consisting of serous and mucous acini) and sublinguals (consisting of mostly mucous acini) are compound acinar glands (see Figure 23.4).

Examine prepared slides of the three different types of salivary glands and compare your observations with Figure 23.4 on page 458.

3. Teeth

Teeth (dentes) are located in the sockets of the alveolar processes of the mandible and maxillae. The alveolar processes are covered by *gingivae* (jin-JĪ-vē), or gums, which extend slightly into each socket. The sockets are lined by a dense fibrous connective tissue called a *periodontal ligament,* which anchors the teeth in position and acts as a shock absorber during chewing.

The parts of a tooth follow:

a. *Crown*—Exposed portion above the level of the gums.

b. *Root*—One to three projections embedded in the socket.

c. *Neck*—Constricted junction line of the crown and root near the gum line.

d. *Dentin*—Calcified connective tissue that gives teeth their basic shape.

e. *Pulp cavity*—Enlarged part of the cavity in the crown within dentin.

f. *Pulp*—Connective tissue containing blood vessels, lymphatic vessels, and nerves.

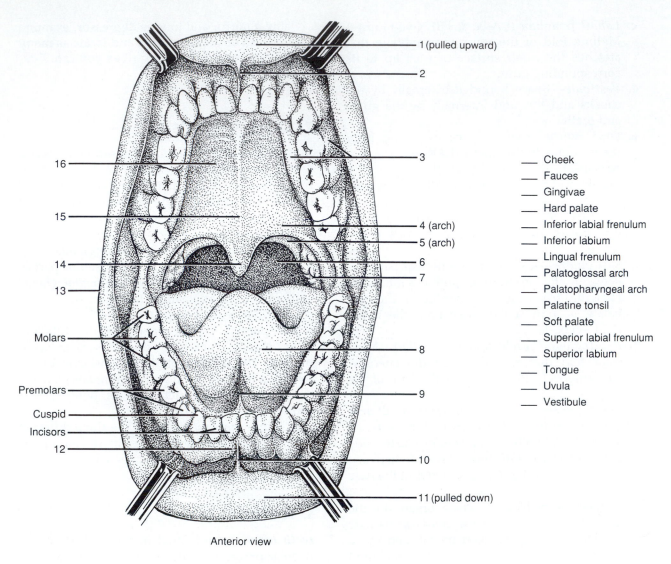

1 (pulled upward)
2
3
4 (arch)
5 (arch)
6
7
8
9
10
11 (pulled down)

16
15
14
13
Molars
Premolars
Cuspid
Incisors
12

Anterior view

____ Cheek
____ Fauces
____ Gingivae
____ Hard palate
____ Inferior labial frenulum
____ Inferior labium
____ Lingual frenulum
____ Palatoglossal arch
____ Palatopharyngeal arch
____ Palatine tonsil
____ Soft palate
____ Superior labial frenulum
____ Superior labium
____ Tongue
____ Uvula
____ Vestibule

FIGURE 23.2   Mouth (oral cavity).

g. *Root canal*—Narrow extension of the pulp cavity in the root.

h. *Apical foramen*—Opening in the base of the root canal through which blood vessels, lymphatic vessels, and nerves enter the tooth.

i. *Enamel*—Covering of the crown that consists primarily of calcium phosphate and calcium carbonate.

j. *Cementum*—Bonelike substance that covers and attaches the root to the periodontal ligament.

With the aid of your textbook, label the parts of a tooth shown in Figure 23.5 on page 459.

4. Dentitions

Dentitions (sets of teeth) are of two types: *deciduous* (baby) and *permanent.* Deciduous teeth begin to erupt at about 6 months of age, and one pair appears at about each month thereafter, until all 20 are present. The deciduous teeth follow:

a. *Incisors*—Central incisor closest to the midline, with a lateral incisor on either side. Incisors are chisel-shaped, adapted for cutting into food, and have only one root.

b. *Cuspids (canines)*—Posterior to the incisors, cuspids have pointed surfaces (cusps) for

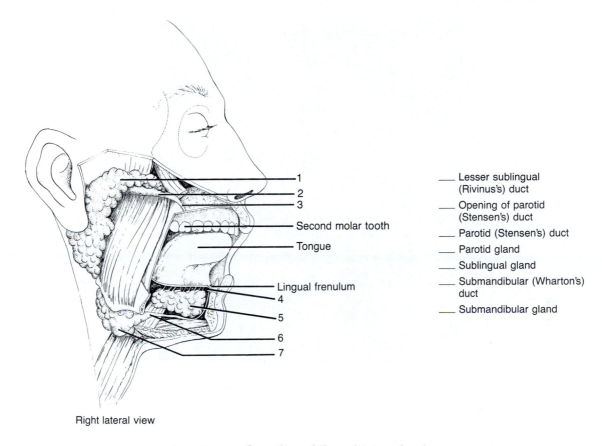

Right lateral view

1 _____
2 _____
3 _____
Second molar tooth
Tongue
Lingual frenulum
4 _____
5 _____
6 _____
7 _____

_____ Lesser sublingual
       (Rivinus's) duct
_____ Opening of parotid
       (Stensen's) duct
_____ Parotid (Stensen's) duct
_____ Parotid gland
_____ Sublingual gland
_____ Submandibular (Wharton's)
       duct
_____ Submandibular gland

FIGURE 23.3   Location of the salivary glands.

tearing and shredding food and have only one root.

c. *Molars*—First and second molars posterior to canines. Molars crush and grind food. Upper molars have four cusps and three roots; lower molars have four cusps and two roots.

All deciduous teeth are usually lost between 6 and 12 years of age and replaced by permanent dentition consisting of 32 teeth that appear between age 6 and adulthood. The permanent teeth follow:

a. *Incisors*—Central incisor and lateral incisor replace those of deciduous dentition.
b. *Cuspids (canines)*—Replace the deciduous dentition.
c. *Premolars (bicuspids)*—First and second premolars replace deciduous molars. Premolars crush and grind food and have two cusps and one root (upper first premolars have two roots).

d. *Molars*—Erupt behind premolars as the jaw grows to accommodate them; they do not replace any deciduous teeth. First molars erupt at age 6, second at age 12, and third (wisdom teeth) after age 18.

Using a mirror and tongue depressor, examine your mouth or your partner's and locate as many teeth of the permanent dentition as you can.

With the aid of your textbook, label the deciduous and permanent dentitions in Figure 23.6 on page 460.

5. Esophagus

The *esophagus* (e-SOF-a-gus) is a muscular, collapsible tube posterior to the trachea. The structure is 23 to 25 cm (10 in.) long and extends from the laryngopharynx through the mediastinum and esophageal hiatus in the diaphragm and terminates in the superior portion of the stomach. The esophagus conveys food from the pharynx to the stomach by peristalsis.

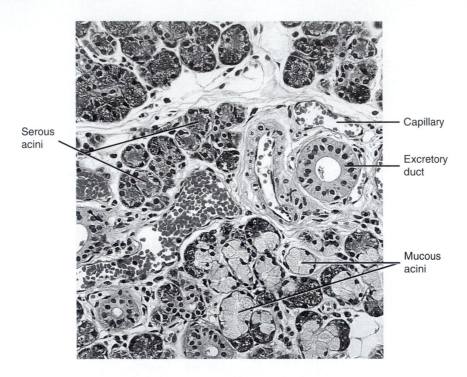

Serous acini

Capillary

Excretory duct

Mucous acini

FIGURE 23.4 Histology of the salivary glands.

Histologically, the esophagus consists of a *mucosa* (nonkeratinized stratified squamous epithelium, lamina propria, muscularis mucosae), *submucosa* (connective tissue, blood vessels, mucous glands), *muscularis* (upper third striated, middle third striated, and smooth, lower third smooth), and *adventitia* (ad-ven-TISH-ya). The esophagus is not covered by peritoneum.

Examine a prepared slide of a cross section of the esophagus that shows its various coats. With the aid of your textbook, label Figure 23.7 on page 460.

6. Stomach

The *stomach* is a J-shaped enlargement of the gastrointestinal tract under the diaphragm in the epigastric, umbilical, and left hypochondriac regions of the abdomen (see Figure 23.1). The superior part is connected to the esophagus; the inferior part empties into the duodenum, the first portion of the small intestine. The stomach is divided into four areas: cardia, fundus, body, and pylorus. The *cardia* surrounds the lower esophageal sphincter, a physiological sphincter in the esophagus just above the diaphragm. The rounded portion above and to the left of the cardia is the *fundus.* Below the fundus, the large

central portion of the stomach is called the *body.* The narrow, inferior region is the *pylorus.* The pylorus consists of a *pyloric antrum,* which is closer to the body of the stomach, and a *pyloric canal,* which is closer to the duodenum. The concave medial border of the stomach is called the *lesser curvature,* and the convex lateral border is the *greater curvature.* The pylorus communicates with the duodenum of the small intestine via a sphincter called the *pyloric sphincter (valve).* The main chemical activity of the stomach is to begin the digestion of proteins.

Label Figure 23.8 on page 461.

The *mucosa* of the stomach contains large folds called *rugae* (ROO-jē). The columnar epithelium of the mucosa contains many narrow openings that extend down into the lamina propria. These openings are called *gastric glands (pits)* and are lined with four kinds of cells: (1) *chief (zymogenic) cells,* which secrete inactive pepsinogen, which is converted to active pepsin, a protein-digesting enzyme; (2) *parietal (oxyntic) cells,* which secrete hydrochloric acid and intrinsic factor; (3) *mucous cells,* which secrete mucus; and (4) *G cells,* which secrete the hormone gastrin. The *submucosa* consists of areolar connective tissue. The *muscularis* has three layers of smooth muscle—outer longitudinal, middle circular, and inner oblique.

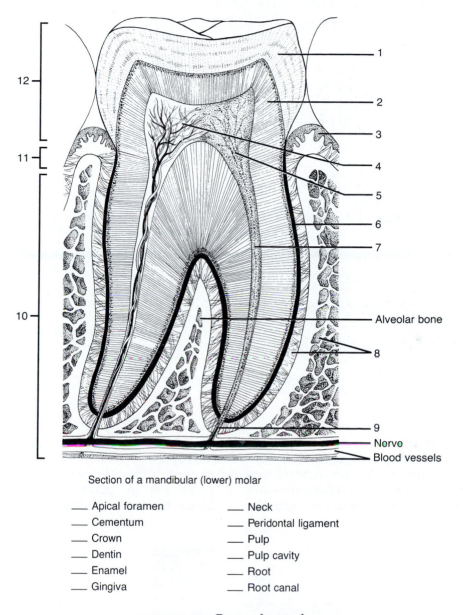

Section of a mandibular (lower) molar

___ Apical foramen	___ Neck
___ Cementum	___ Peridontal ligament
___ Crown	___ Pulp
___ Dentin	___ Pulp cavity
___ Enamel	___ Root
___ Gingiva	___ Root canal

FIGURE 23.5 Parts of a tooth.

The oblique layer is limited chiefly to the body of the stomach. The *serosa* is part of the visceral peritoneum.

Examine a prepared slide of a section of the stomach that shows its various layers. With the aid of your textbook, label Figure 23.9 on page 461.

7. Pancreas

The *pancreas* is a soft, oblong, tubuloacinar gland posterior to the greater curvature of the stomach (see Figure 23.1). The gland consists of a *head* (expanded portion near duodenum), *body* (central portion), and *tail* (terminal tapering portion).

Histologically, the pancreas consists of *pancreatic islets (islets of Langerhans)* that contain (1) glucagon-producing *alpha cells,* (2) insulin-producing *beta cells,* (3) somatostatin-producing *delta cells,* and pancreatic polypeptide-producing *F cells* (see Figure 15.5). The pancreas also consists of *acini,* which produce pancreatic juice (see Figure 15.5). Pancreatic juice contains enzymes that assist in the chemical breakdown of carbohydrates, proteins, lipids, and nucleic acids.

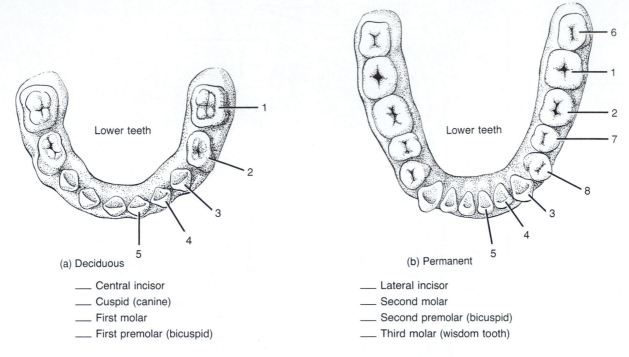

Lower teeth

(a) Deciduous

____ Central incisor
____ Cuspid (canine)
____ First molar
____ First premolar (bicuspid)

Lower teeth

(b) Permanent

____ Lateral incisor
____ Second molar
____ Second premolar (bicuspid)
____ Third molar (wisdom tooth)

FIGURE 23.6   Dentitions.

Pancreatic juice is delivered from the pancreas to the duodenum by a large main tube, the *pancreatic duct (duct of Wirsung).* This duct unites with the common bile duct from the liver and pancreas and enters the duodenum in a common duct called the *hepatopancreatic am-* *pulla (ampulla of Vater).* The ampulla opens on an elevation of the duodenal mucosa, the *duodenal papilla.* An *accessory pancreatic duct (duct of Santorini)* may also lead from the pancreas and empty into the duodenum about 2.5 cm (1 in.) above the hepatopancreatic ampulla. With the

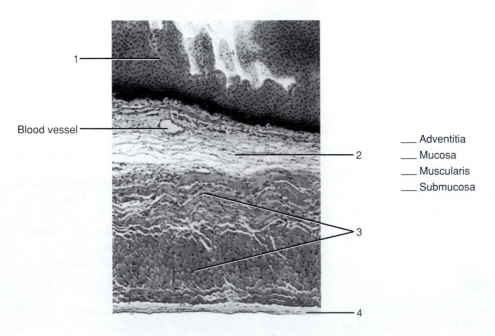

Blood vessel

____ Adventitia
____ Mucosa
____ Muscularis
____ Submucosa

FIGURE 23.7   Histology of the esophagus.

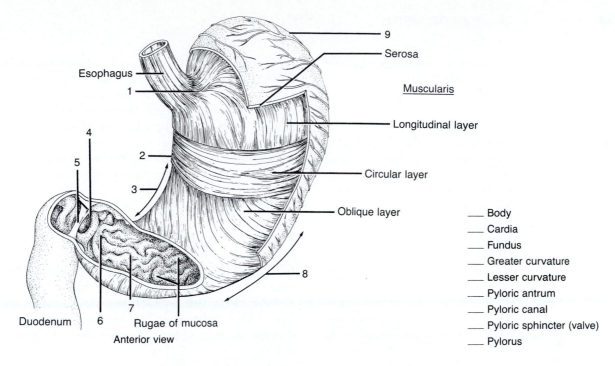

Esophagus

1

4

5

3

2

9

Serosa

<u>Muscularis</u>

Longitudinal layer

Circular layer

Oblique layer

___ Body
___ Cardia
___ Fundus
___ Greater curvature
___ Lesser curvature
___ Pyloric antrum
___ Pyloric canal
___ Pyloric sphincter (valve)
___ Pylorus

8

7

6

Duodenum

Rugae of mucosa

Anterior view

FIGURE 23.8   Stomach. External and internal anatomy.

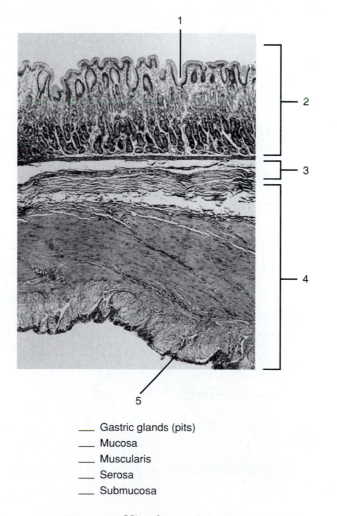

1

2

3

4

5

___ Gastric glands (pits)
___ Mucosa
___ Muscularis
___ Serosa
___ Submucosa

FIGURE 23.9   Histology of the stomach.

aid of your textbook, label the structures associated with the pancreas in Fig 23.10.

8. Liver

The *liver* is located inferior to the diaphragm, occupying most of the right hypochondriac and part of the epigastric regions of the abdomen (see Figure 23.1). The gland is divided into two principal lobes, the *right lobe* and *left lobe,* separated by the *falciform ligament.* The falciform ligament attaches the liver to the anterior abdominal wall and diaphragm. The right lobe also has associated with it an inferior *quadrate lobe* and a posterior *caudate lobe.*

Each lobe is composed of microscopic functional units called *lobules.* Among the structures in a lobule are cords of *hepatic (liver) cells*

arranged in a radial pattern around a *central vein; sinusoids,* endothelial lined spaces between hepatic cells through which blood flows; and *stellate reticuloendothelial (Kupffer) cells,* which destroy bacteria and worn-out blood cells by phagocytosis.

Examine a prepared slide of several liver lobules. Compare your observations with Figure 23.11.

Bile is manufactured by hepatic cells and functions in the emulsification of fats in the small intestine. The liquid is passed to the small intestine as follows: Hepatic cells secrete bile into *bile capillaries* that empty into small ducts. The small ducts merge into larger *right* and *left hepatic ducts,* one in each principal lobe of the liver. The right and left hepatic ducts unite outside the liver to form a single *common hepatic*

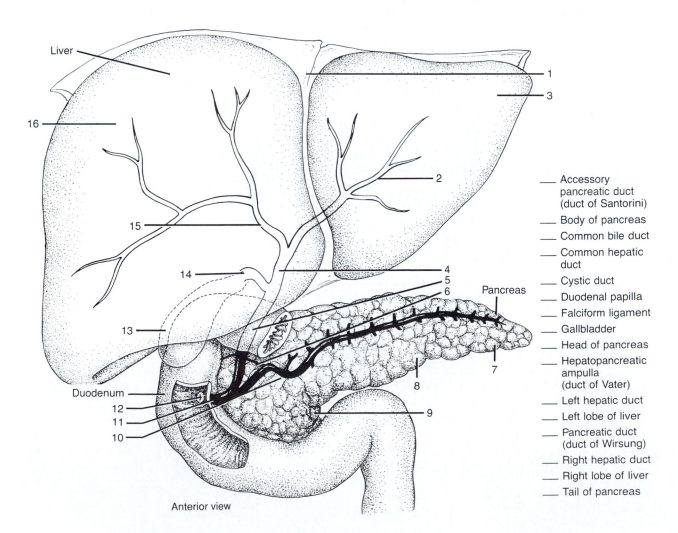

___ Accessory pancreatic duct (duct of Santorini)

___ Body of pancreas

___ Common bile duct

___ Common hepatic duct

___ Cystic duct

___ Duodenal papilla

___ Falciform ligament

___ Gallbladder

___ Head of pancreas

___ Hepatopancreatic ampulla (duct of Vater)

___ Left hepatic duct

___ Left lobe of liver

___ Pancreatic duct (duct of Wirsung)

___ Right hepatic duct

___ Right lobe of liver

___ Tail of pancreas

FIGURE 23.10   Relations of the liver, gallbladder, duodenum, and pancreas.

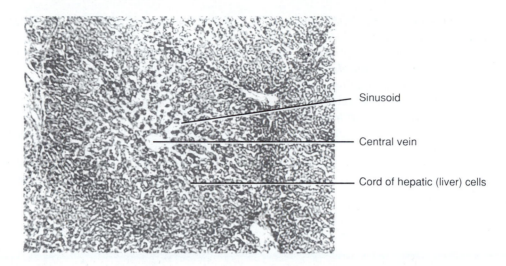

Sinusoid

Central vein

Cord of hepatic (liver) cells

FIGURE 23.11   Photomicrograph of a liver lobule.

duct. This duct joins the *cystic duct* from the gallbladder to become the *common bile duct,* which empties into the duodenum at the hepatopancreatic ampulla (ampulla of Vater). When fats are not being digested, a valve around the hepatopancreatic ampulla (ampulla of Vater), the *sphincter of the hepatopancreatic ampulla (sphincter of Oddi),* closes, and bile backs up into the gallbladder via the cystic duct.

With the aid of your textbook, label the structures associated with the liver in Figure 23.10.

9. Gallbladder (GB)

The *gallbladder (GB)* is a pear-shaped sac in a fossa along the undersurface of the liver (see Figure 23.1). The gallbladder stores and concentrates bile. The cystic duct of the gallbladder and common hepatic duct of the liver merge to form the common bile duct. The *mucosa* of the gallbladder contains rugae. The *muscularis* consists of *bile capillaries* that empty into small ducts. The small ducts merge into larger *right* and *left hepatic ducts,* one in each principal lobe of the liver. The right and left hepatic ducts unite outside the liver to form a single *common hepatic duct.* This duct joins the *cystic duct* from the gallbladder to become the *common bile duct,* which empties into the duodenum at the hepatopancreatic ampulla (ampulla of Vater). When fats are not being digested, a valve around the hepatopancreatic ampulla, the *sphincter of the hepatopancreatic ampulla (sphincter of Oddi),* closes, and bile backs up into the gallbladder via the cystic duct.

With the aid of your textbook, label the structures associated with the gallbladder in Figure 23.10.

10. Small Intestine

The bulk of digestion and absorption occurs in the *small intestine,* which begins at the pyloric sphincter (valve) of the stomach, coils through the central and lower part of the abdomen, and joins the large intestine at the ileocecal sphincter (see Figure 23.1). The mesentery attaches the small intestine to the posterior abdominal wall. The small intestine is about 6.35 m (21 ft) long and is divided into three segments: *duodenum* (doo'-ō-DĒ-num), which begins at the stomach; *jejunum* (jē-JOO-num), the middle segment; and *ileum* (IL-ē-um), which terminates at the large intestine.

Histologically, the *mucosa* contains many pits lined with glandular epithelium called *intestinal glands (crypts of Lieberkühn);* they secrete enzymes that digest carbohydrates, proteins, and nucleic acids (see Figure 23.12). Some of the simple columnar cells of the mucosa are *goblet cells* that secrete mucus (see Figure 23.12); others contain *microvilli* to increase the surface area for absorption (see Figure 3.1). The mucosa contains a series of fingerlike projections, the *villi* (see Figure 23.12). Each villus contains a blood capillary and lymphatic capillary called a *lacteal;* these absorb digested nutrients. The 4 million to 5 million villi in the small intestine greatly increase the surface area for absorption. The mucosa and *submucosa* also contain deep permanent folds, the *circular folds,* or *plicae circulares* (PLĪ-kē SER-kyoo-lar-es), which also help to increase the surface area for absorption. In the submucosa of the duodenum are *duodenal (Brunner's) glands,* which secrete an alkaline mucus to protect the mucosa from excess acid and the action of digestive enzymes. The *muscularis* of the small intestine consists of an outer longitudinal layer and an inner circular layer of smooth muscle. Except for a major portion of the duodenum, the *serosa* (visceral peritoneum) completely covers the small intestine.

Obtain prepared slides of the small intestine (section through its tunics and villi) and identify as many structures as you can, using Figure 23.12 and your textbook as references.

11. Large Intestine

The *large intestine* functions in the completion of the absorption of water that leads to the formation of feces and in the expulsion of feces from the body. Bacteria residing in the large intestine manufacture certain vitamins. The large intestine

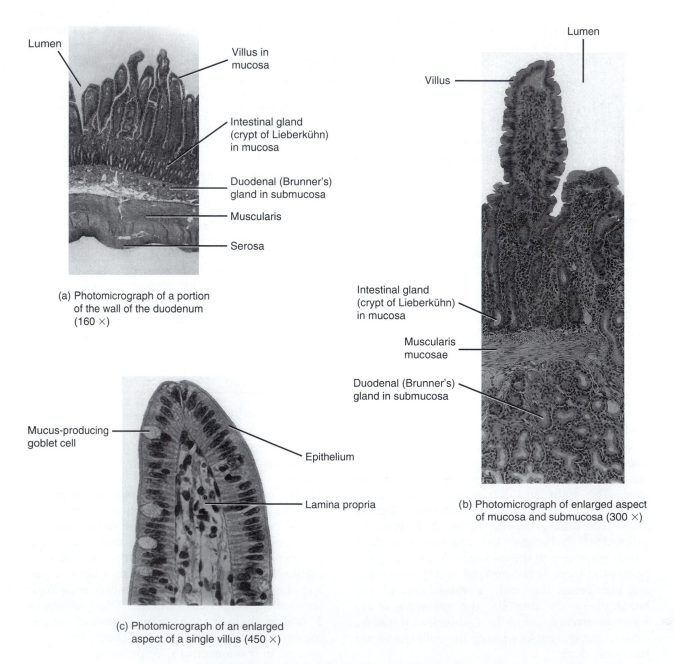

Lumen

Villus in mucosa

Intestinal gland (crypt of Lieberkühn) in mucosa

Duodenal (Brunner's) gland in submucosa

Muscularis

Serosa

(a) Photomicrograph of a portion of the wall of the duodenum (160 ×)

Lumen

Villus

Intestinal gland (crypt of Lieberkühn) in mucosa

Muscularis mucosae

Duodenal (Brunner's) gland in submucosa

(b) Photomicrograph of enlarged aspect of mucosa and submucosa (300 ×)

Mucus-producing goblet cell

Epithelium

Lamina propria

(c) Photomicrograph of an enlarged aspect of a single villus (450 ×)

FIGURE 23.12   Histology of the small intestine.

is about 1½ m (5 ft) long and extends from the ileum to the anus (see Figure 23.1). It is attached to the posterior abdominal wall by an extension of visceral peritoneum called mesocolon. The large intestine is divided into four principal regions: cecum, colon, rectum, and anal canal.

The opening from the ileum into the large intestine is guarded by a fold of mucous membrane, the *ileocecal sphincter (valve)*. Hanging below the valve is a blind pouch, the *cecum*, to which is attached the *vermiform appendix* by an extension of visceral peritoneum called the *mesoappendix*. Inflammation of the vermiform appendix is called *appendicitis*. The open end of the cecum merges with the *colon*. The first division of the colon is the *ascending colon*, which ascends on the right side of the abdomen and turns abruptly to the left at the undersurface of the liver *(right colic [hepatic] flexure)*. The *transverse colon* continues across the abdomen, curves be-

neath the spleen *(left colic [splenic] flexure)*, and passes down the left side of the abdomen as the *descending colon.* The *sigmoid colon* begins near the iliac crest, projects inward toward the midline, and terminates at the rectum at the level of the third sacral vertebra. The *rectum* is the last 20 cm (7 to 8 in.) of the gastrointestinal tract. Its terminal 2 to 3 cm (1 in.) is known as the *anal canal.* The opening of the anal canal to the exterior is the *anus.*

With the aid of your textbook, label the parts of the large intestine in Figure 23.13.

The *mucosa* of the large intestine consists of simple columnar epithelium with numerous *goblet cells* (Figure 23.14). The *submucosa* is similar to that in the rest of the gastrointestinal tract. The *muscularis* consists of an outer longitudinal layer and an inner circular layer of smooth muscle. However, the longitudinal layer is not continuous; it is broken up into three flat bands, the

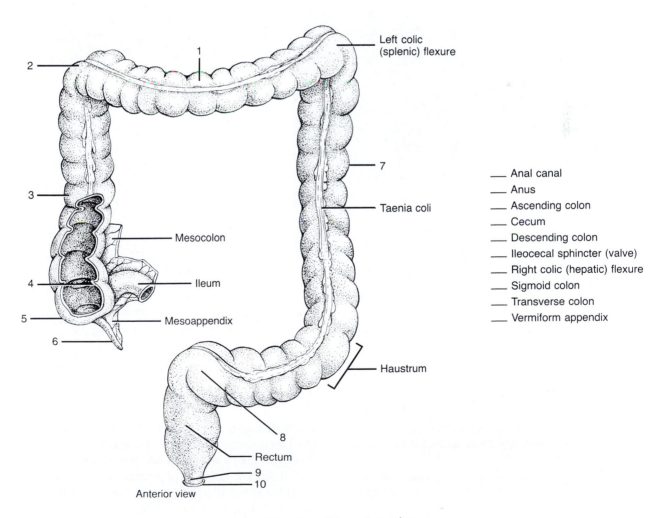

___ Anal canal
___ Anus
___ Ascending colon
___ Cecum
___ Descending colon
___ Ileocecal sphincter (valve)
___ Right colic (hepatic) flexure
___ Sigmoid colon
___ Transverse colon
___ Vermiform appendix

FIGURE 23.13 Large intestine.

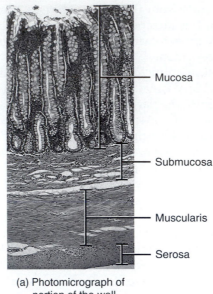

Mucosa

Submucosa

Muscularis

Serosa

(a) Photomicrograph of
portion of the wall

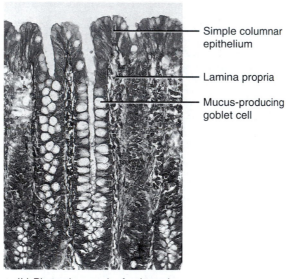

Simple columnar
epithelium

Lamina propria

Mucus-producing
goblet cell

(b) Photomicrograph of enlarged
aspect of mucosa

FIGURE 23.14 Histology of the large intestine.

taeniae coli (TĒ-nē-a KŌ-lī), which gather the colon into a series of pouches called *haustra* (HAWS-tra). Singular is *haustrum.* (see Figure 23.13). The *serosa* of the large intestine is visceral peritoneum.

Examine a prepared slide of the large intestine showing its tunics. Compare your observations to Figure 23.14.

C. DEGLUTITION

Swallowing, or *deglutition* (dē-gloo-TISH-un), is the mechanism that moves food from the mouth through the esophagus to the stomach. It is facilitated by saliva and mucus and involves a complex series of actions by the mouth, pharynx, and esophagus. Swallowing can be initiated voluntarily, but the remainder of the activity is almost completely under reflex control. Swallowing is a very orderly sequence of events involving the movement of food from the mouth to the stomach. It simultaneously involves the inhibition of respiratory activity and the prevention of the entrance of foreign particles into the trachea.

Swallowing is divided into three phases: (1) the *oral,* or *voluntary, phase,* when a *bolus* (soft flexible mass of food) is moved into the oropharynx; (2) the *pharyngeal phase,* involving the movement of the bolus through the pharynx into the esophagus; and (3) the *esophageal phase,* which involves the movement of the bolus from the esophagus, through the lower esophageal (gastroesophageal) sphincter, and into the stomach. Both the pharyngeal and esophageal phases are involuntary.

In the mouth, food is mixed with saliva and is shaped into a bolus as a result of the chewing action of the teeth and the muscular movements of the tongue. The oral, or voluntary, phase of swallowing starts when the tip of the tongue separates a bolus from the larger food mass. The bolus is then forced to the back of the mouth cavity and into the oropharynx by the movement of the tongue against the hard palate. Once the bolus is forced into the oropharynx, tactile receptors are stimulated that initiate the swallowing reflex.

The reflex portion of swallowing is controlled by a *swallowing (deglutition) center* found within the medulla and the lower portion of the pons. Tactile receptors within the pharynx initiate action potentials that are transmitted to the swallowing center. Motor impulses are then transmitted to the muscular portions of the pharynx and esophagus.

The *pharyngeal phase* of swallowing involves an orderly sequence of events, starting with the pulling of the soft palate and uvula superiorly. In addition, the palatopharyngeal folds move closer toward one another, thereby preventing the movement of food into the nasopharynx. Next, the epiglottis covers the opening of the larynx as a result of the larynx moving anteriorly and superiorly, thereby preventing the movement of food into the larynx and trachea. The movement of the larynx also pulls the vocal folds together, closing the rima glottidis, further sealing off the respiratory tract and widening the opening between the laryngopharynx and esophagus. As the bolus of food moves more posteriorly in the oropharynx, the superior con-

strictor muscles of the pharynx relax, thereby allowing food to pass from the oropharynx into the pharynx. As the bolus enters the pharynx, rhythmic contractions within the muscular layers of the pharynx propel the food inferiorly and down the pharynx. As the bolus passes through the laryngopharynx and enters the esophagus, the respiratory passageways reopen and breathing resumes.

As food enters the esophagus and the *esophageal phase* is initiated, peristaltic contractions occur within the esophagus, thereby propelling food inferiorly toward the stomach. *Peristalsis* (per'-i-STAL-sis) consists of involuntary muscular movements and is a function of the muscularis. It is controlled by the swallowing center. In the section of the esophagus lying just above and around the top of the bolus, the inner circular muscle fibers contract. The contraction constricts the esophageal wall and squeezes the bolus downward. Meanwhile, the longitudinal muscle fibers lying around the bottom of and just below the bolus also begin to contract. Contraction of the longitudinal fibers shortens this lower section, pushing its walls outward so it can receive the bolus. The contractions are repeated in a wave that moves down the esophagus, pushing the food toward the stomach. Passage of the bolus is further facilitated by glands secreting mucus. The passage of solid or semisolid food from the mouth to the stomach takes 4 to 8 sec. Very soft foods and liquids pass through in about 1 sec.

PROCEDURE

1. Chew a cracker and note the movements of your tongue as you begin to swallow. Chew another cracker, lean over your chair, placing your mouth below the level of your pharynx, and attempt to swallow again.
2. Note the movements of your laboratory partner's larynx during the pharyngeal stage of deglutition.
3. *Clean the earpieces of a stethoscope with alcohol.* Using the stethoscope, listen to the sounds that occur when your partner swallows water. Place the stethoscope just to the left of the midline at the level of the sixth rib. You should hear the sound of water as it makes contact with the lower esophageal sphincter (valve), followed by the sound of water passing through the sphincter after the sphincter relaxes. The sphincter is at the junction of the esophagus and stomach.

4. Record your observations in Section C of the LABORATORY REPORT RESULTS at the end of the exercise. At what point can deglutition be stopped voluntarily? _____

_____

D. CHEMISTRY OF DIGESTION

In order to understand how food is digested, let us examine the various processes that occur between absorption and the final utilization of nutrients by the body's cells.

Nutrients are chemical substances in food that provide energy, act as building blocks to form new body components, or assist body processes. The six major classes of nutrients are carbohydrates, lipids, proteins, minerals, vitamins, and water. Cells break down carbohydrates, lipids, and proteins to release energy, or use them to build new structures and new regulatory substances, such as hormones and enzymes.

Enzymes are produced by living cells to catalyze or speed up many of the reactions in the body. Enzymes consist of proteins and act as *catalysts* to speed up reactions without being permanently altered by the reaction. Digestive enzymes function as catalysts to speed chemical reactions in the gastrointestinal tract.

Because digestive enzymes function outside the cells that produce them, they are capable of also reacting within a test tube and therefore provide an excellent means of studying enzyme activity.

CAUTION! *Please reread Section A, "General Safety Precautions and Procedures," on page xiii, and Section C, "Precautions Related to Working with Reagents," on page xiv, at the beginning of the laboratory manual, before you begin any of the following experiments. Also, read the experiments before you perform them to be sure that you understand all the procedures and safety precautions.*

1. Positive Tests for Sugar and Starch

Salivary amylase is an enzyme produced by the salivary glands. This enzyme starts starch digestion and *hydrolyzes* (splits using water) it into maltose (a disaccharide), maltotriose (a trisaccharide) and α-dextrines. We will measure the amount of starch and sugar present before and after enzymatic activity. It is expected that the amount of starch will *decrease* and the sugar level will *increase* as a result of salivary amylase activity.

a. TEST FOR SUGAR
Benedict's test is commonly used to detect sugars. Glucose (monosaccharide), maltose (disaccharide), or any other reducing sugars react with *Benedict's solution*, forming insoluble red cuprous oxide. The precipitate of cuprous oxide can usually be seen in the bottom of the tube when standing. Benedict's solution turns green, yellow, orange, or red, depending on the amount of reducing sugar present, according to the following scale:

blue (−)
green (+)
yellow (++)
orange (+++)
red (++++)

Test for the presence of sugar (maltose) as follows:

PROCEDURE
1. Using separate medicine droppers, place 2 ml of maltose and 2 ml of Benedict's solution in a Pyrex test tube.
2. *Using a test tube holder,* place the test tube in a boiling water bath and heat for 5 min.

CAUTION! *Make sure that the mouth of the test tube is pointed away from you and everyone else in the area.* Note the color change.

3. *Using a test tube holder,* remove the test tube from the water bath.
4. Repeat the same procedure using a starch solution instead of maltose. Notice that because the solution contains no sugar, the color does not change. Now you have a method for detecting sugar.

b. TEST FOR STARCH
Lugol's solution is a brown-colored iodine solution used to test certain polysaccharides, especially starch. Starch, for example, gives a *deep blue* to *black* color with Lugol's solution (the black is really a concentrated blue color). Cellulose, monosaccharides, and disaccharides do not react. A negative test is indicated by a yellow to

brown color of the solution itself, or possibly some other color (other than blue to black), a result of the pigments present in the substance being tested.

PROCEDURE

1. Using separate medicine droppers, place a drop of starch solution on a spot plate and then add a drop of Lugol's solution to it. Notice the black color that forms as the starch-iodine complex develops.
2. Repeat the test using a maltose solution in place of the starch solution. Note that there is no color change because Lugol's solution and maltose do not combine. Now you have a method for detecting starch.

C. DIGESTION OF STARCH

PROCEDURE

1. Using separate medicine droppers, transfer 3 ml of a starch solution to a small beaker. Add a fresh enzyme solution consisting of a pinch of amylase powder in 3 ml of water. Mix thoroughly with a glass rod.
2. Wait for 1 min, record the time, remove one drop of the mixture with a glass rod to the depression of a spot plate, and then test for starch with Lugol's solution.
3. At 1-min intervals, test 1-drop samples of the mixture until you no longer note a positive test for starch. Keep the glass rod in the mixture, stirring it from time to time.
4. After the starch test is seen to be negative, test the remaining mixture for the presence of glucose. Do this by adding 2 ml of the mixture with a medicine dropper to 2 ml of the Benedict's solution in a Pyrex test tube. Heat it in a boiling water bath as per the procedure outlined in Step a., "Test for Sugar." Answer questions (a) through (c) in Section D.1 of the LABORATORY REPORT RESULTS at the end of the exercise.

2. Effect of Temperature on Starch Digestion

In the following procedure, you will test starch digestion at five different temperatures to determine how temperature influences enzyme activity. Lugol's solution is again used for the presence or absence of starch.

Again use a fresh enzyme solution consisting of a pinch of amylase powder in 3 ml of water.

Five constant temperature water baths should be available. Starting with the lowest temperature, these are 0°C or cooler, 10°C, 40°C, 60°C, and boiling.

PROCEDURE

1. Prepare 10 test Pyrex tubes, 5 containing 1 ml each of enzyme solution and 5 containing 1 ml each of starch solution.
2. Using rubber bands, pair the tubes (i.e., a tube containing enzyme with one containing starch) and use a test tube holder to place one pair into each water bath.

CAUTION! *Make sure that the mouth of the test tube is pointed away from you and, everyone else in the area.*

3. Permit the tubes to adapt to the bath temperatures for about 5 min, then mix the enzyme and starch solutions of each pair together. *Using a test tube holder,* place the single test tube in its respective water bath.
4. After 30 sec, *using a test tube holder,* remove the test tubes from the water baths and test all five tubes for starch on a spot plate, using Lugol's solution.
5. Repeat every 30 secs until you have determined the time required for the starch to disappear (i.e., to be digested).
6. Record and graph your results in Section D.2 of the LABORATORY REPORT RESULTS at the end of the exercise.

3. Effect of pH on Starch Digestion

You can demonstrate the effectiveness of salivary amylase digestion at different pH readings.

PROCEDURE

1. Prepare three buffer solutions as follows:

    Solution A   pH 4.0
    Solution B   pH 7.0
    Solution C   pH 9.0

2. Once again, use a fresh enzyme solution consisting of a pinch of amylase powder in 3 ml of water.
3. Using medicine droppers, mix 4 ml of a starch solution with 2 ml of buffer solution A in a Pyrex test tube.
4. Repeat this procedure with buffer solutions B and C.

5. You now have three test tubes of a starch-buffer solution, each at a different pH (4.0, 7.0, and 9.0).
6. Using separate medicine droppers, place one drop of starch-buffer solution A on a spot plate and immediately add one drop of the saliva.
7. Test for starch disappearance using Lugol's solution. Record the time when the starch first disappears completely.
8. Repeat this test for the other two starch buffers (solutions B and C) and record the time when starch is no longer present at each pH.
9. Record and explain your results in Section D.3 of the LABORATORY REPORT RESULTS at the end of the exercise.

4. Action of Bile on Fats

Bile is important in the process of digestion because of its emulsifying effects (breaking down of large globules to smaller, uniformly distributed particles) on fats and oils. *Bile does not contain any enzymes.* Emulsification of lipids by means of bile salts serves to increase the surface area of the lipid that will be exposed to the action of the lipase.

PROCEDURE

1. Place 5 ml of water into one Pyrex test tube and 5 ml of bile solution into a second one.
2. Using a medicine dropper, add 1 drop of vegetable oil that has been colored with a fat-soluble dye, such as Sudan B, into each tube.
3. Place a stopper in each tube. Shake them *vigorously*, and then let them stand in a test tube rack, undisturbed, for 10 min. Fat or oil that is broken into sufficiently small droplets will remain suspended in water in the form of an *emulsion*. If emulsification does not occurr, the fat or oil will lie on the surface of the water.
4. Answer the questions in Section D.4 of the LABORATORY REPORT RESULTS at the end of the exercise.

5. Digestion of Fats

You can demonstrate the effect of pancreatic juice on fat by the use of pancreatin, which contains all the enzymes present in pancreatic juice. Because the optimum pH of the pancreatic enzymes ranges from 7.0 to 8.8, the pancreatin is prepared in sodium carbonate. The enzyme used in this test is *pancreatic lipase,* which digests fat to fatty

acids and glycerol. The fatty acid produced changes the color of *blue* litmus to *red.*

PROCEDURE

1. Using a medicine dropper, place 5 ml of litmus cream (heavy cream to which powdered litmus has been added to give it a blue color) in a Pyrex test tube. *Using a test tube holder*, place the test tube in a 40°C water bath.
2. Repeat the procedure with another 5-ml portion in a second tube, but put it in an ice bath.
3. When the tubes have adapted to their respective temperatures (in about 5 min), use a medicine dropper to add 5 ml of pancreatin to each tube. *Using a test tube holder*, replace them in their water baths until a color change occurs in one tube.
4. Summarize your results and your explanation in Section D.5 of the LABORATORY REPORT RESULTS at the end of the exercise.

6. Digestion of Protein

Now you will demonstrate the effect of pepsin on protein and the factors affecting the rate of action of pepsin. *Pepsin,* a proteolytic enzyme, is secreted in inactive form (pepsinogen) by the chief (zymogenic) cells in the lining of the stomach. Pepsin digests proteins (fibrin in this experiment) to peptides. The efficiency of pepsin activity depends on the pH of the solution, the optimum being 1.5 to 2.5. Pepsin is almost completely inactive in neutral or alkaline solutions.

PROCEDURE

1. Prepare and number the following five Pyrex test tubes. For this test, the quantity of each solution must be *measured carefully.*

Tube 1: 5 ml of 0.5% pepsin; 5 ml of 0.8% HCl
Tube 2: 5 ml of pepsin; 5 ml of water
Tube 3: 5 ml of pepsin, boiled for 10 min in a water bath; 5 ml of 0.8% HCl

CAUTION! *Make sure that the mouth of the test tube is pointed away from you and everyone else in the area. Using a test tube holder, remove the test tube from the water bath.*

Tube 4: 5 ml of pepsin; 5 ml of 0.5% NaOH
Tube 5: 5 ml of water; 5 ml of 0.8% HCl

2. First determine the approximate pH of each test tube using Hydrion paper (range 1 to 11).

Enter the values in the table provided in Section D.6 of the LABORATORY REPORT RESULTS at the end of the exercise.

3. Using forceps, place a small amount of fibrin (the protein) in each test tube. An amount near the size of a pea will be sufficient. *Using a test tube holder,* put the tubes in a 40°C water bath and *carefully* shake them occasionally. Maintain the 40°C temperature closely. The tubes should remain in the water bath for *at least 1½ hr.*

4. Watch the changes the fibrin undergoes. Do not confuse the swelling that occurs in some tubes with digestion. Digested fibrin becomes transparent and disappears (dissolves) as the protein is digested to soluble peptides.

5. Finish the experiment when the fibrin is digested in one of the five tubes.

6. Record all your observations in the table provided in Section D.6 of the LABORATORY REPORT RESULTS at the end of the exercise.

ANSWER THE LABORATORY REPORT QUESTIONS AT THE END OF THE EXERCISE.

Digestive System

STUDENT _____   DATE _____

LABORATORY SECTION _____   SCORE/GRADE _____

SECTION C. DEGLUTITION

1. Describe the tongue movement at the beginning of deglutition: _____

_____

2. Describe swallowing with the head below the pharynx: _____

_____

3. Describe laryngeal movements during pharyngeal deglutition: _____

_____

4. What is the time lapse between two sounds of deglutition? _____

_____

SECTION D. CHEMISTRY OF DIGESTION

1. Digestion of Starch

a. How long did it take for the starch to be digested? _____

b. Did your observation indicate the presence of maltose? _____

c. What is the meaning of this result? _____

2. Effect of Temperature on Starch Digestion

d. Record the time required for starch to disappear at the temperatures tested:

_____ 0°C   _____ 40°C   _____ Boiling

_____ 10°C   _____ 60°C

e. What is the optimum temperature for starch digestion? _____

f. What happens to amylase when it is boiled? _____

g. Is the effect of boiling reversible or irreversible? _____

h. Is the effect of freezing reversible or irreversible? _____

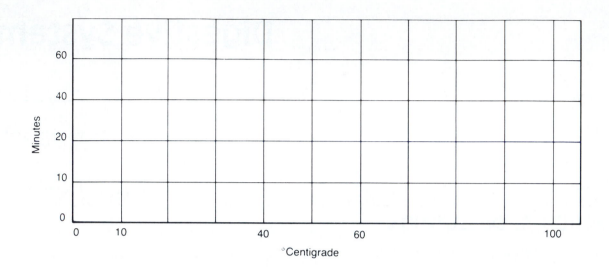

3. Effect of pH on Starch Digestion

i. Record the time required for starch to disappear at the pH readings tested:

Solution A: pH 4.0 _____

Solution B: pH 7.0 _____

Solution C: pH 9.0 _____

j. Explain your results

A: pH 4.0 _____

B: pH 7.0 _____

C: pH 9.0 _____

k. What is the optimum pH for the action of salivary amylase? _____

_____

4. Action of Bile on Fats

l. What difference can you detect in the appearance of the mixtures in the two test tubes? _____

_____

m. How does emulsification of lipids by means of bile aid in the digestion of fat? _____

_____

5. Digestion of Fats

Record the results of the digestion of fats by pancreatic lipase in pancreatic juice in the following table.

Tube no.	Temperature of water bath	Change in pH (color)	Explanation of results
1			
2			

6. Digestion of Protein

Record the results of the digestion of protein by pepsin in the following table:

Tube no.	Tube contents	pH	Digestion observed (yes or no)	Explanation of results
1				
2				
3				
4				
5				

Digestive System

STUDENT _____   DATE _____

LABORATORY SECTION _____   SCORE/GRADE _____

PART 1. Multiple Choice

_____ 1. If an incision has to be made in the small intestine to remove an obstruction, the first layer of tissue to be cut is the (a) muscularis (b) mucosa (c) serosa (d) submucosa

_____ 2. Mesentery, lesser omentum, and greater omentum are all directly associated with the (a) peritoneum (b) liver (c) esophagus (d) mucosa of the gastrointestinal tract

_____ 3. Chemical digestion of carbohydrates is initiated in the (a) stomach (b) small intestine (c) mouth (d) large intestine

_____ 4. A tumor of the villi and circular folds would interfere most directly with the body's ability to carry on (a) absorption (b) deglutition (c) mastication (d) peristalsis

_____ 5. The main chemical activity of the stomach is to begin the digestion of (a) fats (b) proteins (c) carbohydrates (d) all of the above

_____ 6. Surgical cutting of the lingual frenulum would occur in which part of the body? (a) salivary glands (b) esophagus (c) nasal cavity (d) tongue

_____ 7. To free the small intestine from the posterior abdominal wall, which of the following would have to be cut? (a) mesocolon (b) mesentery (c) lesser omentum (d) falciform ligament

_____ 8. The cells of gastric glands that produce secretions directly involved in chemical digestion are the (a) mucous (b) parietal (c) chief (d) pancreatic islets (islets of Langerhans)

_____ 9. An obstruction in the hepatopancreatic ampulla (ampulla of Vater) would affect the ability to transport (a) bile and pancreatic juice (b) gastric juice (c) salivary amylase (d) intestinal juice

_____ 10. The terminal portion of the small intestine is known as the (a) duodenum (b) ileum (c) jejunum (d) pyloric sphincter (valve)

_____ 11. The portion of the large intestine closest to the liver is the (a) right colic flexure (b) rectum (c) sigmoid colon (d) left colic flexure

_____ 12. The lamina propria is found in which coat? (a) serosa (b) muscularis (c) submucosa (d) mucosa

_____ 13. Which structure attaches the liver to the anterior abdominal wall and diaphragm? (a) lesser omentum (b) greater omentum (c) mesocolon (d) falciform ligament

_____ 14. The opening between the oral cavity and pharynx is called (a) vermilion border (b) fauces (c) vestibule (d) lingual frenulum

_____ 15. All of the following are parts of a tooth *except* the (a) crown (b) root (c) cervix (d) papilla

_____ **16.** Cells of the liver that destroy worn-out white and red blood cells and bacteria are termed (a) hepatic cells (b) stellate reticuloendothelial (Kupffer) cells (c) alpha cells (d) beta cells

_____ **17.** Bile is manufactured by which cells? (a) alpha (b) beta (c) hepatic (d) stellate reticuloendothelial (Kupffer)

_____ **18.** Which part of the small intestine secretes the intestinal digestive enzymes? (a) intestinal glands (b) duodenal (Brunner's) glands (c) lacteals (d) microvilli

_____ **19.** Structures that give the colon a puckered appearance are called (a) taenia coli (b) villi (c) rugae (d) haustra

PART 2. Completion

20. An acute inflammation of the serous membrane lining the abdominal cavity and covering the abdominal viscera is referred to as _____.

21. The _____ a sphincter (valve) between the ileum and large intestine.

22. The portion of the small intestine that is attached to the stomach is the _____.

23. The portion of the stomach closest to the esophagus is the _____.

24. The _____ forms the floor of the oral cavity and is composed of skeletal muscle covered with mucous membrane.

25. The convex lateral border of the stomach is called the _____.

26. The three special structures found in the wall of the small intestine that increase its efficiency in absorbing nutrients are the villi, circular folds, and _____.

27. The three pairs of salivary glands are the parotids, submandibulars, and _____.

28. The small intestine is divided into three segments: duodenum, ileum, and _____.

29. The large intestine is divided into four main regions: the cecum, colon, rectum, and _____.

30. The enzyme that is present in saliva is called _____.

31. The transition of the lips where the outer skin and inner mucous membrane meet is called the _____.

32. The _____ papillae are arranged in the form of an inverted V on the posterior surface of the tongue.

33. The portion of a tooth containing blood vessels, lymphatic vessels, and nerves is the _____.

34. The teeth present in permanent dentition, but not in deciduous dentition, that replace the deciduous molars are the _____.

35. The portion of the gastrointestinal tract that conveys food from the pharynx to the stomach is the _____.

36. The inferior region of the stomach connected to the small intestine is the _____.

37. The clusters of cells in the pancreas that secrete digestive enzyme are called _____.

38. The caudate lobe, quadrate lobe, and central vein are all associated with the

    _____.

39. The common bile duct is formed by the union of the common hepatic duct and

    _____.

40. The pear-shaped sac that stores bile is the _____.

41. _____ glands of the small intestine secrete an alkaline substance to protect the mucosa from excess acid.

42. The _____ attaches the large intestine to the posterior abdominal wall.

43. The _____ is the last 20 cm (7 to 8 in.) of the gastrointestinal tract.

44. A midline fold of mucous membrane that attaches the inner surface of each lip to its corresponding

    gum is the _____.

45. The bonelike substance that gives teeth their basic shape is called _____.

46. The _____ anchors teeth in position and helps to dissipate chewing forces.

47. The portion of the colon that terminates at the rectum is the _____ colon.

48. The palatine tonsils are between the palatoglossal and _____ arches.

49. The teeth closest to the midline are the _____.

50. The vermiform appendix is attached to the _____.

PART 3. Matching

_____ 51. Pancreatic lipase

_____ 52. Benedict's solution

_____ 53. Lugol's solution

_____ 54. Salivary amylase

_____ 55. Pepsin

A. Commonly used solution in the test for starch

B. Capable of digesting starch

C. Capable of digesting protein

D. Commonly used solution for detecting reducing sugars

E. Digests fat to fatty acids and glycerol and changes the color of blue litmus to red

Urinary System

The *urinary system* functions to keep the body in homeostasis by controlling the composition and volume of the blood. The system accomplishes these functions by removing and restoring selected amounts of water and various solutes. The kidneys also excrete selected amounts of various wastes, assume a role in erythropoiesis by forming renal erythropoietic factor, help control blood pH, help regulate blood pressure by secreting renin (which activates the renin-angiotensin pathway), and participate in the activation of vitamin D.

The urinary system consists of two kidneys, two ureters, one urinary bladder, and a single urethra (see Figure 24.1). Other systems that help in waste elimination are the respiratory, integumentary, and digestive systems.

Using your textbook, charts, or models for reference, label Figure 24.1.

A. ORGANS OF URINARY SYSTEM

1. Kidneys

The paired *kidneys,* which resemble kidney beans in shape, are found just above the waist between the parietal peritoneum and the posterior wall of the abdomen. Because they are external to the peritoneal lining of the abdominal cavity, they are referred to as *retroperitoneal* (re′-trō-per-i-tō-NĒ-al). The kidneys are positioned between T12 and L3, with the right kidney slightly lower than the left because of the position

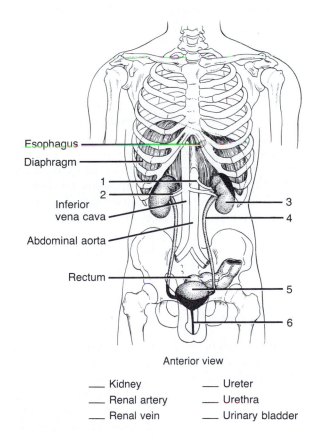

Esophagus
Diaphragm
1
2
Inferior vena cava
Abdominal aorta
Rectum
3
4
5
6

Anterior view

___ Kidney   ___ Ureter
___ Renal artery   ___ Urethra
___ Renal vein   ___ Urinary bladder

FIGURE 24.1 Organs of the male urinary system and associated structures.

of the liver. Near the center of the kidney's concave border, which faces the vertebral column, is a notch called the *hilus,* through which the ureter leaves the kidney and blood and lymphatic vessels and nerves enter and exit the

481

kidney. The hilus is the entrance to a cavity in the kidney called the *renal sinus.* Three layers of tissue surround each kidney: the *renal capsule, adipose capsule,* and *renal fascia.* They function to protect the kidney and hold it firmly in place.

If a coronal (frontal) section is made through a kidney, the following structures can be seen:

a. *Cortex*—Outer, narrow, reddish area.
b. *Medulla* —Inner, wide, reddish-brown area.
c. *Renal (medullary) pyramids*—Striated triangular structures, 8 to 18 in number, in the medulla. The bases of the pyramids face the cortex, and the apices, called *renal papillae,* are directed toward the center of the kidney.
d. *Renal columns*—Cortical substance between the renal pyramids.
e. *Renal pelvis*—Large cavity in the renal sinus, enlarged proximal portion of the ureter.
f. *Major calyces* (KĀ-li-sēz)—Consist of 2 or 3 cuplike extensions of the renal pelvis.
g. *Minor calyces*—Consist of 8 to 18 cuplike extensions of the major calyces.

Within the cortex and renal pyramids of each kidney are more than one million microscopic units called nephrons, the functional units of the kidneys (described shortly). As a result of their activity in regulating the volume and chemistry of the blood, they produce urine. Urine passes from the nephrons to the minor calyces, major calyces, renal pelvis, ureter, urinary bladder, and urethra.

Examine a specimen, model, or chart of the kidney and with the aid of your textbook, label Figure 24.2.

2. Nephrons

Basically, a *nephron* (NEF-ron) consits of a renal corpuscle (KOR-pus-sul) where fluid is filtered and a renal tubule into which the filtered fluid passes. A *renal corpuscle* has two components: a tuft (knot) of capillaries called a *glomerulus* (glō-MER-voo-lus) surrounded by a double-walled epithelial cup, called a *glomerular (Bow-man's) capsule,* lying in the cortex of the kidney. The inner wall of the capsule, the *visceral layer,* consists of epithelial cells called *podocytes* and surrounds the glomerulus. A space separates the visceral layer from the outer wall of the capsule, the *parietal layer,* which is composed of simple squamous epithelium.

The visceral layer of the glomerular (Bow-

man's) capsule and endothelium of the glomerulus form an *endothelial-capsular membrane,* a very effective filter. Electron microscopy has determined that the membrane consists of the following components, given in the order in which substances are filtered (Figure 24.3 on page 484).

a. *Endothelium of glomerulus*—Single layer of endothelium with pores averaging 50 to 100 nm in diameter. It restricts the passage of blood cells.
b. *Basement membrane of glomerulus*—Extracellular membrane that consists of fibrils in a glycoprotein matrix. It restricts the passage of large sized proteins.
c. *Epithelium of visceral layer of the glomerular (Bowman's) capsule*—Epithelial cells are called *podocytes* and consist of footlike structures called *pedicels* (PED-i-sels) arranged parallel to the circumference of the glomerulus. Pedicels cover the basement membrane of the glomerulus, except for spaces between them called *filtration slits (slit pores).* Pedicels contain numerous thin, contractile filaments that are believed to regulate the passage of substances through the filtration slits. In addition, another factor that helps to regulate the passage of substances through the filtration slits is a thin membrane, the *slit membrane,* that extends between filtration slits. This membrane restricts the passage of intermediate sized proteins.

The endothelial-capsular membrane filters blood passing through the kidney. Blood cells and large molecules, such as proteins, are retained by the filter and eventually are recycled into the blood. The filtered substances pass through the membrane and into the space between the parietal and visceral layers of the glomerular (Bowman's) capsule and then enter the renal tubule (described shortly).

As the filtered fluid (filtrate) passes through the remaining parts of a nephron, substances are selectively added and removed. The end product of these activities is urine. Nephrons are frequently classified into two kinds. A *cortical nephron* usually has its glomerulus in the outer cortical zone, and the remainder of the nephron rarely penetrates the medulla. A *juxtamedullary nephron* usually has its glomerulus close to the corticomedullary junction, and other parts of the nephron penetrate deeply into the medulla

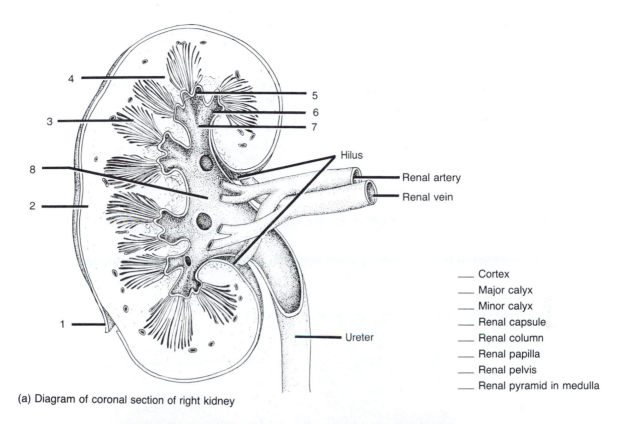

4
5
6
7
3
8
2
1

Hilus

Renal artery

Renal vein

Ureter

____ Cortex
____ Major calyx
____ Minor calyx
____ Renal capsule
____ Renal column
____ Renal papilla
____ Renal pelvis
____ Renal pyramid in medulla

(a) Diagram of coronal section of right kidney

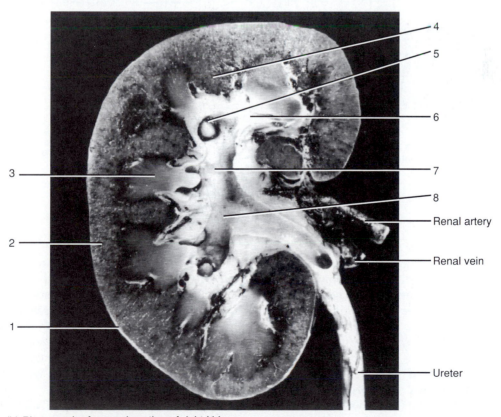

4
5
6
7
8
3
2
1

Renal artery

Renal vein

Ureter

(b) Photograph of coronal section of right kidney

FIGURE 24.2   Kidney.

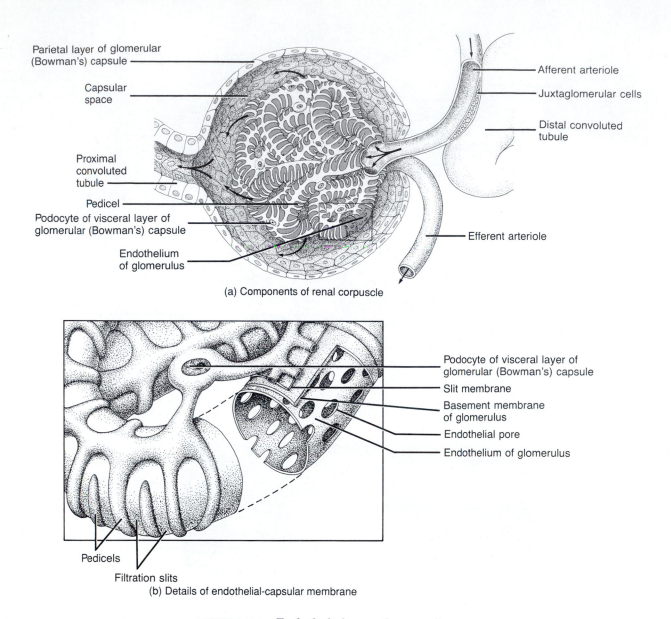

Parietal layer of glomerular
(Bowman's) capsule

Capsular
space

Proximal
convoluted
tubule

Pedicel

Podocyte of visceral layer of
glomerular (Bowman's) capsule

Endothelium
of glomerulus

Afferent arteriole

Juxtaglomerular cells

Distal convoluted
tubule

Efferent arteriole

(a) Components of renal corpuscle

Podocyte of visceral layer of
glomerular (Bowman's) capsule

Slit membrane

Basement membrane
of glomerulus

Endothelial pore

Endothelium of glomerulus

Pedicels

Filtration slits

(b) Details of endothelial-capsular membrane

FIGURE 24.3   Endothelial-capsular membrane.

(see Figure 24.4 on page 485). The following description of the remaining components of a nephron applies to juxtamedullary nephrons.

After the filtrate leaves the glomerular (Bowman's) capsule, it passes through the following parts of a renal tubule:

a. **Proximal convoluted tubule (PCT)**—Coiled tubule in the cortex that originates at the glomerular (Bowman's) capsule; consists of simple cuboidal epithelium with microvilli.

b. **Loop of Henle (nephron loop)**—U-shaped tubule that connects the proximal and distal convoluted tubules. It consists of a *descending*

limb of the loop of Henle, an extension of the proximal convoluted tubule that dips down into the medulla and consists of simple squamous epithelium and an *ascending limb of the loop of Henle* that ascends in the medulla and approaches the cortex and consists of simple squamous, cuboidal, and columnar epithelium; the ascending limb is wider in diameter than the descending limb.

c. **Distal convoluted tubule (DCT)**—Coiled extension of the ascending limb in the cortex; consists of simple cuboidal epithelium with fewer microvilli than in proximal convoluted tubules.

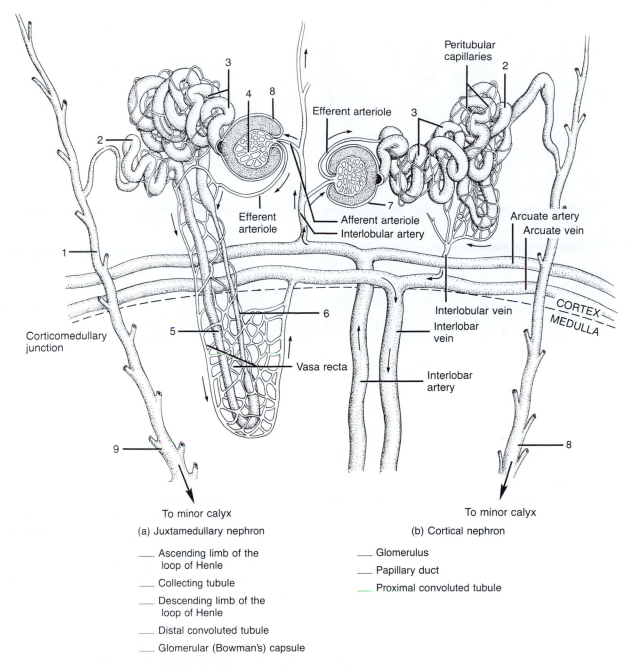

FIGURE 24.4 Nephrons.

Distal convoluted tubules terminate by merging with straight *collecting tubules.* In the medulla, collecting tubules receive distal convoluted tubules from several nephrons, pass through the renal pyramids, and open at the renal papillae into minor calyces through about 30 large *papillary ducts.* The processed filtrate, called urine, passes from the collecting tubules to papillary ducts, minor calyces, major calyces, renal pelvis, ureter, urinary bladder, and urethra.

With the aid of your textbook, label the parts of a nephron and associated structures in Figure 24.4.

Examine prepared slides of various components of nephrons and compare your observations with Figure 24.5.

3. Blood and Nerve Supply

Nephrons are abundantly supplied with blood vessels, and the kidneys actually receive around one-fourth the total cardiac output, or approxi-

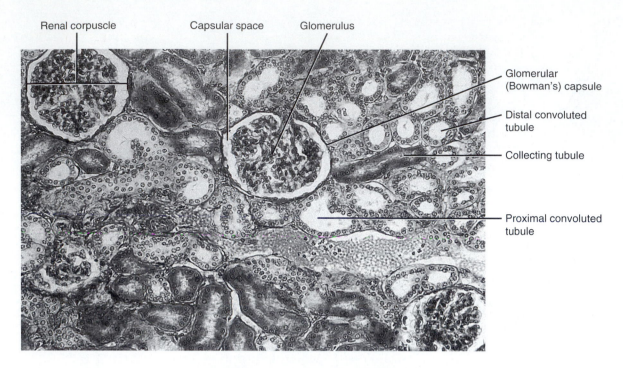

Renal corpuscle   Capsular space   Glomerulus

Glomerular (Bowman's) capsule

Distal convoluted tubule

Collecting tubule

Proximal convoluted tubule

FIGURE 24.5   Histology of nephrons.

mately 1200 ml, every minute. The blood supply originates in each kidney with the *renal artery,* which divides into many branches, eventually supplying the nephron and its complete tubule.

Before or immediately after entering the hilus, the renal artery divides into a larger anterior branch and a smaller posterior branch. From these branches, five *segmental arteries* originate. Each gives off several branches, the *interlobar arteries,* which pass between the renal pyramids in the renal columns. At the bases of the pyramids, the interlobar arteries arch between the medulla and cortex and here are known as *arcuate arteries.* Branches of the arcuate arteries, called *interlobular arteries,* enter the cortex. *Afferent arterioles,* branches of the interlobular arteries, are distributed to the *glomeruli.* Blood leaves the glomeruli via *efferent arterioles.*

The next sequence of blood vessels depends on the type of nephron. Around convoluted tubules, efferent arterioles of cortical nephrons divide to form capillary networks called *peritubular capillaries.* Efferent arterioles of juxtamedullary nephrons also form peritubular capillaries and, in

addition, form long loops of blood vessels around medullary structures called *vasa recta.* Peritubular capillaries eventually reunite to form *interlobular veins.* Blood then drains into *arcuate veins, interlobar veins,* and *segmental veins.* Blood leaves the kidneys through the *renal vein* that exits at the hilus. (The vasa recta pass blood into the interlobular veins, arcuate veins, interlobar veins, and renal veins.)

As the afferent arteriole approaches the renal corpuscle, the smooth muscle fibers (cells) of the tunica media become more rounded and granular and are known as *juxtaglomerular cells.* The cells of the distal convoluted tubule adjacent to the afferent and efferent arterioles become narrower and taller, and are known as the *macula densa.* Together the juxtaglomerular cells and macula densa constitute the *juxtaglomerular apparatus (JGA).* The juxtaglomerular apparatus responds to low renal blood pressure by secreting a substance (renin) that begins a sequence of responses that raises renal pressure back to normal.

Using Figures 24.4 and 24.6 as guides, trace a drop of blood from its entrance into the renal artery to its exit through the renal vein. As you do so, name in sequence each blood vessel through which blood passes for both cortical and juxtaglomerular nephrons.

Label Figure 24.7.

The nerve supply to the kidneys comes from the *renal plexus* of the autonomic system. The nerves are vasomotor because they regulate the circulation of blood in the kidney by regulating the diameters of the arterioles.

4. Ureters

The body has two retroperitoneal *ureters* (YOO-re-ters), one for each kidney; each ureter is a continuation of the renal pelvis and runs to the urinary bladder (see Figure 24.1). Urine is carried through the ureters mostly by peristaltic contractions of the muscular layer of the ureters. Each ureter extends 25 to 30 cm (10 to 12 in.) and enters the urinary bladder at the superior lateral angle of its base.

Histologically, the ureters consist of an inner *mucosa* of transitional epithelium and connective tissue, a middle *muscularis* (inner longitudinal and outer circular smooth muscle), and an outer *fibrous coat*.

Examine a prepared slide of the wall of the ureter showing its various layers. With the aid of your textbook, label Figure 24.8 on page 489.

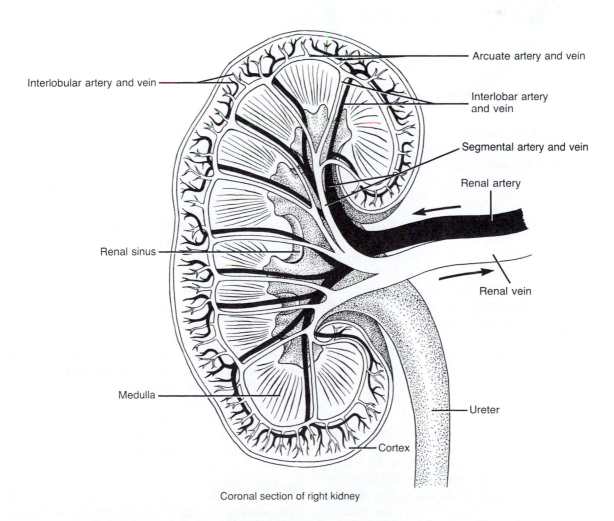

Interlobular artery and vein

Arcuate artery and vein

Interlobar artery and vein

Segmental artery and vein

Renal artery

Renal sinus

Renal vein

Medulla

Ureter

Cortex

Coronal section of right kidney

FIGURE 24.6 Maroscopic blood vessels of the kidney. Macroscopic and microscopic blood vessels are shown in Figure 24.4.

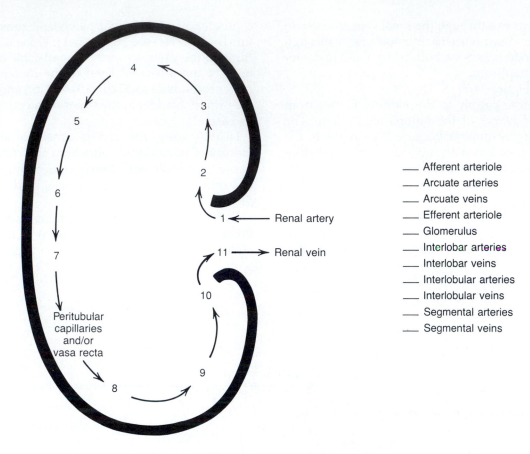

FIGURE 24.7   Blood supply of the right kidney. This view is designed to show the *sequence* of blood flow, not the anatomical location of blood vessels, which is shown in Figure 24.6.

Renal artery

Renal vein

Peritubular capillaries and/or vasa recta

___ Afferent arteriole
___ Arcuate arteries
___ Arcuate veins
___ Efferent arteriole
___ Glomerulus
___ Interlobar arteries
___ Interlobar veins
___ Interlobular arteries
___ Interlobular veins
___ Segmental arteries
___ Segmental veins

5. Urinary Bladder

The *urinary bladder* is a hollow muscular organ located in the pelvic cavity posterior to the pubic symphysis (see Figure 24.1). In the male, the bladder is directly anterior to the rectum; in the female, it is anterior to the vagina and inferior to the uterus.

At the base of the interior of the urinary bladder is the *trigone* (TRĪ-gōn), a triangular area bounded by the opening to the urethra and ureteral openings into the bladder. The *mucosa* of the urinary bladder consists of transitional epithelium and connective tissue that form *rugae.* The *muscularis,* also called the *detrusor* (de-TROO-ser) *muscle,* consists of three layers of smooth muscle: inner longitudinal, middle circular, and outer longitudinal. In the region around the opening to the urethra, the circular muscle fibers form an *internal urethral sphincter.* Below this is the *external urethral sphincter* composed of skeletal muscle. The *serosa* is formed by visceral peritoneum and covers the superior surface of the urinary bladder.

Urine is expelled from the bladder by an act called *micturition* (mik′-too-RISH-un), commonly known as urination or voiding. The average capacity of the urinary bladder is 700 to 800 ml.

Using your textbook as a guide, label the external urethral sphincter, internal urethral sphincter, ureteral openings, and ureters in Figure 24.9 on page 490.

Examine a prepared slide of the wall of the urinary bladder. With the aid of your textbook, label Figure 24.10 on page 491.

6. Urethra

The *urethra* is a small tube leading from the floor of the urinary bladder to the exterior of the body.

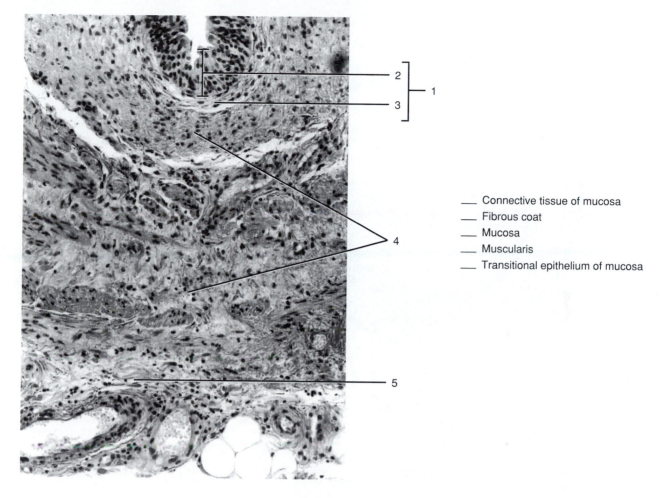

Connective tissue of mucosa
Fibrous coat
Mucosa
Muscularis
Transitional epithelium of mucosa

FIGURE 24.8   Histology of the ureter.

In females, this tube is posterior to the pubic symphysis and is embedded in the anterior wall of the vagina; its length is approximately 3.8 cm (1½ in.). The opening of the urethra to the exterior, the *external urethral orifice,* is between the clitoris and vaginal orifice. In males, its length is around 20 cm (8 in.); it follows a route different from that of the female. Immediately below the urinary bladder, the urethra passes through the prostate gland, pierces the urogenital diaphragm, and traverses the penis. The urethra is the terminal portion of the urinary system and serves as the passageway for discharging urine from the body. In addition, in the male, the urethra serves as the duct through which reproductive fluid (semen) is discharged from the body.

Label the urethra and urethral orifice in Figure 24.9.

B. DISSECTION OF SHEEP (OR PIG) KIDNEY

The sheep kidney is very similar to both human and cat kidney. You may use Figure 24.2 as a reference for this dissection.

CAUTION! *Please reread Section D, "Precautions Related to Dissection," at the beginning of the laboratory manual, on page xv, before you begin your dissection.*

PROCEDURE

1. Examine the intact kidney and notice the hilus and fatty tissue that normally surround the kidney. Strip away the fat.
2. As you peel the fat off, look carefully for the **adrenal (suprarenal) gland.** This gland is usu-

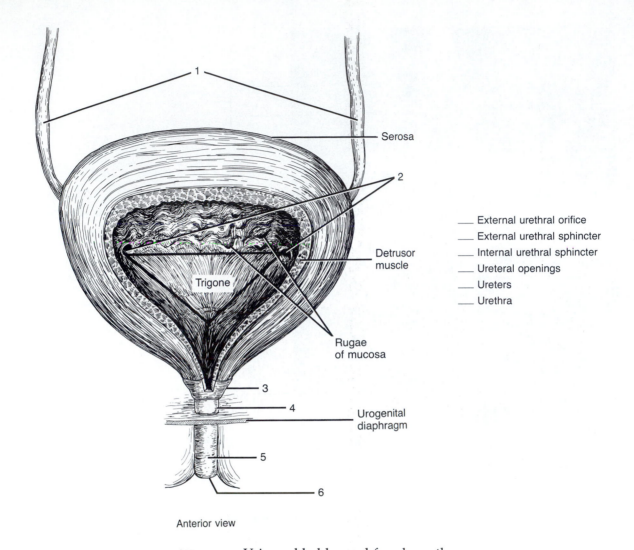

1

Serosa

2

Detrusor
muscle

Trigone

Rugae
of mucosa

3

4

Urogenital
diaphragm

5

6

Anterior view

____ External urethral orifice
____ External urethral sphincter
____ Internal urethral sphincter
____ Ureteral openings
____ Ureters
____ Urethra

FIGURE 24.9   Urinary bladder and female urethra.

ally found attached to the superior surface of the kidney, as it is in the human. Most preserved kidneys do not have this gland. If it is present, remove it, cut it in half, and note its distinct outer *cortex* and inner *medulla.*

3. Look at the *hilus,* which is the concave area of the kidney. From here the *ureter, renal artery,* and *renal vein* enter and exit.

4. Differentiate these blood vessels by examining the thickness of their walls. Which vessel has the thicker wall?

5. With a sharp scalpel, *carefully* make a longitudinal (coronal) section through the kidney.

6. Identify the *renal capsule* as a thin, tough layer of connective tissue completely surrounding the kidney.

7. Immediately beneath this capsule is an outer, light-colored area called the *renal cortex.* The inner dark-colored area is the *renal medulla.*

8. The *renal pelvis* is the large chamber formed by the expansion of the ureter inside the kidney. This renal pelvis divides into many smaller areas called *renal calyces,* each of which has a dark tuft of kidney tissue called a *renal pyramid.*

9. The bases of these pyramids face the cortical area. Their apices, called *renal papillae,* are directed toward the center of the kidney.

10. The calyces collect urine from collecting ducts and drain it into the renal pelvis and out through the ureter.

11. The renal artery divides into several branches

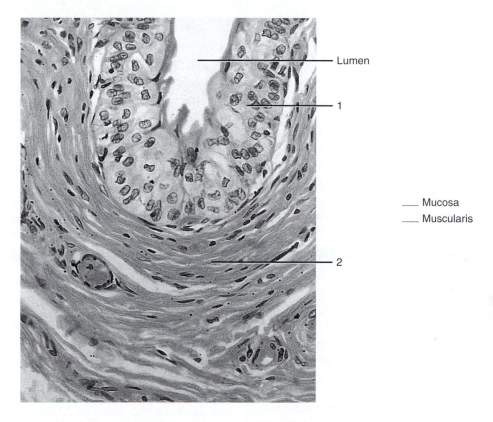

FIGURE 24.10   Histology of the urinary bladder.

that pass between the renal pyramids. These vessels are small and delicate and may be too difficult to dissect and trace through the renal medulla.

C. URINE

The kidneys function to maintain bodily homeostasis. This is accomplished by three processes: (1) filtration of the blood by the glomeruli, (2) tubular reabsorption, and (3) tubular secretion. As a result of these three functions, *urine* is formed and eliminated from the body. Urine contains a high concentration of solutes. In a healthy person, the volume, pH, and solute concentration of urine will vary with the needs of the internal environment. In certain pathological conditions, the characteristics of urine may change drastically. An analysis of the volume and the physical and chemical properties of urine tells us much about the state of the body.

1. Physical Characteristics

Normal urine usually varies between a straw yellow and an amber transparent color and possesses a characteristic odor. Urine color varies considerably according to the ratio of solutes to water and according to an individual's diet.

Cloudy urine sometimes reflects the secretion of mucin from the urinary tract lining and is not necessarily an indicator of a pathological condition. The normal pH of urine is 7.4 but may range between 4.5 and 8.0. The pH of urine is also strongly affected by diet, with a high-protein diet lowering the pH, and a mostly vegetable diet increasing the pH of the urine.

Specific gravity is the ratio of the weight of a volume of a substance to the weight of an equal volume of distilled water. Water has a specific gravity of 1.000. The specific gravity of urine depends on the amount of solids in solution and normally ranges from 1.008 to 1.030. The greater the concentration of solutes, the higher the specific gravity. In certain conditions, such as diabetes mellitus, specific gravity is high because of the high glucose content.

2. Abnormal Constituents

When the body's metabolism becomes abnormal, many substances not normally found in urine may appear in varying amounts, while normal constituents may appear in abnormal amounts. *Urinalysis* is the analysis of the physical and chemical properties of urine and is a vital tool in diagnosing pathological conditions.

a. *Albumin*—Normally absent from a urine sample because the particles are too large to be filtered out of the blood through the endothelial-capsular membrane. When albumin is found in the urine, the condition is called *albuminuria.*

b. *Glucose*—Urine normally contains such small amounts of *glucose* that clinically glucose is considered to be absent from urine samples. Its presence in significant amounts is called *glycosuria,* and the most common cause is a high blood sugar (glucose) level seen in certain diseases, such as diabetes mellitus.

c. *Erythrocytes—Hematuria* is the term utilized to describe the presence of red blood cells in a urine sample. Hematuria usually indicates the presence of a pathological condition within the kidneys.

d. *Leucocytes—Pyuria* is the condition that occurs when white blood cells and other components of pus are found in the urine; this usually indicates a pathological condition.

e. *Ketone bodies*—Normal urine contains a small amount of *ketone (acetone) bodies.* Their appearance in large quantities in the urine produces the condition called *ketosis (acetonuria)* and may indicate physiological abnormalities.

f. *Casts*—Tiny masses of various substances that have hardened and assumed the shape of the lumens of the nephron tubules. They are microscopic in size and are composed of many different substances.

g. *Calculi*—Insoluble *calculi* (stones) are various salts that have solidified in the urinary tract. They may be found anywhere from the kidney tubules to the external opening. Their presence causes considerable pain as they attempt to pass through the various lumens of the urinary system.

D. URINALYSIS

In this exercise you will determine some of the characteristics of urine and perform tests for some abnormal constituents that may be present in urine. Some of these tests are used to determine unknowns in urine specimens.

CAUTION! *Please reread Section A, "General Safety Precautions and Procedures," on page xiii, and Section D, "Precautions Related to Working with Reagents," on page xiv, at the beginning of the laboratory manual before you begin any of the following experiments. Read the experiments before you perform them, to be sure that you understand all the procedures and safety precautions.*

When working with urine, avoid any kind of contact with an open sore, cut, or wound. Wear tight-fitting surgical gloves and safety goggles.

Work with your own urine only. After you have completed your experiments, place all glassware in a fresh household bleach solution or other comparable disinfectant, wash the laboratory tabletop with a fresh household bleach solution or comparable disinfectant, and dispose of the gloves and Chemstrips® in the appropriate biohazard container provided by your instructor.

1. Urine Collection

PROCEDURE

1. A specimen of urine can be collected at any time for routine tests; urine voided within 3 hr after meals, however, may contain abnormal constituents. For this reason, the first voiding in the morning is preferred.

2. Both males and females should collect a midstream sample of urine in a sterile container. A midstream sample is essential to avoid contamination from the external genitalia and to avoid the presence of pus cells and bacteria normally found in the urethra.

3. If not examined immediately, the specimen should be refrigerated to prevent unnecessary bacterial growth.

4. Before testing, *always* mix urine by swirling, inverting the container, or stirring with a wooden swab stick.

5. *Keep all containers clean!*

6. Wrap all papers and sticks and dispose of them in the garbage pail.

7. Rinse all test tubes and glass containers carefully, with *cold water,* after they have cooled.

8. Flush sinks well with cold water.
9. Obtain either your own freshly voided urine sample or one that has been provided.

2. Chemstrip® Testing

Alternate methods can be employed for several of the tests you are about to perform. One method uses plastic strips to which are attached paper squares impregnated with various reagents. These strips display a color reaction when dipped into urine with any abnormal constituents.

PROCEDURE

1. At this point, take a Chemstrip® and test your urine sample for the following: pH, protein, glucose, ketones, bilirubin, and blood (hemoglobin). Record your results below.
2. Determine which component(s) of urine are measured by the Chemstrip® you are using. If you are using a strip that tests for multiple substances, determine which squares measure which substances. Locate the color chart used to read the results. Note the appropriate time for reading each test.
3. Remove a test strip from the vial and *replace the cap.* Dip the test strip into your urine sample for no longer than 1 sec, being sure that all reagents on the strip are immersed.
4. Remove any excess urine from the strip by drawing the edge of the strip along the rim of the container holding your urine sample.
5. After the appropriate time, as indicated on the Chemstrip® vial, hold the strip close to the color blocks on the vial.
6. Make sure that the strip blocks are properly lined up with the color chart on the vial.

Test	Chemstrip® result
pH	_____
Protein	_____
Glucose	_____
Ketones	_____
Bilirubin	_____
Blood	_____

3. Physical Analysis

a. COLOR

Normal urine varies in color from straw yellow to amber because of the pigment *urochrome,* a byproduct of hemoglobin destruction. Observe the color of your urine sample. Some abnormal colors follow:

Color	Possible cause
Silvery, milky	Pus, bacteria, epithelial cells
Smoky brown, rust	Blood
Orange, green, blue, red	Medications or liver disease

Record the color of your urine in Section D.3.a of the LABORATORY REPORT RESULTS at the end of the exercise.
How would sickle-cell anemia affect urine color?

_____

_____

b. TRANSPARENCY

A fresh urine sample should be clear. Cloudy urine may be due to substances such as mucin, phosphates, urates, fat, pus, mucus, microbes, crystals, or epithelial cells.

PROCEDURE

1. To determine transparency, cover the container, shake your urine sample, and observe the degree of cloudiness.
2. Record your observations in Section D.3.b of the LABORATORY REPORT RESULTS at the end of the exercise.

c. PH

The pH of urine varies with several factors already indicated. You can test the pH of your urine by using either a Chemstrip® or pH paper. Because you have already determined the pH of your urine by using a Chemstrip®, you might want to verify the results using pH paper.

PROCEDURE

1. Place a strip of pH paper into your urine sample three consecutive times.
2. Shake off any excess urine.

3. Let the pH paper sit for 1 min and then compare it to the color chart provided.
4. Record your observations in Section D.3.c of the LABORATORY REPORT RESULTS at the end of the exercise.

How would the consumption of antacid (sodium bicarbonate) affect urine pH? _____

_____

_____

d. SPECIFIC GRAVITY

Specific gravity is easily determined using a urinometer (hydrometer). The urinometer is a float with a numbered scale near the top that indicates specific gravity directly.

PROCEDURE

1. Familiarize yourself with the scale on the urinometer neck. Determine the change in specific gravity represented by each calibration.
2. Allow your urine to reach room temperature. Urinometers are calibrated to read the specific gravity at 15°C (69°F). If the temperature differs, add or subtract 0.001 for each 3°C above or below 15°C.
3. Fill the cylinder ¾ full of urine. Insert the urinometer and make sure that it is free floating; if not, spin the neck gently.
4. Read the scale at the bottom of the meniscus when the urinometer is at rest.
5. Record the specific gravity in Section D.3.d of the LABORATORY REPORT RESULTS at the end of the exercise.
6. Rinse the urinometer and cylinder. *Follow your instructor's directions for cleaning them.*

How would dehydration affect the specific gravity of urine? _____

_____

_____

4. Chemical Analysis

a. CHLORIDE AND SODIUM CHLORIDE

Most of the sodium chloride (NaCl) present in the renal filtrate is reabsorbed; a small amount remains as a normal component of urine. The normal value for chloride (Cl^-) is 476 mg/100 ml; for sodium (Na^+) it is 294 mg/100 ml.

PROCEDURE

1. Using a medicine dropper, place 10 drops of urine into a Pyrex test tube.
2. Using a medicine dropper, add 1 drop of 20% potassium chromate and *gently* agitate the tube. It should be a yellow color.
3. Using a medicine dropper, add 2.9% silver nitrate solution *one drop at a time,* counting the drops and *gently* agitating the test tube during the time the solution is being added.
4. Count the number of drops needed to change the color of the solution from bright yellow to brown.
5. Determine the chloride and sodium chloride concentrations of the sample and record your results in Section D.4.a of the LABORATORY REPORT RESULTS at the end of the exercise. Each drop of silver nitrate added in Step 4 is equivalent to (1) 61 mg of Cl^- per 100 ml of urine and (2) 100 mg of NaCl per 100 ml of urine.

Thus, to determine the chloride concentration of the sample, multiply the number of drops times 61 to obtain the number of mg of Cl^- per 100 ml of urine.

To determine the sodium chloride concentration of the sample, multiply the number of drops times 100 to obtain the number of mg of NaCl per 100 ml of urine.

What would a high NaCl content in the urine indicate? _____

_____

_____

b. GLUCOSE

Glycosuria (the presence of glucose in the urine) occurs in patients with diabetes mellitus and other disorders. Traces of glucose may occur in normal urine, but detection of these small amounts requires special tests.

(1) Clinitest® Reagent Method

An alternative method is to use Clinitest tablets.

PROCEDURE

1. Using medicine droppers, place 10 drops of water and 5 drops of urine in a Pyrex test tube.

CAUTION! *Place the test tube in a test tube rack because it will become too hot to handle.*

2. With forceps, add one Clinitest tablet®.

CAUTION! *The concentrated sodium hydroxide in the tablet generates enough heat to make the liquid in the test tube boil. Be sure to point the mouth of the test tube away from you and everyone else in the area.*

3. Fifteen seconds after boiling has stopped, shake the test tube *gently* and compare the color obtained to the color chart. Disregard any color change that occurs after 15 sec.

CAUTION! *Make sure the mouth of the test tube is pointed away from you and others.*

4. Record your results in Section D.4.b of the LABORATORY REPORT RESULTS at the end of the exercise.

How would a high-sugar diet affect the urine?

Explain. _____

_____

_____

(2) Chemstrip® Method

Record your results in Section D.4.b of the LABORATORY REPORT RESULTS at the end of the exercise.

C. PROTEIN

Normal urine contains traces of proteins that are hard to detect through regular laboratory procedures. Albumin is the most abundant serum protein and is the one usually detected. Because tests for albumin are determined by precipitating the protein either by heat (coagulation) or by adding a reagent, the urine sample should either be filtered or centrifuged (see Figure 24.11). The test for protein will be done by the Albutest® reagent method and the Chemstrip® method.

(1) Albutest® Reagent Method

An alternative method is to use Albutest® reagent tablets containing bromphenol blue.

PROCEDURE

1. Using forceps, place an Albutest® tablet on a clean, dry paper towel and add 1 drop of urine.

2. After the drop has been absorbed, add 2 drops of water and allow these to penetrate before reading the color change.

3. Compare the color (in daylight or fluorescent light) on the top of the tablet with the color chart provided in lab.

4. If albumin is present in the urine, a *blue-green* spot will remain on the surface of the tablet after the water is added. The amount of protein is indicated by the intensity of the blue-green color.

5. If the test is negative, the original color of the tablet will not be changed at the completion of the test.

6. Record your results in Section D.4.c of the LABORATORY REPORT RESULTS at the end of the exercise.

Why is protein not normally found in urine?

_____

_____

(2) Chemstrip® Method

Record your results in Section D.4.c of the LABORATORY REPORT RESULTS at the end of the exercise.

d. KETONE (ACETONE) BODIES

The presence of ketone (acetone) bodies in urine is a result of abnormal fat catabolism. Reagents such as sodium nitroprusside, ammonium sulfate, and ammonium hydroxide are available in the form of tablets. Ketones turn purple when added to these chemicals.

(1) Acetest® Tablet Method

PROCEDURE

1. Using forceps, place an Acetest® tablet on a clean, dry paper towel. Using a medicine dropper, place 1 drop of urine on the tablet.

2. If acetone or ketone is present, a *lavender-purple* color develops within 30 sec. If the tablet becomes cream-colored with wetting, the results are negative. Compare the results with the color chart that comes with the reagent.

3. Record your results in Section D.4.d of the LABORATORY REPORT RESULTS at the end of the exercise.

Why would starvation cause ketones? _____

_____

_____

(2) Chemstrip® Method

Record your results in Section D.4.d of the LABORATORY REPORT RESULTS at the end of the exercise.

e. BILE PIGMENTS

Bile pigments (biliverdin and bilirubin) are not normally present in urine. The presence of large quantities of bilirubin in the extracellular fluids produces jaundice, a yellowish tint to the body tissues, including yellowness of the skin and deep tissues.

(1) Ictotest® for Bilirubin

PROCEDURE

1. Using a medicine dropper, place a drop of urine on one square of the special mat provided in the Ictotest® kit.
2. Using forceps, place one Ictotest® reagent tablet in the center of the moistened area.

CAUTION! *Do not touch the tablets with your fingers. Recap the bottle.*

3. Add 1 drop of water directly to the tablet; after 5 sec, add another drop of water to the tablet so that the water runs off onto the mat. Observe the color of the mat around the tablet at 60 sec. The presence of bilirubin will turn the mat *blue* or *purple*. A slight *pink* or *red* color is negative for bilirubin.
4. Record your results in Section D.2.e of the LABORATORY REPORT RESULTS at the end of the exercise.

(2) Chemstrip® Method

Record your results in Section D.2.e of the LABORATORY REPORT RESULTS at the end of the exercise.

f. HEMOGLOBIN

Hemoglobin is not normally found in urine.

(1) Chemstrip® Method

Record your results in Section D.4.f of the LABORATORY REPORT RESULTS at the end of the exercise.

5. Microscopic Analysis

If you allow a urine specimen to stand undisturbed for a few hours, many suspended materials will settle to the bottom. A much faster method is to centrifuge a urine sample.

PROCEDURE

1. Place 5 ml of fresh urine in a centrifuge tube (Figure 24.11). Follow the directions of your

FIGURE 24.11   Tabletop centrifuge used for spinning urine samples to obtain sediment for microscopic analysis.

instructor to centrifuge the tube for 5 min at a slow speed [1500 revolutions per minute (rpm)].
2. Dispose of the supernatant (the clear urine) and mix the sediment by shaking the test tube.
3. Using a long medicine dropper or a Pasteur pipette with a bulb, place a small drop of sediment on a clean glass slide, add 1 drop of Sedi-stain® or methylene blue, and place a cover glass over the specimen.
4. Using low power and reduced light, examine the sediment for any of the microscopic elements pictured in Figure 24.12.
5. Increase the light and use high power or oil immersion to look for crystals.
6. Crystals can be identified as follows (Figure 24.12):
   a. *Calcium oxalate*—Dumbbell and octahedral shapes.
   b. *Calcium phosphate*—Very pointed, wedge-shaped formations that may occur as individual crystals or grouped together to form rosettes.
   c. *Cholesterol*—Spherical crystals with a crosslike configuration inside.
   d. *Hippuric acid*—Long, needlelike crystals.
   e. *Triple phosphates*—Prisms or feathery forms.
   f. *Uric acid*—Rhombic prisms, wedges, dumbbells, rosettes, irregular crystals. These are pigmented in sediment; the color varies from yellow to dark reddish brown.

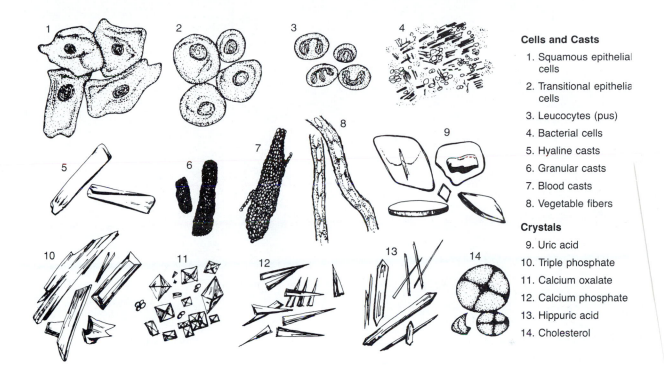

Cells and Casts

1. Squamous epithelial cells
2. Transitional epithelial cells
3. Leucocytes (pus)
4. Bacterial cells
5. Hyaline casts
6. Granular casts
7. Blood casts
8. Vegetable fibers

Crystals

9. Uric acid
10. Triple phosphate
11. Calcium oxalate
12. Calcium phosphate
13. Hippuric acid
14. Cholesterol

FIGURE 24.12   Diagram of microscopic elements in urine.

Draw the results of your observations in Section D.5 of the LABORATORY REPORT RESULTS at the end of the exercise.

Why might some red blood cells in urinary sediment be crenated? _____

_____

Why might yeast be seen in the urine of a person with diabetes mellitus? _____

_____

6. Unknown Specimens

a. UNKNOWNS PREPARED BY INSTRUCTOR

When the composition of a substance has not been defined, it is called an *unknown.* In this exercise, the unknowns will be urine specimens to which the instructor has added glucose, albumin, or any other detectable substance. Each unknown contains only one added substance.

Perform the previously outlined tests until you identify the substance in your unknown.

Record your results in Section D.6.a of the LABORATORY REPORT RESULTS at the end of the exercise.

b. UNKNOWNS PREPARED BY CLASS (OPTIONAL)

The class should be divided into two groups. Each group adds certain substances—such as glucose, protein, starch, or fat—to normal, freshly voided urine, keeping accurate records as to what was added to each sample. The two groups then exchange samples, and each group does the basic chemical tests on urine to detect which substances were added. Each student should add one substance to one urine sample and see if another student can detect what was added. Record your results in Section D.6.b of the LABORATORY REPORT RESULTS at the end of the exercise.

ANSWER THE LABORATORY REPORT QUESTIONS AT THE END OF THE EXERCISE.

Urinary System

SECTION D. URINALYSIS

3. Physical Analysis

Characteristic	Normal	Your sample
a. Color	Straw yellow to amber	
b. Sediment	None	
c. pH	5.0–7.8	
d. Specific gravity	1.008–1.030	

4. Chemical Analysis

Record your results in the spaces provided.

a. CHLORIDE AND SODIUM CHLORIDE

Chloride concentration _____ mg/100 ml

Sodium chloride concentration _____ mg/100 ml

b. GLUCOSE

Clinitest® tablet _____

Chemstrip® _____

c. PROTEIN

Albutest® reagent tablets _____

Chemstrip®_____

d. KETONE (ACETONE) BODIES

Acetest® tablet _____

Chemstrip® _____

e. BILE PIGMENTS

Ictotest® (bilirubin) _____

Chemstrip® _____

f. HEMOGLOBIN

Chemstrip® _____

5. Microscopic Analysis

Draw some of the substances (types of cells, types of crystals, or other elements) that you found in the microscopic examination of urinary sediment.

6. Unknown Specimens

a. What substance did you find in the unknown specimen prepared by your instructor?

_____

b. What substance did you find in the unknown specimen prepared by other students?

_____

Urinary System

STUDENT _____   DATE _____

LABORATORY SECTION _____   SCORE/GRADE _____

PART 1. Multiple Choice

_____ 1. Beginning at the innermost layer and moving toward the outermost layer, identify the order of tissue layers surrounding the kidney. (a) renal capsule, renal fascia, adipose capsule (b) renal fascia, adipose capsule, renal capsule (c) adipose capsule, renal capsule, renal fascia (d) renal capsule, adipose capsule, renal fascia

_____ 2. The functional unit of the kidney is the (a) nephron (b) ureter (c) urethra (d) hilus

_____ 3. Substances filtered by the kidney must pass through the endothelial-capsular membrane, which is composed of several parts. Which of the following choices lists the correct order of the parts as substances pass through the membrane? (a) epithelium of the visceral layer of glomerular (Bowman's) capsule, endothelium of the glomerulus, basement membrane of the glomerulus (b) endothelium of the glomerulus, basement membrane of the glomerulus, epithelium of the visceral layer of glomerular (Bowman's) capsule (c) basement membrane of the glomerulus, endothelium of the glomerulus, epithelium of the visceral layer of the glomerular (Bowman's) capsule (d) epithelium of the visceral layer of the glomerular (Bowman's) capsule, basement membrane of the glomerulus, endothelium of the glomerulus.

_____ 4. In the glomerular (Bowman's) capsule, the afferent arteriole divides into a capillary network called a(n) (a) glomerulus (b) interlobular artery (c) peritubular capillary (d) efferent arteriole

_____ 5. Transport of urine from the renal pelvis into the urinary bladder is the function of the (a) urethra (b) calculi (c) casts (d) ureters

_____ 6. The terminal portion of the urinary system is the (a) urethra (b) urinary bladder (c) ureter (d) nephron

_____ 7. Damage to the renal medulla would interfere first with the functioning of which parts of a juxtamedullary nephron? (a) glomerular (Bowman's) capsule (b) distal convoluted tubule (c) collecting tubules (d) proximal convoluted tubules

_____ 8. An obstruction in the glomerulus would affect the flow of blood into the (a) renal artery (b) efferent arteriole (c) afferent arteriole (d) intralobular artery

_____ 9. Urine that leaves the distal convoluted tubule passes through the following structures in which sequence? (a) collecting tubule, hilus, calyces, ureter (b) collecting tubule, calyces, pelvis, ureter (c) calyces, collecting tubule, pelvis, ureter (d) calyces, hilus, pelvis, ureter

_____ 10. The position of the kidneys behind the peritoneal lining of the abdominal cavity is described by the term (a) retroperitoneal (b) anteroperitoneal (c) ptosis (d) inferoperitoneal

_____ 11. Of the following structures, the one to receive filtrate *last* as it passes through the nephron is the (a) proximal convoluted tubule (b) ascending limb of the loop of Henle nephron (c) glomerulus (d) collecting tubule

_____ 12. Peristalsis of the ureter is a function of the (a) serosa (b) mucosa (c) submucosa (d) muscularis

_____ 13. The trigone and the detrusor muscle are associated with the (a) kidney (b) urinary bladder (c) urethra (d) ureters

_____ 14. The notch on the medial surface of the kidney through which blood vessels enter and exit is called the (a) medulla (b) major calyx (c) hilus (d) renal column

_____ 15. Blood is drained from the kidneys by the (a) renal arteries (b) interlobar arteries (c) interlobular veins (d) renal veins

_____ 16. The epithelium of the urinary bladder that permits distension is (a) stratified squamous (b) transitional (c) simple squamous (d) pseudostratified

_____ 17. How many times a day is the entire volume of blood in the body filtered by the kidneys? (a) 100 times (b) 5 times (c) 30 times (d) 60 times

_____ 18. The average urine capacity of the urinary bladder is (a) 1000 to 1200 ml (b) 50 to 100 ml (c) 700 to 800 ml (d) 200 to 300 ml

_____ 19. The normal pH of urine is between (a) 5.0 and 7.8 (b) 2.0 and 4.8 (c) 10.0 and 12.0 (d) none of the above

_____ 20. Normal urine has a specific gravity of approximately (a) 1.008 to 1.030 (b) 1.030 to 1.080 (c) 1.100 to 1.200 (d) none of the above

_____ 21. The special chemical that can be used to detect glucose in the urine is (a) sulfosalicylic acid (b) Benedict's solution (c) Lugol's solution (d) nitric acid

PART 2. Completion

22. In addition to the urinary system, other systems that help eliminate wastes are the respiratory, integumentary, and _____ systems.

23. The double-walled cup found in a nephron is called a(n) _____.

24. The special capillary network found inside of this double-walled cup is the

_____.

25. The major blood vessel that enters each kidney is the _____.

26. The nerve supply to the kidneys comes from the autonomic nervous system and is called the

_____.

27. Urine is expelled from the urinary bladder by an act called urination, voiding, or

_____.

28. The small tube in the urinary system that leads from the floor of the urinary bladder to the outside is the _____.

29. The apices of renal pyramids are referred to as renal _____.

30. The cortical substance between renal pyramids is called a renal _____.

31. Cuplike extensions of the renal pelvis, usually two or three in number, are referred to as

_____.

32. Epithelial cells of the visceral layer of the glomerular (Bowman's) capsule are called

_____.

33. Distal convoluted tubules terminate by merging with _____.

34. Long loops of blood vessels around the medullary structures of juxtamedullary nephrons are called

_____.

35. Which blood vessel comes next in this sequence? Interlobar artery, arcuate artery, interlobular

artery, _____.

36. The abnormal condition when red blood cells are found in the urine in appreciable amounts is called

_____.

37. Various salts that solidify in the urinary tract are called _____.

38. Various substances that have hardened and assumed the shape of the lumens of the nephron

tubules are the _____.

39. The pH of urine in individuals on high-protein diets tends to be _____ than normal.

40. The greater the concentration of solutes in urine, the greater will be its _____.

pH and Acid-Base Balance

A. THE CONCEPT OF PH

When molecules of inorganic acids, bases, or salts dissolve in water, they undergo *ionization* (ī'-on-i-ZĀ-shun) or *dissociation* (dis'-sō-sē-Ā-shun); that is, they separate into ions.

An *acid* can be defined as a substance that dissociates into one or more *hydrogen ions (H⁺)* and one or more *anions* (negative ions). Because H^+ is a single proton with a charge of +1, an acid can also be defined as a proton donor. A *base,* by contrast, dissociates into one or more *hydroxide ions (OH⁻)* and one or more *cations* (positive ions). A base can also be viewed as a proton acceptor. Hydroxide ions have a strong attraction for protons. A *salt,* when dissolved in water, dissociates into cations and anions, neither of which is H^+ or OH^-. Acids and bases react with one another to form salts. Body fluids must constantly contain balanced quantities of acids and bases. In solutions such as those found inside or outside body cells, acids dissociate into hydrogen ions (H^+) and anions. Bases, on the other hand, dissociate into hydroxide ions (OH^-) and cations. The more hydrogen ions that exist in a solution, the more acidic the solution; conversely, the more hydroxide ions, the more basic (alkaline) the solution.

Biochemical reactions—those that occur in living systems—are very sensitive to even small changes in acidity or alkalinity. Any departure from the narrow limits of normal H^+ and OH^- concentrations may greatly modify cell functions and disrupt homeostasis. For this reason, the acids and bases that are constantly formed in the body must be kept in balance.

A solution's acidity or alkalinity is expressed on the *pH scale,* which runs from 0 to 14. This scale is based on the concentration of H^+ in a solution. The midpoint in the scale is 7, where the concentrations of H^+ and OH^- are equal. A substance with a pH of 7, such as distilled (pure) water, is neutral. A solution that has more H^+ than OH^- in an *acidic solution* and has a pH below 7. A solution that has more OH^- than H^+ is a *basic (alkaline) solution* and has a pH above 7. A change of one whole number on the pH scale represents a 10-fold change from the previous concentration. A pH of 1 denotes 10 times more H^+ than a pH of 2. A pH of 3 indicates 10 times fewer H^+ than a pH of 2 and 100 times fewer H^+ than a pH of 1.

B. MEASURING PH

1. Using Litmus Paper

PROCEDURE

CAUTION! *Please reread Section A, "General Safety Precautions and Procedures," on page xiii, and Section C, "Precautions Related to Working with Reagents," on page xiv at the beginning of the laboratory manual, before you begin any of the following experiments. Read the experiments before you perform them, to be sure that you understand all the procedures and safety precautions.*

1. Before you begin, it is important to know that *an acid solution will turn blue litmus paper red and a basic (alkaline) solution will turn red litmus paper blue.*

2. Using forceps, dip a strip of red litmus paper into each of the solutions to be tested. Use a *new strip* for each solution. Record your observations in the Table in Section B.1 of the LABORATORY REPORT RESULTS at the end of the exercise.

3. Using forceps, dip a strip of blue litmus paper into each of the solutions to be tested. Use a *new strip* for each solution. Record your observations in the Table in Section B.1 of the LABORATORY REPORT RESULTS at the end of the exercise.

4. Using your textbook as a guide, determine the pH of the following body fluids:

Body Fluid	pH
Bile	
Saliva	
Gastric juice	
Blood	
Pancreatic juice	
Semen	
Urine	

2. Using pH Paper

PROCEDURE

CAUTION! *Please reread Section A, "General Safety Precautions and Procedures," on page xiii, and Section C, "Precautions Related to Working with Reagents," on page xiv, at the beginning of the laboratory manual, before you begin any of the following experiments. Read the experiments before you perform them, to be sure that you understand all the procedures and safety precautions.*

1. Using forceps, dip a strip of pH paper into each of the solutions to be tested. Use a *new strip* for each solution. Use wide-range pH paper to determine the approximate pH and narrow-range pH paper to determine a more accurate pH.

2. Compare the color of the strip of pH paper to the color chart on the pH paper container.

3. Record your observations in the Table in Section B.2 of the LABORATORY REPORT RESULTS at the end of the exercise.

3. Using a pH Meter

Because there are different types of pH meters, your instructor will demonstrate how to use the pH meter in your laboratory.

PROCEDURE

CAUTION! *Please reread Section A, "General Safety Precautions and Procedures," on page xiii, and Section C, "Precautions Related to Working with Reagents," on page xiv, at the beginning of the laboratory manual, before you begin any of the following experiments. Read the experiments before you perform them, to be sure that you understand all the procedures and safety precautions.*

1. Examine the pH meter that has been made available to you and identify the following parts: (1) electrodes, (2) pH dial or digital display, (3) temperature control, and (4) calibration control.

2. Plug in the pH meter and turn it on. (*NOTE:* Some models take up to one-half hour to warm up.)

3. Using the temperature control knob, adjust the pH meter for the temperature of the solutions to be tested.

4. To calibrate the pH meter, place the electrode(s) in a beaker that contains a pH 7 buffer solution. The electrode(s) should be immersed at least 1 in. into the solution. Adjust the pH meter with the appropriate controls so that the meter will show a pH value of 7. Now the instrument is calibrated.

5. Depress the standby button and remove the electrode(s) from the buffer solution. *The electrode(s) should not touch anything.* Rinse the electrode(s) with distilled water, using a wash bottle. The rinse water can be collected in an empty beaker.

6. Immerse the electrode(s) into the first solution to be tested. Release the standby button and note the pH. Record the pH in the Table in Section B.3 of the LABORATORY REPORT RESULTS at the end of the exercise.

7. Depress the standby button, remove the electrode(s) from the solution, and rinse with

distilled water. Test the pH of the remaining solutions and record each pH in the Table in Section B.3 of the LABORATORY REPORT RESULTS at the end of the exercise.

C. ACID-BASE BALANCE

A very important electrolyte in terms of the body's acid–base balance is the hydrogen ion (H^+). Although some hydrogen ions enter the body in ingested foods, most are produced as a result of the cellular metabolism of substances such as glucose, fatty acids, and amino acids. One of the major challenges to homeostasis is keeping the hydrogen ion concentration at an appropriate level to maintain proper acid–base balance.

The balance of acids and bases is maintained by controlling the H^+ concentration of body fluids, particularly extracellular fluid. In a healthy person, the pH of the extracellular fluid remains between 7.35 and 7.45. Metabolism typically produces a significant excess of H^+. If there were no mechanisms for disposal of acids, the rising concentration of H^+ in body fluids would quickly lead to death. Homeostasis of H^+ concentration within a narrow pH range is essential to survival and depends on three major mechanisms.

1. *Buffer systems*—Buffers act quickly to bind H^+ temporarily, which removes excess H^+ from solution but not from the body.
2. *Exhalation of carbon dioxide*—By increasing the rate and depth of breathing, more carbon dioxide can be exhaled. This reduces the level of carbonic acid and is effective within minutes.
3. *Kidney excretion*—The slowest mechanism, taking hours or days, but the only way to eliminate acids other than carbonic acid is through their passage into urine and their excretion by the kidneys.

In the following experiments, you will note the relationship between buffers and the exhalation of carbon dioxide to pH.

1. Buffers and pH

Most *buffer systems* of the body consist of a weak acid and the salt of that acid, which functions as a weak base. Buffers function to prevent rapid,

drastic changes in the pH of a body fluid by changing strong acids and bases into weak acids and bases. Buffers work within fractions of a second. A strong acid dissociates into H^+ more easily than does a weak acid. Strong acids therefore lower pH more than weak ones because strong acids contribute more H^+. Similarly, strong bases raise pH more than weak ones because strong bases dissociate more easily into hydroxide ions (OH^-). The principal buffer systems of the body fluids are the carbonic acid–bicarbonate system, the phosphate system, and the protein buffer system.

a. CARBONIC ACID-BICARBONATE BUFFER SYSTEM

The *carbonic acid–bicarbonate buffer system* is based on the bicarbonate ion (HCO_3^-), which can act as a weak base, and carbonic acid (H_2CO_3), which can act as a weak acid. Thus, the buffer system can compensate for either an excess or a shortage of H^+. For example, if there is an excess of H^+ (an acid condition), HCO_3^- can function as a weak base and remove the excess H^+ as follows:

$$\underset{\substack{\text{Hydrogen} \\ \text{ion}}}{H^+} + \underset{\substack{\text{Bicarbonate} \\ \text{ion} \\ \text{(weak base)}}}{HCO_3^-} \rightarrow \underset{\substack{\text{Carbonic} \\ \text{acid}}}{H_2CO_3} \rightarrow \underset{\text{Water}}{H_2O} + \underset{\substack{\text{Carbon} \\ \text{dioxide}}}{CO_2}$$

On the other hand, if there is a shortage of H^+ ions (an alkaline condition), H_2CO_3 can function as a weak acid and provide H^+ as follows:

$$\underset{\substack{\text{Carbonic} \\ \text{acid} \\ \text{(weak acid)}}}{H_2CO_3} \rightarrow \underset{\text{Hydrogen ion}}{H^+} + \underset{\text{Bicarbonate ion}}{HCO_3^-}$$

A typical bicarbonate buffer system consists of a mixture of carbonic acid (H_2CO_3) and its salt, sodium bicarbonate ($NaHCO_3$). The carbonic acid–bicarbonate buffer system is an important regulator of blood pH. When a strong acid, such as hydrochloric acid (HCl) is added to a buffer solution containing sodium bicarbonate, which behaves like a weak base, the following reactions occurs:

$$\underset{\substack{\text{Hydrochloric} \\ \text{acid} \\ \text{(strong acid)}}}{HCl} + \underset{\substack{\text{Sodium} \\ \text{bicarbonate} \\ \text{(weak base)}}}{NaHCO_3} \rightarrow \underset{\substack{\text{Sodium} \\ \text{chloride}}}{NaCl} + \underset{\substack{\text{Carbonic} \\ \text{acid} \\ \text{(weak acid)}}}{H_2CO_3}$$

If a strong base, such as sodium hydroxide (NaOH), is added to a buffer solution containing a weak acid, such as carbonic acid, the following reaction occurs:

$$NaOH + H_2CO_3 \rightarrow H_2O + NaHCO_3$$

Sodium hydroxide (strong base)   Carbonic acid (weak acid)   Water   Sodium bicarbonate (weak base)

Normal metabolism produces more acids than bases and thus tends to acidify the blood rather than make it more alkaline. Accordingly, the body needs more bicarbonate salt than it needs carbonic acid. Bicarbonate molecules outnumber carbonic acid molecules 20:1.

b. PHOSPHATE BUFFER SYSTEM

The *phosphate buffer system* acts in the same manner as the carbonic acid–bicarbonate buffer system. The components of the phosphate buffer system are the sodium salts of dihydrogen phosphate and sodium monohydrogen phosphate ions. The dihydrogen phosphate ion acts as the weak acid and is capable of buffering strong bases.

$$NaOH + NaH_2PO_4 \rightarrow H_2O + Na_2HPO_4$$

Sodium hydroxide (strong base)   Sodium dihydrogen phosphate (weak acid)   Water   Sodium monohydrogen phosphate (weak base)

The monohydrogen phosphate ion acts as the weak base and is capable of buffering strong acids.

$$HCl + Na_2HPO_4 \rightarrow NaCl + NaH_2PO_4$$

Hydrochloric acid (strong acid)   Sodium monohydrogen phosphate (weak base)   Sodium chloride (salt)   Sodium dihydrogen phosphate (weak acid)

Because the phosphate concentration is highest in intracellular fluid, the phosphate buffer system is an important regulator of pH in the cytosol. It also is present at a lower level in extracellular fluids and acts to buffer acids in urine. NaH_2PO_4 is formed when excess H^+ in the kidney tubules combines with Na_2HPO_4. In this reaction, Na^+ released from Na_2HPO_4 forms sodium bicarbonate ($NaHCO_3$) and passes into the blood. The H^+ that replaces Na^+ becomes part of the NaH_2PO_4 that passes into the urine. This reaction is one of the mechanisms by which the kidneys help maintain pH by the acidification of urine.

c. PROTEIN BUFFER SYSTEM

The *protein buffer system* is the most abundant buffer in body cells and plasma. Inside red blood cells the protein hemoglobin is an especially good buffer. Proteins are composed of amino acids. An amino acid is an organic compound that contains at least one carboxyl group (COOH) and at least one amine group (NH$_2$). The free carboxyl group at one end of a protein acts like an acid by releasing hydrogen ions (H$^+$) ions when pH rises and can dissociate in this way:

$$NH_2 - \overset{\overset{R}{|}}{\underset{\underset{H}{|}}{C}} - COOH \rightarrow NH_2 - \overset{\overset{R}{|}}{\underset{\underset{H}{|}}{C}} - COO + H^+$$

The H$^+$ is then able react with any excess hydroxide ion (OH$^-$) in the solution to form water.

The free amine group at the other end of a protein can act as a base by combining with hydrogen ions when pH falls as follows:

$$COOH - \overset{\overset{R}{|}}{\underset{\underset{H}{|}}{C}} - NH_2 + H^+ \rightarrow COOH - \overset{\overset{R}{|}}{\underset{\underset{H}{|}}{C}} - NH_3^+$$

Thus, proteins act as both acidic and basic buffers.

The following exercise will demonstrate how buffers resist changes in pH.

PROCEDURE

CAUTION! *Please reread Section A, "General Safety Precautions and Procedures," on page xiii, and Section C, "Precautions Related to Working with Reagents," on page xiv, at the beginning of the laboratory manual, before you begin any of the following experiments. Read the experiments before you perform them, to be sure that you understand all the procedures and safety precautions.*

1. Using a pH meter, immerse the electrode(s) in a beaker containing distilled water and determine the pH. _____
2. Drop by drop, slowing add 0.05 M hydrochloric acid (HCl) to the distilled water. Gently swirl the beaker after each drop is added. Note how many drops it takes for the pH of the solution to change one whole number. _____

   What is the pH of the solution? _____
3. Remove the electrode(s) from the solution and rinse with distilled water.

4. Now immerse the electrode(s) in a pH 7 buffer solution. Drop by drop, slowly add 0.05 M HCl to the solution. Gently swirl the beaker after each drop is added. Note how many drops it takes for the pH of the solution to change one whole number. _____
What conclusion can you draw from this observation? _____
_____
_____

5. Remove the electrodes from the solution and rinse with distilled water.
6. Immerse the electrode(s) in a beaker of fresh distilled water and determine the pH. _____
_____

7. Drop by drop, slowly add 0.05 M sodium hydroxide (NaOH) to the distilled water. Gently swirl the beaker after each drop is added. Note how many drops it takes for the pH of the solution to change one whole number.

_____

What is the pH of the solution? _____
8. Remove the electrode(s) from the solution and rinse with distilled water.
9. Now immerse the electrode(s) in a pH 7 buffer solution. Drop by drop, slowly add 0.05 M NaOH to the solution. Gently swirl the beaker after each drop is added. Note how many drops it takes for the pH of the solution to change one whole number. _____.
What conclusion can you draw from this observation? _____
_____
_____

2. Respirations and pH

Breathing also plays a role in maintaining the pH of the body. An increase in the carbon dioxide (CO_2) concentration in body fluids increases H^+ concentration and thus lowers the pH (makes it more acidic). This is illustrated by the following reactions:

$$CO_2 + H_2O \rightleftharpoons H_2CO_3 \rightleftharpoons H^+ + HCO_3^-$$

Conversely, a decrease in the CO_2 concentration of body fluids raises the pH (makes it more basic).

The pH of body fluids can be adjusted, usually in 1 to 3 mins, by a change in the rate and depth of breathing. If the rate and depth of breathing increase, more CO_2 is exhaled, the reaction just given is driven to the left, H^+ concentration falls, and the blood pH rises. Because carbonic acid can be eliminated by exhaling CO_2, it is called a *volatile acid.* If the rate of respiration slows down, less carbon dioxide is exhaled, and the blood pH falls. Doubling the breathing rate increases the pH by about 0.23, from 7.4 to 7.63. Reducing the breathing rate to one-quarter its normal rate lowers the pH by 0.4, from 7.4 to 7.0. These examples show the powerful effect of alterations in breathing on pH of body fluids.

The pH of body fluids, in turn, affects the rate of breathing. If, for example, the blood becomes more acidic, the increase in hydrogen ions is detected by chemoreceptors that stimulate the inspiratory center in the medulla. As a result, the diaphragm and other muscles of respiration contract more forcefully and frequently—the rate and depth of breathing increase.

The same effect is achieved if the blood level of CO_2 increases. The increased rate and depth of respiration remove more CO_2 from blood to reduce the H^+ concentration, and blood pH increases. On the other hand, if the pH of the blood increases, the respiratory center is inhibited and respirations decrease. A decrease in the CO_2 concentration of blood has the same effect. The decreased rate and depth of respirations cause CO_2 to accumulate in blood and the H^+ concentration increases. The respiratory mechanism normally can eliminate more acid or base than can all the buffers combined, but it is limited to eliminating only the single volatile acid, carbonic acid.

The following exercise will demonstrate the relationship of exhalation of carbon dioxide to pH.

PROCEDURE

CAUTION! *Please reread Section A, "General Safety Precautions and Procedures," on page xiii, and Section C, "Precautions Related to Working with Reagents," on page xiv, at the beginning of the laboratory manual, before you begin any of the following experiments. Read the experiments before you perform them, to be sure that you understand all the procedures and safety precautions.*

1. Fill a large beaker with 100 ml of distilled water.
2. Add 5 ml of 0.10 normal sodium hydroxide (NaOH) solution and 5 drops of phenol red.
3. Phenol red is a pH indicator. It remains red in a basic solution, changes to orange in a neutral solution, and changes to yellow in an acidic solution.
4. While at rest, exhale through a straw into the solution. Your partner should determine how long it takes for the solution to change from

   orange to yellow. _____

CAUTION! *Perform Step 5 only if you have no known or apparent cardiac or other health problems and are capable of such an activity.*

5. *Run in place for about 100 steps.*
6. Exhale through a straw into a fresh solution as prepared in steps 1 and 2. Your partner should determine how long it takes for the solution to change from orange to yellow.
7. Change places with your partner, and repeat the experiment. Explain the difference in time it took for the solution to change from orange

   to yellow at rest and following exercise. _____

   _____

   _____

   _____

D. ACID-BASE IMBALANCES

The normal blood pH range is 7.35 to 7.45. *Acidosis* (or *acidemia*) is a condition in which blood pH is below 7.35. *Alkalosis* (or *alkalemia*) is a condition in which blood pH is higher than 7.45.

A change in blood pH that leads to acidosis or alkalosis can be compensated to return pH to normal. *Compensation* refers to the physiological response to an acid–base imbalance. If a person has an altered pH due to metabolic causes, respiratory mechanisms (hyperventilation or hypoventilation) can help compensate for the alteration. Respiratory compensation occurs within minutes and is maximized within hours. On the other hand, if a person has an altered pH due to respiratory causes, metabolic mechanisms (kidney excretion) can compensate for the alteration.

Metabolic compensation may begin in minutes but takes days to reach a maximum.

1. Physiological Effects

The principal physiological effect of acidosis is depression of the CNS through depression of synaptic transmission. If the blood pH falls below 7, depression of the nervous system is so severe that the individual becomes disoriented and comatose and dies. Patients with severe acidosis usually die in a state of coma. On the other hand, the major physiological effect of alkalosis is overexcitability in both the CNS and peripheral nerves. Nerves conduct impulses repetitively, even when not stimulated by normal stimuli, resulting in nervousness, muscle spasms, and even convulsions and death.

In the discussion that follows, note that both respiratory acidosis and alkalosis are primary disorders of blood pCO_2 (normal range 35 to 45 mm Hg). On the other hand, both metabolic acidosis and alkalosis are primary disorders of bicarbonate (HCO_3^-) concentration (normal range 22 to 26 mEq/liter).

2. Respiratory Acidosis

The hallmark of *respiratory acidosis* is an elevated pCO_2 of arterial blood (above 45 mm Hg). Inadequate exhalation of CO_2 decreases the blood pH. It occurs as a result of any condition that decreases the movement of CO_2 from the blood to the alveoli of the lungs to the atmosphere and therefore causes a buildup of carbon dioxide, carbonic acid, and hydrogen ions. Such conditions include emphysema, pulmonary edema, injury to the respiratory center of the medulla, airway obstruction, or disorders of the muscles involved in breathing. Metabolic compensation involves increased excretion of H^+ and increased reabsorption of HCO_3^- by the kidneys. Treatment of respiratory acidosis aims to increase the exhalation of CO_2. Excessive secretions can be suctioned out of the respiratory tract, and artificial respiration can be given. In addition, intravenous administration of bicarbonate and ventilation therapy to remove excessive carbon dioxide can be used.

3. Respiratory Alkalosis

In *respiratory alkalosis* arterial blood pCO_2 is decreased (below 35 mm Hg). Hyperventilation causes the pH to increase. It occurs in conditions that stimulate the respiratory center. Such conditions include oxygen deficiency due to high altitude or pulmonary disease, cerebrovascular accident (CVA), severe anxiety, and aspirin overdose. The kidneys attempt to compensate by decreasing excretion of H^+ and decreasing reabsorption of HCO_3^-. Treatment of respiratory alkalosis is aimed at increasing the level of CO_2 in the body. One simple measure is to have the person breathe into a paper bag and then rebreathe the exhaled mixture of CO_2 and oxygen from the bag.

4. Metabolic Acidosis

In *metabolic acidosis* there is a decrease in HCO_3^- concentration (below 22 mEq/liter). The decrease in pH is caused by loss of bicarbonate, such as may occur with severe diarrhea or renal dysfunction; accumulation of an acid, other than carbonic acid, as may occur in ketosis; or failure of the kidneys to excrete H^+ derived from metabolism of dietary proteins. Compensation is respiratory by hyperventilation. Treatment of metabolic acidosis consists of intravenous solutions of sodium bicarbonate and correcting the cause of acidosis.

5. Metabolic Alkalosis

In *metabolic alkalosis* HCO_3^- concentration is elevated (above 26 mEq/liter). A nonrespiratory loss of acid by the body or excessive intake of alkaline drugs causes the pH to increase. Excessive vomiting of gastric contents results in a substantial loss of hydrochloric acid and is prob-

ably the most frequent cause of metabolic alkalosis. Other causes of metabolic alkalosis include gastric suctioning, use of certain diuretics, endocrine disorders, and administration of alkali. Compensation is respiratory by hypoventilation. Treatment of metabolic alkalosis consists of fluid therapy to replace chloride, potassium, and other electrolyte deficiencies and correcting the cause of alkalosis.

A summary of acidosis and alkalosis is presented in Table 25.1 on page 512.

Based on an analysis of respiratory gases, you can determine if a person has acidosis or alkalosis and whether the acidosis or alkalosis is respiratory or metabolic, as reflected by the change in pH.

In the following table (bottom of page), note the normal ranges for pH, pCO_2, and HCO_3^-. Also note how values above and below normal relate to acidosis and alkalosis.

If a change in pH is a result of an abnormal pCO_2 value, then the condition is respiratory in nature. If, instead, a change in pH is a result of an abnormal HCO_3^- value, the condition is metabolic in nature.

Problems related to acid-base imbalance are reflected in each of the following conditions. Determine whether each is (1) acidosis or alkalosis and (2) metabolic or respiratory:

a. pH = 7.32
   $HCO_3^- = 10$ mEq/liter
b. pH = 7.48
   $pCO_2 = 32$ mmHg
c. pH = 7.52
   $HCO_3^- = 28$ mEq/liter
d. pH = 7.30
   $pCO_2 = 48$ mmHg

	pH	**pCO_2**	**HCO_3^-**
Normal range	7.35-7.45	35-45 mmHg	22-26 mEq/liter
Acidosis	Below 7.35	Above 45 mmHg	Below 22 mEq/liter
Alkalosis	Above 7.45	Below 35 mmHg	Above 26 mEq/liter

TABLE 25.1

Summary of Acidosis and Alkalosis

Condition	Definition	Common cause	Compensatory mechanism
Respiratory acidosis	Increase pCO_2 (above 45 mm Hg) and decreased pH (below 7.35) if there is no compensation.	Hypoventilation due to emphysema, pulmonary edema, trauma to respiratory center, airway obstructions, dysfunction of muscles of respiration.	Renal: increased excretion of H^+; increased reabsorption of HCO_3^-. If compensation is complete, pH will be within normal range, but pCO_2 will be high.
Respiratory alkalosis	Decreased pCO_2 (below 35 mm Hg) and increased pH (above 7.45) if there is no compensation.	Hyperventilation due to oxygen deficiency, pulmonary disease, cerebrovascular accident (CVA), anxiety, or aspirin overdose.	Renal: decreased excretion of H^+; decreased reabsorption of HCO_3^-. If compensation is complete, pH will be within normal range, but pCO_2 will be low.
Metabolic acidosis	Decreased bicarbonate (below 22 mEq/liter) and decreased pH (below 7.35) if there is no compensation.	Loss of bicarbonate due to diarrhea, accumulation of acid (ketosis), renal dysfunction.	Respiratory: hyperventilation, which increases loss of CO_2. If compensation is complete, pH will be within normal range but HCO_3^- will be low.
Metabolic alkalosis	Increased bicarbonate (above 26 mEq/liter) and increased pH (above 7.45) if there is no compensation.	Loss of acid or excessive intake of alkaline drugs; due to vomiting, gastric suctioning, use of certain diuretics, and administration of alkali.	Respiratory: hypoventilation, which slows loss of CO_2. If compensation is complete, pH will be within normal range, but HCO_3^- will be high.

pH and Acid–Base Balance

STUDENT _____   DATE _____

LABORATORY SECTION _____   SCORE/GRADE _____

SECTION B. MEASURING PH
1. Using Litmus Paper

Solution	Color of red litmus paper	Color of blue litmus paper	Is the solution acidic or basic?
Milk of magnesia			
Vinegar			
Coffee			
Carbonated soft drink			
Orange juice			
Distilled water			
Baking soda			
Lemon juice			

2. Using pH Paper

Solution	pH
Milk of magnesia	
Vinegar	
Coffee	
Carbonated soft drink	
Orange juice	
Distilled water	
Baking soda	
Lemon juice	

3. Using a pH Meter

Solution	pH
Milk of magnesia	
Vinegar	
Coffee	
Carbonated soft drink	
Orange juice	
Distilled water	
Baking soda	
Lemon juice	

pH and Acid–Base Balance

STUDENT _____   DATE _____

LABORATORY SECTION _____   SCORE/GRADE _____

PART 1. Multiple Choice

_____ 1. An acid is a substance that dissociates into (a) OH^- (b) H^+ (c) HCO_3^- (d) Na^+

_____ 2. Which of the following pHs is more acidic? (a) 6.89 (b) 6.91 (c) 7.00 (d) 6.83

_____ 3. The pH of bile is (a) 4.2 (b) 6.35 to 6.85 (c) 7.6 to 8.6 (d) 3.0 to 3.5

_____ 4. Which mechanism is quickest to restore pH? (a) exhalation of CO_2 (b) buffers (c) kidney excretion (d) inhalation of oxygen

_____ 5. In the carbonic acid-bicarbonate buffer system, which substance functions to buffer a strong base? (a) NaOH (b) $NaHCO_3$ (c) H_2CO_3 (d) NaCl

_____ 6. The most abundant buffer in body cells and plasma is the (a) protein buffer (b) phosphate buffer (c) hemoglobin buffer (d) carbonic acid-bicarbonate buffer

_____ 7. Doubling the breathing rate increases pH by about (a) 0.75 (b) 2.21 (c) 1.86 (d) 0.23

PART 2. Completion

8. Bases dissociate into _____ ions and cations.

9. A change of one whole number on the pH scale represents a _____-fold change from the previous concentration.

10. A(n) _____ solution will turn red litmus paper blue.

11. The pH of blood is _____.

12. Most H^+ in the body is produced as a result of _____.

13. In the protein buffer system, the carboxyl group acts as a(n) _____.

14. If the rate and depth of respiration increase, pH will _____.

15. If blood becomes more basic, the rate and depth of respiration will _____.

16. A pH higher than 7.45 is referred to as _____.

17. Metabolic acidosis and alkalosis are disorders of _____ concentration.

18. The principal physiological effect of _____ is depression of the CNS.

19. _____ acidosis is characterized by a pH below 7.35 and an elevated pCO_2.

20. Compensation for metabolic acidosis is _____.

26

Reproductive Systems

In this exercise you will study the structure of the male and female reproductive organs and associated structures.

A. ORGANS OF MALE REPRODUCTIVE SYSTEM

The *male reproductive system* includes (1) the testes, or male gonads (*gonos* = seed), which produce sperm; (2) ducts that either transport, receive, or store sperm; (3) accessory glands whose secretions contribute to semen; and (4) several supporting structures, including the penis.

1. Testes

The *testes,* or *testicles,* are paired oval glands that lie in the pelvic cavity for most of fetal life. They usually begin to enter the scrotum during the latter half of the seventh month of fetal development; full descent is not complete until just before birth. If the testes do not descend, the condition is called *cryptorchidism* (krip-TOR-ki-dizm). Cryptorchidism results in sterility because the cells that stimulate the initial development of sperm cells are destroyed by the higher temperature of the pelvic cavity.

Each testis is covered by a dense white fibrous capsule, the *tunica albuginea* (al'-byoo-JIN-ē-a), which extends inward and divides the testis into a series of 200 to 300 internal compartments called *lobules.* Each lobule contains one to three tightly coiled *seminiferous tubules* where sperm production *(spermatogenesis)* occurs.

Label the structures associated with the testes in Figure 26.1.

Examine a microscope slide of a testis in cross section. Using oil immersion, note the developing reproductive cells in the seminiferous tubules. From the edge of the tubules toward the lumen, these cells are *spermatogonia, primary spermatocytes, secondary spermatocytes, spermatids,* and mature tailed *spermatozoa (sperm cells).* Distinguish between these cells and the *sustentacular (Sertoli) cells* that nourish the reproductive cells, phagocytize degenerating reproductive cells, and secrete the hormone inhibin. Locate, in the tissue surrounding the seminiferous tubules, the clusters of *interstitial endocrinocytes (interstitial cells of Leydig),* which secrete the male hormone testosterone, the most important androgen. Because they produce both sperm and hormones, the testes are both exocrine and endocrine glands.

Using your textbook, charts, or models as reference, label Figure 26.2 on page 519.

Spermatozoa (sper'-ma-tō-ZŌ-a) are produced at the rate of about 300 million per day. Once ejaculated, they usually live about 48 hr in the female reproductive tract. The parts of a spermatozoon (sper-mat'-ō-ZŌ-on) follow:

a. *Head*—Contains the *nucleus* and *acrosome* (produces hyaluronic acid and proteinases to effect penetration of secondary oocyte).

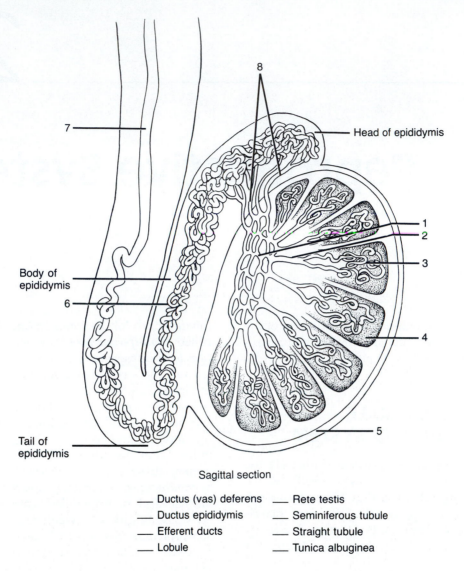

Sagittal section

___ Ductus (vas) deferens   ___ Rete testis
___ Ductus epididymis   ___ Seminiferous tubule
___ Efferent ducts   ___ Straight tubule
___ Lobule   ___ Tunica albuginea

FIGURE 26.1   Testis showing its system of ducts.

b. *Midpiece*—Contains numerous mitochondria in which the energy for locomotion is generated.

c. *Tail*—Typical flagellum used for locomotion.

With the aid of your textbook, label Figure 26.3 on page 520.

2. Ducts

As spermatozoa mature, they are moved through seminiferous tubules into tubes called **straight tubules,** from which they are transported into a network of ducts, the *rete* (RĒ-tē) *testis.* The spermatozoa are next transported out of the testes through a series of coiled *efferent ducts* that empty into a single *ductus epididymis* (ep'-i-DID-i-mis; *epi* = above, *didymos* = testis). From there, they are passed into the *ductus (vas) deferens,* which ascends along the posterior border of the testis, penetrates the inguinal canal, enters the pelvic cavity, and loops over the side and down the posterior surface of the urinary bladder. The ductus (vas) deferens and the duct from the seminal vesicle (gland) together form the *ejaculatory* (e-JAK-yoo-la-tō'-rē) *duct,* which propels the spermatozoa into the *urethra,* the terminal duct of the system. The male urethra is divisible into (1) a *prostatic portion,* which passes through the prostate gland; (2) a *membranous portion,* which passes through the urogenital diaphragm; and (3) a *spongy (cavernous) portion,* which

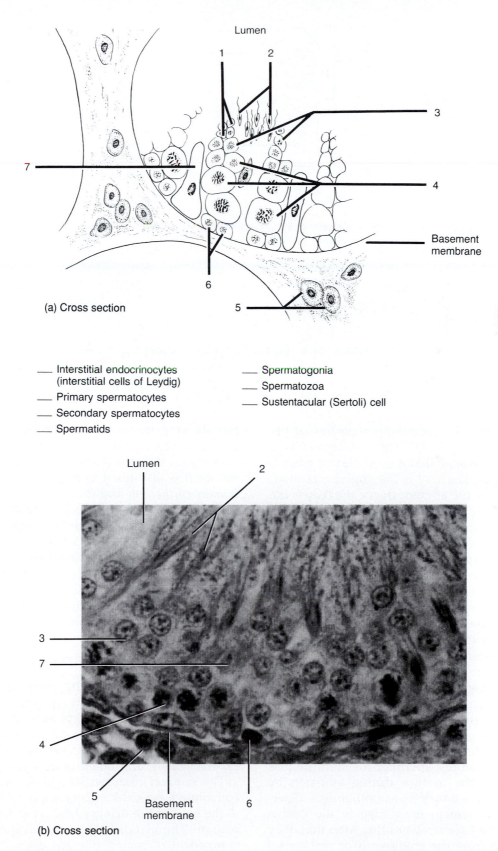

Lumen

3

7

4

Basement
membrane

6

(a) Cross section

5

___ Interstitial endocrinocytes
   (interstitial cells of Leydig)
___ Primary spermatocytes
___ Secondary spermatocytes
___ Spermatids

___ Spermatogonia
___ Spermatozoa
___ Sustentacular (Sertoli) cell

Lumen

2

3

7

4

5

Basement
membrane

6

(b) Cross section

FIGURE 26.2 Seminiferous tubules showing various stages of spermatogenesis.
(a) Diagram. (b) Photomicrograph.

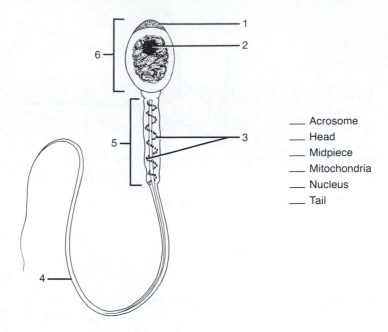

___	Acrosome
___	Head
___	Midpiece
___	Mitochondria
___	Nucleus
___	Tail

FIGURE 26.3   Parts of a spermatozoon.

passes through the corpus spongiosum of the penis (see Figure 26.6). The *epididymis* is a comma-shaped organ that is divisible into a head, body, and tail. The *head* is the superior portion, which contains the efferent ducts; the *body* is the middle portion, which contains the ductus epididymis; and the *tail* is the inferior portion, in which the ductus epididymis continues as the ductus (vas) deferens.

Label the various ducts of the male reproductive system in Figure 26.1.

The ductus epididymis is lined with *pseudostratified columnar epithelium.* The free surfaces of the cells contain long, branching microvilli called *stereocilia.* The muscularis deep to the epithelium consists of smooth muscle. Functionally, the ductus epididymis is the site of sperm maturation (increased motility and fertility potential). They require between 10 and 14 days to complete their maturation—that is, to become capable of fertilizing an ovum. The ductus epididymis also stores spermatozoa and propels them toward the urethra during emission by peristaltic contraction of its smooth muscle. Spermatozoa may remain in storage in the ductus epididymis up to several months. After that, they are expelled from the epididymis or reabsorbed in the epididymis.

Obtain a prepared slide of the ductus epididy-

mis showing its mucosa and muscularis. Compare your observations to Figure 26.4.

Histologically, the *ductus (vas) deferens (seminal duct)* is also lined with *pseudostratified columnar epithelium* and its muscularis consists of three layers of smooth muscle. Peristaltic contractions of the muscularis propel spermatozoa toward the urethra during ejaculation. One method of sterilization in males, *vasectomy,* involves removal of a portion of each ductus (vas) deferens.

Obtain a prepared slide of the ductus (vas) deferens showing its mucosa and muscularis. Compare your observations to Figure 26.5.

3. Accessory Sex Glands

Whereas the ducts of the male reproductive system store or transport sperm, a series of *accessory sex glands* secrete most of the liquid portion of *semen.* Semen is a mixture of spermatozoa and the secretions of the seminal vesicles, prostate gland, and bulbourethral glands.

The *seminal vesicles* (VES-i-kuls) are paired, convoluted, pouchlike structures posterior to and at the base of the urinary bladder in front of the rectum. The glands secrete the alkaline viscous component of semen into the ejaculatory duct. The seminal vesicles contribute about 60% of the volume of semen.

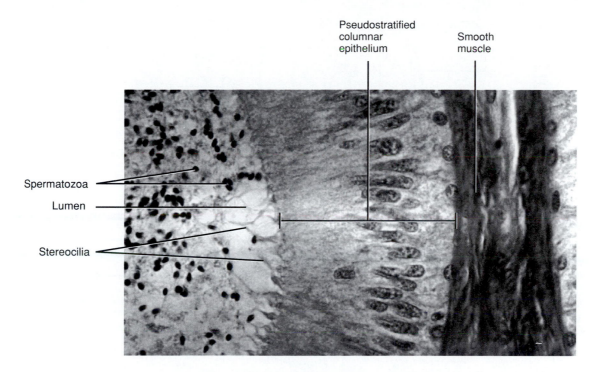

FIGURE 26.4   Histology of the ductus epididymis.

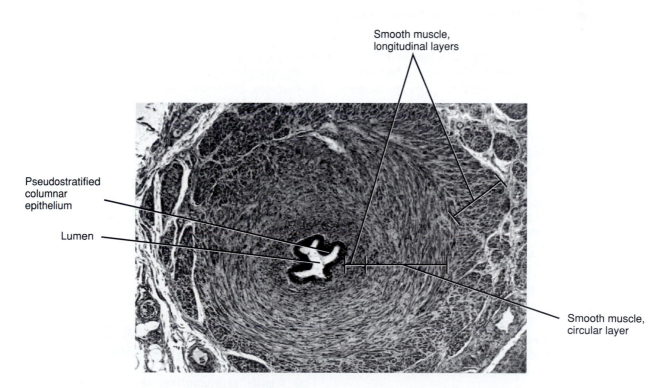

FIGURE 26.6   Histology of the ductus (vas) deferens.

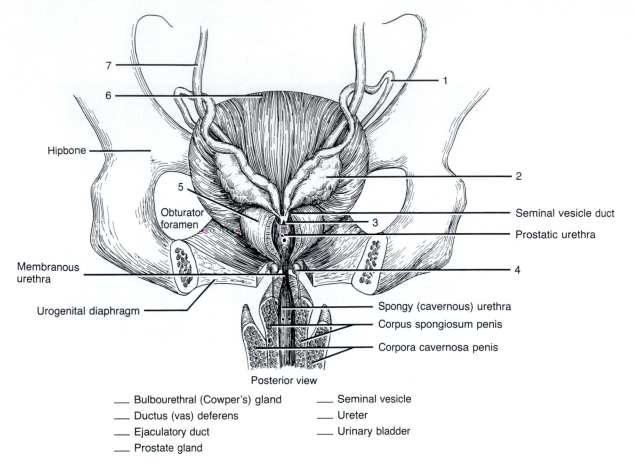

Hipbone

7

6

1

Obturator
foramen

5

2

Seminal vesicle duct

Prostatic urethra

3

4

Membranous
urethra

Urogenital diaphragm

Spongy (cavernous) urethra

Corpus spongiosum penis

Corpora cavernosa penis

Posterior view

____ Bulbourethral (Cowper's) gland   ____ Seminal vesicle
____ Ductus (vas) deferens   ____ Ureter
____ Ejaculatory duct   ____ Urinary bladder
____ Prostate gland

FIGURE 26.6   Relationships between some male reproductive organs.

The **prostate** (PROS-tāt) **gland,** a single dough-nut-shaped gland inferior to the urinary bladder, surrounds the prostatic urethra. The prostate secretes a slightly acidic fluid into the prostatic urethra. The prostatic secretion constitutes about 25% of the total semen produced.

The paired **bulbourethral** (bul'-bō-yoo-RĒ-thral), or **Cowper's, glands,** located inferior to the prostate on either side of the membranous urethra, are about the size of peas. They secrete an alkaline substance through ducts that open into the spongy (cavernous) urethra.

With the aid of your textbook, label the accessory glands and associated structures in Figure 26.6.

4. Penis

The **penis** conveys urine to the exterior and introduces spermatozoa into the vagina during copulation. Its principal parts follow:

a. *Glans penis*—Slightly enlarged distal end.
b. *Corona*—Margin of the glans penis.

c. *Prepuce*—Foreskin; loosely fitting skin covering the glans penis.
d. *Corpora cavernosa penis*—Two dorsolateral masses of erectile tissue.
e. *Corpus spongiosum penis*—Midventral mass of erectile tissue that contains the spongy (cavernous) urethra.
f. *External urethral orifice*—Opening of the spongy (cavernous) urethra to the exterior.

With the aid of your textbook, label the parts of the penis in Figure 26.7.

Now that you have completed your study of the organs of the male reproductive system, label Figure 26.8 on page 524.

B. ORGANS OF FEMALE REPRODUCTIVE SYSTEM

The **female reproductive system** includes the female gonads (ovaries), which produce ova; uterine (Fallopian) tubes, or oviducts, which

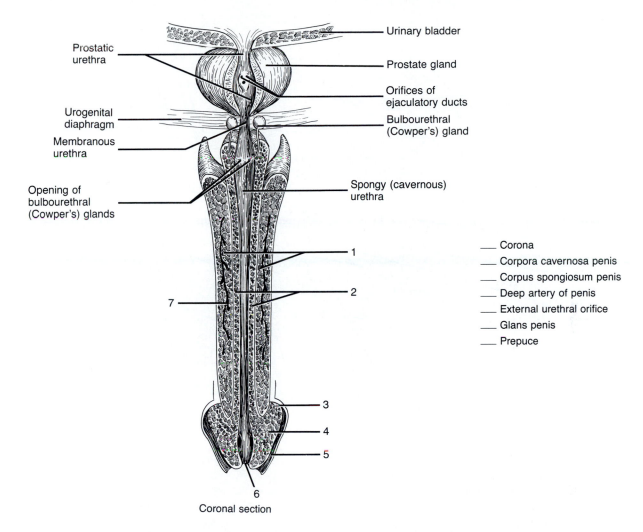

Prostatic urethra

Urogenital diaphragm

Membranous urethra

Opening of bulbourethral (Cowper's) glands

Urinary bladder

Prostate gland

Orifices of ejaculatory ducts

Bulbourethral (Cowper's) gland

Spongy (cavernous) urethra

1

2

7

3

4

5

6

Coronal section

___ Corona
___ Corpora cavernosa penis
___ Corpus spongiosum penis
___ Deep artery of penis
___ External urethral orifice
___ Glans penis
___ Prepuce

FIGURE 26.7   Internal structure of the penis viewed from the floor of the penis.

transport ova to the uterus; vagina; external organs that compose the vulva; and the mammary glands.

1. Ovaries

The *ovaries* are paired glands that resemble almonds in size and shape. Functionally, the ovaries produce ova (eggs), discharge them about once a month by a process called ovulation, and secrete female sex hormones (estrogens, progesterone, relaxin, and inhibin). The point of entrance for blood vessels and nerves is the *hilus*. The ovaries are positioned in the upper pelvic cavity, one on each side of the uterus, by a series of ligaments:

a. *Mesovarium*—Double-layered fold of the peritoneum that attaches the ovaries to the broad ligament of the uterus.

b. *Ovarian ligament*—Anchors the ovary to the uterus.

c. *Suspensory ligament*—Attaches the ovary to the pelvic wall.

With the aid of your textbook, label the ovarian ligaments in Figure 26.9 on page 525.

Histologically, the ovaries consist of the following parts:

a. *Germinal epithelium*—Layer of simple squamous or cuboidal epithelium covering the free surface of ovary.

b. *Tunica albuginea*—Collagenous connective tissue capsule immediately deep to the germinal epithelium.

c. *Stroma*—Region of connective tissue deep to the tunica albuginea. The outer region, called the *cortex*, contains the ovarian follicles; the inner region is the *medulla*.

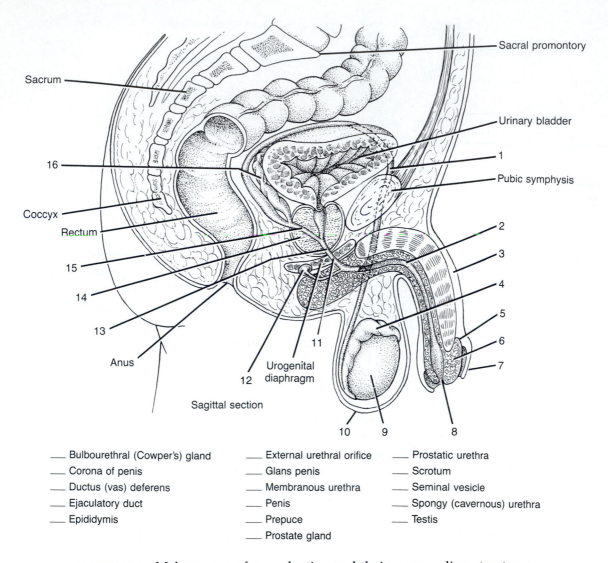

Sacral promontory

Sacrum

Urinary bladder

1

Pubic symphysis

16

2

Coccyx

3

Rectum

4

15

5

14

6

13

7

Anus

11

Urogenital
diaphragm

12

Sagittal section

10   9   8

___ Bulbourethral (Cowper's) gland   ___ External urethral orifice   ___ Prostatic urethra
___ Corona of penis   ___ Glans penis   ___ Scrotum
___ Ductus (vas) deferens   ___ Membranous urethra   ___ Seminal vesicle
___ Ejaculatory duct   ___ Penis   ___ Spongy (cavernous) urethra
___ Epididymis   ___ Prepuce   ___ Testis
   ___ Prostate gland

FIGURE 26.8 Male organs of reproduction and their surrounding structures.

d. *Ovarian follicles*—Oocytes (immature ova) and their surrounding tissue in various stages of development.

e. *Vesicular ovarian (Graafian) follicle*—Relatively large fluid-filled follicle that contains immature ovum.

f. *Corpus luteum*—Mature vesicular ovarian follicle that has ruptured to expel a secondary oocyte (potential ovum), a process called ovulation; after changes occur in the ruptured follicle, it secretes estrogens, progesterone, relaxin, and inhibin.

With the aid of your textbook, label the parts of an ovary in Figure 26.10 on page 526.

Obtain prepared slides of the ovary, examine them, and compare your observations with Figure 26.11 on page 527.

2. Uterine (Fallopian) Tubes

The *uterine (Fallopian) tubes,* or *oviducts,* extend laterally from the uterus and transport the potential ovum from the ovaries to the uterus. Fertilization normally occurs in the uterine tubes. The tubes are positioned between folds of the broad ligaments of the uterus. The funnel-shaped, open distal end of each uterine tube, called the *infundibulum,* is surrounded by a fringe of finger-like projections called *fimbriae* (FIM-brē-ē). The *ampulla* (am-POOL-la) of the uterine tube is the widest, longest portion, constituting about two-thirds of its length. The *isthmus* (IS-mus) is the short, narrow, thick-walled portion that joins to the uterus.

With the aid of your textbook, label the parts of the uterine tubes in Figure 26.9.

Histologically, the mucosa of the uterine tubes consists of ciliated columnar cells and secretory cells. The muscularis is composed of inner circular and outer longitudinal layers of smooth muscle. Wavelike contractions of the muscularis help move the ovum down into the uterus. The serosa is the outer covering.

Examine a prepared slide of the wall of the uterine tube, and compare your observations with Figure 26.12 on page 528.

3. Uterus

The *uterus* is the site of menstruation, implantation of a fertilized ovum, development of the fetus during pregnancy, and labor. Located between the urinary bladder and the rectum, the organ is shaped like an inverted pear. The uterus is subdivided into the following regions:

a. *Fundus*—Dome-shaped portion above the uterine tubes.
b. *Body*—Major, tapering portion.
c. *Cervix*—Inferior narrow opening into the vagina.
d. *Isthmus*—Constricted region between the body and the cervix.
e. *Uterine cavity*—Interior of the body.
f. *Cervical canal*—Interior of the cervix.
g. *Internal os*—Junction of the isthmus and cervical canal.
h. *External os*—Site where the cervix opens into the vagina.

Label these structures in Figure 26.9.

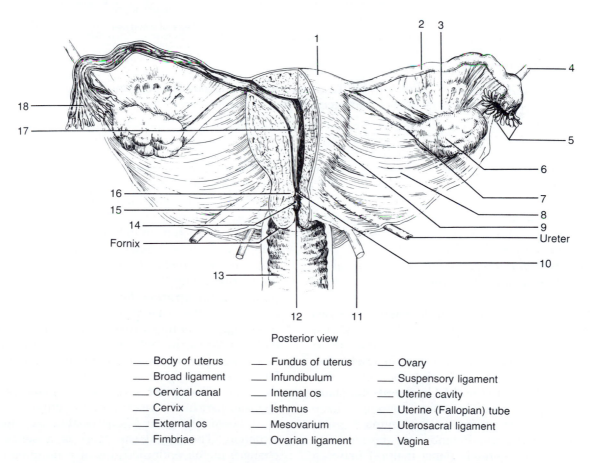

Posterior view

___ Body of uterus	___ Fundus of uterus	___ Ovary
___ Broad ligament	___ Infundibulum	___ Suspensory ligament
___ Cervical canal	___ Internal os	___ Uterine cavity
___ Cervix	___ Isthmus	___ Uterine (Fallopian) tube
___ External os	___ Mesovarium	___ Uterosacral ligament
___ Fimbriae	___ Ovarian ligament	___ Vagina

FIGURE 26.9 Uterus and associated female reproductive structures. The left side of the figure is sectioned to show internal structures.

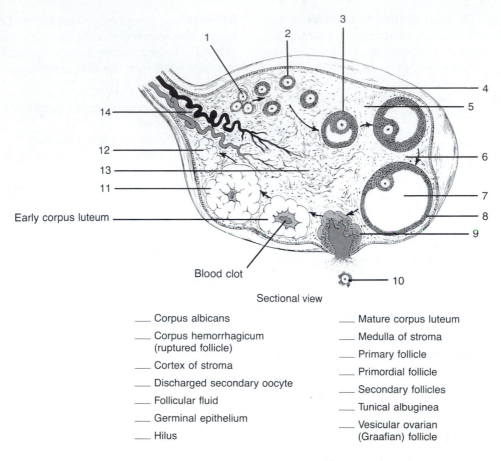

Early corpus luteum

Blood clot

Sectional view

___ Corpus albicans

___ Corpus hemorrhagicum
(ruptured follicle)

___ Cortex of stroma

___ Discharged secondary oocyte

___ Follicular fluid

___ Germinal epithelium

___ Hilus

___ Mature corpus luteum

___ Medulla of stroma

___ Primary follicle

___ Primordial follicle

___ Secondary follicles

___ Tunical albuginea

___ Vesicular ovarian
(Graafian) follicle

FIGURE 26.10   Histology of the ovary. Arrows indicate the sequence of developmental stages that occur as part of the ovarian cycle.

The uterus is maintained in position by the following ligaments:

a. *Broad ligaments*—Double folds of the parietal peritoneum that anchor the uterus to either side of the pelvic cavity.

b. *Uterosacral ligaments*—Parietal peritoneal extensions that connect the uterus to the sacrum.

c. *Cardinal (lateral cervical) ligaments*—Tissues containing smooth muscle, uterine blood vessels, and nerves. These ligaments extend below the bases of the broad ligaments between the pelvic wall and the cervix and vagina. They are the chief ligaments maintaining the position of the uterus, helping to keep it from dropping into the the vagina.

d. *Round ligaments*—Extend from the uterus to the external genitals (labia majora) between the folds of the broad ligaments.

Label the uterine ligaments in Figure 26.9.

Histologically, the uterus consists of three principal layers: endometrium, myometrium, and perimetrium (serosa). The inner ***endometrium*** is a mucous membrane with numerous glands and blood vessels and consists of two layers: (1) ***stratum functionalis,*** the layer closer to the uterine cavity that is shed during menstruation; and (2) ***stratum basalis*** (ba-SAL-is), the permanent layer that produces a new stratum functionalis after menstruation. The middle ***myometrium*** forms the bulk of the uterine wall and consists of three layers of smooth muscle. During labor, its coordinated contractions help to expel the fetus. The outer layer is the ***perimetrium (serosa),*** part of the visceral peritoneum.

As noted, the uterus is associated with menstruation. The ***menstrual cycle*** is a series of changes in the endometrium of a nonpregnant female that prepares the endometrium to receive

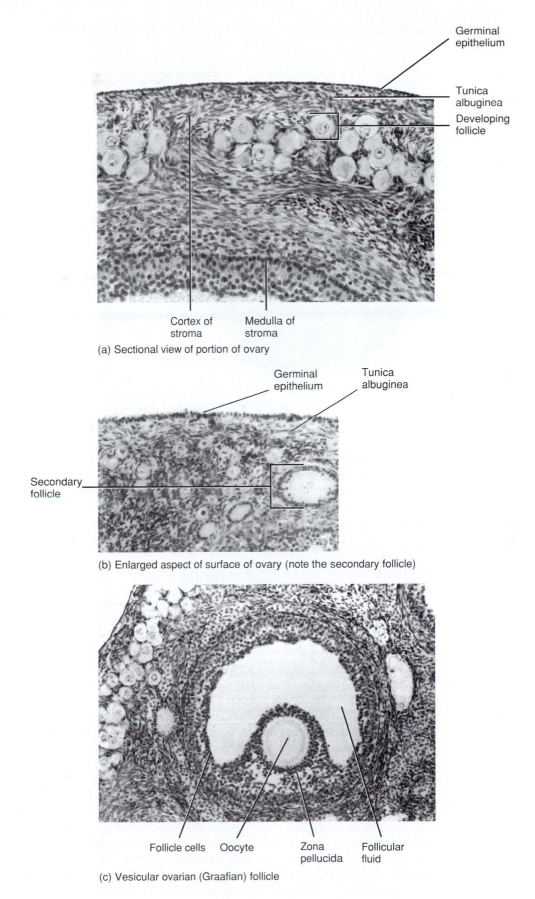

Germinal
epithelium

Tunica
albuginea

Developing
follicle

Cortex of
stroma

Medulla of
stroma

(a) Sectional view of portion of ovary

Germinal
epithelium

Tunica
albuginea

Secondary
follicle

(b) Enlarged aspect of surface of ovary (note the secondary follicle)

Follicle cells   Oocyte   Zona   Follicular
pellucida   fluid

(c) Vesicular ovarian (Graafian) follicle

FIGURE 26.11   Histology of ovary.

527

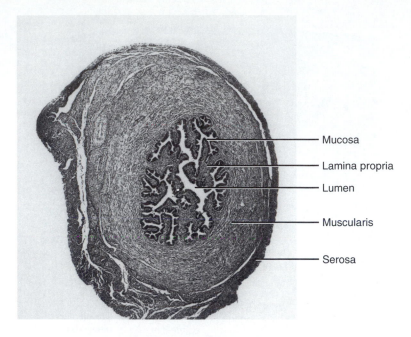

FIGURE 26.12   Histology of the uterine (Fallopian) tube.

a fertilized ovum each month (Figure 26.13). The cycle is controlled by estrogens, progesterone, follicle-stimulating hormone (FSH), luteinizing hormone (LH), and gonadotropin-releasing hormone (GnRH). The menstrual cycle is divided into the following stages:

1. *Menstrual phase (menstruation)*—The periodic discharge of 25 to 65 ml of blood, fluid, mucus, and epithelial cells due to the sudden drop in estrogens and progesterone. It lasts for about 5 days in an average 28-day cycle. During this phase, the stratum functionalis of the endometrium is sloughed off.
2. *Preovulatory (proliferative) phase*—Lasts from about day 6 through day 13 in a 28-day cycle. Rises in levels of estrogens stimulate endometrial repair and the proliferation of endometrial glands and blood vessels and result in the thickening of the endometrium. *Ovulation,* the discharge of a secondary oocyte, occurs on the 14th day.
3. *Postovulatory (secretory) phase*—Lasts from about day 15 through day 28 in a 28-day cycle. Mainly under the influence of progesterone, the vascular supply of the endometrium increases even more. Also, the endometrial glands further increase in size. These changes are greatest about one week after ovulation.

Obtain microscope slides of the endometrium showing the menstrual, preovulatory, and postovulatory phases of the menstrual cycle. See if you can note the differences in thickness of the endometrium, distribution of blood vessels, and distribution and size of the endometrial glands.

4. Vagina

A muscular, tubular organ lined with a mucous membrane, the *vagina* is the passageway for menstrual flow, the receptacle for the penis during copulation, and the inferior portion of the birth canal. The vagina is situated between the urinary bladder and rectum and extends from the cervix of the uterus to the vestibule of the vulva. Recesses called *fornices* (FOR-ni-sēz'; *fornix* = arch or vault) surround the vaginal attachment to the cervix (see Figure 26.9) and make possible the use of contraceptive diaphragms. The opening of the vagina to the exterior, the *vaginal orifice,* may be bordered by a thin fold of vascularized membrane, the *hymen.*

Label the vagina in Figure 26.9.

Histologically, the mucosa of the vagina consists of nonkeratinized stratified squamous epithelium and connective tissue that lies in a series of transverse folds, the *rugae.* The muscularis is composed of a longitudinal layer of smooth muscle.

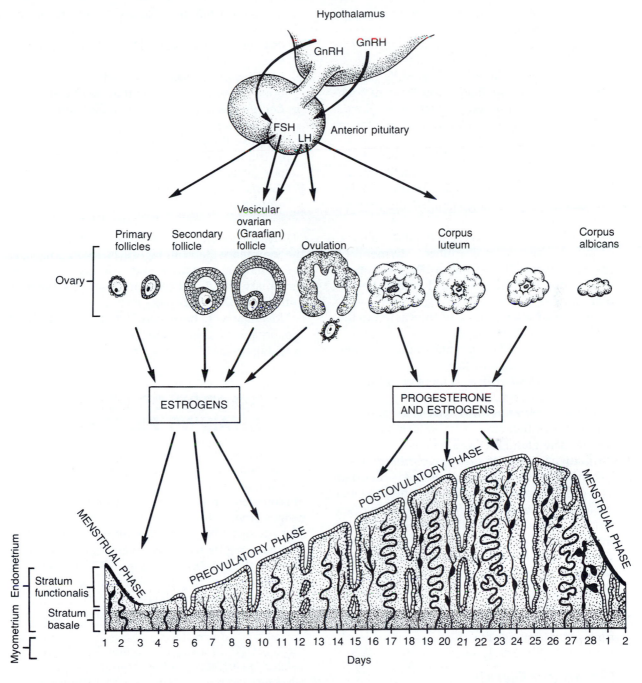

FIGURE 26.13   Menstrual cycle.

5. Vulva

The vulva (VUL-va; *volvere* = to wrap around), or ***pudendum*** (pyoo-DEN-dum), is a collective term for the external genitals of the female. It consists of the following parts:

a. ***Mons pubis***—Elevation of adipose tissue over the pubic symphysis covered by skin and pubic hair.

b. ***Labia majora*** (LĀ-bē-a ma-JŌ-ra)—Two longitudinal folds of skin that extend inferiorly and posteriorly from the mons pubis. The folds, covered by pubic hair on their superior lateral surfaces, contain abundant adipose tissue and sebaceous (oil) and sudoriferous (sweat) glands.

c. *Labia minora* (MĪ-nō-ra)—Two folds of mucous membrane medial to the labia majora. The folds have numerous sebaceous glands but few sudoriferous glands and no fat or pubic hair.

d. *Clitoris* (KLĪ-to-ris)—Small cylindrical mass of erectile tissue at the anterior junction of the labia minora. The exposed portion is called the *glans;* the covering is called the *prepuce* (foreskin).

e. *Vestibule*—Cleft between the labia minora; contains the vaginal orifice, hymen (if present), external urethral orifice, and openings of ducts of the lesser and greater vestibular glands.

f. *Vaginal orifice*—Opening of the vagina to the exterior.

g. *Hymen*—Thin fold of vascularized membrane that borders vaginal orifice.

h. *External urethral orifice*—Opening of the urethra to the exterior.

i. *Orifices of paraurethral (Skene's) glands*—Located on either side of the external urethral orifice. The glands secrete mucus.

j. *Orifices of ducts of greater vestibular* (ves-TIB-yoo-lar), or *Bartholin's, glands*—Located in a groove between the hymen and the labia minora. These glands produce a mucoid secretion that supplements lubrication during intercourse.

k. *Orifices of ducts of lesser vestibular glands*—Microscopic orifices opening into the vestibule.

Using your textbook as an aid, label the parts of the vulva in Figure 26.14.

6. Mammary Glands

The *mammary glands* are modified sweat glands (branched tubuloalveolar glands) that lie over the pectoralis major muscles and are attached to them by a layer of connective tissue. They consist of the following structures:

a. *Lobes*—About 15 to 20 compartments separated by adipose tissue.

b. *Lobules*—Small compartments in lobes that contain clusters of milk-secreting cells called *alveoli.*

c. *Secondary tubules*—Receive milk from the alveoli.

d. *Mammary ducts*—Receive milk from the secondary tubules.

e. *Lactiferous sinuses*—Expanded distal portions of mammary ducts that store milk.

f. *Lactiferous ducts*—Receive milk from lactiferous sinuses.

g. *Nipple*—Projection on the anterior the surface of the mammary gland that contains lactiferous ducts.

h. *Areola* (a-RĒ-ō-la)—Circular pigmented skin around the nipple.

With the aid of your textbook, label the parts of the mammary gland in Figure 26.15 on page 532.

Examine a prepared slide of alveoli of the mammary gland and compare your observations with Figure 26.16 on page 532.

Now that you have completed your study of the organs of the female reproductive system, label Figure 26.17 on page 533.

C. DISSECTION OF FETUS-CONTAINING PIG UTERUS

Examination of the uterus of a pregnant pig reveals that the fetuses are equally spaced in the two uterine horns. Each fetus produces a local enlargement of the horn. The litter size normally ranges from 6 to 12. Your instructor may have you dissect the fetus-containing uterus of a pregnant pig or have one available as a demonstration (Figure 26.18 on page 537). If you do a dissection, use the following directions. Also examine a chart or model of a human fetus and of a pregnant uterus if one is available.

CAUTION! *Please reread Section D, ''Precautions Related to Dissection,'' at the beginning of the laboratory manual, on page xv, before you begin your dissection.*

PROCEDURE

1. Using a sharp scissors, cut open one of the enlargements of the horn. You will see that each fetus is enclosed together with an elongated, sausage-shaped *chorionic vesicle.*

2. You will also notice many round bumps called *areolae* located over the chorionic surface.

3. The lining of the uterus together and the wall of the chorionic vesicle form the *placenta.*

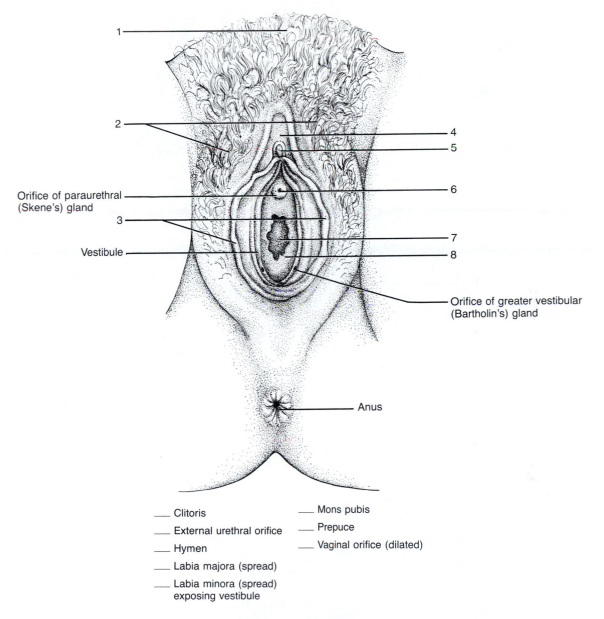

1

2

4

5

6

Orifice of paraurethral
(Skene's) gland

3

7

Vestibule

8

Orifice of greater vestibular
(Bartholin's) gland

Anus

___ Clitoris   ___ Mons pubis

___ External urethral orifice   ___ Prepuce

___ Hymen   ___ Vaginal orifice (dilated)

___ Labia majora (spread)

___ Labia minora (spread)
exposing vestibule

FIGURE 26.14   Vulva.

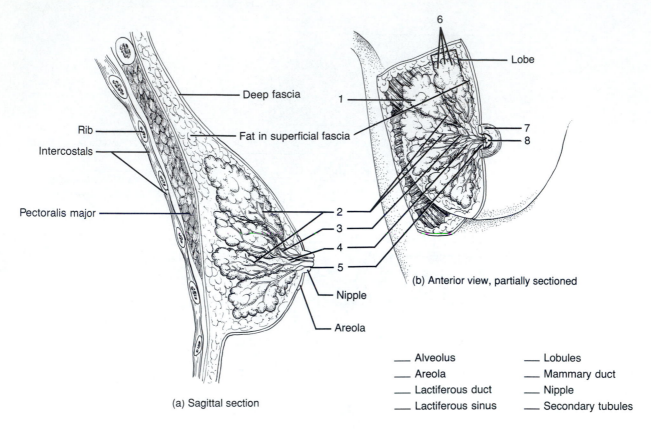

Deep fascia

Rib

Intercostals

Pectoralis major

Fat in superficial fascia

1

2

3

4

5

Nipple

Areola

6

Lobe

7

8

(a) Sagittal section

(b) Anterior view, partially sectioned

___ Alveolus   ___ Lobules

___ Areola   ___ Mammary duct

___ Lactiferous duct   ___ Nipple

___ Lactiferous sinus   ___ Secondary tubules

FIGURE 26.15   Mammary glands.

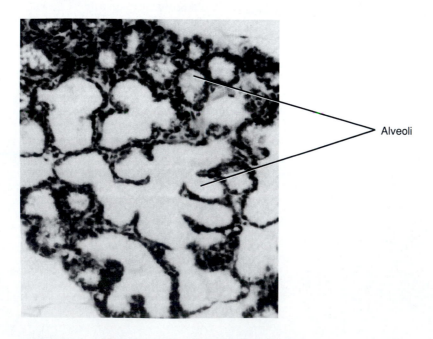

Alveoli

FIGURE 26.16   Histology of the mammary gland showing alveoli.

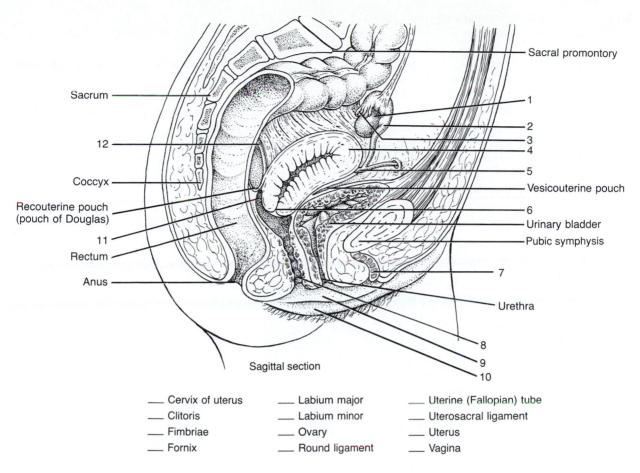

Sacral promontory

Sacrum

12

Coccyx

Recouterine pouch
(pouch of Douglas)

11

Rectum

Anus

1
2
3
4
5

Vesicouterine pouch

6

Urinary bladder

Pubic symphysis

7

Urethra

8
9
10

Sagittal section

___ Cervix of uterus   ___ Labium major   ___ Uterine (Fallopian) tube

___ Clitoris   ___ Labium minor   ___ Uterosacral ligament

___ Fimbriae   ___ Ovary   ___ Uterus

___ Fornix   ___ Round ligament   ___ Vagina

FIGURE 26.17   Female organs of reproduction and surrounding structures.

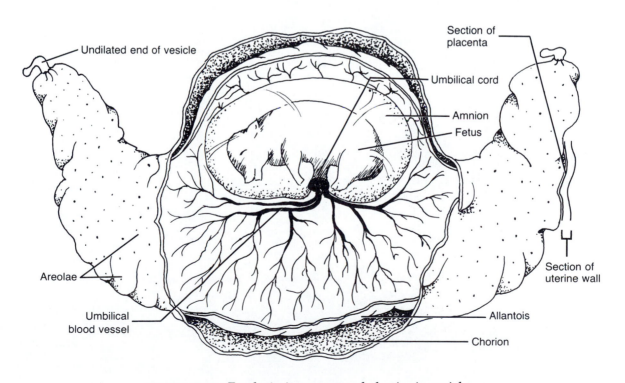

Section of placenta

Undilated end of vesicle

Umbilical cord

Amnion

Fetus

Areolae

Section of uterine wall

Umbilical blood vessel

Allantois

Chorion

FIGURE 26.18   Fetal pig in an opened chorionic vesicle.

4. Carefully cut open the chorionic vesicle. Avoid cutting or breaking the second sac surrounding the fetus itself.

5. The vesicle wall is the fusion of two extraembryonic membranes, the outer *chorion* (KOR-ē-on) and the inner *allantois* (a-LAN-tō-is). The allantois is the large sac growing out from the fetus; the umbilical cord contains its stalk.

6. The *umbilical blood vessels* are seen in the allantoic wall spreading out in all directions and are also seen entering the *umbilical cord.*

7. A thin-walled nonvascular *amnion* surrounds the fetus. This membrane is filled with *amniotic fluid,* which acts as a protective water cushion and prevents adherence of the fetus and membranes.

ANSWER THE LABORATORY REPORT QUESTIONS AT THE END OF THE EXERCISE.

Reproductive Systems

STUDENT _____   DATE _____

LABORATORY SECTION _____   SCORE/GRADE _____

PART 1. Multiple Choice

_____ 1. Structures of the male reproductive system responsible for producing sperm are the (a) efferent ducts (b) seminiferous tubules (c) seminal vesicles (d) rete testis

_____ 2. The superior portion of the male urethra is encircled by the (a) epididymis (b) testes (c) prostate gland (d) seminal vesicles

_____ 3. Cryptorchidism is a condition associated with the (a) prostate gland (b) testes (c) seminal vesicles (d) bulbourethral (Cowper's) glands

_____ 4. Weakening of the suspensory ligament would directly affect the position of the (a) mammary glands (b) uterus (c) uterine (Fallopian) tubes (d) ovaries

_____ 5. Organs in the female reproductive system responsible for transporting ova from the ovaries to the uterus are the (a) uterine (Fallopian) tubes (b) seminal vesicles (c) inguinal canals (d) none of the above

_____ 6. The name of the process that is responsible for the actual production of sperm is called (a) cryptorchidism (b) oogenesis (c) spermatogenesis (d) spermatogonia

_____ 7. Fertilization normally occurs in the (a) uterine (Fallopian) tubes (b) vagina (c) uterus (d) ovaries

_____ 8. The portion of the uterus that assumes an active role during labor is the (a) serosa (b) endometrium (c) peritoneum (d) myometrium

_____ 9. Stereocilia are associated with the (a) ductus (vas) deferens (b) oviduct (c) epididymis (d) rete testis

_____ 10. The major portion of the volume of semen is contributed by the (a) bulbourethral (Cowper's) glands (b) testes (c) prostate gland (d) seminal vesicles

_____ 11. The chief ligament supporting the uterus and keeping it from dropping into the vagina is the (a) cardinal ligament (b) round ligament (c) broad ligament (d) ovarian ligament

_____ 12. Which sequence, from inside to outside, best represents the histology of the uterus? (a) stratum basalis, stratum functionalis, myometrium, perimetrium (b) myometrium, perimetrium, stratum functionalis, stratum basalis (c) stratum functionalis, stratum basalis, myometrium, perimetrium (d) stratum basalis, stratum functionalis, perimetrium, myometrium

_____ 13. The white fibrous capsule that divides the testis into lobules is called the (a) dartos (b) raphe (c) tunica albuginea (d) germinal epithelium

_____ 14. Which sequence best represents the course taken by spermatozoa from their site of origin to the exterior? (a) seminiferous tubules, efferent ducts, epididymis, ductus (vas)

deferens, ejaculatory duct, urethra (b) seminiferous tubules, efferent ducts, epididymis, ductus (vas) deferens, urethra, ejaculatory duct (c) seminiferous tubules, efferent ducts, ductus (vas) deferens, epididymis, ejaculatory duct, urethra (d) seminiferous tubules, epididymis, efferent ducts, ductus (vas) deferens, ejaculatory duct, urethra

———— 15. The ovaries are anchored to the uterus by the (a) ovarian ligament (b) broad ligament (c) suspensory ligament (d) mesovarium

———— 16. The terminal duct for the male reproductive system is the (a) urethra (b) ductus (vas) deferens (c) inguinal canal (d) ejaculatory duct

———— 17. The site of sperm maturation is the (a) ductus (vas) deferens (b) spermatic cord (c) epididymis (d) testes

———— 18. Which of the following is the site of menstruation, implantation of a fertilized ovum, development of the fetus during pregnancy, and labor? (a) uterus (b) uterine (Fallopian) tubes (c) vagina (d) cervix

———— 19. Branched tubuloalveolar glands lying over the pectoralis major muscles are (a) lesser vestibular glands (b) adrenal (suprarenal) glands (c) mammary glands (d) greater vestibular (Bartholin's) glands

PART 2. Completion

20. Discharge of a secondary oocyte from the ovary about once each month is a process referred to as

————————————.

21. The inferior, narrow portion of the uterus that opens into the vagina is the

————————————.

22. The clusters of milk-secreting cells of the mammary glands are referred to as

————————————.

23. The distal end of the penis is a slightly enlarged region called the ————————————.

24. Covering the slightly enlarged region of the penis is a loosely fitting skin called the

————————————.

25. The circular pigmented area surrounding each nipple of the mammary glands is the

————————————.

26. After a secondary oocyte leaves the ovary, it enters the open, funnel-shaped distal end of the uterine

(Fallopian) tube called the ————————————.

27. The portion of a spermatozoon that contains the nucleus and acrosome is the

————————————.

28. Vasectomy refers to removal of a portion of the ————————————.

29. The mass of erectile tissue in the penis that contains the spongy urethra is the

————————————.

30. Both the vesicular ovarian (Graafian) follicle and ———————————— of the ovary secrete hormones.

31. The superior dome-shaped portion of the uterus is called the ————————————.

32. The ———————————— anchor the uterus to either side of the pelvic cavity.

33. The passageway for menstrual flow and inferior portion of the birth canal is the

————————————.

34. Two longitudinal folds of skin that extend inferiorly and posteriorly from the mons pubis and are covered with pubic hair are the _____.

35. The _____ is a small mass of erectile tissue at the anterior junction of the labia minora.

36. The thin fold of vascularized membrane that borders the vaginal orifice is the _____.

37. Complete the following sequence for the passage of milk: alveoli, secondary tubules, lactiferous sinuses, _____, lactiferous ducts, nipple.

38. The layer of simple epithelium covering the free surface of the ovary is the _____.

39. The phase of the menstrual cycle between days 6 and 13 during which endometrial repair occurs is the _____ phase.

40. During menstruation, the stratum _____ the endometrium is sloughed off.

27

Development

Development refers to the sequence of events starting with fertilization of a secondary oocyte and ending with the formation of a complete organism. Consideration will be given to how reproductive cells are produced and to a few developmental events associated with pregnancy.

A. SPERMATOGENESIS

The process by which the testes produce spermatozoa involves several phases, including meiosis, and is called *spermatogenesis* (sper'-ma-tō-JEN-e-sis). In order to understand spermatogenesis, review the following concepts.

1. In sexual reproduction, a new organism is produced by the union and fusion of sex cells called *gametes* (*gameto* = to marry). Male gametes, produced in the testes, are called sperm cells, and female gametes, produced in the ovaries, are called ova.
2. The cell resulting from the union and fusion of gametes, called a *zygote* (*zygo* = joined), contains two full sets of chromosomes (DNA), one from each parent. Through repeated mitotic cell divisions, a zygote develops into a new organism.
3. Gametes differ from all other body cells (somatic cells) in that they contain the *haploid* (one-half) *chromosome number,* symbolized as *n*. In humans, this number is 23, which composes a single set of chromosomes. Uninucleated somatic cells contain the *diploid chromosome number,* symbolized as 2*n*. In

humans, this number is 46, which composes two sets of chromosomes.

4. In a diploid cell, two chromosomes that belong to a pair are called *homologous* (*homo* = same) *chromosomes (homologues).* In human diploid cells, the members of 22 of the 23 pairs of chromosomes are morphologically similar and are called *autosomes.* The other pair constitutes the *sex chromosomes,* designated as X and Y. In the female, the homologous pair of sex chromosomes consists of two X chromosomes; in the male, the pair consists of an X and a Y chromosome.
5. If gametes were diploid (2*n*), like somatic cells, the zygote would contain twice the diploid number (4*n*). With every succeeding generation the chromosome number would continue to double, and normal development could not occur.
6. This continual doubling of the chromosome number does not occur because of *meiosis,* a process of cell division by which gametes produced in the testes and ovaries receive the haploid chromosome number. Thus, when haploid (*n*) gametes fuse, the zygote contains the diploid chromosome number (2 *n*) and can undergo normal development.

In humans, spermatogenesis takes about 74 days. The seminiferous tubules are lined with immature cells called *spermatogonia* (sper'-ma-tō-GŌ-nē-a; *sperm* = seed, *gonium* = generation, or offspring), or sperm mother cells (see Figure 26.2). Singular is *spermotogonium.* These cells develop from *primordial* (*primordialis* = primitive, or early form) *germ cells* that arise from yolk sac endoderm and enter the testes early in development. In the embryonic testes, the primordial germ cells differentiate into spermatogonia but remain dormant until they begin to undergo mitotic proliferation at puberty. Spermatogonia contain the diploid (2n) chromosome number and represent a heterogenous group of cells in which three subtypes can be distinguished. These are referred to as Pale Type A, Dark Type A, and Type B and are distinguished by the appearance of their nuclear chromatin. Pale Type A spermatogonia remain relatively undifferentiated and capable of extensive mitotic division. Following division, some of the daughter cells remain undifferentiated and serve as a reservoir of precursor cells to prevent depletion of the stem cell population. Such cells remain near the basement membrane. The remainder of the daughter cells differentiate into Type B spermatogonia. These cells lose contact with the basement membrane of the seminiferous tubule, undergo certain developmental changes, and become known as *primary spermatocytes* (SPER-ma-tō-sītz'). Primary spermatocytes, like spermatogonia, are diploid (2n)—that is, they have 46 chromosomes. Dark Type A spermatogonia are believed to represent reserve stem cells, only becoming activated if Pale Type A cells become critically depleted.

1. Reduction Division (Meiosis I)

Each primary spermatocyte enlarges before dividing. Then two nuclear divisions take place as part of meiosis. In the first, DNA is replicated and 46 chromosomes (each made up of two chromatids) form and move toward the equatorial plane of the nucleus. There they line up by homologous pairs so that there are 23 pairs of duplicated chromosomes in the center of the nucleus. This pairing of homologous chromosomes is called *synapsis.* The four chromatids of each homologous pair then become associated with each other to form a *tetrad.* In a tetrad, portions of one

chromatid may be exchanged with portions of another. This process, called *crossing over,* permits an exchange of genes among chromatids (Figure 27.1) that results in the recombination of genes. Thus, the spermatozoa eventually produced are genetically unlike each other and unlike the cell that produced them—one reason for the great variation among humans. Next, the meiotic spindle forms and the kinetochore microtubules organized by the centromeres extend toward the poles of the cell. As the pairs separate, one member of each pair migrates to opposite poles of the dividing nucleus. The random arrangement of chromosome pairs on the spindle is another reason for variation among humans. The cells formed by the first nuclear division (reduction division) are called *secondary spermatocytes.* Each cell has 23 chromosomes—the haploid number. Each chromosome of the secondary spermatocytes, however, is made up of two chromatids. Moreover, the genes of the chromosomes of secondary spermatocytes may be rearranged as a result of crossing over.

2. Equatorial Division (Meiosis II)

The second nuclear division of meiosis is *equatorial division.* There is no replication of DNA. The chromosomes (each composed of two chromatids) line up in single file around the equatorial plane, and the chromatids of each chromosome separate from each other. The cells formed from the equatorial division are called *spermatids.* Each contains half the original chromosome number, or 23 chromosomes, and is haploid. Each primary spermatocyte therefore produces four spermatids by meiosis (reduction division and equatorial division). Spermatids lie close to the lumen of the seminiferous tubule.

3. Spermiogenesis

The final stage of spermatogensis, called *spermiogenesis* (sper'-mē-ō-JEN-e-sis), involves the

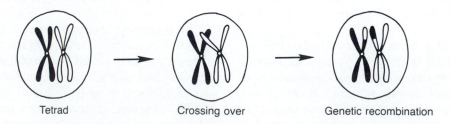

Tetrad   Crossing over   Genetic recombination

FIGURE 27.1 Crossing over within a tetrad, resulting in genetic recombination.

maturation of spermatids into spermatozoa. Each spermatid embeds in a sustentacular (Sertoli) cell and develops a head with an acrosome (described shortly) and a flagellum (tail). Sustentacular cells extend from the basement membrane to the lumen of the seminiferous tubule, where they nourish the developing spermatids. Because there is no cell division in spermiogenesis, each spermatid develops into a single *spermatozoon (sperm cell).* The release of a spermatozoon from a sustentacular cell is known as *spermiation.*

Spermatozoa enter the lumen of the seminiferous tubule and migrate to the ductus epididymis, where in 10 to 14 days they complete their maturation and become capable of fertilizing an ovum. Spermatozoa are also stored in the ductus (vas) deferens. Here, they can retain their fertility for up to several months.

With the aid of your textbook, label Figure 27.2.

B. OOGENESIS

The formation of haploid (*n*) secondary oocytes in the ovary involves several phases, including meiosis, and is referred to as *oogenesis* (ō'-ō-JEN-e-sis). With some important exceptions, oogenesis occurs in essentially the same manner as spermatogenesis.

1. Reduction Division (Meiosis I)

During early fetal development, primordial germ cells migrate from the endoderm of the yolk sac to the ovaries. There, germ cells differentiate into *oogonia* (ō'-ō-GŌ-nē-a; *oo* = egg), cells that can give rise to other cells that develop into ova (Figure 27.3 on page 543). Singular is *oogonium.* Oogonia are diploid (2*n*) cells that divide mitotically to produce a large population of cells. At about the third month of prenatal development, oogonia divide and develop into larger diploid (2*n*) cells called *primary oocytes* (Ō'-ō-sitz). These cells enter prophase of reduction division (meiosis I) but do not complete it until after the female reaches puberty. Each primary oocyte is surrounded by a single layer of flattened epithelial cells (follicular) and the entire structure is called a *primordial follicle.* Primordial follicles do not begin further development until they are stimulated by follicle-stimulating hormone (FSH) from the anterior pituitary gland, which has responded to gonadotropin-releasing hormone (GnRH) from the hypothalamus. Many primor-

dial follicles degenerate (atresia) before birth and are known as *atretic follicles.*

Starting with puberty, several primary follicles respond each month to the rising level of FSH and form *primary follicles,* which are surrounded by a single layer of cuboidal-shaped follicular cells. As the preovulatory phase of the menstrual cycle proceeds and lutzinizing hormone (LH) is secreted from the anterior pituitary, one of the primary follicles reaches a stage in which meiosis resumes and the diploid primary oocyte completes reduction division (meiosis I). Synapsis, tetrad formation, and crossing over occur, and two cells of unequal size, both with 23 chromosomes (*n*) of two chromatids each, are produced. The smaller cell, called the *first polar body,* is essentially a packet of discarded nuclear material. The larger cell, known as the *secondary oocyte,* receives most of the cytoplasm. Each secondary oocyte is surrounded by several layers of cuboidal then columnar epithelial cells. The entire structure is called a *secondary (growing) follicle.* Once a secondary oocyte is formed, it proceeds to the metaphase of equatorial division (meiosis II) and then stops at this stage. The equatorial division (meiosis II) is completed following ovulation and fertilization.

2. Equatorial Division (Meiosis II)

At ovulation, the secondary oocyte, with its polar body and some surrounding supporting cells, is discharged. The discharged secondary oocyte enters the uterine (Fallopian) tube and, if spermatozoa are present and fertilization occurs, the second division, the equatorial division (meiosis II), is completed.

3. Maturation

The secondary oocyte produces two cells of unequal size, both of them haploid (*n*). The larger cell eventually develops into an *ovum,* or mature egg; the smaller is the *second polar body.*

The first polar body may undergo another division to produce two polar bodies. If it does, meiosis of the primary oocyte results in a single haploid (*n*) secondary oocyte and three haploid (*n*) polar bodies. In any event, all polar bodies disintegrate. Thus, each oogonium produces a single secondary oocyte, whereas each spermatocyte produces four spermatozoa. Spermatogenesis and oogenesis differ in other ways as well. Spermatogenesis is a continuous process that begins in puberty and continues throughout life; oogenesis begins at menstruation and ends at

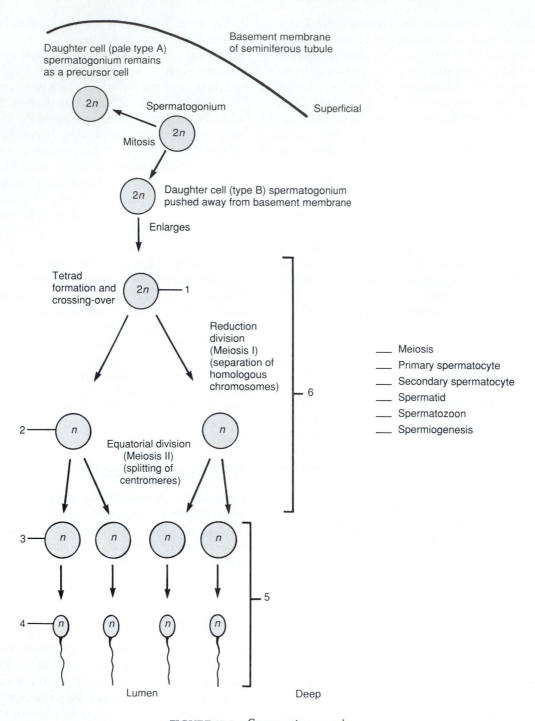

FIGURE 27.2 Spermatogenesis.

menopause. Also, spermatozoa are quite small, have flagella for locomotion, and contain few nutrients; a secondary oocyte is larger, lacks flagella, and contains more nutrients for nourishment until implantation occurs in the uterus.

With the aid of your textbook, label Figure 27.3.

C. EMBRYONIC PERIOD

The *embryonic period* is the first 2 months of development. The developing human is called an *embryo.*

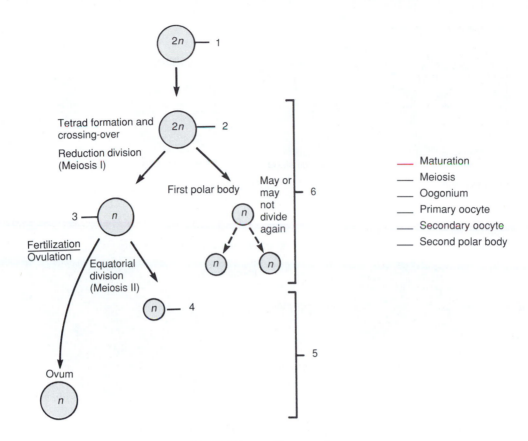

FIGURE 27.3 Oogenesis.

1. Fertilization

The term *fertilization* refers to the penetration of a secondary oocyte by a spermatozoon and the subsequent union of the sperm nucleus and the nucleus of the oocyte (Figure 27.4a). Fertilization normally occurs in the uterine (Fallopian) tube when the oocyte is about one-third of the way down the tube, usually within 24 hr after ovulation. Peristaltic contractions and the action of cilia transport the oocyte through the uterine tube.

In addition to assisting in the transport of sperm, the female reproductive tract also confers on sperm the capacity to fertilize a secondary oocyte. Although sperm undergo maturation in the epididymis, they are still not able to fertilize an oocyte until they have remained in the female reproductive tract for about 10 hr. The functional changes that sperm undergo in the female reproductive tract that allow them to fertilize a secondary oocyte are referred to as *capacitation* (ka'-pas'-i'-TĀ-shun). During this process, secreted substances in the female reproductive tract cause the membrane around the acrosome to become fragile and to release three enzymes: hyaluroni-

dase, acrosin, and neuraminidase. The enzymes help sperm penetrate the *corona radiata,* several layers of follicular cells around the oocyte, and a gelantinous glycoprotein layer internal to the corona radiata called the *zona pellucida* (pe-LOO-si-da) (Figure 27.4a and b). Spermatozoa bind to receptors in the zona pellucida. Once this is accomplished, normally only one spermatozoon enters and fertilizes a secondary oocyte. Once union is achieved, the ionic changes in the plasma membrane of the oocyte block the entry of other sperm. Also, enzymes produced by the fertilized ovum (egg) alter receptor sites so that sperm already bound are detached and others are prevented from binding. In this way, *polyspermy,* fertilization by more than one spermatozoon, is prevented. Once a spermatozoon has entered a secondary oocyte, the oocyte completes equatorial division (meiosis II). The secondary oocyte divides into a larger ovum (mature egg) and a smaller second polar body that fragments and disintegrates.

When a spermatozoon has entered a secondary oocyte, the tail is shed and the nucleus in the head develops into a structure called the *male*

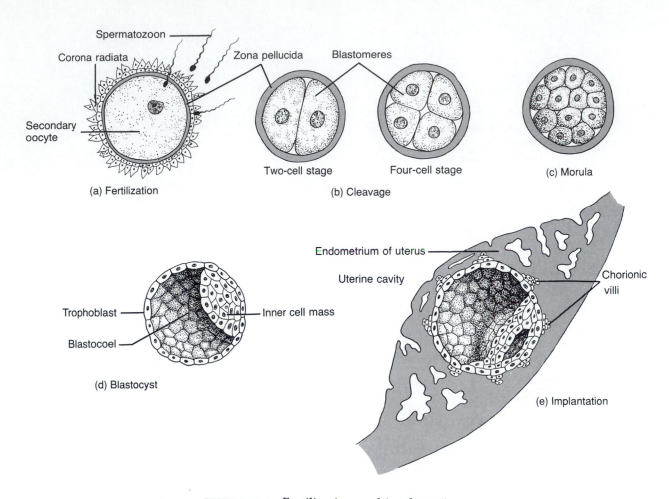

FIGURE 27.4   Fertilization and implantation.

pronucleus. The nucleus of the oocyte develops into a *female pronucleus.* After the pronuclei are formed, they fuse to produce a *segmentation nucleus.* The segmentation nucleus contains 23 chromosomes *(n)* from the male pronucleus and 23 chromosomes *(n)* from the female pronucleus. Thus, the fusion of the haploid *(n)* pronuclei restores the diploid number *(2n).* The fertilized ovum, consisting of a segmentation nucleus, cytoplasm, and zona pellucida, is called a *zygote* (*zygotos* = yoked).

2. Formation of the Morula

Immediately after fertilization, rapid cell division of the zygote takes place. This early division of the zygote is called *cleavage* (Figure 27.4b). During this time, the dividing cells are contained by the zona pellucida. The progressively smaller cells produced by cleavage are called *blastomeres*

(BLAS-tō-mērz). A few days after fertilization, the successive cleavages have produced a solid mass of cells, the *morula* (MOR-yoo-la), or mulberry, which is about the same size as the original zygote (Figure 27.4c).

3. Development of the Blastocyst

As the number of cells in the morula increases, it moves from the original site of fertilization down through the ciliated uterine (Fallopian) tube toward the uterus and enters the uterine cavity. By this time, the dense cluster of cells is altered to form a hollow ball of cells. The mass is now referred to as a *blastocyst* (Figure 27.4d). The blastocyst is differentiated into an outer covering of cells called the *trophoblast* (TRŌF-ō-blast; *troph* = nourish), an *inner cell mass (embryoblast)*, and an internal fluid-filled cavity called the *blastocoel* (BLAS-tō-sēl). The trophoblast ultimately forms part of the membranes composing the fetal portion of the placenta; the inner cell mass develops into the embryo.

4. Implantation

The blastocyst remains free within the cavity of the uterus from 2 to 4 days before it actually attaches to the uterine wall. During this time, nourishment is provided by secretions of the endometrium, sometimes called uterine milk. The attachment of the blastocyst to the endometrium occurs 7 to 8 days after fertilization and is called *implantation* (Figure 27.4e).

PROCEDURE

1. Obtain prepared slides of the embryonic development of the sea urchin or starfish. First try to find a zygote. This will appear as a single cell surrounded by an inner fertilization membrane and an outer, jellylike membrane. Draw a zygote in the spaces provided.
2. Now find several cleavage stages. See if you can isolate 2-cell, 4-cell, 8-cell, and 16-cell stages. Draw the various stages in the spaces provided.
3. Try to find a blastula (called a blastocyst in humans), a hollow ball of cells with a lighter center created by the presence of the blastocoel. Draw a blastula in the space provided.

4-cell stage

8-cell stage

16-cell stage

Blastula

Zygote

2-cell stage

5. Primary Germ Layers

Following implantation, the inner cell mass of the blastocyst differentiates into three *primary germ layers: ectoderm, endoderm,* and *mesoderm.* The primary germ layers are embryonic tissues from which all tissues and organs of the body will develop. Various movements of groups of cells leading to establishment of primary germ layers are called *gastrulation*. Cells of the inner cell mass divide by mitosis and form two cavities: the *amniotic cavity* and *extraembryonic coelom* (Figure 27.5). In gastrulation, the layer of cells nearest the amniotic cavity is the ectoderm; the layer nearest the extraembryonic coelom is the endoderm. Cells that form mesoderm develop between the ectoderm and endoderm.

As the embryo develops, the endoderm becomes the epithelium that lines most of the gastrointestinal tract, urinary bladder, gallbladder, liver, pharynx, larynx, trachea, bronchi, lungs, vagina, urethra, and thyroid, parathyroid, and thymus glands, among other structures. The mesoderm develops into muscle; cartilage, bone and other connective tissues; bone marrow, lymphoid tissue, endothelium of blood and lymphatic vessels, gonads, dermis of the skin, and other structures. The ectoderm develops into the entire nervous system, epidermis of skin, epidermal derivatives of the skin, and portions of the eye and other sense organs.

6. Embryonic Membranes

During the embryonic period, the *embryonic membranes* form outside the embryo. Collectively, they function to protect and nourish the embryo and later the fetus. These membranes include the *yolk sac, amnion, chorion* (KŌ-rē-on), and *allantois* (a-LAN-tō-is; *allas* = sausage) (see Figures 27.5 and 27.6).

The *placenta* (pla-SEN-ta) is a mostly vascular organ formed by the chorion of the embryo and a portion of the endometrium (decidua basalis) of the mother (Figure 27.6). It functions to exchange nutrients and wastes between fetus and mother and secretes the hormones necessary to maintain pregnancy. When the baby is delivered, the placenta detaches from the uterus and is called the afterbirth. The scar that marks the site of the entry of the fetal umbilical cord into the abdomen is called the *umbilicus*. The umbilical cord contains blood vessels that (1) deliver fetal blood containing CO_2 and wastes to the placenta and

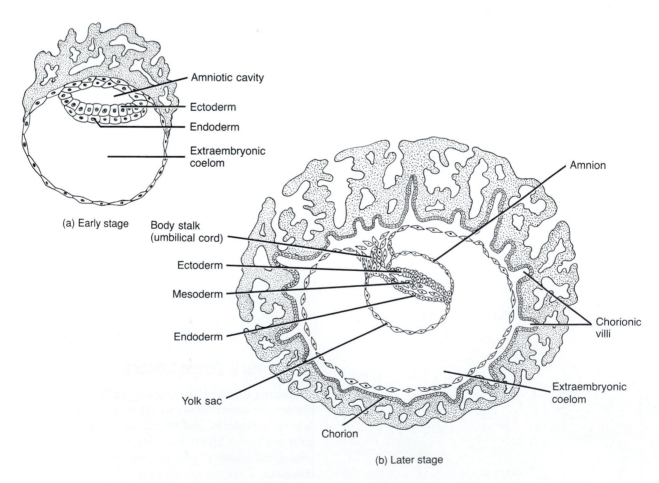

(a) Early stage

(b) Later stage

FIGURE 27.5 Formation of the three primary germ layers and embryonic membranes.

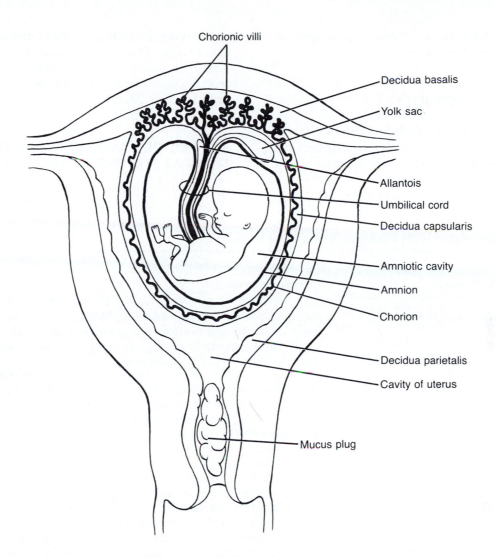

FIGURE 27.6   Embryonic membranes.

(2) return fetal blood containing oxygen and nutrients from the placenta.

D. FETAL PERIOD

During the *fetal period,* the months of development after the second month, all the organs of the body grow rapidly from the original primary germ layers, and the organism takes on a human appearance. During this time, the developing human is called a *fetus.* Some of the principal changes associated with fetal growth are summarized in Table 27.1.

ANSWER THE LABORATORY REPORT QUESTIONS AT THE END OF THE EXERCISE.

TABLE 27.1

Changes Associated with Embryonic and Fetal Growth

End of month	Approximate size and weight	Representative changes
1	0.6 cm (3/16 in.)	Eyes, nose, and ears not yet visible. Backbone and vertebral canal form. Small buds that will develop into upper and lower extremities form. Heart forms and starts beating. Body systems begin to form. CNS appears at the start of the third week.
2	(3 cm (1¼ in.) 1 g (⅟30 oz)	Eyes far apart, eyelids fused, nose flat. Ossification begins. Limbs become distinct as upper and lower extremities. Digits well formed. Major blood vessels form. Many internal organs continue to develop.
3	7.5 cm (3 in.) 28 g (1 oz)	Eyes almost fully developed but eyelids still fused; nose develops bridge and external ears are present. Ossification continues. Extremities are fully formed and nails develop. Heartbeat can be detected. Urine starts to form. Fetus begins to move, but movement cannot be detected by mother. Body systems continue to develop.
4	18 cm (6½–7 in.) 113 g (4 oz)	Head large in proportion to rest of body. Face takes on human features and hair appears on head. Skin bright pink. Many bones ossify, joints begin to form. Rapid development of body systems.
5	25–30 cm (10–12 in.) 227–454 (½–1 lb)	Head less disproportionate to rest of body. Fine hair (lanugo) covers body. Skin still bright pink. Brown fat forms, the site of heat production. Fetal movements commonly felt by mother. Rapid development of body systems.
6	27–35 cm (11–14 in.) 567–681 g (1¼–1½ lb)	Head become even less disproportionate to rest of body. Eyelids separate and eyelashes form. Substantial weight gain. Skin wrinkled and pink. Type-II alveolar cells begin to produce surfactant (surface-active agent).
7	32–42 (13–17 in.) 1,135–1,362 g (2½–3 lb)	Head and body more proportionate. Skin wrinkled and pink. Seven-month fetus (premature baby) capable of survival because lungs can breathe air and CNS can control respiration and body temperature. Fetus assumes upside-down position.
8	41–45 cm (16½–18 in.) 2,043–2,270 g (4½–5 lb)	Subcutaneous fat deposited. Skin less wrinkled. Testes descend into scrotum. Bones of head soft. Changes of survival much greater at end of 8th month.
9	50 cm (20 in.) 3,178–3,405 g (7–7½ lb)	Additional subcutaneous fat accumulates. Lanugo shed. Nails extend to tips of fingers and maybe even beyond.

Development

STUDENT _____   DATE _____

LABORATORY SECTION _____   SCORE/GRADE _____

PART 1. Multiple Choice

_____ 1. The basic difference between spermatogenesis and oogenesis is that (a) two more polar bodies are produced in spermatogenesis (b) the secondary oocyte contains the haploid chromosome number, whereas the mature sperm contains the diploid number (c) in oogenesis, one secondary oocyte is produced, and in spermatogenesis four mature sperm are produced (d) both mitosis and meiosis occur in spermatogenesis, but only meiosis occurs in oogenesis

_____ 2. The union of a sperm nucleus and a secondary oocyte nucleus resulting in formation of a zygote is referred to as (a) implantation (b) fertilization (c) gestation (d) parturition

_____ 3. The most advanced stage of development for these stages is the (a) morula (b) zygote (c) ovum (d) blastocyst

_____ 4. Damage to the mesoderm during embryological development would directly affect the formation of (a) muscle tissue (b) the nervous system (c) the epidermis of the skin (d) hair, nails, and skin glands

_____ 5. The placenta, the organ of exchange between mother and fetus, is formed by union of the endometrium with the (a) yolk sac (b) amnion (c) chorion (d) umbilicus

_____ 6. One oogonium produces (a) one ovum and three polar bodies (b) two ova and two polar bodies (c) three ova and one polar body (d) four ova

_____ 7. Implantation is defined as (a) attachment of the blastocyst to the uterine (Fallopian) tube (b) attachment of the blastocyst to the endometrium (c) attachment of the embryo to the endometrium (d) attachment of the morula to the endometrium

_____ 8. Epithelium lining most of the gastrointestinal tract and a number of other organs is derived from (a) ectoderm (b) mesoderm (c) endoderm (d) mesophyll

_____ 9. The nervous system is derived from the (a) ectoderm (b) mesoderm (c) endoderm (d) mesophyll

_____ 10. Which of the following is *not* an embryonic membrane? (a) amnion (b) placenta (c) chorion (d) allantois

PART 2. Completion

11. A normal human sperm cell, as a result of meiosis, contains _____ chromosomes.
12. The process that permits an exchange of genes resulting in their recombination and a part of the

variation among humans is called _____.

13. The result of meiosis in spermatogenesis is that each primary spermatocyte produces four

_____.

14. The stage of spermatogenesis that results in maturation of spermatids into spermatozoa is called

_____.

15. The afterbirth expelled in the final stage of delivery is the _____.

16. After the second month, the developing human is referred to as a(n) _____.

17. Embryonic tissues from which all tissues and organs of the body develop are called the

_____.

18. The cells of the inner cell mass divide to form two cavities: amniotic cavity and

_____.

19. Somatic cells that contain two sets of chromosomes are referred to as _____.

20. At the end of the _____ month of development, a heartbeat can be detected.

28

Genetics

Genetics is the branch of biology that studies inheritance. *Inheritance* is the passage of hereditary traits from one generation to another. It is through the passage of hereditary traits that we acquire our characteristics from our parents and transmit our characteristics to our children. If all individuals were brown-eyed, we could learn nothing of the hereditary basis of eye color. However, because some people are blue-eyed and marry brown-eyed people, we can gain some knowledge of how hereditary traits are transmitted. We constantly analyze the genetic bases of the *differences* between individuals. Some of these differences occur normally, such as differences in eye color, blood groups, or the ability to taste PTC (phenylthiocarbamide). Other differences are abnormal: physical abnormalities and abnormalities in the processes of metabolism.

A. GENOTYPE AND PHENOTYPE

The vast majority of human cells, except gametes, contain 23 pairs of chromosomes (diploid number) in their nuclei. One chromosome from each pair comes from the mother, and the other comes from the father. The two chromosomes that belong to a pair are called *homologous* (hō-MOL-ō-gus) *chromosomes,* and these homologues contain genes that control the same traits. The homologue of a chromosome that contains a gene for height also contains a gene for height.

The relationship of genes to heredity can be illustrated by the disorder called *phenylketon-*

uria, or *PKU* (see Figure 28.1). People with PKU are unable to manufacture the enzyme phenylalanine hydroxylase. Current belief is that PKU results from the presence of an abnormal gene symbolized as *p*. The normal gene is symbolized as *P*. *P* and *p* are said to be alleles. An *allele* is one of many alternative forms of a gene, occupying the same *locus* (position of a gene on a chromosome) in homologous chromosomes. The chromosome that has the gene that directs phenylalanine hydroxylase production will have either *p* or *P* on it. Its homologue will also have either *p* or *P*. Thus, every individual will have one of the following genetic makeups, or *genotypes* (JĒ-nō-tīps): *PP, Pp,* or *pp*. Although people with genotypes of *Pp* have the abnormal gene, only those with genotype *pp* suffer from the disorder because the normal gene masks the abnormal one. A gene that masks the expression of its allele is called the *dominant gene,* and the trait expressed is said to be a dominant trait. The homologous gene that is masked is called the *recessive gene.* The trait expressed when two recessive genes are present is called the recessive trait. Several dominant and recessive traits inherited in human beings are listed in Table 28.1.

Traditionally, the dominant gene is symbolized with a capital letter and the recessive one with a lowercase letter. When the same genes appear on homologous chromosomes, as in *PP* or *pp,* the person is said to be *homozygous* for a trait. When the genes on homologous chromosomes are different, however, as in *Pp,* the person is said to be *heterozygous* for the trait. *Phenotype*

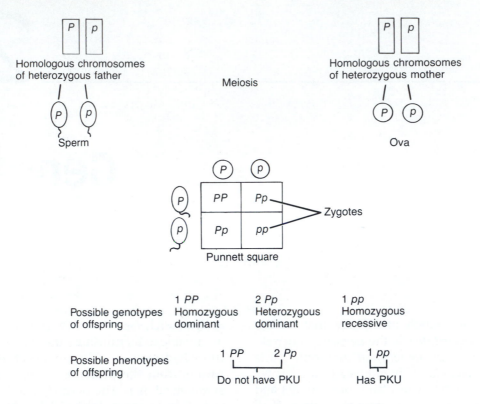

FIGURE 28.1   Inheritance of phenylketonuria (PKU).

(FĒ-nō-tīp; *pheno* = showing) refers to how the genetic composition is expressed in the body. An individual with *Pp* has a different genotype than one with *PP*, but both have the same phenotype—which in this case is normal production of phenylalanine hydroxylase.

Table 28.1
Hereditary Traits in Human Beings

Dominant	Recessive
Curly hair	Straight hair
Dark brown hair	All other colors
Normal skin pigmentation	Albinism
Nearsightedness or farsightedness	Normal vision
Normal color vision	Color blindness
Syndactylism (webbed digits)	Normal digits
Diabetes insipidus	Normal excretion
Huntington's chorea	Normal nervous system
A or B blood factor	O blood factor
Rh blood factor	No Rh blood factor
Normalcy	Phenylketonuria (PKU)

B. PUNNETT SQUARES

To determine how gametes containing haploid chromosomes unite to form diploid fertilized eggs, special charts called *Punnett squares* are used. The Punnett square is merely a device that helps one visualize all the possible combinations of male and female gametes and is invaluable as a learning exercise in genetics. Usually, the male gametes (sperm cells) are placed at the side of the square and the female gametes (ova) at the top (Figure 28.1). The spaces in the chart represent the possible combinations of male and female gametes that could form fertilized eggs. Possible combinations are determined simply by dropping the female gamete on the left into the two boxes below it and dropping the female gamete on the right into the two spaces under it. The upper male gamete is then moved across to the two spaces in line with it, and the lower male gamete is moved across to the two spaces in line with it.

C. SEX INHERITANCE

Lining up human chromosomes in pairs reveals that the last pair (the twenty-third pair) differs in males and in females (Figure 28.2a). In females, the pair consists of two rod-shaped chromosomes designated as *X* chromosomes. One *X* chromosome is also present in males, but its mate is hook-shaped and called a *Y* chromosome. The *XX* pair in the female and the *XY* pair in the male are called the *sex chromosomes,* and all other pairs of chromosomes are called *autosomes.*

The sex of an individual is determined by the sex chromosomes (Figure 28.2b). When a spermatocyte undergoes meiosis to reduce its chromosome number from diploid to haploid, one daughter cell will contain the X chromosome and the other will contain the Y chromosome. When the secondary oocyte is fertilized by an X-bearing sperm, the offspring normally will be a female (XX). Fertilization by a Y sperm normally produces a male (XY).

Sometimes chromosomes fail to move toward opposite poles of a cell in meiotic anaphase. This is called *nondisjunction* and results in one sex cell having two members of a chromosome pair while the other receives none. Thus, eggs can contain

two Xs or no X (symbolized as 0), and sperm can contain both an X and a Y chromosome, two Xs or two Ys, or no sex chromosomes at all.

Because the X chromosome contains so many genes unrelated to sex that are necessary for development, a zygote must contain at least one X chromosome to survive. Thus, Y0 and YY zygotes do not develop. However, other zygotes with sex chromosome anomalies do develop. Examples are Turner's and Kleinfelter's syndromes.

"Extra" X chromosomes (more than two in the female and more than one in the male) have a surprisingly minor effect on the individual, compared with the significant effect of other additional chromosomes. Studies show that only one X chromosome is active in any cell. Any additional X chromosomes are randomly inactivated early in development and do not express the genes contained on them.

These inactivated X chromosomes remain tightly coiled against the cell membrane and can be seen as what are called *Barr bodies*

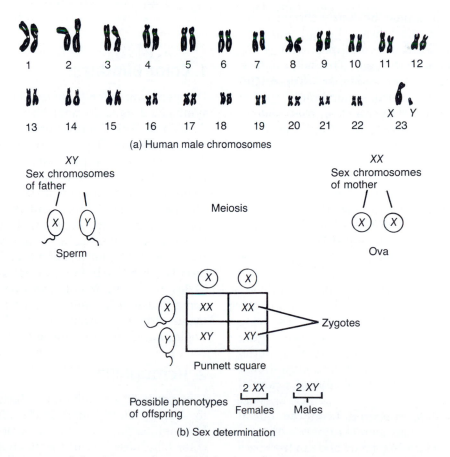

(a) Human male chromosomes

(b) Sex determination

FIGURE 28.2 Inheritance of sex. In (a), note the sex chromosomes, X and Y.

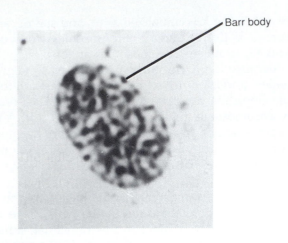

FIGURE 28.3 Barr body in a cell from the buccal mucosa of a human female. Feulgen stain; magnification 300×.

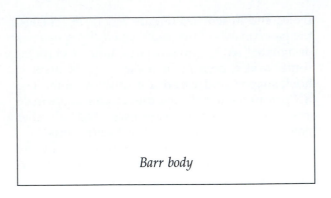

Barr body

(Figure 28.3). Because males do not have inactivated X chromosomes, no Barr bodies will be seen in normal male cells.

PROCEDURE

1. Make a buccal smear by *gently* scraping the inside of your cheek with the flat end of a toothpick. Discard the toothpick and *gently* scrape the same area again. This will produce more live cells from a deeper layer of the epithelium of the mucous membrane.
2. Spread the material over a clean glass slide.
3. Promptly place the slide in a Coplin jar filled with fixative for 1 min.
4. Remove the slide and wash it gently under running tap water.
5. Place the slide in a Coplin jar filled with Giemsa stain for 10 to 20 min. Fresh solutions of stain prepared within 2 hr require 10 min; older stains require a longer time.
6. Wash the slide *gently* under running water and air-dry it.
7. Examine the slide under high power and look for interphase nuclei. Identify Barr bodies, small disc-shaped chromatin bodies lying against the nuclear membrane (see Figure 28.3). Depending on the position of the nuclei and the staining technique, Barr bodies should be seen in 30 to 70 % of the cells of a normal female.
8. Examine a slide prepared from the buccal epithelium of a class member not of your sex.
9. Draw a cell containing a Barr body in the space provided here.

D. SEX-LINKED INHERITANCE

The sex chromosomes contain genes that are responsible for the transmission of a number of nonsexual traits, as do the other 22 pairs of chromosomes. Genes for these traits appear on X chromosomes, but many of these genes are absent from Y chromosomes. Traits transmitted by genes on the X chromosome are called *sex-linked (X-linked) traits.* This pattern of heredity differs from the pattern described earlier. About 150 X-linked traits are known in humans. Examples of sex-linked traits are color blindness and hemophilia.

1. Color Blindness

The gene for *color blindness* is a recessive one symbolized as *c*. Normal vision, symbolized *C*, dominates. The C and c genes are located on the X chromosome. The Y chromosome, however, does not contain the segment of DNA that programs color vision. Thus, the ability to see colors depends entirely on the X chromosome. The genetic possibilities and the inheritance of color blindness are shown in Figure 28.4. Only females who have two X^c chromosomes are color blind. In $X^C X^c$ females, the trait for color blindness is inhibited by the normal dominant gene. Males, however, do not have a second X chromosome that would inhibit the trait. Therefore, all males with an X^c chromosome will be color blind. Their pair of sex chromosomes is $X^c Y$.

2. Hemophilia

Hemophilia is a condition in which the blood fails to clot or clots very slowly after an injury. Hemophilia is a much more serious defect than color blindness because individuals with severe hemophilia can bleed to death from even a small

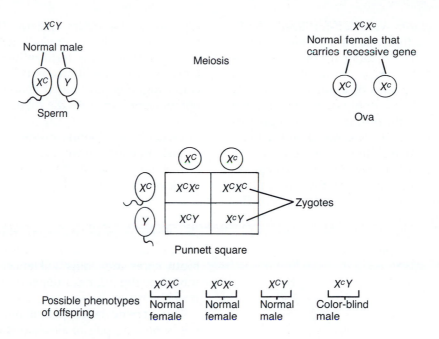

FIGURE 28.4 Inheritance of color blindness.

cut. Hemophilia is caused by a recessive gene, as is color blindness. If H represents normal clotting and h represents abnormal clotting, X^hX^h females will be hemophiliacs. Males with X^HY will be normal and males with X^hY will be hemophiliacs. Other sex-linked traits in human beings are certain forms of diabetes, night blindness, juvenile glaucoma, and juvenile muscular dystrophy.

E. MENDELIAN LAWS

In any genetic cross, all the offspring in the first (that is, the parental, or P_1) generation are symbolized as F_1. The F is from the Latin word *filial*, which means progeny. The second generation is symbolized as F_2, the third as F_3, and so on. The recognized "father" of genetics is Gregor Mendel, whose basic experiments were performed on garden peas. As a result of his tests, Mendel postulated what are now called *Mendelian Laws*, or *Mendelian Principles*. The *First Mendelian Law*, or the *Law of Segregation*, asserts that, in cells of individuals, genes occur in pairs, and that when those individuals produce germ cells, each germ cell receives only one member of the pair.

This law applies equally to pollen grains (or sperm) and to ova. The genetic cross is represented here:

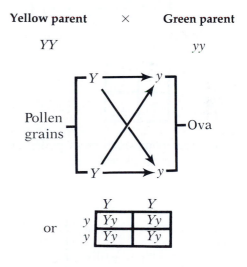

All possible combinations of pollen grains and ova are indicated by the arrows. Notice that all combinations yield the genotype Yy. All these F_1 seeds were yellow, yellow being dominant to green, or, in genetic terms, Y being dominant to y. These F_1 individuals resembled the yellow parent in phenotype (being yellow) but not in genotype (Yy as opposed to YY). Both parents were homozygous. Both members of that pair of alleles were the same. The yellow parent was homozygous for Y and the green parent for y. The F_1 individuals were heterozygous, having one Y and one y.

When the F_1 plants were self-fertilized, the F_2 seeds appeared in the ratio of 3 yellow:1 green. Mendel found similar 3:1 ratios for the other traits he studied, and this type of result has been reported in many species of animals and plants for a variety of traits. Not only does the recessive trait reappear in the F_2, but, in a definite proportion of the individuals, in one-fourth of the total. If the sample is small, the ratio may deviate considerably from 3:1. However, as the progeny, or sampling numbers, get larger, the ratio usually comes closer and closer to an exact 3:1 ratio. The reason is that the ratio depends on the random union of gametes. The result is a 3:1 phenotypic ratio or a 1:2:1 genotypic ratio.

F_1 Parents	**Yellow**	$\times$	**Yellow**
genotype	Yy		Yy

Thus, the four combinations of pollen and ova are expected to occur as follows:

$$\frac{1}{4}YY = \text{yellow}$$
$$\frac{1}{4}Yy = \text{yellow}$$
$$\frac{1}{4}Yy = \text{yellow} \quad \Big\} \; 3/4$$
$$\frac{1}{4}yy = \text{green} \quad \Big\} \; 1/4$$

It is important to realize that these fractions depend on the operation of the laws of probability. A model using coins will emphasize the point. This model consists of two coins, a nickel and a penny, tossed at the same time. The penny represents the pollen (male parent). At any given toss, the chances are equal that the penny will come up "heads" or that it will come up "tails." Similarly, at any given fertilization, the chances are equal that a Y-bearing pollen grain or that a y-bearing one will be transmitted. The nickel represents the ovum. Again, the chances are equal for "heads" or "tails," just as the chances are equal that in any fertilization a Y-bearing or a y-bearing ovum will take part. If we toss the two coins together and do it many times, we will obtain approximately the following:

¼ nickel heads; penny heads	$(= YY)$
¼ nickel heads; penny tails	$(= Yy)$
¼ nickel tails; penny heads	$(= yY)$
¼ nickel tails; penny tails	$(= yy)$

If we assume "heads" as dominant, we find that three-quarters of the time there is at least one "head" and one-fourth of the time no "heads" (both coins are "tails"). Hence, this gives us a model of the 3:1 ratio dependent on the laws of probability.

The genetic cross just demonstrated considered only one pair of alleles (yellow versus green or heads versus tails) and is therefore called a ***monohybrid cross.*** Mendel's second principle applied to genetic crosses in which two traits, or two pairs of alleles, were considered. These were ***dihybrid crosses.*** They enabled him to postulate his second principle, the ***Principle of Independent Assortment.*** This principle stated that the segregation of one pair of traits occurred independently of the segregation of a second pair of traits. This is the case only if the traits are caused by genes located on nonhomologous chromosomes.

When Mendel crossed garden peas with round yellow seeds with garden peas with wrinkled green seeds, his F_1 generation showed that yellow and round were dominant. If self-fertilization then occurred, the F_2 generation resulted as follows:

Round yellow	$\frac{3}{4} \times \frac{3}{4} = \frac{9}{16}$
Round green	$\frac{3}{4} \times \frac{1}{4} = \frac{3}{16}$
Wrinkled yellow	$\frac{1}{4} \times \frac{3}{4} = \frac{3}{16}$
Wrinkled green	$\frac{1}{4} \times \frac{1}{4} = \frac{1}{16}$

Therefore, in a dihybrid cross, the expected phenotypic ratio was 9:3:3:1, with 9/16 of the F_2 being doubly dominant and only 1/16 being doubly recessive.

F. MULTIPLE ALLELES

In the genetics examples we have considered to this point, we have discussed only two alleles of each gene. However, many, and possibly all genes, have *multiple alleles*—that is, they exist in more than two allelic forms, even though a diploid cell cannot carry more than two alleles.

One example of multiple alleles in humans involves ABO blood groups (Exercise 16). The four basic blood groups of the ABO system are determined by three alleles: I^A, I^B, and i. Alleles I^A and I^B are not dominant over each other (they are co-dominant), but are dominant over allele i. These three alleles can give rise to six genotypes as follows:

Genotype	Phenotype (blood type)
$I^A I^A$ or $I^A I$	A
$I^B I^B$ or $I^B I$	B
$I^A I^B$	AB
ii	O

Given this information, is it possible for a child with Type O blood to have a mother with Type O blood and a father with Type AB blood?

Explain: _____

_____

_____

If two children in a family have type O blood, the mother has Type B blood and the father has Type A blood, what is the genotype of the father?

_____

What is the genotype of the mother? _____

G. GENETICS EXERCISES

1. Karyotyping

A group of cytogeneticists meeting in Denver, Colorado, in 1960 adopted a system for classifying and identifying human chromosomes. Chromosome *length* and *centromere position* were the bases for classification. The Denver classification has become a standard for human chromosome studies. By the early 1970s, most human chromosomes could be identified microscopically.

Every chromosome pair could not be identified consistently until chromosome *banding tech-niques* finally distinguished all 46 human chromosomes. Bands are defined as parts of chromosomes that appear lighter or darker than adjacent regions with particular staining methods.

A *karyotype* is a chart made from a photograph of the chromosomes in metaphase. The chromosomes are cut out and arranged in matched pairs, according to length (see Figure 28.2a). Their comparative size, shape, and morphology are then examined to determine if they are normal.

Karyotyping helps scientists to visualize chromosomal abnormalities. For example, individuals with Down syndrome typically have 47 chromosomes, instead of the usual 46, with chromosome 21 being represented three times rather than only twice. The syndrome is characterized by mental retardation, retarded physical development, and distinctive facial features (round head, broad skull, slanting eyes, and large tongue). With chronic myelogenous leukemia, part of the long arm of chromosome 22 is missing. The chromosome is referred to as the Philadelphia chromosome, named for the city where it was first detected.

2. PKU Screening

Phenylketonuria (PKU), an inherited metabolic disorder that occurs in approximately 1 in 16,000 births, is transmitted by an autosomal recessive gene (see Figure 28.1). Individuals with this condition do not have the enzyme phenylalanine hydroxylase, which converts the amino acid phenylalanine to tyrosine. As a result, phenylalanine and phenylpyruvic acid accumulate in the blood and urine. These substances are toxic to the CNS and can produce irreversible brain damage. Most states in the United States require routine screening at birth for this disorder. The test is accomplished by a simple color change in treated urine.

The procedure for testing for PKU follows.

PROCEDURE

1. A Phenistix® strip is made specifically for testing urine for phenylpyruvic acid. Dip this test strip in freshly voided urine.
2. Compare the color change with the color chart on the Phenistix® bottle. The test is based on the reaction of ferric ions with phenylpyruvic acid to produce a gray-green color.
3. Record your results in Section G.1 of the LABORATORY REPORT RESULTS at the end of the exercise.

3. PTC Inheritance

The ability to taste the chemical compound known as phenylthiocarbamide, commonly called PTC, is inherited. On the average, 7 out of 10 people, on chewing a small piece of paper treated with PTC, detect a definite bitter or sweet taste. Others do not taste anything.

Individuals who can taste something (bitter or sweet) are called tasters and have the dominant allele *T*, either as *TT* or *Tt*. A nontaster is a homozygous recessive and is designated *tt*.

Determine your phenotype for tasting PTC and record your results in Section G.2 of the LABORATORY REPORT RESULTS at the end of the exercise.

Note: If PTC paper is not available, a 0.5% solution of phenylthiourea (PTT) can be substituted because the capacity to taste PTT is also inherited as a dominant.

4. Corn Genetics

Genetic corn can be purchased and used in this exercise. Each ear of corn represents a family of offspring. Mark a starting row with a pin to avoid repetition. Count the kernels (individuals) for each trait (color, wrinkled, or smooth). Record your results in Section G.3 of the LABORATORY REPORT RESULTS at the end of the exercise.

Develop a ratio by using your lowest number as "1" and dividing it into the others to determine what multiples of it they are. See how close you come to Mendel's ratios. Figure out the probable genotype and phenotype of the parent plants if you can. Monohybrid crosses, test crosses, dihybrid crosses, and trihybrid crosses are available.

5. Color Blindness

Using either Stilling or Ishihara test charts, test the entire class for color blindness. Tests for color blindness depend on the person's ability to distinguish various colors from one another and also on his/her ability to judge correctly the degree of contrast between colors.

Of all men, 2% are color blind to red and 6% to green, so 8% of all men are red-green color-blind. Red-green color blindness is rare in females, occurring in only 1 of every 250 women. Record your results in Section G.4 of the LABORATORY REPORT RESULTS at the end of the exercise.

6. Mendelian Laws of Inheritance

Follow the procedure outlined in the explanation of the Mendelian Law of Segregation, tossing a nickel and a penny simultaneously to demonstrate the law and determine ratios.

PROCEDURE

1. Toss the nickel and the penny together 10 times to get the genotypes of a family of 10. Repeat this procedure for a total of five times to obtain 5 families of 10 offspring. Record all of the results on the chart in Section G.5 of the LABORATORY REPORT RESULTS at the end of the exercise.
   Note: Use the following symbols for the following exercises.

   G = gene for yellow
   g = gene for green
   GG = the genotype of an individual pure (homozygous) for yellow
   gg = the genotype of an individual pure (homozygous) for green
   Gg = the genotype of the hybrid (heterozygous) individual, phenotypically yellow
   ♀ = symbol for female
   ♂ = symbol for male

2. Obeying the Mendelian Law of Segregation and using the Punnett square shown in Section G.6 of the LABORATORY REPORT RESULTS at the end of the exercise, cross yellow garden peas with green garden peas (a monohybrid cross). Show the P_1, F_1, and F_2 generations and all the different phenotypes and genotypes.

3. Obeying the Mendelian Law of Independent Assortment and using the Punnett square shown in Section G.8 of the LABORATORY REPORT RESULTS at the end of the exercise, cross the round yellow seeds with the wrinkled green seeds (a dihybrid cross). Show the P_1, F_1, and F_2 generations and all the different phenotypes and genotypes. The F_2 generation can be generated from a Punnett square comparable to that for the monohybrid cross, but with 16 rather than 4 squares.

7. Observing Phenotypes

The pattern of inheritance of many human traits is complex and involves many genes; the inheritance of other traits is controlled by single genes. Record your phenotype and your genotype, if they can be determined, for several traits controlled by single genes. For example, if you have the phenotypically dominant trait *A*, your genotype will be *A—*, indicating that the allele symbolized as "—" is not known, because you could be homozygous dominant (*AA*) or heterozygous (*Aa*) for the trait. If you have the recessive trait *a—*, your genotype will be *aa*. Record your phenotype and genotype for the following traits in Section G.10 of the LABORATORY REPORT RESULTS at the end of the exercise.

1. *Attached earlobes*—The dominant gene *E* causes earlobes that develop free from the neck; *ee* results in adherent earlobes connected to the cervical skin.
2. *Tongue rolling*—The dominant gene *R* causes the development of muscles that allow the tongue to be rolled into a U shape. The *rr* genotype prohibits such rolling.
3. *Hair whorl direction*—The dominant gene *W* causes the hair whorl on the cranial surface of the scalp to turn in a clockwise direction; the genotype *ww* determines a counterclockwise whorl.
4. *Little-finger bending*—The dominant gene *B* causes the distal segment of the little finger to

bend laterally. The genotype *bb* results in a straight distal segment.
5. *Double-jointed thumbs*—The dominant gene *J* results in loose ligaments that allow the thumb to be bent out of the constricted orientation caused by the recessive genotype *jj*.
6. *Widow's peak*—The dominant gene *W* causes the hairline to extend caudally in the midline of the forehead. The recessive genotype *ww* results in a straight hairline.
7. *Rh factor*—A dominant gene *Rh* results in the presence of the Rh antigen on red blood cells. This antigen is not present with the recessive genotype *rhrh*. Use the results obtained in Exercise 17.H.4 to determine your phenotype. What additional information would you need to determine your complete genotype if you have the dominant phenotype for these traits?

ANSWER THE LABORATORY REPORT QUESTIONS AT THE END OF THE EXERCISE.

Genetics

STUDENT _____   DATE _____

LABORATORY SECTION _____   SCORE/GRADE _____

SECTION G. GENETICS EXERCISES
PKU Screening

1. _____ negative—cream color

   _____ 15 mg%—light green

   _____ 40 mg%—medium green

   _____ 100 mg%—dark green

PTC Inheritance

2. _____ bitter taste

   _____ sweet taste

   _____ negative (no taste)

Corn Genetics

3. Ratio_____ monohybrid cross

   Ratio _____ test cross

   Ratio _____ dihybrid cross

   Ratio_____ trihybrid cross

Color Blindness

	Male students	Female students
4. Red color blind	_____	_____
Green color blind	_____	_____

Mendelian Laws of Inheritance

5. Record the results of tossing a nickel and a penny together 10 times.

	Female (nickel)	Male (penny)	1	2	3	4	5	Total	Class total
Dominant offspring	A Heads	A Heads							
	a Heads	A Tails							
	a Tails	A Heads							
Recessive offspring	a Tails	a Tails							
Ratio dominant to recessive									

6. Complete the following monohybrid cross. Fill in genotypes (within circles) and phenotypes (under circles).

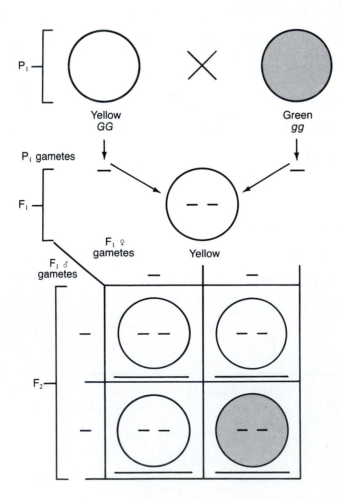

Monohybrid cross in the garden pea *(Pisum sativum):* G = allele for yellow, g = allele for green, P_1 = parental generation, F_1 = first filial generation, F_2 = second filial generation.

7. What is the phenotype ratio of the F_2 generation? _____ yellow/_____ green.

8. Complete the following dihybrid cross. Fill in genotypes (within circles) and phenotypes (under circles).

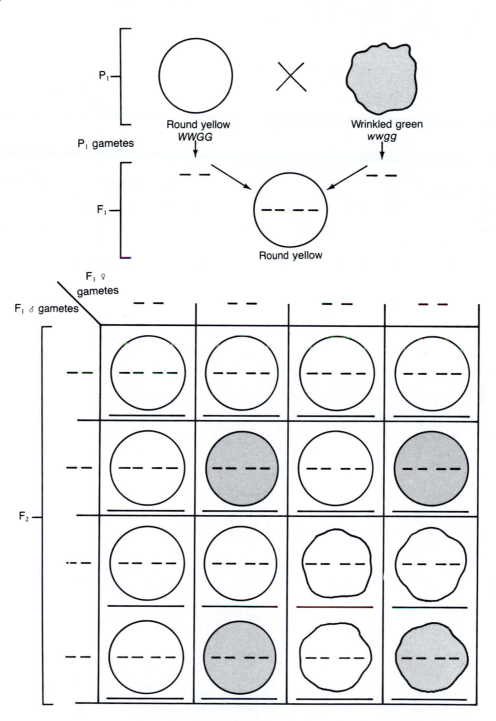

Dihybrid cross in the garden pea *(Pisum sativum):* W = allele for round, *w* = allele for wrinkled, G = allele for yellow, *g* = allele for green.

9. What is the phenotype ratio of the F_2 generation?

_____ round yellow/_____ wrinkled yellow/_____ round green/

_____wrinkled green.

10. Observing phenotypes

Trait	Phenotype	Genotype
Earlobes		
Tongue rolling		
Hair whorl		
Double-jointed thumbs		
Widow's peak		
Rh factor		

Genetics

STUDENT _____   DATE _____

LABORATORY SECTION _____   SCORE/GRADE _____

PART 1. Multiple Choice

_____ 1. Using the symbols *Aa* to represent genes, which of the following is true? (a) the trait is homozygous for the dominant characteristic (b) the trait is homozygous for the recessive characteristic (c) the trait is heterozygous (d) X-linked inheritance is in operation

_____ 2. Which statement concerning the normal inheritance of sex is correct? (a) all zygotes contain a Y chromosome (b) some ova contain a Y chromosome (c) all ova and all sperm contain an X chromosome (d) all ova have an X chromosome, some sperm have an X chromosome, and some sperm have a Y chromosome

_____ 3. The genotype that will express characteristics associated with hemophilia (assume that *H* represents the gene for normal blood) is (a) $X^H X^h$ (b) $X^h Y$ (c) $X^H X^H$ (d) $X^H Y$

_____ 4. The exact position of a gene on a chromosome is called the (a) homologue (b) locus (c) triad (d) allele

PART 2. Completion

5. When the same genes appear on homologous chromosomes, as in *PP* or *pp*, the individual is said to be _____ for the trait.

6. If different genes appear on homologous chromosomes, as in *Pp*, the individual is said to be

   _____ for the trait.

7. Genetic composition expressed in the body or morphologically is called the body's

   _____.

8. The device that helps one visualize all the possible combinations of male and female gametes is

   called the _____.

9. The twenty-third pair of human chromosomes are the sex chromosomes. All of the other pairs of

   chromosomes are called _____.

10. Color blindness is an inherited trait that is specifically called _____ trait.

11. The recognized "father" of genetics is _____.

12. A genetic cross that involves only one pair of alleles (or traits) is called a(n)

    _____ cross.

13. Passage of hereditary traits from one generation to another is called _____.

14. The genetic makeup of an individual is called the person's _____.

15. One of the many alternative forms of a gene is called its _____.

Some Important Units of Measurement

English Units of Measurement

Fundamental or derived unit	Units and Equivalents
Length	12 inches (in.) = 1 foot (ft) = 0.333 yard (yd) 3 ft = 1 yd 1760 yd = 1 mile (mi) 5280 ft = 1 mi
Mass	1 ounce (oz) = 28.35 grams (g); 1 g = 0.0353 oz 1 pound (lb) = 453 g = 16 oz; 1 kilogram (kg) = 2.205 lb 1 ton = 2000 lb = 907 kg
Time	1 second (sec) = 1/86 400 of a mean solar day 1 minute (min) = 60 sec 1 hour (hr) = 60 min = 3600 sec 1 day = 24 hr = 1440 min = 86 400 sec
Volume	1 fluid dram (fl dr) = 0.125 fluid ounce (fl oz) 1 fl oz = 8 fl dr = 0.0625 quart (qt) = 0.008 gallon (gal) 1 qt = 256 fl dr = 32 fl oz = 2 pints (pt) = 0.25 gal 1 gal = 4 qt = 128 fl oz = 1024 fl dr

Metric Units of Length and Some English Equivalents

Metric Unit	Meaning of Prefix	Metric Equivalent	English Equivalent
1 kilometer (km)	kilo = 1000	1000 m	3280.84 ft or 0.62 mi; 1 mi = 1.61 km
1 hectometer (hm)	hecto = 100	100 m	328 ft
1 dekameter (dam)	deka = 10	10 m	32.8 ft
1 meter (m)		Standard unit of length	39.37 in. or 3.28 ft or 1.09 yd
1 decimeter (dm)	deci = $\frac{1}{10}$	0.1 m	3.94 in.
1 centimeter (cm)	centi = $\frac{1}{100}$	0.01 m	0.394 in.; 1 in. = 2.54 cm
1 millimeter (mm)	milli = $\frac{1}{1000}$	0.001 m = $\frac{1}{10}$ cm	0.0394 in.
1 micrometer (μm) [formerly micro (μ)]	micro = $\frac{1}{1,000,000}$	0.0000001 m = $\frac{1}{10000}$ cm	3.94×10^{-5} in.
1 nanometer (nm) [Formerly millimicrons (mμ)]	nano = $\frac{1}{1,000,000,000}$	0.000000001 m = $\frac{1}{10,000,000}$ cm	3.94×10^{-8} in.
1 angstrom (Å)		0.0000000001 m = $\frac{1}{100,000,000}$ cm	3.94×10^{-9} in.

Temperature

Unit	K	°F	°C
1 degree Kelvin (K)	1	$\frac{9}{5}$(K) − 459.7	K + 273.16*
1 degree Fahrenheit (°F)	$\frac{5}{9}$(°F) + 255.4	1	$\frac{5}{9}$(°F − 32)
1 degree Celsius (°C)	°C − 273	$\frac{9}{5}$(°C) + 32	1

Volume

Unit	ml	cm³	qt	oz
1 milliliter (ml)	1	1	1.06×10^{-3}	3.392×10^{-2}
1 cubic centimeter (cm³)	1	1	1.06×1^{-3}	3.392×10^{-2}
1 quart (qt)	943	943	1	32
1 fluid ounce (fl oz)	29.5	29.5	3.125×10^{-2}	1

*Absolute zero (K) = −273.16°C

Periodic Table of the Elements

KEY

6	Atomic Number
C	Symbol
12.01	Atomic Weight
Carbon	Name

1 H 1.0080 Hydrogen																	2 He 4.003 Helium
3 Li 6.940 Lithium	4 Be 9.013 Berilium											5 B 10.82 Boron	6 C 12.011 Carbon	7 N 14.008 Nitrogen	8 O 16.000 Oxygen	9 F 19.00 Fluorine	10 Ne 20.183 Neon
11 Na 22.991 Sodium	12 Mg 24.32 Magnesium											13 Al 26.98 Aluminum	14 Si 28.09 Silicon	15 P 30.975 Phosphorus	16 S 32.066 Sulfur	17 Cl 35.457 Chlorine	18 Ar 39.944 Argon
19 K 39.100 Potassium	20 Ca 40.08 Calcium	21 Sc 44.96 Scandium	22 Ti 47.90 Titanium	23 V 50.95 Vanadium	24 Cr 52.01 Chromium	25 Mn 54.94 Manganese	26 Fe 55.85 Iron	27 Co 58.94 Cobalt	28 Ni 58.71 Nickel	29 Cu 63.54 Copper	30 Zn 65.38 Zinc	31 Ga 69.72 Gallium	32 Ge 72.60 Germanium	33 As 74.91 Arsenic	34 Se 78.96 Selenium	35 Br 79.916 Bromine	36 Kr 83.80 Krypton
37 Rb 85.48 Rubidium	38 Sr 87.63 Strontium	39 Y 88.92 Yttrium	40 Zr 91.22 Zirconium	41 Nb 92.91 Niobium	42 Mo 95.95 Molybdenum	43 Tc (99) Technetium	44 Ru 101.1 Ruthenium	45 Rh 102.91 Rhodium	46 Pd 106.4 Palladium	47 Ag 107.880 Silver	48 Cd 112.41 Cadmium	49 In 114.82 Indium	50 Sn 118.70 Tin	51 Sb 121.76 Antimony	52 Te 127.61 Tellurium	53 I 126.91 Iodine	54 Xe 131.30 Xenon
55 Cs 132.91 Cesium	56 Ba 137.36 Barium	57 La 138.92 Lanthanum	72 Hf 178.50 Hafnium	73 Ta 180.95 Tantalum	74 W 183.86 Wolfram	75 Re 186.22 Rhenium	76 Os 190.2 Osmium	77 Ir 192.2 Iridium	78 Pt 195.09 Platinum	79 Au 197.0 Gold	80 Hg 200.61 Mercury	81 Tl 204.39 Thallium	82 Pb 207.21 Lead	83 Bi 209.00 Bismuth	84 Po (210) Polonium	85 At (210) Astatine	86 Rn (222) Radon
87 Fr (223) Francium	88 Ra (226) Radium	89 Ac (227) Actinium	104 (Russian Proposal Unofficial)														

Lanthanide series

58 Ce 140.13 Cerium	59 Pr 140.92 Praseodymium	60 Nd 144.27 Neodymium	61 Pm (147) Promethium	62 Sm 150.35 Samarium	63 Eu 152.0 Europium	64 Gd 157.26 Gadolinium	65 Tb 158.93 Terbium	66 Dy 162.51 Dysprosium	67 Ho 164.94 Holmium	68 Er 167.27 Erbium	69 Tm 168.94 Thulium	70 Yb 173.04 Ytterbium	71 Lu 174.99 Lutetium

Actinide series

90 Th (232) Thorium	91 Pa (231) Protactinium	92 U 238.07 Uranium	93 Np (237) Neptunium	94 Pu (242) Plutonium	95 Am (243) Americium	96 Cm (247) Curium	97 Bk (249) Berkelium	98 Cf (251) Californium	99 Es (254) Einstenium	100 Fm (253) Fermium	101 Md (256) Mendelevium	102 No (253) Nobelium	103 Lw 257 Lawrencium

Eponyms Used in This Laboratory Manual

Eponym	Current Terminology
Achilles tendon	calcaneal tendon
Adam's apple	thyroid cartilage
ampulla of Vater (VA-ter)	hepatopancreatic ampulla
Bartholin's (BAR-tō-linz) gland	greater vestibular gland
Billroth's (BIL-rōtz) cord	splenic cord
Bowman's (BŌ-manz) capsule	glomerular capsule
Bowman's (BŌ-manz) gland	olfactory gland
Broca's (BRŌ-kaz) area	motor speech area
Brunner's (BRUN-erz) gland	duodenal gland
bundle of His (HISS)	atrioventricular (AV) bundle
canal of Schlemm (SHLEM)	scleral venous sinus
circle of Willis (WIL-is)	cerebral arterial circle
Cooper's (KOO-perz) ligament	suspensory ligament of the breast
Cowper's (KOW-perz) gland	bulbourethral gland
crypt of Lieberkühn (LĒ-ber-kyoon)	intestinal gland
duct of Rivinus (ri-VĒ-nus)	lesser sublingual duct
duct of Santorini (san'-tō-RĒ-nē)	accessory duct
duct of Wirsung (VĒR-sung)	pancreatic duct
end organ of Ruffini (roo-FĒ-nē)	type-II cutaneous mechanoreceptor
Eustachian (yoo-STĀ-kē-an) tube	auditory tube
Fallopian (fal-LŌ-pē-an) tube	uterine tube
gland of Zeis (ZĪS)	sebaceous ciliary gland
Golgi (GOL-jē) tendon organ	tendon organ
Graafian (GRAF-ē-an) follicle	vesicular ovarian follicle
Hassall's (HAS-alz) corpuscle	thymic corpuscle
Haversian (ha-VĒR-shun) canal	central canal
Haversian (ha-VĒR-shun) system	osteon
interstitial cell of Leydig (LĪ-dig)	interstitial endocrinocyte
islet of Langerhans (LANG-er-hanz)	pancreatic islet
Kupffer's (KOOP-ferz) cells	stellate reticuloendothelial cell
loop of Henle (HEN-lē)	loop of the nephron
Malpighian (mal-PIG-ē-an) corpuscle	splenic nodule

Eponym	Current Terminology
Meibomian (mī-BŌ-mē-an) gland	tarsal gland
Meissner's (MĪS-nerz) corpuscle	corpuscle of touch
Merkel's (MER-kelz) disc	tactile disc
Müller's (MIL-erz) duct	paramesonephric duct
Nissl (NISS-l) bodies	chromatophilic substance
node of Ranvier (ron-VĒ-ā)	neurofibral node
organ of Corti (KOR-tē)	spiral organ
Pacinian (pa-SIN-ē-an) corpuscle	lamellated corpuscle
Peyer's (PĪ-erz) patches	aggregated lymphatic follicles
plexus of Auerbach (OW-er-bak)	myenteric plexus
plexus of Meissner (MĪS-ner)	submucous plexus
pouch of Douglas	rectouterine pouch
Purkinje (pur-KIN-jē) fiber	conduction myofiber
Rathke's (rath-KĒZ) pouch	hypophyseal pouch
Schwann (SCHVON) cell	neurolemmocyte
Sertoli (ser-TŌ-lē) cell	sustentacular cell
Skene's (SKĒNZ) gland	paraurethral gland
sphincter of Oddi (OD-dē)	sphincter of the hepatopancreatic ampulla
Stensen's (STEN-senz) duct	parotid duct
Volkmann's (FŌLK-manz) canal	perforating canal
Wharton's (HWAR-tunz) duct	submandibular duct
Wharton's (HWAR-tunz) jelly	mucous connective tissue
Wormian (WER-mē-an) bone	sutural bone

Figure Credits

1.1 Visuals Unlimited.

3.5 (a)–(f) Carolina Biological Supply Company/ Phototake.

4.1 (a) Biophoto Associates/Science Source/ Photo Researchers, Inc. (b) Walker/Photo Researchers, Inc. (c) Biophoto Associates/Science Source/Photo Researchers, Inc. (d) Biological Photo Service. (e) Biophoto Associates/Science Source/Photo Researchers, Inc. (f) Ed Reschke. (g) Ed Reschke.

4.3 (a) Photo Researchers, Inc. (b) Biophoto Associates/Science Source/Photo Researchers, Inc. (c) Robert Brons/Biological Photo Service. (d) Biophoto Associates/Science Source/Photo Researchers, Inc. (e) Ed Reschke. (f) Bruce Iverson/ Visuals Unlimited. (g) Fred Hossler/Visuals Unlimited. (h) Frederick C. Skvara. (j) C. W. Brown/ Photo Researchers, Inc.

5.2 From *Tissues and Organs: A Text-Atlas of Scanning Electron Microscopy* by Richard G. Kessel and Randy H. Kardon. Copyright © 1979, by W. H. Freeman and Company. Reprinted by permission.

6.1 (b) From *The Musculoskeletal System in Health and Disease* by Cornelius Rosse and D. Kay Clawson. Reprinted by permission of the author. (c), (d) Courtesy of Fisher Scientific.

8.2 (a)–(c), (e) Copyright © 1983 by Gerard J. Tortora, courtesy of Matt Iacobino and Lynne Borghesi. (d), (i), (j) Copyright © 1991 by Evan J. Colella. (f), (g), (h) Courtesy of Matt Iacobino.

8.4 © By John Eads.

9.1 By James R. Smail and Russell A. Whitehead, Macalester College.

9.3 (a) Biological Photo Service. (b) Alfred Owczarzak/Biological Photo Service.

9.4 (a) Johannes Rhodin, Don Fawcett/Visuals Unlimited. (b) Don W. Fawcett/Visuals Unlimited.

11.2–11.8, 11.9 (a), **11.10–11.12** Biomedical Graphics Department, University of Minnesota Hospitals.

11.9 (b) Courtesy of O. Richard Johnson.

12.1 Carolina Biological Supply Co./Phototake.

15.2 By Andrew Kuntzman, Wright State University.

15.3 (b) Cabisco/Visuals Unlimited.

15.6 Biomedical Graphics Department, University of Minnesota Hospitals.

15.13 (a) John Cunningham/Visuals Unlimited. (b) Photo Researchers, Inc.

15.14 Copyright © 1983 by Gerard J. Tortora, courtesy of James Borghesi.

16.2 By James R. Smail and Russell A. Whitehead, Macalester College.

16.3 By James R. Smail, Macalester College.

16.4 Biological Photo Service.

16.5 Don W. Fawcett/Visuals Unlimited.

16.6 By Andrew Kuntzman, Wright State University.

17.3 Courtesy of Becton-Dickinson, Division of Becton, Dickenson and Company.

17.4 Modified from *RBC Determination Methods* and *WBC Determination for Manual Methods*, Becton-Dickenson, Division of Becton, Dickinson and Company.

17.6, 17.7 Courtesy of Fisher Scientific.

17.8 Courtesy of Leica, Inc., Buffalo, NY.

Index